Process Design and Engineering Practice

Donald R. Woods

McMaster University

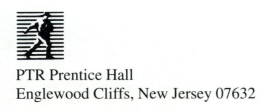

PTR Prentice Hall
Englewood Cliffs, New Jersey 07632

Library of Congress Cataloging-in-Publication Data

Woods, Donald R., 1954–
    Process design and engineering practice / Donald R. Woods.
      p. cm.
    Includes index.
    ISBN 0–13–805755–9
    1. Chemical processes.   I. Title.
TP155.7.W66   1995
660′281—dc20   93–5526
                                   CIP

Cover design: *Oysterpond Press*
Acquisitions editor: *Betty Sun*
Manufacturing manager: *Alexis A. Heydt*

©1995 by PTR Prentice Hall
Prentice-Hall, Inc.
A Paramount Communications Company
Englewood Cliffs, New Jersey 07632

The publisher offers discounts on this book when ordered in bulk quantities.
For more information, contact:

    Corporate Sales Department
    PTR Prentice Hall
    113 Sylvan Avenue
    Englewood Cliffs, NJ 07632

    Phone: 201–592–2863
    Fax: 201–592–2249

Printed in the United States of America
10 9 8 7 6 5 4 3 2 1

ISBN 0-13-805755-9

Prentice-Hall International (UK) Limited, *London*
Prentice-Hall of Australia Pty. Limited, *Sydney*
Prentice-Hall Canada Inc., *Toronto*
Prentice-Hall Hispanoamericana, S.A., *Mexico*
Prentice-Hall of India Private Limited, *New Delhi*
Prentice-Hall of Japan, Inc., *Tokyo*
Simon & Schuster Asia Pte. Ltd., *Singapore*
Editora Prentice-Hall do Brasil, Ltda., *Rio de Janeiro*

Contents

Chapter 2
Transportation   2-1

Chapter 3
Energy Exchange: Mechanical—Electrical—Thermal   3-1

Minichapter 4
Selecting Options for the Separation of Components from Homogeneous Phases   4-1

Minichapter 5
Selecting Options for the Separation of Components from Heterogeneous Systems
5–1

Minichapter 6
Selecting Options for Reactions and Storage   *6–1*

Appendices   *A–1*

Preface

Our professional skills are needed to build and design "smarter." We need to be able to visualize a problem, identify about five possible options, select maybe two of these options for more detailed study, and thus synthesize innovative processes that will keep our companies at the leading edge. We need to be able to bring in perspectives from mining, metallurgical, chemical, petroleum, bioprocessing, food, pulp and paper, and environmental disciplines.

Crucial to these tasks are the ability to apply the principles of optimum sloppiness and successive approximation, the skill to do back-of-the-envelope calculations and the possession of a rich set of practical know-how. Most of us acquire these skills from practical experience after we graduate. This book offers a vehicle to be used to help nurture these skills in the undergraduate programs of chemical, environmental, metallurgical, mechanical, and mining engineering. It also guides the young (and older) professional in structuring our experience knowledge.

This book provides quickie selection and sizing procedures that are directly applicable in capstone design and synthesis courses. Indeed, since I have taught these courses for over twenty-five years, this text evolved to provide practical, explicit guidelines about "experience" and how to use it to consider *many*—not just one—optional configurations for processes or for trouble-shooting. However, as our engineering disciplines have grown and expanded in terms of the possible unit operations that can be considered, we have seen the need to better prepare our students for the capstone courses. The design-synthesis activities, the consideration of numerous options, and the practical application of the fundamentals all need to be integrated from our freshman courses through to our senior courses. But how? One option is to provide the big picture about the excitement of the engineering profession and then, through a variety of processes and examples, to illustrate why we study mass and energy balances, why and how to apply thermo, why and how to use fluid mechanics and heat transfer. Such an approach would offer concrete examples throughout the curriculum—examples that integrated the application, that introduced simplified PIDs and taught how to read and use them. It would teach how to look at a PID and, through the use of fundamentals, extract the keys to different process operations.

How will you benefit from this text and how might you use it? The book provides practical applications to the fundamentals of process engineering. It shows how it is used. It empowers you to be able to do order-of-magnitude calculations, to start memorizing and putting together your own set of experience knowledge. It develops your confidence that you can look at an idea and judge whether it is reasonable or not.

This book is meant to be used. It provides about two hundred worked examples, extensive collections of previously unavailable data, and over five hundred charts and tables. The Appendices add needed enrichment in the subtlety of choosing the materials of construction, and of control and operability of an option. To provide feedback, in Appendix III are given practical installation configurations used for some of the problems listed at the end of the Chapters in this book. The companion text "Data for Process Design and Engineering Practice" (referred to as **Data** in this text) offers a rare consolidation of data needed by engineers. The cross referencing to Tables in the Data book uses the symbol T. Thus, **Data** TC–12 refers to Table C–12 in Data for Process Design and Engineering Practice.

For students in a freshman design/synthesis course,

Chapter 1 provides a background introduction to our profession; the extensive index will then lead to the key selection/sizing procedure needed for your project.

For students in mass and energy balance and thermodynamics courses, the variety of PID diagrams, the diameters of pipes and the size of solids conveyors (from Chapter 2), and the section on selecting reactors (from Chapter 6) provide the applications and enrichment. Indeed, in problem-based courses, you might start with a PID and questions related to why and how it is hooked up the way it is. This provides motivation and understanding as to why these fundamentals are so vital. For thermodynamics, for example, the hydrodealkylation reactor provides a very good vehicle for understanding why thermo is so powerful a subject.

For the fluid dynamics course, Chapter 2—excluding the enrichment sections—should be studied at the beginning and the end of the course. Depending on the focus in the fluids course, the sections on filtration, centrifugation, and settling (from Chapter 5) are pertinent. Similarly, Chapters 3 and 4 provide overviews and concrete applications for courses in heat transfer and for mass transfer. These can be studied profitably at the beginning and end of such courses. The PIDs can be used throughout to illustrate the application of the fundamentals.

Chapter 6 provides the overview for a reaction kinetics and reactor design course.

Chapter 5 is a critical, but often neglected, portion of our curriculum and so may not directly enrich a particular course other than the design course.

Other volumes in this series give additional resources for the capstone equipment sizing and selection and for the design course.

Practicing engineers will find this book extremely valuable as a starting place for most problems in retrofitting, process improvement, process analysis and synthesis, and even trouble-shooting. It helps us to keep options open, to explore ideas, yet it is quantitative. For example, several of us used this book to rough-size and cost seven optional routes to synthesize a product in three hours. From this we could then decide which two options might hold the most promise.

Chemists have found this material helpful. They can see how engineers take the syntheses from the lab and create large-scale processes. This gives them an easy introduction to the questions asked and information needed by engineers to solve process problems.

Throughout the book, I have used charts, pictures, and graphs extensively and refrained from providing computer AI programs. Although those developing artificial intelligence and expert systems will find this collection of "tacit knowledge" or engineering experience an invaluable source around which they can develop the programs, my focus is on providing you with the hands-on experience of how professionals make choices.

To people like the late Don Ormston, Distiller's Company Ltd., Saltend, UK; Chuck Watson and the late Ed Crosby, Chemical Engineering Dept, University of Wisconsin, Madison; Stan A. Chodkiewicz, Engineering, Polysar Ltd., Sarnia; Reg H. Clark, Chemical Engineering Dept., Queen's University, Kingston, Ontario; Roger Butler, Imperial Oil Ltd., Sarnia; and J. Mike F. Drake, British Geon Ltd., Barry, South Wales, I owe much because they started me on my search for sound estimating procedures by willingly sharing their experience with me and being superb engineers. Many others have helped me with different sections of the text; indeed, because of the scope of this work, I could not have formulated the procedures without their guidance. I thank my colleagues at McMaster University (and especially Terry W. Hoffman, Joe D. Wright, Andy Hrymak, Tom Marlin, Jim Dickson, Phil E. Wood, Archie E. Hamielec, Les W. Shemilt, Marios Tsezos, Neil R. Bayes, Malcolm Baird, John Brash, and John Vlachopulos), Ken D. Hester of Rio Algom Ltd. and K. D. Hester and Associates, Consultants, Toronto; Bob McAndrew and Glen Dobby, University of Toronto; G. S. Peter Castle and Maurice Bergognou, University of Western Ontario, London; R. E. Edmondson, of Canadian Industries Ltd., Courtright, Ontario; V. I. Lakshmanan, Ontario Research Foundation, Sheridan Park, Ontario; Gordon Agar, INCO Research, Sheridan Park, Ontario; Peter Silveston, Chemical Engineering Dept., University of Waterloo, Ontario; Ed Capes, National Research Council of Canada, Ottawa; Peter Clark, of Epstein Engineering, Chicago, Illinois; Paul Belter, The Upjohn Co., Kalamazoo, Michigan; Jim Couper, Chemical Engineering Dept., University of Arkansas; Gary Powers and Art Westerberg, Carnegie Mellon University, Pittsburgh; Jim Douglas, University of Massachusetts, Amherst; Bill Cotton, DuPont of Canada Ltd., Kingston; Emil Nenninger, Hatch and Associates, Toronto; Keith Murphy, IEC, Toronto; Graham Davies, Manchester Institute of Technology, Manchester, England; and Don Dahlstrom, Chemical Engineering Dept., University of Utah, Salt Lake City for helping me with this text.

I am grateful to the Natural Sciences and Engineering Research Council of Canada, which financially supported much of the research required to develop the materials in this book.

My students in the design and communication skills courses did literature searches and gave me feedback about the text. Roberto Narbaitz, Valerie Meng, Helen Hadcock, and Evan Diamadopoulos read sections of the text and made suggestions for improvement.

I thank Laura Honda, from McMaster University's Word Processing Centre, Cynthia J. Woods, and my wife, Diane Elliott Woods, for their assistance.

Donald R. Woods
Waterdown, Ontario

The PD Symbols List

Symbols used in this volume and in subsequent volumes of Process Design and Engineering Practice

a	activity
a	effective packing area per unit volume
a	constant in the Langmuir isotherm
a	constant in pressure drop calculation, dimensionless (value between 0.3 to 0.5) [Eq 2–22]
A	code for characterizing solids: particle size, size less than 100 Mesh, "very fine"
A	area, m^2
A	heat transfer area
A^+	dimensionless area of a drum = $A/\pi/4\ D^2$)
A_b	area of the blades of a paddle reel
A_1	correction factor for batch distillation
A_i	abrasion index, dimensionless
A	hydraulic permeability through a membrane, $kg/s.\ m^2.\ MPa$
A_g'	reduction in solvent permeability because of the gel layer in membrane modeling, $kg/s.\ m^2.\ MPa.$
A subscript	solvent
A_w	liquid to wet solids underflow ratio in a leaching operation
Ar	dimensionless number, Archimedes no., = Galileo no. × density ratio $(\rho^2\ D_{p3}\ g/\ \mu^2)((\rho_s - \rho_g)/\ \rho_g)$
b	correlating parameter in the BET isotherm
b	correlating exponent in model of

	coalescence in a band of drops in a decanter (value usually = 5)
b	distance between discs in a disc-bowl centrifuge
Bi	dimensionless group: Bird number, the mass percentage of solids within a ±0.1 density variation from the critical or cut density ρ_{50}
Bi*	modified Bird number: Bi/(1 less (mass fraction of feed with density greater than 2.0))
B	code for characterizing solids: particle size, size less than 0.3 cm, "fine"
B	magnetic induction, T
B	solute B permeability of a membrane, $\mu m/s$
B^*	sodium chloride permeability of a membrane corrected for the membrane type, $\mu m/s$
BOD	acronym for Biological Oxygen Demand
c_p	heat capacity, $kJ/kg.K$
c	mass or mol/ unit volume depending on the superscript
c^+	dimensionless concentration
c_{B2}	concentration of solute B at exit stream 2
c_{B1}	concentration of solute B in the feed stream 1
c_o	initial concentration
C	code for characterizing solids: particle size, size less than 1.3 cm, "granular"

C_D	drag coefficient for a moving blade in a flocculator.
C_f	packing factor
Ca_D	Capillary drainage number, 1/m
COD	acronym for Chemical Oxygen Demand
CMC	abbreviation for critical micelle concentration
D	code for characterizing solids: particle size, size greater than 1.3 cm and less than 5 cm, "pebble"
D	diameter
D	diameter of the impeller
D	internal diameter of a pipe, m
D	diameter of the pan for agglomerators
D	diameter of a roll briquetter
d	diameter of a briquette
d	dimensionless thickness of a briquette d/D
d_o	diameter of nozzle, mm
D_p	diameter of a particle, μm
D_{pl}	limiting diameter of a particle; mm
D_{p50}	diameter of a particle that has a 50% chance of reporting to either of the two exit streams. (see also "cut" diameter), μm
d_h	hydraulic radius = 4 (cross-sectional area)/wetted perimeter, cm
$\mathscr{D}$	diffusivity, usually cm^2/s
$\mathscr{D}_f$	diffusivity in the fluid, cm^2/s
DAF	acronym for dissolved air flotation
DMS	acronym for dense media separation
E	code for characterizing solids: particle size, "fibrous and stringy"
f	correction factor for concentration driving force
f	correction factor for the Log Mean Temperature Driving Force [Eq 3–10]
f	correction factor of Johanson for briquetting
f	mass of solution that goes with the solid underflow from stage to stage in leaching
f_T	correction factor of Johanson for power for briquetting
f_T	correction factor for velocity to account for turbulence
f_c	correction factor for particle separation in a hydrocyclone, dimensionless
f_{ss}	correction factor for particle velocity to account for short circuiting
f	correction factor to convert batch

	coalescence data to continuous data, $= 1.3 \times 10^5$
f	Selker and Sleicher correlating parameter to predict the dispersed phase
f	correlating parameter defined as $A^+/h^+(0.85\text{-}A^+)$
f	friction or drag coefficient; dimensionless
f	friction factor for flow inside pipes, dimensionless
F	code for characterizing solids: corrosivity, non-corrosive. pH > 7
F	force on briquetting rolls
F	force on the particle, N
F	flowrate
$\hat{F}$	mass flowrate, kg/s
$\check{F}$	volumetric flowrate, L/s
$\tilde{F}$	molar flowrate, mol/s
F_L	liquid flowrate
F_G	gas flowrate
F_m	fictitious flow of solute through the membrane
Fr	dimensionless ratio of convective to gravitational forces, Froude number ($v^2/ D_p \, \mathbf{g}$)
Fo_M	Fourier number for mass transfer, dimensionless $= \mathscr{D}_{eff} \, t/ (D_p/2)^2$
$\mathbf{g}$	gravitational force of acceleration, 9.8 m/ s^2
G	code for characterizing solids: corrosivity, mildly corrosive. pH 5 to 7
G	Gibbs free energy
G	velocity gradient in flocculation, s^{-1}
Ga	dimensionless number, Galileo number ($\rho^2 D_{p2} \, \mathbf{g} / \mu^2$)
h	wall thickness of a vessel, cm
h_c	thickness allowance for corrosion, cm
h_p	thickness of wall needed from hoop stress equation
h	thickness of the film separating two approaching drops about to coalesce
h^+	dimensionless height = h/D
h_2	height (or depth) of the overflow liquid layer in decanter
h_3	height (or depth) of the underflow liquid layer
h	depth of bed, height of a bed, m
h	length of a pore
h	individual heat transfer coefficient, kW/m^2 K [Eq 3–11 and 3–12]

h_i	individual heat transfer coefficient relative to the inside of the tubes [Eq 3–12]	$\tilde{J}$	molar flux relative to other components, $kmol/m^2.s$
h_{id}	"dirty" individual heat transfer coefficients based on the inside fluid, kW/m^2 K		see N for fluxes relative to fixed coordinates
h_{od}	"dirty" individual heat transfer coefficients based on the outside fluid, kW/m^2 K	J	code for characterizing solids: handling characteristics, "fluidizes"
h	height of liquid in a bed	JTU	unit of measurement for turbidity, called the Jackson Turbidity Unit
h	vertical head, m [defined Eq 2–2 and 2–3]	k	correlating parameter for torque
h_c	height of porous bed where the pores are filled with liquid by capillary action, m	k	correlating parameter for gas-liquid separation
h_c	height of the continuous phase in a decanter	k	correlating parameter for gas-liquid separation in foams
h_d	height of the discontinuous phase in a decanter	k	correlating ratio of the water velocity to the blade velocity
h_L	height of a liquid in a drum	k	correlating parameter in the Freundlich isotherm
h_G	height of the gas phase in a drum	k	correlating exponent on the density ratio
h_G	height of a transfer unit in the gas phase	k	correlating parameter for coalescence
h_o	mass ratio of pure solvent to inert solid in the feed to a leaching operation, dimensionless	k_T	correction factor if the operating temperature is above the critical
ΔH_{react}	heat of reaction	k	mass % of the feed that is misplaced into the concentrate
h/D	height to diameter ratio; also defined as "r" for vertical vessels	k	thermal conductivity, mW/m.K
H	code for characterizing solids: corrosivity, very corrosive. pH 1.5 to 5	k	mass transfer coefficient
H	total enthalpy	k^+	dimensionless mass transfer coefficient = k/B
H	humidity	k_G	gas phase mass transfer coefft
$\mathbb{H}$	Henry's law constant. kPa/mol ratio	k_s	mass transfer coefficient through the gel layer
H	applied magnetic field, T	k_L	liquid phase mass transfer coefft
HETP	height equivalent theoretical plate	k_x	local mass transfer coefft [Eq 3–22]
HTU	height of a transfer unit	k_T	thermal diffusivity
HTU	heat transfer unit UA/$\check{F}$ρ $\hat{c}_p$ [defined Eq 3–7]	k_p	pore transport controlling factor
HGMS	acronym for high gradient magnetic separator	k_s	coefficient for collision energy to cause coalescence
i	correlating exponent representing migration behavior of drops in coalescing bands	k	rate constant, min^{-1}
IX	ion exchange	K_o	mass transfer coefft
j	correlating exponent for coalescence	K_{1-2}	mass transfer coefficient to transfer solute B from the feed stream bulk condition to the surface of the membrane
j_H	Chilton Colburn dimensionless j-factor for heat transfer, = (h $Pr^{2/3}$ / ρ $\hat{c}_p$ v) [defined Eq 3–21]	K	distribution coefficient for species between two phases or selectivity
j_D	Chilton Colburn dimensionless j-factor for mass transfer, = (k_x $Sc^{2/3}$/ $\tilde{c}_p$ v) [defined Eq 3–22]	K	solubility in the phase
J	flux	K_D	solute distribution coefficient
$\hat{J}$	mass flux of solids relative to other components, $kg/m^2. s$	K_s	solute distribution coefficient between inside the membrane and the local concentration in solution on the high pressure side of the membrane
$\check{J}$	volumetric flux of liquids relative to other components, $L/s.m^2$		

K	permeability of a bed of solid particles, μm^2
K	dialysis mass transfer coefficient
K	code for characterizing solids: handling characteristics, "cakes, builds up and hardens"
script l	length of the briquette in the direction of rotation
L/D	length to diameter ratio: sometimes given symbol "r"
L	flowrate of reflux to column
L	length of a blade on a reel for flocculation
L	length of the centrifuge bowl
L	height or depth of a porous bed, m
L	length of the magnetic matrix bed in the gap, m
L	length of the filter in the direction of flow
L	length of pipe, m [Eq 2–3]
L	code for characterizing solids: handling characteristics, "light and fluffy"
Le	dimensionless number, Lewis number, = Sc/Pr
LGMS	acronym for low gradient magnetic separator
LMTD	log mean temperature difference [Eq 3–9; see also ΔT_{im} Eq 3–22]
m	slope of the equilibrium line, y = mx, dimensionless
$\mathbb{M}$	molar mass, kg/mol
M	magnetization, T
M_o	reference magnetization for ferro- and ferri-magnetic materials, T
M	code for characterizing solids: handling characteristics, "dusty"
MTD	mean temperature difference, °C [Eq 3–10]
n	number of blades for a flocculator reel, dimensionless
n	number of spaces between the discs in a disc bowl centrifuge, dimensionless
n	counter for the "number of stages"
n	number of theoretical stages
n_t	total number of theoretical stages; see also NTS
n_m	minimum number of theoretical stages
n	degree of wetting, dimensionless
n	exponent in the Freundlich isotherm, dimensionless
	exponent for the variation in viscosity with temperature

	exponent for the variation in capital cost with capacity
N	rotational speed, rpm
N_s	rotational speed of a shaft, rpm
N_c	critical rotational speed of a trommel
$\hat{N}$	mass flux of species relative to fixed coordinates, $kg/m^2.s$
$\tilde{N}$	molar solute flux relative to fixed coordinates,
Nm^3	unit of measurement, "standard cubic metres"
Nu	dimensionless number, Nusselt number, hD/k
Nu_{AB}	dimensionless number, Nusselt number for mass transfer: convective transfer/diffusive transfer
NTU	number of transfer units
NTU_G	Number of transfer units based on the gas phase
NTU_L	number of transfer units based on the liquid phase
NTS	number of theoretical stages; see also n_t
N	code for characterizing solids: handling characteristics, "sticky"
O	code for characterizing solids: handling characteristics, "hygroscopic"
O	liquid in the overflow stream per ton of solids in the CCD operation
p	total pressure, kPa
p	compression pressure, kPa
p	internal pressure, kPa
p	pressure on a briquetter, kPa or MPa
p_B	partial pressure of component B, kPa
P_i^o	ideal single component partial pressure, kPa
p^*	partial pressure of a gas as defined by Henry's law, kPa
p	total column pressure, kPa
p^+	dimensionless pressure = π/p
$\mathbf{P_B}$	membrane permeability coefficient, $Nm^3.\mu m/s.m^3.MPa$
Pe	dimensionless number; Peclet number for flow: convective flow/ diffusion
PDI	potential (or surface charge) determining ions.
P	power, kW
P	power input, kW
P	code for characterizing solids: handling characteristics, "becomes plastic"

q	heat flow per unit time
q_s	sensible heat flow per unit time
q_c	latent heat flow per unit time
q_r	heat flow because of reaction per unit time
Q	code for characterizing solids: handling characteristics, "packs under pressure"
r	reduction ratio for crushing
r	vessel length to diameter ratio: L/D
r_i	resistances to heat transfer for inside conditions because of fouling
r_m	resistance to heat transfer because of the solid material
r	$m\ F_G/\ F_L$ for gas/liquid $F_L/\ f.\ F_S$ for leaching
r	outer radius
r_i	radius of a rotating shaft
Re	dimensionless number: Reynolds number: convection/viscous effects
R_c	radius of the cage of a reel
R_b	radius of the centrifuge bowl
R_l	radius of the inner part of centrifuge bowl
$\mathcal{R}$ R(script)	fraction of dissolved materials recovered in the washing stream and present in the pregnant liquor
$\mathcal{R}$ R(script)	mass fraction of feed solute that is recovered in the exit extract or leachate solution for leaching
$\mathcal{R}$ R(script)	mass recovery = mass of component recovered/feed mass
$\mathcal{R}$ R(script)	mass recovery of magnetics
R	recycle ratio: retentate recycled/retentate as exit product, dimensionless
R	molar reflux ratio, dimensionless
R_m	minimum reflux ratio, dimensionless
R	rejection by a membrane (conc solute in feed—concn solute in permeate)/ (concn of solute in feed), dimensionless
$\mathbb{R}$	ideal gas law constant
RO	acronym for reverse osmosis
R	code for characterizing solids: handling characteristics, "interlocks, mats, or agglomerates"
S	code for characterizing solids: safety-hazard, "generates static electricity"
S	volume fraction of voids filled with liquid, dimensionless
S_c	spreading coefficient
S_i	spreading coefficient

Sc	dimensionless number: viscous effects/diffusive effects; Schmidt number = $\mu/\rho\ D_B$
Sh	dimensionless number, Sherwood number. = Nusselt number for mass transfer Nu_{AB}. = $k_s\ D/\ D_B$
S	concentration ratio, dimensionless
S	design stress for a material
SX	acronym for solvent extraction
t	time, usually s
t^+	dimensionless time = $t/\ t_c$
t^+	dimensionless time = $4t/\ 10^3\ \pi\ D^2$
t^*	dimensionless time = $4t/10^3\ \pi\ r(1-A^+)$
t_c	time to load the bed (in ion exchange or adsorption)
t_c	residence time of the continuous phase in a decanter
t_b	coalescence time or break time of the dispersion band
T	temperature
ΔT	temperature difference
T	code for characterizing solids: safety-hazard, "toxic fumes given off"
U_v	volumetric heat transfer coefficient
U	overall heat transfer coefficient
UF	acronym for ultrafiltration
U	code for characterizing solids: safety-hazard, "flammability"
v_o	superficial velocity, usually $L/\ m^2.s$
v	velocity, usually m/s
v	drift velocity, usually cm/s
v_d	velocity of the dispersed phase, usually m/s
v_m	minimum fluidization velocity, usually m/s
v_f	design fluidization velocity, usually m/s
v_t	terminal settling velocity of a particle, usually m/s
v_F	flooding velocity, usually m/s
v_H	hindered settling velocity of groups of particles through fluids, usually m/s
v_{Hl}	settling velocity of the limiting diameter of particle in a classifier, usually m/s
v_{oo}	settling velocity of a single particle in an infinite fluid, usually m/s
v_x	velocity in the "x"-direction, usually m/s
V	total volume, m^3
V^+	reduced volume = V/V_{mm}

V^+	reduced, corrected volume of gas = V, saturated gas/ V, inlet gas conditions	Y	code for characterizing solids: product considerations, "prevent decomposition"
V	code for characterizing solids: safety-hazard, "explosive dust"	z	vertical height/distance measure
V	liquid in the underflow stream per ton of dry solids in a CCD	z	valence
		z	ratio of valence
w	width of a briquetter roll	z	diffusion path length
w	width of a blade in a reel	zpc	conditions when the surface charge is zero, similar to "iep" the isoelectric point
w	width of the vessel	α	code for characterizing solids: particle size, size less than 250 Mesh, "silt"
W	mass of solute transferred, g/s	α	separation factor
W	mass or amount, kg. or mols depending on the superscript	α	parameter
W	volumetric wash rate per ton of dry solids in a CCD operation	α	correlating parameter for drum-area
W_i	work index, kWh. $\mu m^{0.5}$/Mg	β	correction factor
W	work per unit solids = Power/solids throughput, kWh/Mg	β	exponent
		β	correlating parameter for drum-area
W	code for characterizing solids: product considerations, "prevent contamination"	δ	code for characterizing solids: particle size, size greater than 5 cm and less than 10 cm, "small cobble"
x	mass or mol fraction depending on the superscript, usually refers to the underflow or liquid phase in separation systems	δ	boundary layer thickness
$\hat{x}$	mass fraction	δ	nominal membrane thickness
$\tilde{x}$	mol fraction	δ	Hildebrand solubility parameters, $(J/cm^3)^{1/2}$
$\check{x}$	volume fraction	δ_d	Hildebrand solubility parameter accounting for the dispersion forces, $(J/cm^3)^{1/2}$
$\hat{x}'$	mass of solute per mass of inert solids feed	δ_p	Hildebrand solubility parameter accounting for the polar forces, $(J/cm^3)^{1/2}$
$\hat{x}^+$	dimensionless mass fraction relative to the feed mass fraction = x/x_1	δ_h	Hildebrand solubility parameter accounting for the hydrogen bond forces, $(J/cm^3)^{1/2}$
$\hat{x}^+_2$	dimensionless mass fraction in the overflow relative to the feed mass fraction = x_2/x_1 = upgrading ratio or upgrading factor	Δ	code for characterizing solids: particle size, size greater than 10 cm and less than 15 cm, "medium cobble"
$\hat{x}^+\bullet$	dimensionless mass fraction of solute in the permeate at the exit of the membrane to the value in the permeate at the inlet of the membrane = $x^+_{B3\ exit}/x^+_{B3\ inlet}$	Δ^+	separation ratio for batch distillation = F_3/F_1 = W_3/W_1, dimensionless
x_g	gel concentration for ultrafiltration	Δ	separation for membranes = F_2/F_1 = W_2/W_1 or mass retentate recovery, or the reciprocal of the "ratio of concentration," dimensionless
$\check{x}_d$	feed volume fraction of the dispersed phase (see also ϵ and ϕ)		
$\tilde{x}_{LK}$	mol fraction of the light key component	Δ	difference between two conditions
$\tilde{x}_{HK}$	mol fraction of the heavy key component	ϵ	volume fraction vapor
X	code for characterizing solids: product considerations, "prevent degradation"	ϵ	interstitial volume fraction liquid in liquid/solid mixture; void volume, voidage, dimensionless
y	mass, mol or vol fraction depending on the superscript; usually refers to the overflow or gas phase in a separation system	ϵ	heat transfer effectiveness [Eq 3–29]
		ϵ_r	electrical conductivity, relative permittivity
y	distance or linear dimension	γ	surface tension, mN/m
y	correlating exponent for coalescence = (j/i) +1	γ_c	critical surface tension of a solid at which the wetting properties of the solid change, mN/m

γ	correction factor for solute
γ^{o}	shear stress at the wall
Γ	surface concentration, kg/kg solid or resin
Γ^{+}	dimensionless surface concentration
Γ_{T}	total ion exchange capacity of an ion exchange bed
Γ_{B}^{+*}	equilibrium exchange concentration
λ	latent heat of evaporation
λ	correlating parameter for collision frequency for coalescence
η	tray efficiency, number of ideal trays/number of actual trays, dimensionless
η	efficiency
η	efficiency of separation
η	particle collection efficiency, number (mass) of particles collected/total number (mass) of particles present, dimensionless (care is needed to identify whether this is the "number" or "mass")
η	particle collection efficiency
η^{*}	packing efficiency for drops at the interface, usually $= \pi/4$
ω	angular velocity
ϕ	volume fraction solids
ϕ	mobility of the surfaces during coalescence
Ψ	induction energy for solvent
Ψ_{s}	sphericity of the particles
Ψ_{p}	accounting for pore diffusion, dimensionless
π	osmotic pressure
Θ	osmotic pressure coefficient MPa. mol/mol
θ	contact angle between a gas/liquid/solid system
θ^{+}	dimensionless solute to solvent flux ratio
θ	angle that discs make with the axis in a disc-bowl centrifuge

ρ	density
ρ_{G}	vapor density
ρ_{sg}	mass density of soluble gas
ρ_{L}	mass density of a liquid or of the light liquid phase
ρ_{H}	mass density of the heavy phase
ρ_{c}	mass density of the continuous phase
ρ_{d}	mass density of the dispersed phase
ρ_{50}	mass density at which 50% of the mass goes overhead and 50% of the mass goes as underflow
ρ_{s}	mass density of a solid, Mg/m^{3}
ρ_{bs}	bulk mass density of a mass of solid particles, $Mg/m^{3} = \varepsilon\,\rho_{G} + (1-\varepsilon)\,\rho_{s}$ approximately $= (1-\varepsilon)\,\rho_{s}$ or approximately $0.5\,\rho_{s}$
μ	viscosity
μ_{L}	viscosity of the light phase
μ_{H}	viscosity of the heavy phase
μ_{c}	viscosity of the continuous phase
μ_{d}	viscosity of the dispersed phase
μ^{+}	electrical mobility
μ^{+}	dimensionless flux ratios
μ_{s}	surface shear coefficient of viscosity
υ	kinematic viscosity $= \mu/\rho$
σ	Staverman coefficient, dimensionless
σ_{G}	geometric standard deviation
Σ	capacity factor or equivalent horizontal cross-sectional area for settling, m^{2}
τ	coalescence time, s
κ	rationalized magnetic susceptibility, dimensionless (some are based on the volume, m^{3}, and some on the mass, m^{3}/kg)

Chapter 1

Engineering and Selecting Process Equipment

Engineers solve problems to give us a better world in which to live. Whatever means they use to solve those problems, some processing equipment is usually needed.

Energy cannot be converted from one form into another without some piece of equipment. Contamination cannot be removed from drinking water without some piece of equipment. Actually, it would be better to replace the words *piece of equipment* with *system*, of which equipment is an integral part. How does an engineer select equipment as he or she devises systems, solves problems, improves existing processes, and troubleshoots faults?

We will consider first the "system" contexts in which engineers select process equipment, then the general "thinking" principles engineers use as they perform these tasks. Then we will highlight the role the information in this book plays and how to use the book.

1.1 THE EXCITING ENGINEERING WORLD

The world of engineering is multifaceted. Engineers synthesize new systems to convert materials and energy into useful products. They analyze situations to uncover methods for improvement and discover new knowledge so as to perform their function better. Engineers are also called upon to troubleshoot difficulties in a plant.

1.1-1 The Creator and Synthesizer

Engineers develop new processes to make pharmaceuticals, foods, paper, cosmetics, perfumes, materials of construction, transistors, and fibers. Some of these products are used directly by the consumer and some are intermediates. To see what is involved, consider the task of making 180,000 Mg/a of ethylene, one of the most important organic chemical intermediates. We have ethane available as a raw material. Figure 1–1 illustrates some of the dimensions of the engineering creative process. First the engineer considers the various "tasks" required to ensure that the product is produced *and* that the overall system is safe, is cost and energy efficient, is easy to operate and control, and has flexibility to exploit new feedstocks and new technical developments.

Just consider the first task of "react." Which equipment should be selected to do this task? The engineer probably has about a dozen equipment options for this task alone. Thus, we see a need to systematically consider a sequence of tasks, to decide among numerous equipment options for each task, and to maintain an overall concern that the "system" satisfy a wide range of constraints and criteria. Quite a challenge!

Also shown in Figure 1–1 is one team of engineers' creation, called a flowsheet, which shows the equipment

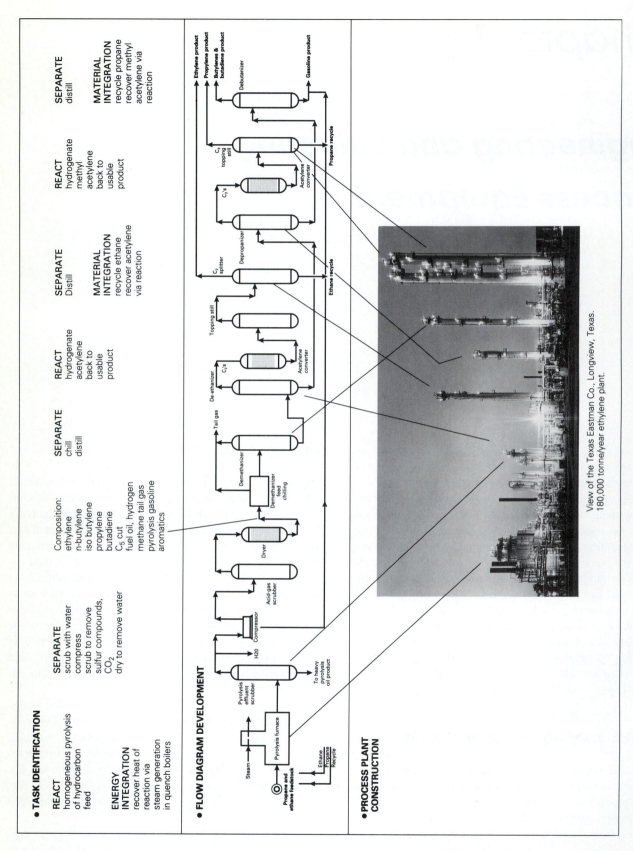

Figure 1-1 Creating a Process To Produce Ethylene from Ethane (Courtesy of The M. W. Kellogg Co., a subsidiary of Dresser Industries, Inc.)

they selected, the sequencing and interconnects, and the overall system. Since engineering results in actual construction and operation of a process, a photograph of the M. W. Kellogg ethylene plant is shown.

1.1-2 The Analyst and Improver

Once process plants are constructed, the equipment physically lasts about twenty years; the process technology lasts about five years. Thus, engineers are continually on the search for ways and means of improving and upgrading existing processes. Figure 1–2 is an engineer's "think board"—the Process Instrument Diagram. Often this will be computerized, but here we show a hard copy of part of an ethylene plant. This diagram shows equipment sizing, interconnects, and instrumentation-control.

The engineer—the analyst—will scrutinize such a diagram to understand the secrets of the process as it was designed and constructed a number of years ago. Where should he or she start to probe? How does the probing occur? How does one recognize the areas for improvement? Armed with an economic analysis of how and where the process is making money, an engineer does order-of-magnitude estimates of how the different types of equipment are functioning. He or she asks a whole series of "What if?" questions. What if the distillation column is replaced with a bank of membranes? What if the feed stream is split and goes to six locations along the reactor? Again the challenge is to be very familiar with options, to be astute in the choice of what if? questions to ask, and to be able to do many order-of-magnitude calculations. We want to be able to poke the system and see how it responds.

1.1-3 The Troubleshooter and Sleuth

Here is a scenario where the engineer becomes a troubleshooter.

"Trouble on the ethylene plant!"

At low flowrates of 70 Mg/d of C_2H_4, we encounter no difficulties with production. Recently, we have encountered excessive pressure drops across the third column when we worked as usual, but high rates of 150 Mg/d. At these rates, the pressure drop was so large that we could not operate satisfactorily. The lost capacity is worth $20,000/day. We cannot shut this plant down because the rest of the plant uses C_2H_4 as a raw material and the current inventories are very low. Indeed, C_2H_4 is the feedstock for the rest of the plant site. Usual operating conditions are, for the columns in Figure 1–3:

column 3	−40°C,	3.2 MPa
columns 1 and 2	0 to −10°C,	3.2 MPa

Get our inventories up so that the rest of this plant site can function properly!

This case problem was supplied courtesy of John Gates, B.Eng., McMaster University (1968).

1.1-4 The Researcher and Explorer

For the ethylene plant, we want to improve it by recovering some of the heat from the pyrolysis effluent scrubber or quencher. The function of this piece of equipment in the system is to quench the pyrolysis reaction that occurs at about 850°C. We want to stop it cold! The current design sprays liquid in and we depend on the latent heat of vaporization to sop up the heat. But the result is low-grade heat in the quench liquid that is difficult to recover. Can't we use this quench operation to generate steam or some other form of "high-grade" energy? Let's explore the type of steam boiler that we would have to design to change the temperature from 850°C to about 400°C in milliseconds— one that has a very small pressure drop and does not plug from coke and tar formation. Engineers would do some order-of-magnitude calculations to screen out the poor ideas, ask some what if? questions, and do theoretical and fundamental research to create a new concept, a new piece of equipment. Again, the challenge is to identify the most profitable areas to explore!

1.1-5 The Specialist and Consultant

Engineers are generalists and yet each becomes a specialist. Some might specialize in pumps, in packings and seals, in pressure-relief valves, in estimating physical properties, in marketing. As specialists, some are technical sales representatives or licensers for a particular type of equipment, technology or know-how. Their tasks are to consider a client's unique position, to consider numerous options, to see how the particular specialty fits into those options from the client's point of view. Order-of-magnitude estimates of options are a daily occurrence.

1.1-6 The Predictor and Crystal-Ball Gazer

Engineers also have to anticipate the future. How would we operate this plant if the cost of energy were ten times what it is now? if the environmental constraints were three times more stringent than they are now? What happens if we wish to change the feed material from ethane to naphtha? or to soybean oil? Should we be continuing to develop this type of technology or should we be considering microbiological production of ethylene? To fulfill this role, the engineer needs to be able to do order-of-magnitude estimates, to be able to make numerous clever assumptions so that many different future scenarios can be explored. True, computer simulations will play a dominant role. But the programming of the simulations will be based on the approximations and the procedures that are outlined in this book.

1.1-7 The Selector of Process Equipment

Throughout all of these above tasks, engineers will be continually selecting and evaluating process equipment—equip-

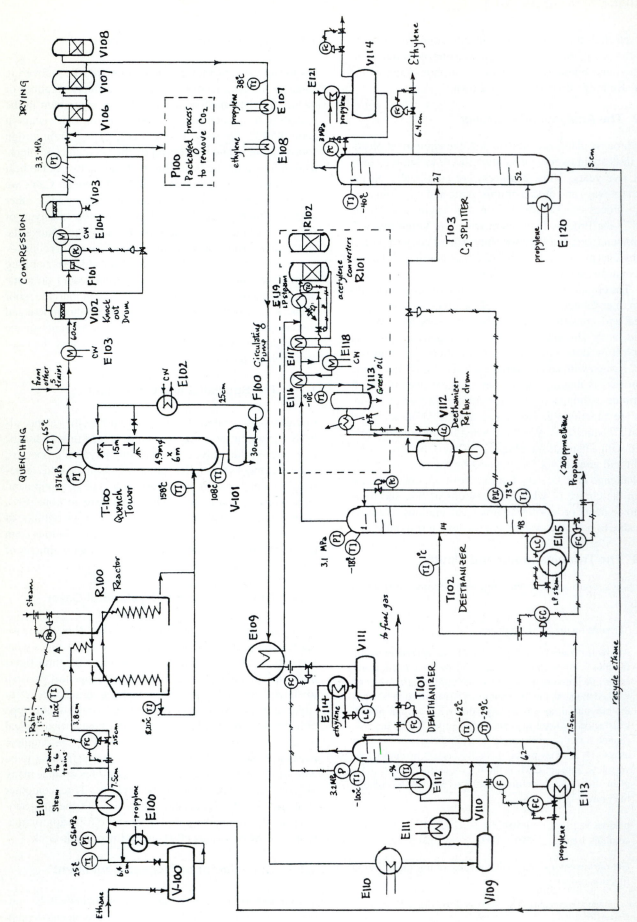

Figure 1–2 A Process and Instrumentation Diagram from an Ethylene Plant

1-4

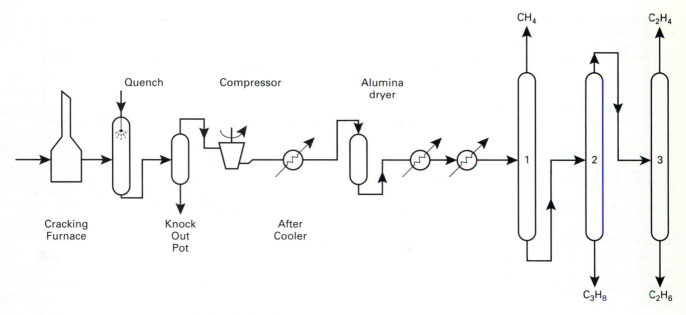

Figure 1–3 A Section of an Ethylene Plant Where Trouble Developed

ment that fits into a system, but individual pieces of equipment nevertheless. Even this is a challenge. The equipment selected must technically solve the problem, be financially feasible and economically attractive, be environmentally acceptable, be reliable, be safe to operate, be available for purchase, be serviceable, be operable and controllable, and be from a reputable supplier. Consider each in turn.

A. Technical Feasibility.
The piece of equipment must do the job. It makes no sense to purchase a "Superseparator" that will not separate the desired components. The name of the equipment or prestige of the supplier cannot do the technical impossibility.

B. The Money.
There are two vital money matters to consider: the capital cost and the operating cost. The capital cost tells you how much you have to pay to buy the equipment in the first place. This tells us how much money we need. We may have a fantastically good piece of equipment, but if it costs $30 million and we can only raise $10 million, this equipment is not financially feasible.

However, *if* the money is available, we select the equipment with the smallest annual operating cost or that gives us the most attractive return on our investment. For example, if a Gasflex costs $16,000 with an operating cost of $3,000/year and a Flexgas costs $13,000 with an operating cost of $2,800/year, then naturally we would purchase the Flexgas. The more usual situation is, however, that the equipment whose initial purchase price is higher has a smaller annual operating cost. For this situation, more detailed calculations need to be done as described, for example, in Chapter 5 of Woods (1975). The main point, however, is that we do not select equipment on its capital cost or its purchase cost. A vital consideration is annual operating cost.

Included in the financial analysis of the *annual operating cost* will be such factors as the energy usage, the quality and amount of operating labor required, the resale value of the equipment, the amount and expertise (of labor and of type of tool needs) required for the maintenance, the implications of the installation of one particular type of equipment on the upstream and downstream processing equipment, the weight of the equipment, and the space required in the equipment layout.

As an aside, one might consider buying used equipment as opposed to brand-new equipment. An example of the FOB price ratios for different types of equipment is given in Table 1–1. However, the FOB price when one buys equipment does not include the cost of installing a working module of equipment. A brand-new pump costing $1,000 might require an additional $3,000 to convert it into a working module on a new installation. A second-hand pump may cost $300 and also cost $3,000 to convert into a working module on a new installation. Unlike buying a new versus a used car, the initial cost of equipment must include whatever costs are required to put in all the necessary concrete, pipework, structural steel, etc., to convert the purchased equipment into a working reality. This is discussed in detail in Chapter 1 of *Cost Estimation for the Process Industries* (Woods, 1986). Used equipment may be purchased "as is" on location, rebuilt and guaranteed, or on approval. Detailed suggestions are given by Epstein (1965).

TABLE 1-1. Comparing delivered cost of old versus new equipment (From Epstein, 1965, and used with permission of the American Institute of Chemical Engineers © 1965 AIChE)

Rotary kiln	0.22
Mill	0.44
Mixer, double arm	0.12
Complete packaged hydrogen plant	0.20

C. The Environment. The equipment must fit into the environment safely. This means that there is no noise pollution, mechanical hazard, fire hazard, etc. For example, for most process plants where much of the construction is outside, Totally Enclosed Fan Cooled, or TEFC motors are often used because the motor is well protected from the weather. Large-size compressors or blowers are usually steam turbine-driven instead of motor-driven because the former are quieter.

D. Reliability. Downtime or time when the whole plant is sitting idle because of the malfunction of one piece of equipment is costly. Plants operate twenty-five days of the month to clear expenses and the last five days to make a profit. If the plant is down for six days in a month, the company is operating at a loss; if it is down three days of the month, the investors in the company should have invested their money in savings bonds. In selecting equipment, we should know the reliability quantitatively. Some illustrative reliability numbers are given in Table 1–2.

E. Availability. The equipment must be available for the conditions specified. This has five aspects. First, is the equipment applicable for the combination of conditions specified? We must know the general region of applications for each piece of equipment. For example, we may want to separate a mixture of 0.1% solids of 10 μm diameter using a solids-retaining filtering centrifuge. However, we will discover that the usual range of application is about 2 to 8% solids for this type of centrifuge. We may calculate that we need a 10,000 kW electric motor to drive our grinders. Second, do they make electric motors that big? Hence, we select equipment based on the availability of the size we want.

Third, are the equipment specifications standard or the usual off-the-shelf models or must we build a special unit just for our needs? For example, shell-and-tube heat exchangers usually have 4.9 m bundles of tubes; diffuser pipes for aeration basins are usually 4.9 m long. This requires that we learn something about the usual sizes of manufactured equipment.

Fourth, is the equipment available here? Because of licensing regulations and import restrictions, a desired type of equipment may not be available.

Fifth, will the equipment be available in time? Perhaps the time when a given piece of equipment is available and the time when it is needed differ so much that for practical purposes, that type of equipment is unavailable. Some indications of possible delivery delays are given in Table 1–3.

F. Serviceability. When machines break down, we want to repair them quickly and easily, so we usually keep spare parts on hand. To minimize the amount of money tied up in spare parts, we try to buy parts that are interchangeable. Companies usually standardize on one brand line. Naturally, in selecting that brand, we expect that the supplier will have the full range of spare parts available, together with whatever instructions may be needed to make the repairs.

We also expect that the equipment is designed so that it is easy to take apart and repair. If the seal breaks on a pump, you do not want to have to completely disassemble the pump to make the repair.

G. Operability. As engineers, we want the plants we design to be easy to operate. If a particular piece of equipment is awkward to start up, control, and operate, the morale of the operators will be low. For example, the valves and switches should be in convenient locations.

H. Reputable Supplier. The equipment supplier must stand behind his or her product. If he or she does not, then you want to discover that *before* you buy. Such information as to who is a reliable supplier is not easy to obtain. However, we should try to get some idea of the supplier's performance through such questions as:

How long have you been in business?

How long have you handled this product line?

What specifically do you do if the equipment does not meet our agreed-upon specifications?

What service manuals do you provide?

What sort of maintenance record does this equipment have?

What spare parts do you recommend that we buy?

Who has bought your product recently? Do you mind if I contact them?

TABLE 1-2. Reliability of equipment (Reprinted from "Reliability Revisited" by C. L. Cornett and J. L. Jones, Chemical Engineering Progress, Vol. 66, No. 12, pp. 29-33 [1970 reproduced by permission of the American Institute of Chemical Engineers. © 1970 AIChE. All rights reserved]; Anykora, 1971, "The Chemical Engineer"; and Pekrul, 1975, McGraw Hill, Inc.)

Type	Failures per annum	Mean time to repair, h/a
turbine	0.876	2000
pumps, condenser	0.569	54
pumps, cooling tower	0.569	54
pumps, boiler feed	0.569	74
pumps, recirculating	0.569	256
utilities	0.37	
furnaces	0.23	
piping	0.18	
towers and reactors	0.15	
exchangers	0.11	
reactor core internals	0.254	2000
nuclear core rod assembly	0.175	50
valves, control	0.053	32
valves, steam	0.035	32
valves, control	0.57	
valve, solenoid	0.30	
valve positioner	0.41	
current/pressure transducer	0.54	
pressure measurement	0.97	
flow measurement (fluid)	1.09	
differential pressure transducer	1.86	
transmitting variable area meter	1.01	
indicating variable area meter	0.34	
level measurements (liquids)	1.55	
differential pressure transducer	1.77	
float-type level transducer	1.64	
capacitance-type level transducer	0.22	
temperature measurement	0.29	
thermocouple	0.4	
resistance thermometer	0.32	
mercury in steel thermometer	0.027	
temperature transducer	0.85	
controller	0.26	
pressure switch	0.3	
flame failure detector	1.37	
analyzer	6.17	
pH meter	4.27	
gas-liquid chromatograph	20.9	
O_2 analyzer	7.	
CO_2 analyzer	10.5	
IR liquid analyzer	1.4	
impulse lines	0.91	
purge systems	1.00	

In summary, we try to discover a quality product (reliable, designed for easy maintenance) backed by a reliable supplier (who knows and stands behind the product).

1.1-8 Challenges and Opportunities

Whichever of the numerous roles engineers play from one day to the next, each is characterized by a need to consider zillions of options, to astutely select the most productive avenues to spend time on, and to continually monitor progress to ensure that what is being done is "reasonable."

Consider a few scenarios.

Sam looked at the design calculations of the young engineer and said, "I think you've made a mistake; that answer doesn't look right." How did Sam know? Experience!

"Perhaps ion exchange is something we should consider here," suggested Michelle, when there was a lull in the de-

TABLE 1-3. Delivery Times for Different Types of Equipment (Reprinted with permission from Chemical Engineering, **84**, 13, 78, 1977)

Types of equipment	Lag in months between placement and shipment of order, U.S. conditions								
	1977	1976	1975	1974	1973	1972	1971	1970	1969
Compressors	7.1	7.5	4.1	8.3	4.8	4.7	4.6	4.1	4.2
Conveyors	2.8	7.6	7.8	7.2	3.7	4.6	4.4	4.2	NA
Crushers	5.5	5.4	6.9	5.5	6.6	3.9	3.9	4.3	4.2
Engines	6.7	9.8	4.8	5.6	5.9	6.3	7.6	7.7	8.0
Grinding machinery	8.7	13.4	18.5	7.5	8.5	8.1	6.5	NA	NA
Heating and cooling equipment	5.1	4.9	6.7	9.0	3.9	3.8	3.3	3.4	5.0
Mineral classifying, separating, cleaning	5.5	4.2	9.0	5.8	3.7	NA	NA	NA	NA
Mixers	6.1	4.6	4.8	4.2	3.6	3.1	2.8	4.2	3.5
Motors, 7 kW	2.0	2.0	3.4	6.2	2.2	2.0	0.5	2.1	1.9
Motors, 350 kW	4.9	2.4	5.8	7.4	3.1	2.5	1.3	4.7	2.8
Motor controls	3.6	1.4	2.8	4.2	1.8	1.9	2.0	3.0	1.3
Process instruments and controls	6.9	3.5	4.5	2.8	1.4	2.4	1.2	NA	NA
Pumps	5.8	5.4	3.2	6.1	2.7	2.6	2.4	2.9	2.8
Screening equipment	2.2	0.5	3.6	6.0	1.8	NA	NA	NA	NA
Turbines	11.5	13.5	17.1	19.5	11.6	10.8	11.2	10.8	11.5
Valves	3.9	0.5	1.8	2.1	0.8	1.9	1.0	1.0	2.3
Water pollution control equipment	9.0	9.0	10.4	8.2	6.6	8.3	8.3	8.3	NA
General industrial machinery	10.0	9.9	7.8	9.7	4.4	3.9	2.58	3.0	4.6
Specialized industrial machinery	6.0	5.1	5.4	5.8	4.9	3.5	4.3	4.7	5.6

sign group meeting. How was Michelle able to come up with this idea? Experience!

This brings us to the focus of this book. Successful engineers have experience; but more than this, we can document and classify that experience that is vital to them. Successful engineers:

1. use the thinking principles of "optimum sloppiness" and "successive approximation"; they know when to spend two minutes on a calculation and when to spend six hours. They know when they must be sloppy!
2. have a broad understanding of the equipment and

processing options. They have a well-developed catalogue of equipment.
3. have an extensive mental bank of systematically organized experience factors and rules of thumb.
4. have a good intuitive feeling, based on fundamentals, of when a given piece of equipment could and should be used. They have "ballpark charts" that tell them when to use what.
5. have numerous short-cut design methods available so that they can rough-size equipment at a flick of the wrist.
6. always consider money as the bottom line. Thus, calculations and selections, troubleshooting and research, future considerations and process analyses are all done in the context of a "cost."

1.2 OPTIMUM SLOPPINESS AND SUCCESSIVE APPROXIMATION: OUR MAJOR THINKING TOOLS

Two of our most important engineering laws are:

- the Law of Optimum Sloppiness, which says that we are only as accurate as we need be; or we match the accuracy of our answer with the resources available.
- the Law of Successive Approximation, which says that we do simple, sloppy calculations first; then if it seems as though we are headed in the correct direction, we do more accurate calculations.

Consider an example to illustrate the law of optimum sloppiness.

Reg and Sue were given the task of determining, in five minutes, the number of tennis balls needed to fill a VW camper. The key constraint is *five minutes*. Reg said he needed a week before he could give an answer. He wanted to purchase a VW full of tennis balls and count the number of balls. He would have a very accurate answer, perhaps within five or six balls, depending on how you squeezed them in. But it would take a week and the money to arrange for the camper and purchase the balls. Sue, on the other hand, realized that she knew enough that she could approximate the situation and come up with an answer in five minutes. She knew the approximate size of tennis balls (about 6 cm diameter). She recalled the shape of tennis balls (they were spherical, and she recalled the formula for the volume of a sphere), the total volume occupied by spheres (she realized that when they "fit together" there was some air space between the balls and she estimated that the total space occupied was the volume of the balls plus air around it, which was probably about 40% air to 60% balls). Then she thought about a VW camper. She recalled that its shape was box-like. She estimated the volume and so in about three minutes Sue had an answer. The answer was sloppy! It was probably ± 50%.

In this scenario, Sue applied the law of optimum sloppiness. Reg could not come up with an answer in the five minutes. We note also that for Sue to apply the law, she had to recall a lot of "experience" knowledge about tennis balls and VW campers.

Here is another example.

Estimate how many tubes of 150 mL tubes of toothpaste a drugstore will sell per week if it is the only drugstore in an isolated community of 2,000 people?

In trying to solve this problem, we are trying to discover whether it is 10, 100, 1,000, or 10,000 tubes a week. We can tackle this estimation problem several ways. One approach would be to estimate that a tube of toothpaste lasts me about two months. Thus, in one week, one-eighth

of the population would need to purchase a new supply. Therefore a reasonable answer might be 250 tubes/week.

Another approach would be to estimate the amount of toothpaste used per brushing. A cylinder 5 mm diameter ×10 mm long or $\pi/4(0.5)^2 \times 1$ cm $\simeq 0.25$ mL. Hence, for a tube of 150 mL, the number of brushings would be 150 mL/0.25 mL $\propto$ 600. At three brushings per day, that would be 200 days or approximately six months. If six months is the time that the toothpaste would last, then the number per week would be 2,000 people purchase 200 tubes every 24 weeks, or 80 tubes.

Now, how do we resolve the two-month estimate in the previous estimate with the six-month estimate we get here?

We really do not. We are estimating that maybe 50 to 250 tubes per week would be sold. It is unlikely that it is 10 or 1,000.

Activity 1.1: You might want to estimate the number of meters of sutures needed per week for a hospital operating room that functions twelve hours per day six days per week. Or, the amount of ice cream needed for a picnic for 200 people on the July 4 weekend.

Think about how you were sloppy, what experience information you needed, and what you might do if a more accurate answer were needed.

In our engineering professional world, an example application of the law of optimum sloppiness is when you are given the mass balances for seven flowsheets and asked to rough-size and estimate the capital costs for all seven in an afternoon. To be able to complete this type of assignment, the engineer has to have at his or her fingertips a host of experience knowledge. Actually, for all engineering functions, we need to recognize and apply the law of optimum sloppiness and to have the set of experience knowledge that allows us to do this.

Where does successive approximation come in? Because engineers must consider so many options and factors, we should always start any task with a very sloppy calculation "just to see that we are going in the right direction." Then, if we are, we use a more accurate approach, invest resources, and do the next level of calculations. So it proceeds, successively investing more resources to obtain a more accurate answer.

1.3 TYPES OF EXPERIENCE KNOWLEDGE WE NEED

The experience knowledge required by engineers consists of eleven different types of information and procedures; indeed, some of the experience knowledge consists of "how to do sloppy calculations." This book is an integrated consideration of experience knowledge as it is applied to *selecting* the type of equipment that might be appropriate and in *rough-sizing,* or approximately sizing, the equipment. Before we plunge into the text itself, let us draw back and consider the types of experience knowledge that are important. It is useful to classify the different sources of experience information separate in our minds because different types of information change with different conditions.

1.3-1 Generalized Properties

Physical, health, safety, thermal, and corrosive properties of materials are vital information to all process equipment that we select and size. Indeed, many equipment decisions have gone amiss because of an inadequate use of the property information. **Data** Parts C and D present a collection of approximations to the properties for a variety of compounds. Despite these precautions about the importance of material properties, they also form a key set of experience knowledge upon which to build. Although every compound is unique, we often can generalize the different properties for classes of compounds. Before we start this task, we must choose a unit system in which to work. In this text, the SI system will be used.

In general, the compounds of interest are water, air, organics, metals, and inorganics. We need to keep separate the properties for solids, liquids, and gases. Thus, we extract that reasonable values might be as follows:

heat capacity	liquid water	4.2 kJ/kg °C
	gas water	2 kJ/kg °C
	air	1 kJ/kg °C

and so on for bulk and transport properties.

It also helps to identify important ratios: latent heat of steam/organic is 5:1. Also included in this should be the heats of reaction.

This information is sensitive to temperature, pressure, and composition. Hence, we need to memorize whether these memorized values remain about the same, increase, or decrease, when these state variables change.

1.3-2 Generalized Phenomena

In section 1.3-1, the physical, thermal, and chemical properties of substances (and reactions) were generalized. However, other phenomena important to engineers can be generalized as well. These do not change with developments nor do they depend much on the device. They include:

friction factors inside pipes for turbulent flow: approximately 0.005.

crushing strength of limestone: Work Index 12.7.

the temperature difference for a boiler to ensure that it is boiling in the right regime.

These values are independent of temperature, pressure, and composition but depend on conditions that should be memorized along with the information.

1.3-3 Generalized Operating Conditions

This class of experience information depends on the device, but we can generalize it. The class includes:

heat transfer coefficients for different services: liquid-liquid heat transfer 1 kW/m² °C.

efficiencies of contacting devices: tray efficiencies 60%.

efficiencies of drives and motors: pumps 40 to 60%; compressors 60 to 80%.

This is actually a very large set of information because it should include typical input and output conditions, general efficiencies. These are relatively independent of temperature, pressure, and composition, but depend on the configuration of the equipment. Often this will be memorized for "current" equipment. These numbers evolve as developments occur. They represent usual practice.

1.3-4 Information Derived from Fundamentals

From our knowledge of fundamentals, and based on the data from the three previous sections, we can calculate more experience information.

For example, based on turbulent flow inside pipes and a friction factor of 0.005, we can use the definition of friction factor and pressure drop to estimate that one "velocity head loss" occurs for every 50 to 65 length-to-diameter ratios. Thus, for a 5 cm diameter pipe there is a velocity head loss every 250 to 325 cm length of pipe.

Similarly, we can estimate that we need 15 L of cooling water at the top of a distillation column for every kilogram of steam used to heat the reboiler.

Alternatively, we can focus primarily on the fundamentals. For example, what effect on the overall heat transfer coefficient would it have if we blocked off 10% of the tubes, or if we increased the flowrate inside the tubes or if we changed the tube diameter? Here, we might have at our fingertips the Nusselt equation:

$$\frac{hD}{k} = \left(\frac{\rho v D}{\mu}\right)^{0.8} = \left(\frac{4FD}{\pi D^2 \mu}\right)^{0.8} \qquad (1\text{-}1)$$

From this we can see the various relationships. Of course, we need to remember that this is for one of the heat transfer coefficients and not for the overall coefficient.

This extension of ideas builds on previous approximations and on fundamentals, which are sensitive to the same variables as are used to create them. One can see as generalization builds upon generalization how important it is to try to keep straight in one's mind the limitations of all the components used.

1.3-5 How Big Is Big?

Just as we memorized magnitudes for properties of materials, so we need to memorize and visualize the size of

a 5 μm particle relative to the diameter of hair, for example.
a 20 kW motor.
a 50 Mg/h solids processing plant.

These all have practical implications. For example, a manhole has a minimum diameter of 45 cm; otherwise a person cannot get through it. Thus, if we want a column to have a manhole, the diameter of the column should be greater than 45 cm. Furthermore, the plate spacing should be more than 45 cm and the plates should be located relative to the manhole so that there is access.

Some of this we just memorize and relate a numeral to a size we see. For other pieces of equipment we need to tabulate the size of the unit. For example, we can rough-cost an electrostatic precipitator by knowing the volumetric gas flowrate. But we also should be able to visualize how big that thing is. Will it fill this room? fit over in a corner? take up the whole gymnasium? or fill a football field?

These values change relatively little with time and they usually do not depend on the conditions. Perhaps as technology develops the size of the biggest that is available will increase. However, most of these numbers are relatively robust.

1.3-6 Constraints and Their Implications

Many constraints restrict the options we can select. They can arise from safety or environmental issues, insurance policies, government and state regulations about transportation, or taxes. They include such considerations as the maximum width of objects that can be transported by train or on the highways without a police escort, or the fact that vessels that are connected directly to a steam line fall into a different insurance classification than if the direct connection is made by a hot-water pipe. This form of experience information varies with government policy and standards.

1.3-7 Optimization Rules of Thumb

We can often write economic balances to account for the various tradeoffs made when we size or design. Usually we can generalize such balances, solve them and create optimization rules of thumb. Indeed, many of the rules of

thumb that engineers carry around in their heads are of this type.

The general procedure is:

1. Choose an objective function (usually the annual operating cost).
2. Write out the components that contribute to the costs (and check to see that one component increases as another one decreases; if this does not occur, then there is not an optimization problem. We choose where the objective function cuts a constraint.)
3. Substitute technical estimates for the cost contributions. Try to express all these in terms of *one* sizing or design variable.
4. Solve to determine the optimum.

Happel (1958) gives an excellent description and illustration of how these balance equations are set up and solved.

The results of such analyses yield such experience information as:

- the economic velocity for pumping liquids is about 1 to 2 m/s;
- try distillation first when separating a homogeneous phase;
- the economic approach temperature for shell and tube heat exchanger design is 5°C;
- the optimum reflux ratio is about 1.2 to 2 times the minimum;
- the optimum size reduction operation is to maximize the crush and minimize the grind;
- for optimization, the conditions when you use "direct search" methods as opposed to "dynamic programming" have been outlined by Rudd and Watson (1967).

Research develops this experience information for individual types of equipment and for systems. Much research has been done in this area, especially by Douglas (1984, 1988), Westerberg (1985), Rudd, et al. (1973), and King (1971). Peters and Timmerhaus (1980), Baasel (1980), Ulrich (1984), Walas (1987), Chauvel, et al. (1981), and Jordan (1968) present summaries of different rules of thumb.

The characteristics of this form of experience information are that as the relative costs of utilities, labor, capital equipment, and depreciation change, so will the rule of thumb change.

1.3-8 Good Practice

For any piece of equipment to operate efficiently and safely, some commonsense suggestions are usually put forward by the supplier and are gathered from operating experience. For example, always put in, and allow for, sufficient intermediate storage. Or always put in a feeder between a hopper and a belt conveyor to ensure smooth operation. Another example might be the interaction between erosion and corrosion from a flowing liquid and the temperature. From this we have a constraint that we do not let the cooling water get hotter than 50°C or we will get excessive erosion. Or, some suggest that we do not let a centrifugal pump operate at less than 20% of its rated capacity.

This type of experience is difficult to obtain. It is passed along from experienced design and process engineers. Some of the suggestions are very personal and may be ultraconservative. Nevertheless, we should attempt to gather and tabulate such information and the rationale for it. As new design procedures evolve so that we need less "safety factors" built into the design, and as equipment changes, so does this type of experience knowledge.

1.3-9 Key Equipment Characteristics for Costing

One major use of experience information is to allow an engineer to rough-cost—or screen—many alternatives very quickly. This means that the sizing parameter that we come up with must lead us to a capital and operating cost. For some equipment this poses no difficulty: the sizing and costing parameters for heat exchangers are the surface area, pressure, and type; for distillation, diameter, number, and type of trays, pressure, column height.

However, for rotary drum vacuum filters we may need to size and cost the drive and the vacuum system. Hence, we need this additional information (namely, the drive and vacuum requirements in terms of filter area). For centrifuges, we size based on sigma, defined as the equivalent surface area of a gravity settler; we cost in terms of diameter of the device. Again we need to know the relationship between sigma and diameter.

This type of data varies slightly with time, although, again, the tabulation or plotting of such data provides a convenient place for recording and comparing new information. (A companion text tabulates cost data for the process equipment discussed in this text.)

All of the above experience knowledge is used in two major functions: to select and rough-size equipment. In turn, the selection and rough-sizing procedures become experience knowledge.

1.3-10 Ballpark Selection or Regions of Application

A very important set of experience information records when a given device is applicable. The typical question might be "What device should I use to separate solids from liquids if the particle size is 10 μm and the concentration is 0.5%?"

To answer this type of question we must be familiar with the options available, the unique characteristics of each device, and the characteristics of the output/input conditions. In addition, we must be very clear on the functional and fundamental principles of operation of pieces of equipment.

Data in this category are sensitive to economics, and to the types of devices available and their development. Thus, these charts need to be continually updated. Although this text has numerous charts, I hope that you will continually update the charts and the tables with data as you discover it. This text is meant to be a working text.

1.3-11 Shortcut or Rough-Sizing Procedures

In addition, we should have rapid—but inaccurate—methods of sizing process equipment without having to do bench tests or detailed calculations. By this we mean "values that are within a factor of 10 (and hopefully within a factor of 2 to 5) of the accurate answer." Thus, we want to know if the screen is likely to be 0.4m^2? 4m^2? 40 m^2? or 400 m^2?

1.4 THIS BOOK AND YOU

This book is meant to be used to have your personal data added to tables and figures. It is a collection of experience knowledge organized to help engineers select equipment, rough-size equipment, and obtain the key parameters needed to estimate the capital and operating costs.

This book complements and enriches the fundamentals of process engineering. Where possible, I have tried to provide enrichment references.

Process equipment is classified into four general classes according to function:

1. Transport . . .   Chapter 2
                 gases
                 liquids
                 solids
2. Change Conditions . . . exchange energy   Chapter 3
                 mechanical
                 electrical
                 thermal
       . . . mix
                 liquids
                 solids
       . . . change size and shape
3. Separate . . . concentrate
       starting with a homogeneous mixture   Chapter 4
       starting with a heterogeneous mixture   Chapter 5
4. React/contain in vessels   Chapter 6

This classification is summarized in Figure 1–4. Because the information, procedures, and tables of data are

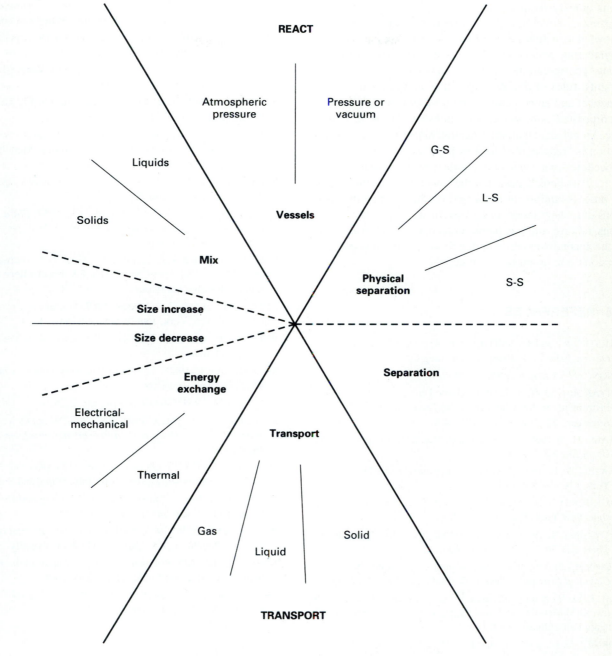

Figure 1–4   Classification of Equipment and Unit Operations

extensive, the text focuses on transportation, energy exchange with a detailed look at selecting options for separations and reactions.

1.5 SUMMARY

Engineers create, analyze, sleuth, specialize, explore, and anticipate ways to improve our world. Their products are ideas, which eventually get translated into the sizing and selection of pieces of process equipment that work in integrated systems. Between the "ideas" and the actual equip-

ment selection, engineers mentally test out an overwhelming number of options.

To facilitate this testing, engineers use two thinking laws: the law of optimum sloppiness and the law of successive approximation. The former says that the accuracy of the answer depends on the resources available; there are times when we must be sloppy. The latter says that we start any option by considering very approximate, sloppy approaches to see if the option has merit for further study. Then we successively invest more resources to obtain better answers to the most promising options.

To be able to apply these laws and to evaluate options quickly requires that engineers have a broad, well-organized, structured set of experience knowledge. That knowledge includes generalized properties of materials and engineering phenomena; familiarity with equipment options, configurations, and size; an appreciation of constraints; rules of thumb based on economics and on good practice; and an economics and cost viewpoint. These are incorporated into two rather extensive classes of experience knowledge: ballpark selection hints (which equipment option is "reasonable" for this situation) and quick sizing procedures (what size of equipment are we considering).

This book focuses on the selection and quick sizing of process equipment organized by the engineering functions of transportation, energy exchange, separations, reactions, mixing, and size increase/decrease. Vital to all of this is the appendix material on the SI system of units and conversions, and properties of materials.

1.6 REFERENCES

AERSTIN, F., and G. SMITH. 1978. *Applied Chemical Process Design.* New York: Plenum Publishing Co.

ANON. 1977. *Chem Eng* **84,** 13: 78.

ANYAKORA, S. N., et al. 1971. "Some Data on the Reliability of Instruments in the Chemical Plant Environment." *The Chem Engineer* 255 (November): 396–402.

ATALLAH, S. 1980. "Assessing and Managing Industrial Risk." *Chem Eng* **87,** 18: 99.

BAASEL, W. D. 1980. *Preliminary Chemical Plant Design.* New York: Elsevier Publishing.

BACKHURST, J. B., and J. H. HARKER. 1973. *Process Plant Design.* New York: Elsevier Publishing.

BROWNING, R. 1969. "Relative Probabilities of Loss Incidents." *Chem Eng* **76,** 27 (December): 135.

CHAUVEL, ALAIN, et al. 1981. *Manual of Economic Analysis of Chemical Processes.* New York: McGraw-Hill Publishing Co.

CIM. 1981. *Workshop on Liquid/Solid Separation: Design Problem.* CIM Metallurgical Society, Hydrometallurgy Section, Niagara Falls, Ontario, Oct. 18–21.

CORNET, C. L., and J. L. JONES. 1970. "Reliability Revisited." *Chem Eng Prog* **66,** 12: 29.

DOUGLAS, J. M. 1984. *Process Design.* Amherst, MA: Dept. of Chemical Engineering, University of Massachusetts.

———. 1988. *Conceptual Design of Chemical Processes.* New York: McGraw-Hill.

EPSTEIN, J. P. 1965. "How to Buy Used Equipment." *Chem Eng Prog* **61,** 11: 111–113.

FAIR, E. W. 1973. "Shall We Buy That New Equipment?" *Chem Eng* **80,** 24: 124–126.

GARCIA-BORRAS, T. 1977. "Research-Project Evaluation." *Hydrocarbon Processing* (Jan): 171.

HAPPEL, JOHN. 1958. *Chemical Process Economics.* New York: John Wiley and Sons.

HAPPEL, JOHN, and D. G. JORDAN. 1975. *Chemical Process Economics,* 2nd ed. New York: Marcel Dekker, Inc.

JORDAN, DONALD G. 1968. *Chemical Process Development.* New York: Interscience Publishers, John Wiley and Sons.

KEARNS, G. D. 1972. "How to Check Process Designs." *Hydrocarbon Processing* (Jan): 100–102.

KING. C. J. 1971. *Separation Processes.* New York: McGraw-Hill Publishing Co.

KLETZ, T. A. 1985. *What Went Wrong?* Houston, TX: Gulf Publishing Co.

MULAR, A. L., and R. B. BHAPPU. 1980. *Mineral Processing Plant Design.* New York: Am. Inst. of Mining, Metallurgical, and Petroleum Engineers.

PEKRUL, P. L. 1975. "Vibration Monitoring Increases Equipment Availability." *Chem Eng* **82,** 17: 109–114.

PETERS, M. S., and K. D. TIMMERHAUS. 1980. *Plant Design and Economics for Chemical Engineers,* 3rd ed. New York: McGraw-Hill Publishing Co.

POWERS, G. J., and S. A. LAPP. 1983. *Fault Tree Analysis.* Carnegie Mellon University, Pittsburgh, PA: Post College Professional Education Shortcourse.

RUDD, D. F., and C. C. WATSON. 1967. *Strategy of Process Engineering.* New York: John Wiley and Sons.

RUDD, D. F., et al. 1973. *Process Synthesis.* New York: John Wiley and Sons.

SHANMUGAN, C. 1980. "Estimating Delivery Times." *Chem Eng* **87,** 22 (November 3): 166.

SOUDERS, M. 1964. *Chem Eng Prog* **60,** 2: 75.

STREET, G. L., and T. E. CORRIGAN. 1967. "Make Quick Evaluation Estimates." *Hydrocarbon Processing/Petroleum Refiner* **46,** 12: 147.

ULRICH, G. D. 1984. *A Guide to Chemical Engineering Process Design and Economics.* New York: John Wiley and Sons.

WALAS, S. M. 1987. "Rule of Thumb: Selecting and Designing Equipment." *Chem Eng* (March 16): 75–81.

WESTERBERG, A. W. 1985. Forthcoming text, personal communication. Pittsburgh, PA: Carnegie Mellon University.

WOODS, D. R. 1975. *Financial Decision Making in the Process Industry.* Englewood Cliffs, NJ: Prentice-Hall.

———. 1986. *Cost Estimation for the Process Industry.* Hamilton, Ont.: McMaster University.

1.7 EXERCISES

1-1 How many Ping-Pong balls would fill the room you are in?

1-2 How many piano tuners are there in the city you are in?

1-3 How much money would a city bus driver collect (in equivalent fares, allowing for passes to be full-fare) in one eight-hour shift?

1-4 How many goose feathers are there in a pillow?

1-5 How many liters of punch should you prepare for a crowd of 200 on a warm, sunny afternoon?

1-6 How many kilograms of 10-cm spikes would a carpenter need to frame a two-story, single-family house?

1-7 How many examination booklets does your college/university have to order for the spring final examinations?

1-8 The density of organic vapors is about _____ .

1-9 The viscosity of an organic liquid is _____ .

1-10 The latent heat of steam is _____ times (bigger/smaller) (**CIRCLE ONE**) than the latent heat of an organic.

1-11 The diffusivity of a species in a vapor (increases/decreases) (**CIRCLE ONE**) with an increase in temperature.

1-12 Draw an H-S *or* p-H *or* pV *or* TS diagram, and on it, show lines of constant p, T, V, H, and S.

1-13 On the diagram in Question 1.12, show:
(a) flow through a valve;
(b) flow through a turbine;
(c) flow through a steam ejector,
(d) flow through a heat exchanger;
(e) flow through an evaporator.

1-14 Complete the following chart:

	organic liquid	water	organic vapor	steam
heat capacity				
latent heat				
thermal conductivity				
viscosity				
diffusivity of CO_2 in				
Prandtl No.				
Schmidt No.				

How do the following properties of pure components vary with temperature and pressure?

	GASES		LIQUIDS	
	temperature	pressure	temperature	pressure
density				
viscosity				
vapor pressure				
heat capacity				
thermal conductivity				

What general statements can we make about the effect of temperature and pressure on:

The solubility of a gas in a liquid: _____

The solubility of an immiscible liquid in another: _____

The reaction equilibrium?

.... $A_{(solid)} + B_{(solid)} \rightarrow C_{(gas)}$ _____

.... $A_s + B_s \rightarrow C_s$ _____

.... $A_{(gas)} + B_{(gas)} \rightarrow D_{(gas)}$ _____

Reasonable velocities for

gas velocities in a pipe _____
liquid velocities in a pipe _____
fluidization velocity _____

Heat transfer coefficients for

boiling liquids _____
condensing organic vapors _____
liquid heating/cooling _____
gas heating/cooling _____

Below are some coordinate axes. On these, draw the saturated liquid and vapor lines and the two-phase region for water. Do not worry about numerical values—concentrate on the shapes. On these plots, indicate (for all regions) lines of constant temperature, pressure, volume, entropy, and enthalpy.

What processes are isentropic $\Delta S = 0$?   _____
                 are isenthalpic $\Delta H = 0$?   _____

Equipment	Size	Comments
pressure vessels:		
horizontal cylinder	max	_____
	min	_____
vertical cylinder	max	_____
	min	_____
distillation towers	max	_____
	min	_____
jacketed reactors	max	_____
	min	_____
vacuum vessels	max	_____
	min	_____
atmospheric vessels:		
processing	max	_____
	min	_____
storage	max	_____
	min	_____
solids hoppers	max	_____
	min	_____

1-15 Trouble on the Ethylene Plant

Background: The ethylene plant shown in Figure 1–3 has two sections pertinent to this case: a drying section to remove moisture from the gas stream and a distillation train to separate the gas stream into the desired component streams.

Drying Section: Three alumina driers are installed as shown in Figure 1–5. One drier is regenerated while two are hooked in series on stream. In the diagram, drier V106 is being regenerated, with V107 and V108 removing the moisture to less than 3 to 4 ppm. Then V107 will be regenerated with V108 and V106 in series, etc. The cycle lasts 12 hours.

During regeneration, towns gas, heated with 2.8 MPa steam, flows though the drier in the direction reverse to normal flow. Once it is through the drier the regeneration effluent gas is cooled and sent to the fuel gas system. During regeneration the drier temperature rises to 190°C and is maintained at this temperature for one hour. Then the towns gas bypasses the heater and is sent directly into the drier to cool it.

The driers are all appropriately manifolded and valved so that any drier can be regenerated, bypassed, or used. The regenerating drier is separated from the line driers by a single gate valve. The details are not shown on the figure.

Separation Section: This is shown in the figure as T101, T102, and T103 and they represent the low temperature distillation units operating at 3.2 MPa.

Problem: The plant manager exclaims, "The low temperature distillation unit is freezing up. This baby is so severe that we are shutting down to de-ice. I don't think this'll do it though, because I think it's something more than a random slug of moisture forming these hydrates. I have nine more hours to finish the de-icing and by that time I'll have lost $10,000 in ethylene and propylene production. In twenty more hours we'll have lost our reserves and we'll have to shut down this whole end of the plant. That's money—$30,000/hour. Get this problem solved." Fix it.

This case problem was supplied courtesy of John Gates, B. Eng., McMaster University (1968).

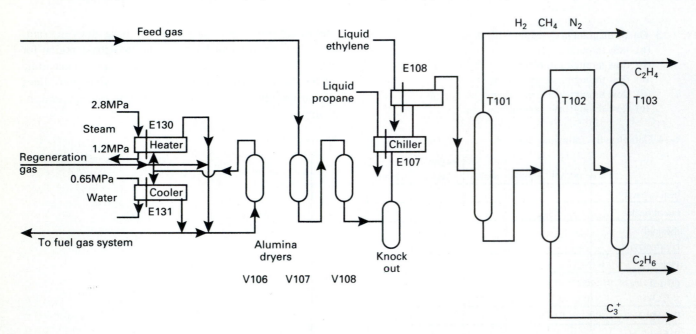

Figure 1–5   The Case of the Frozen Column

1-16 Compare Figure 1–1 with 1–2. What are the most significant implications of having propane as a possible feed instead of having ethane alone?

1-17 An interesting feature of this process to produce ethylene, shown in Figures 1–1, 1–2, 1–3, 1–5, and PID–1A, is that the distillation temperatures are cold, relatively speaking, and the cooling must be done by refrigerants, rather than by cooling water. The temperature at the bottom of the column is the highest temperature. Thus, if the temperature at the top of the column is −100°C, is the temperature at the bottom of the column more likely to be −150°C or −50°C?

1-18 Some commonly used refrigerants are:

propylene: at temperatures of − 6, −30, − 46
and ethylene: at temperatures of −73, −84, −101

Why is there a range of temperatures for each refrigerant? What determines the temperature? Why do we not use propylene for even colder temperatures? or higher temperatures?

1-19 For the process to produce ethylene
   a. How might we get hydrogen sulfide as a contaminant in the reactor gas?
   b. How might we get carbon dioxide as a contaminant?
   c. Why are the hydrogen sulfide and the carbon dioxide removed early in the process?

1-20 PID–1B (see page 1-19) shows a CO_2 removal system based on chemical absorption. What would be the differences in the flowsheet (or PID) if a physical solvent, like Selexol, had been used?

1-21 Identify several ways the process shown in Figure PID–1B could be altered to:
   a. recover as much hydrocarbon as possible in the main process stream entering column T150.
   b. recover as much energy as possible for the process shown in Figure PID–1B.

1.8 PROCESS AND INSTRUMENTATION DIAGRAMS FOR ETHYLENE ILLUSTRATES APPLICATIONS

The results of selecting equipment for process engineering and design are a conceptual description of a network of equipment to achieve a stated purpose. Such diagrams are developed in a variety of detail. Throughout this book are scattered simplified Process and Instrumentation Diagrams (PIDs). These show the arrangement sequencing and interconnections among the pumps, reactors, and separators. Often they provide the sizes and types of equipment. The instrumentation and control systems are usually shown. To complement Chapter 1, a simplified PID for an ethylene plant and a more detailed PID for a "packaged plant" to remove carbon dioxide from a gas stream are included.

PID-1A Ethylene Production

Ethylene is in great demand as a major raw material for the plastics and the petrochemical industry. Some specific products are ethylene oxide (to make antifreeze, ethylene glycol, other polyglycols, and acrylonitrile), ethyl alcohol (and this in turn is a raw material to make acetaldehyde, acetic acid, acetic anhydride), polyethylene, styrene (and subsequently polystyrene), ethyl chloride, and ethylene dichloride (and subsequently PVC or polyvinyl chloride).

A variety of different purities of ethylene can be produced from a wide variety of feedstocks. Feedstocks could include naphtha, n-pentane, ethane, propane, gas oil, or refinery gases. Ethane is the feedstock used in this process. Although one could produce a stream containing 75% ethylene (for example, if we wanted to use this as a feedstock for the alkylation of benzene to produce styrene), in this process the ethylene product is 99.9 mol %. A simplified Process and Instrument Diagram (PID) for such an ethylene plant to produce 90,000 Mg/a of ethylene is given in PID-1A.

The process has basically two sections: reaction and separation. In the **reaction section,** feed ethane is preheated and mixed with steam (in the ratio of 5:1) and fed into the pyrolysis furnace. Here, at temperatures of about 800 to 850°C cracking occurs to yield a gas stream containing about:

	mol%	wt %
hydrogen	32.7	3.52
methane	6.3	5.42
acetylene	0.2	
ethylene	33.8	50.88
ethane	24.9	40.18
propylene	1.0	
propane	0.2	
C_4^+	0.9	
	100	

The **separation section** uses a sequence of separation stages. First, the hot effluent from the cracking furnace is quenched to cool the gas (before it is compressed) and to remove cracking residue from the pyrolysis gas. Fast quenching is done to stop the reaction of the olefins reverting back to paraffins. The gas temperature from the quencher is about 65°C. Now the task is one of separating all of the constituent components and recycling unreacted ethane. Usually a sequence of distillations is used; these operate at high pressures and low temperatures.

We have to remove carbon dioxide and, depending on the contaminants in the feedstock, sulfur-containing compounds and acetylene. Removal of the former is required to prevent CO_2 from freezing and solidifying in the subsequent distillation columns. This is done with a standard absorption package plant that uses a scrubbing liquid such as diethanolamine or dilute caustic. PID–1B gives the

PID 1–A PID for the Ethylene Process

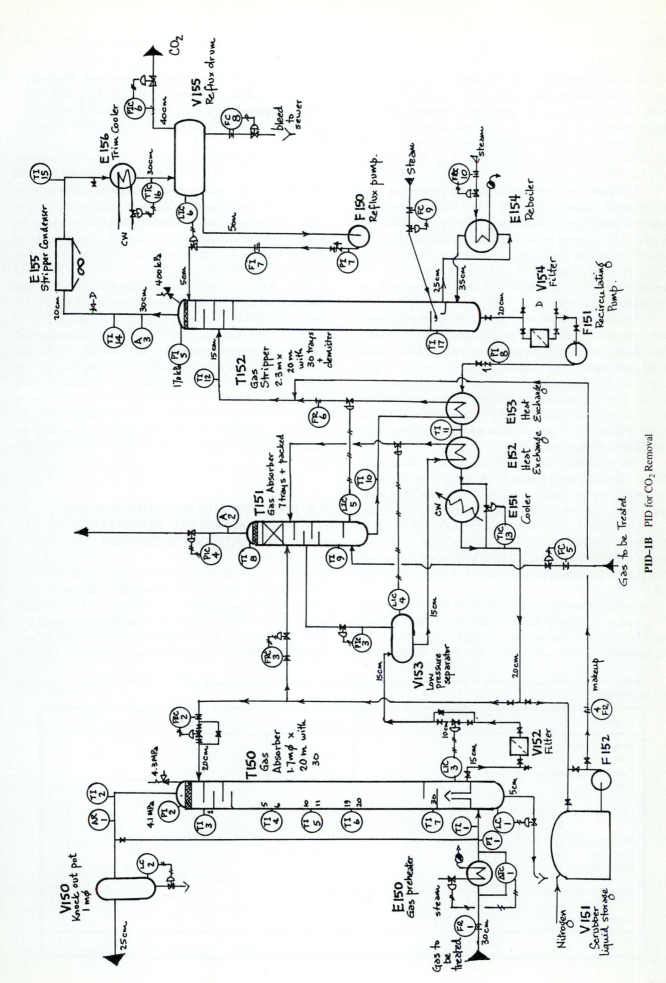

PID-1B PID for CO_2 Removal

1-19

details of this packaged plant. In some processes this is done before compression, some after compression, and some part way through the multistage compression. For ethane feed, this is usually done after the compression.

Hydrogen sulfide must be removed because it would deactivate the palladium catalyst used in the removal process for acetylene.

Similarly, any acetylene that might be present, as in this example, might be removed at this time or it could be removed later from the overhead of the de-ethanizer. The acetylene is removed by a "standard" packaged plant that converts acetylene to ethylene via selective hydrogenation over a palladium catalyst. If acetylene is not removed, it would contaminate the ethylene product since it forms an azeotrope with ethylene.

The gas compression system increases the gas pressure from about atmospheric to 35 atmospheres. Usually a multistage centrifugal compressor is used, although for a small-sized operation a reciprocating device might be preferred. In this diagram, only the first of three stages is shown. The compressed gas is dried (that is, the water is removed) by fixed bed adsorbers in this process. This removal is necessary to prevent the water from freezing in the distillation columns that follow.

The distillation sequence is usually demethanization (that is, removing the methane and hydrogen), de-ethanization (removing the ethane-ethylene stream), and then separating the C_2 species of ethane and ethylene in a C_2-splitter.

(In Figure 1–1, the feedstock can be either ethane or propane. If propane is used, a depropanizer, a C_3-splitter, and a debutanizer are usually added to produce high purity propylene, a C_4 stream, propane recycle, and gasoline product.)

For the ethane feed plant shown in PID-IA the overall capacity is 90,000 Mg/a of ethylene. The reaction-quencher train is actually six parallel units. About 4.4 are operating at any one time with 1.6 on standby. The compression/scrub/dry in done is a single train. The exit gas contains less than 1.2 ppm H_2S, 10 ppm CO_2, and 1 ppm water. Details of the three columns are not given because later PIDs will give details of distillation column configurations.

Other crucial information we must know about a process includes the cost per unit of production and the capital cost. The cost per tonne helps us see which component contributes most significantly to the total cost. The capital cost is the amount of money needed to construct the plant. An estimate of the contributions to the unit cost per Mg of product is given in Table 1A–1. The capital cost for a plant to produce 450,000 Mg/a is about 100 million ± 30% (MS = 1000). MS is the Marshall and Swift construction cost inflation index. Its value for construction in the process industry was 100 in 1926, about 300 in 1970, and is 1000 in the early 1990s. Current values of the index can be found in *Chemical Engineering* magazine. The MS index works well for relating capital costs from one time to another. To estimate the capital cost for the time of interest, multiply the cited cost by the ratio of the MS index at the time of interest/1000. For example, in the first quarter of 1992, the MS index value was 932.9. Thus, the cost of an ethylene plant at that time would be about 932.9/1000 × $100 million.

1.9 PROCESS AND INSTRUMENTATION DIAGRAM FOR CO₂ REMOVAL ILLUSTRATES APPLICATIONS

The PID for the ethylene included a "subplant" to remove carbon dioxide. The use of "packaged" subplants is common to raise steam, produce an inert gas, such as nitrogen, to remove unwanted species, or to transform unwanted species. The example configuration considered here is very flexible. By changing the solvent, we can remove HCN, or

TABLE 1A-1. Cost contributions to the cost per tonne of ethylene product (Based on ethylene from ethane as the feedstock)

Contribution from:	Approximate Unit Cost breakdown, %
Raw material	64
Utilities	24
Catalyst and chemicals	3.2
Labor	3
Credit for byproducts	-14.5
Maintenance	4
Supervision	incl in labor cost
Depreciation	12.5
Indirectly attributable costs	1.3
General expense	2.5
Total	100

H_2S, or SO_2, or various combinations. Figure 1B–1 illustrates the different types of solvent we might use for various mixtures of CO_2 and H_2S.

PID-1B CO_2 Scrubber

Carbon dioxide is a commonly occurring contaminant in gas streams. Indeed, a small amount of CO_2 (and H_2S) are contaminants in the exit from the Quencher T100 for the ethylene plant shown in PID–1A. For the ethylene plant we noted "CO_2 removal by a packaged plant" so that the concentration of CO_2 is less than about 10 ppm and that of H_2S less than about 1 ppm. What kinds of packaged plants remove CO_2?

The properties of carbon dioxide are given in **Data** Part D. It has a molar mass of 44, solidifies at −78°C, is mildly to moderately corrosive when dissolved in liquid (with corrosive ratings of 2 3 0 for copper, carbon steel and 316 stainless steel, respectively; the meanings of these ratings are given in **Data** Part Da). The presence of the H_2S makes the result even more corrosive.

Carbon dioxide can be removed from a gas stream by contacting the gas with a shower of liquid—liquid that absorbs carbon dioxide easily. The liquid can then be withdrawn from contact with the gas. By regenerating this liquid we can recover and reuse the liquid and, hopefully, recover the carbon dioxide as a relatively pure by-product. Naturally, the goal is to choose a solvent that selectively absorbs the target species, and the target species only.

Selecting the Solvent: The process configuration and operating conditions depend on the solvent selected. The solvents illustrated in Table 1B–1 and Figure 1B–1 can be classified depending on how we desorb the absorbed species and thus regenerate the solvent for reuse:

physical solvents, like Selexol, where desorption and regeneration occur by reducing the pressure. (Solvents in this class are known mainly by their trade names: Selexol, Fluor solvent, Rectisol, Purisol, Estosolvan, and Sepasol.)

chemical solvents, like MEA, Benfield, where desorption and regeneration require heat. (Solvents in this class are aqueous solutions of amines [MEA, DEA, Diisopropyl amine, Diethyl amine, Methyl diethanolamine] and alkali carbonates [Benfield, Catacarb].)

A description of some of the solvent options is given in Table 1B–1. The amount of CO_2 absorbed by different physical and chemical solvents is illustrated in Figure 1B–2. The absorption by physical solvents is relatively linear. Thus, over the dilute range, one could use Henry's law to model the absorption.

Henry's law constant is expressed either in terms of the mol fraction of solute per mol of liquid:

$$p^* = H \tilde{x} \qquad (1B\text{-}1)$$

or in terms of mol concentration.

$$p^* = H \tilde{c} \qquad (1B\text{-}2)$$

The data shown in Figure 1B–2 relate the partial pressure to the volume of gas absorbed per volume of liquid. Since the temperature is specified we can divide the abscissa value by the molal volume of 22.4 L per mol × 298/273 to give the molar concentration of CO_2 absorbed. Thus, the Henry's law constant obtained from these data would be of the form given in Equation 1B–2. The procedures outlined in **Data** Part B can be used to convert this to a mol fraction. Thus, data for solvents that physically absorb the solute follow Henry's law.

The absorption isotherm for chemical absorbents is not linear. Indeed, it is highly curved as is illustrated in Figure 1B–2.

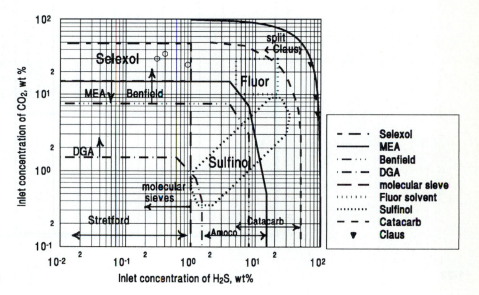

Figure 1B–1 Solvent Choice to Remove H_2S and CO_2 (Reprinted courtesy of the *Can. J. Chem. Eng.* from Woods et al., 1982)

TABLE 1B-1. Characteristics of possible solvents to remove CO_2

Name of process or solvent: physical solvents	Description	Typical CO_2 inlet concentration	Typical CO_2 outlet concentration	Typical operating pressure, MPa	Comments
Water		low			use when $C_3^+ > 6\%$
Dilute caustic wash + water					
Estasolvan	tributyl phosphate				
Fluor solvent	propylene carbonate				use when C_3^+ small
Purisol	n-Methyl pyrrolidone		0.1%		removes H_2S, COS, CS_2
Rectisol	Methanol	> 10%	10 to 100 ppm	>7	use when C_3^+ small; removes H_2S, NH_3, COS
Selexol	dimethyl ether of polyethylene glycol			>7	removes H_2S, COS, CS_2, mercaptan
Sepasolv	mix of polyethylene glycol dialkyl ethers				

TABLE 1B-1 (Continued)

Name of process or solvent: chemical solvents	Description	Typical CO$_2$ inlet concentration	Typical CO$_2$ outlet concentration	Typical operating pressure, MPa	Comments
MEA	10 to 20% w/w in water				O$_2$ and higher pressures promote corrosion
DEA	25 to 35% w/w in water				O$_2$ and higher pressures promote corrosion. Keep <0.4 mol CO$_2$/mol amine to limit corrosion
"Girbitol"	mix of MEA, DEA and TEA				
Diglycol amine; "Econamine"	40 to 60% w/w in water			>6	
Methyl DEA	30 to 50% w/w in water				
Alkazid M	potassium salt of methyl amino propionic acid in water				
Catacarb	activated potassium carbonate	15 to 30% for ammonia plants; 5 to 50% for acid gas	%	3	
GV alkali carbonate	alkali carbonate in water		%	> 0.4 to 0.8	
Benfield	25% w/w activated hot potassium carbonate in water	5 to 50%	1 to 2 %	> 0.4 to 0.8	removes H$_2$S, HCN, and some COS, CS$_2$, and mercaptan
Amisol	DEA, methyl DEA in methanol				
Sulfinol	tetrahydrothiophene dioxide, DEA and water	1 to 45%		> 0.4 to 0.8	use when C$_5^+$ is small and C$_3^+$ is minimal; removes H$_2$S, COS, CS$_2$, and mercaptan.

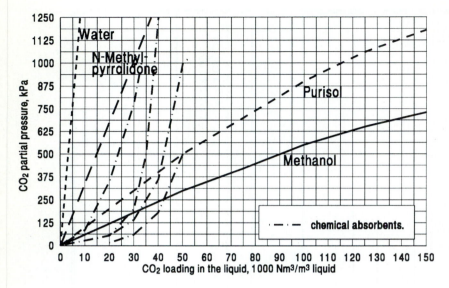

Figure 1B–2   Absorption of CO_2 by Various Scrubber Liquids at Constant Temperature (From Strelzoff [1975] and reprinted courtesy of Chemical Engineering)

So far we have seen example data for the absorption of CO_2. However, an important consideration is the absorption of other nontarget species in the solvent. Table 1B–2 gives some semiquantitative comparison of the absorption of species in the different solvents. For physical solvents where, in general, Henry's law applies, the data give the mol fraction of CO_2 at 25°C and 101 kPa partial pressure. For the chemical solvents, where Henry's law does not apply, the data reported are estimates of the absorption for the same conditions. These are very qualitative. Furthermore, because the data depend on the concentration, the temperature, and pressure, the data should be used as a guide in understanding the relative extent to which other nontarget species would be removed. For example, the amount of heptane absorbed is highest with selexol > purisol > fluor solvent. Thus, if the C_7 concentration is high and is **not** a target, of the three, Fluor solvent would probably be most appropriate.

Case Applications: For the ethylene plant, the amount of contamination present is usually so small that monoethanolamine or even a dilute caustic wash can be used. Often a two- or three-stage dilute wash is used.

The process shown in PID–1B is more applicable to treating "sour gas" or for an ammonia plant. The latter is given as an illustrative application of the principles given in Chapter 4 (in PID–4). For these applications the concentration of CO_2 is about 10 to 30%. This uses chemical absorption with regeneration by heat. Indeed, the system is used to absorb CO_2 from two different gas streams. The absorption occurs at high pressure in columns T150 and T151. The solvent is regenerated in the gas stripper tower, T152. The overhead gas is carbon dioxide. The bottoms from this tower are cooled and then pumped, via exchangers E151, 152, and 153 and recirculating pump

F151, back around the absorption circuit. A continual bleed to sewer, from overhead reflux drum V155, is needed to minimize buildup of undesirable species in the solvent. Fresh solvent is added as makeup via pump F152.

The costs depend markedly on the inlet and outlet compositions, the operating pressures, the type of solvent chosen. Some illustrative operating cost of a gas treatment process using amine is given in Table 1B–3. Table 1B–4 summarizes the capital cost. Since treatment plants are often a component part of a facility to produce ammonia, the estimated cost of a CO_2 removal plant for an ammonia process producing 1000 Mg/d is 6.3 million ± 30% (MS = 1000).

REFERENCES 1B:

HAYDUK, W., and S. C. CHENG. 1970. "The Solubilities of Ethane and other Gases in Paraffin Solvents." *Can J Chem Eng* **48:** 93–99.

LEE, J. C., F. D. OTTO, and A. E. MATHUR. 1974. "The Solubility of H_2S and CO_2 in Aqueous Monoethanolamine Solutions." *Can J Chem Eng* **52:** 803–805.

METCALF & EDDY, INC. 1972. *Wastewater Engineering.* New York: McGraw-Hill.

MOORE, R. G., and F. D. OTTO. 1972. "The Solubility of H_2 in Aliphatic Amines." *Can J Chem Eng* **50:** 355–360.

NEWMAN, S. A. 1985. *Acid and Sour Gas Treating Processes.* Houston, TX: Gulf Publishing Co.

STRELZOFF, S. 1975. "Choosing the Optimum CO_2-removal System." *Chemical Engineering* **82:** 19 (Sept. 15): 115–120.

WOODS D. R., S. J. ANDERSON, and S. L. NORMAN SILLS. 1982. "Evaluation of Capital Cost Data: Onsite Utilities (Industrial Gases)." *Can J Chem Eng* **60:** 173–201.

TABLE 1B-2. Qualitative Comparison of the Absorption of Different Species by Solvents

mol fraction 25C, 100kPa partial pressure	ammonia	Benfield	DEA	Estasol	Fluor	MEA	MDEA	propylamine	1,2-propanediamine	Purisol	Rectisol	Selexol	Sepasolv	TEA	Sulfinol
water					0.04					0.1	0.01998	0.20385			
acetylene															
air	1.4E-05											0			
ammonia											0.14	0.22			
benzene					high							high			
butane					0.023					0.05		0.1			
i-butane					0.015					0.03		0.2			
carbon dioxide	0.0006	0.032	0.02	0.025	0.013	0.03	0.035			0.014	0.006	0.045	0.04	0.04	0.06
carbon disulfide					0.4							high			
carbon monoxide	1.7E-05				0.0003					0.0003	0.0001	0.001			
carbonyl sulfide					0.02					0.04	0.015	0.1	0.1		
cyclohexane					0.61						0.36				
decane					high										
dimethyl sulfide										high					
ethane	3.3E-05				0.002					0.005	0.003	0.02			
ethylene					0.0046					0.008	0.02	0.022			
ethyl mercaptan										high					

Table 1B-2. (Continued)

	1	2	3	4	5	6	7	8	9	10	11	12	13	14	15
hexane						0.18					0.6		0.5		
heptane	1.4E-05					0.38					0.7		high		
hydrogen		8E-05				0.0001			0.00033	0.00014	9E-05		0.0006	0.0002	
hydrogen cyanide											high		high		
hydrogen sulfide	0.0018		0.04	0.14		0.043	0.04	0.02			0.14	0.025	0.4	0.3	0.06 / 0.05
methyl mercaptan						0.35					0.5		high	0.9	
methane	2.4E-05			0.001		0.0005					0.001	0.001	0.003	0.003	
nitrogen	1.2E-05					0.0001						7.2E-05			
NO	3.5E-05														
NO₂						0.22									
octane						0.85									
oxygen	2.3E-05					0.0003					0.0005	0.00012			
propane						0.007					0.015	0.0141	0.05		
pentane						0.065							0.25		
i-pentane						0.046							0.2		
sulfur dioxide						0.89							high		
thiophene													high	high	
water						3.9					high		high		

1-26

TABLE 1B-3. Cost contributions to the cost per 1000 Nm^3 gas treated

Contribution from	Approximate Unit Cost breakdown, %
Raw material	0
Utilities	33
Catalyst and chemicals	2
Labor	15
Credit for byproducts	depends on CO_2 downstream purification
Maintenance	15
Supervision	
Depreciation	27
Indirectly attributable costs	8
General expense	
Total	100

TABLE 1B-4. Capital Cost Distribution for the Cost to construct a battery limits Facility to remove CO_2 (A plant to treat 280,000 Nm^3/d is an estimated $6 million +/- 30% (MS=1000).*)

Contribution from:	Battery Limits Approximate Capital Cost Breakdown, %
Absorber column and trays	23
Lean amine cooler	
Stripping column and trays	11
Stripper reboiler and condensers	42
Flash tanks	4
Pumps	15
Total	100

*MS is the Marshall and Swift construction cost inflation index. Its value for construction in the process industry was 100 in 1926, was about 300 in 1970, and is 1000 in the early 1990s. Current values of the index can be found in *Chemical Engineering* magazine. The MS index works well for relating capital costs from one time to another. To estimate the capital cost for the time of interest, multiply the cited cost by the ratio of the MS index at the time of interest/1000. For example, in the first quarter of 1992, the MS index value was 932.9. Thus, the cost of a CO_2 removal plant at that time would be about 932.9/1000 × $6 million.

Chapter 2

Transportation

Materials need to be transported from one piece of equipment to the next. In general, the materials to be transported are done so as "single" phases such as gases, as liquids or as solids (powders included). Sometimes we want to transport a gas-liquid mixture or a liquid-solid mixture, etc., but usually it is a "single" phase.

Actually, the function of most devices in this section is twofold:

1. to transport
2. to supply energy in the form of pressure.

Hence, although we will be talking about the function of transportation, we should realize that sometimes this equipment is a reactor where we wish to control the gas pressure at 1000 kPa, as illustrated in Figure 2–1. To do this, we measure the pressure of the reactor, use this signal to open a valve on the exit line to control the pressure by letting gas escape if the pressure gets too high, and then we pump in the reactants. The pump supplies new reactants and it supplies pressure. Another example is illustrated in Figure 2–2. Here, we want to maintain a vacuum in a distillation column. The "vacuum" pump is downstream of the condenser. In this way, it does not transport out of the system any of the distilled product; rather, it transports out of the column the air or inerts that have leaked into the column through the flanges and valve stems. Thus, although a

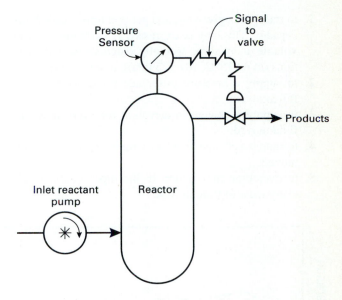

Figure 2–1   Controlling "Pressure" in a Reactor

vacuum pump pumps air out of the system, its main function is the pressure difference that it maintains between inside and outside the column. This dual role of transporting and supplying pressure is illustrated by plotting the pressure difference achieved between the inlet and outlet of the pump versus the amount of material that can be transported. This is illustrated in Figure 2–3, where the two ex-

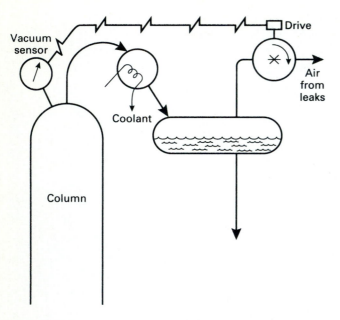

Figure 2–2 Controlling "Vacuum" in a Column

tremes in functions are depicted for different parts of the plot. Such a plot characterizes what the *device* will supply. This must be matched with what the system needs. The needs of a system are as follows:

1. to transport a given amount per unit time—called the capacity. This may be expressed as a volume (L^3/s), a volume under some specially stated standard conditions (m^3/s (15°C, 101.325 kPa)), or a mass (kg/h);
2. to supply a pressure difference for every mass of fluid moved;
3. to supply a change in elevation for every mass of fluid moved;
4. to supply a change in velocity for every mass of fluid moved;
5. to overcome the friction in the pipes of ducts for every mass of fluid moved.

These needs are illustrated in Figure 2–4.

All these "pressure" needs can be summed to yield the total pressure requirements in the system:

$$\text{Total needs} = \left(\begin{array}{c}\text{Pressure} \\ \text{difference}\end{array}\right) + \left(\begin{array}{c}\text{Height} \\ \text{difference}\end{array}\right) + \left(\begin{array}{c}\text{Velocity} \\ \text{difference}\end{array}\right) + \left(\begin{array}{c}\text{Friction} \\ \text{loss}\end{array}\right) \quad (2\text{-}1)$$

It is convenient to express this total need in "units" of pressure that would be independent of the density of the fluid being transported. To do this, we could divide the pressure by the force of gravity acting on the mass contained in a unit volume of the fluid. This result is something having units of distance—vertical distance in this case:

$$\frac{\text{Pressure}}{g\rho} = \text{units of distance called "head"}$$

To understand how this can happen and what the so-called head means, let us explore this idea briefly.

In Figure 2–5 is shown an infinitely tall pipe that is open at the top and is submerged in a large lake. If a pressure of 2 kPa is exerted on the lake, the water in the lake will move up the tube a distance of about 20 cm. If we exert 100 kPa, the liquid will rise up the tube a distance of 10 m. We can calculate this relationship between pressure and head as follows:

$$\text{head, h} = \text{pressure (kPa)}/\rho g \quad (2\text{-}2)$$

where

ρ = density of the fluid; for water this is 1 Mg/m^3

g = local acceleration due to gravity; for earth, this is 9.8 m/s^2

h = "head," m.

Thus, for water, the result is

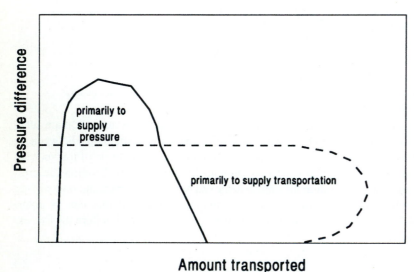

Amount transported

Figure 2–3 Different Functions of "Transportation" Equipment

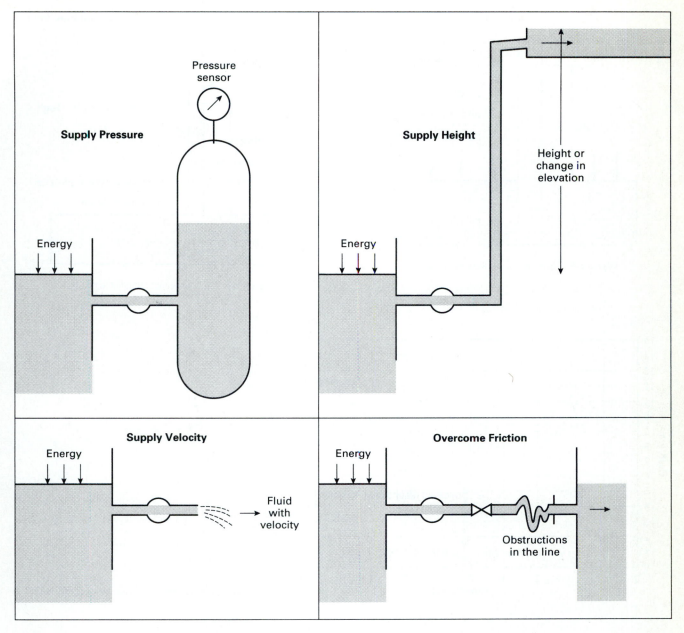

Figure 2–4 The Energy Needs To Supply Pressure, Height, and Velocity and To Overcome Friction

$$h(m) = \frac{kPa}{\rho \times g} = \left(\frac{10^3 \text{ kg}}{s^2 \cdot m}\right) \frac{m^3}{10^3 \text{ kg}} \times \frac{s^2}{9.8 \text{ m}}$$

$$= 0.10 \times \text{the "pressure" value expressed in kPa}$$

Thus, for a pressure of 2 kPa, the "head" is 0.2 m or 20 cm; for 100 kPa, the head is 10 m.

It is interesting to turn this idea around and ask, if I created a "perfect" vacuum, what height of water could I support below the vacuum chamber? The situation is posed in Figure 2–6. Since a perfect vacuum is 0 kPa and since the usual atmospheric pressure is 101 kPa, the answer is "about 10 m" of water.

Thus for any piping-equipment configuration, the transportation "pressure" requirements will be represented as the sum of the components illustrated in Bernouli's

Equation 2–1. Since some of the "pressure" requirements depend on the amount transported, a system requirement curve, such as that shown in Figure 2–7, can be estimated. The pressure needed can be conveniently expressed as:

1. a pressure difference achieved across the device (usually used when discussing gases);
2. an absolute pressure (used when discussing vacuum operations);
3. vertical height or head of fluid (used when discussing liquids).

The pressure-capacity requirements of the "system" are supplied by pumps and blowers that have "performance" curves as shown in Figure 2–8. Thus, in design we

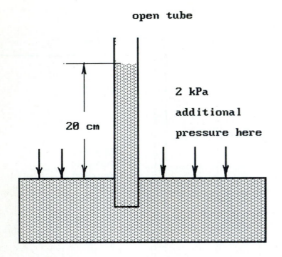

Figure 2–5   How Pressure Is Related to "Head"

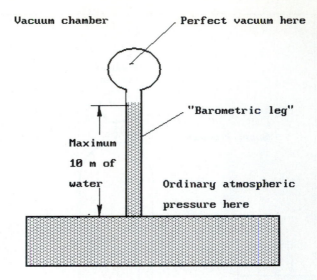

Figure 2–6   How Vacuum Is Related to "Head"

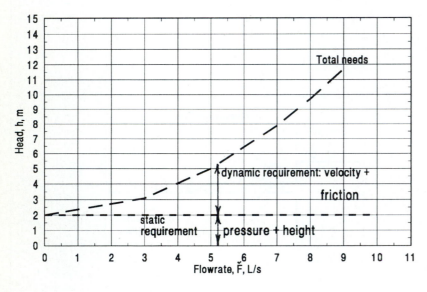

Figure 2–7   System Resistance: Head and Capacity

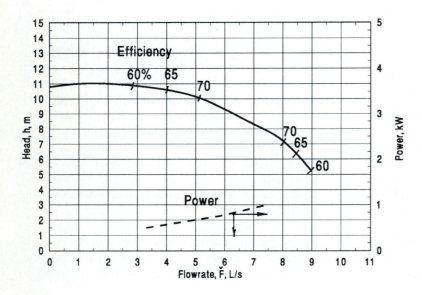

Figure 2–8   Pump Supplies: Head and Capacity

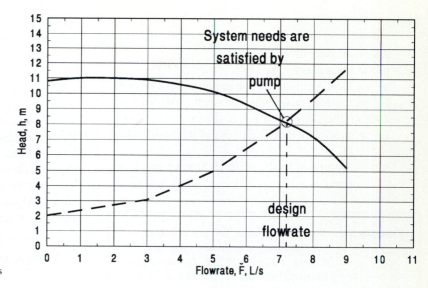

Figure 2–9 Pump Supplies System Requirements

select a device to satisfy the system and so superimpose Figure 2–8 on Figure 2–7 so that the curves intersect close to the design specifications, as illustrated in Figure 2–9.

In summary, transporting fluids means supplying pressure/selecting devices that will transport and supply pressure.

2.1 TRANSPORTING GASES: PRESSURE SERVICE

The devices used are illustrated in Figure 2–10 and are called fans (which transport high volumes at low pressure), blowers (which transport and supply moderate pressures), and compressors (which transport and supply high pressures).

2.1-1 Selecting Devices

The general regions of applications of the different devices are shown in Figure 2–11. Figure 2–12 (see p. 2–9) provides a convenient method of estimating the drive power needed to run the devices. It is common for gas-moving equipment to express the pressure in kPa instead of as head.

Example 2–1: Select a device to convey 10^3 dm^3/s of air and supply a pressure differential of 1000 kPa.

An Answer: From Figure 2–11, this is the usual range for a centrifugal compressor or a two- or three-stage reciprocating compressor. A rotary screw compressor would also satisfy this need. From Figure 2–12, the energy needed would be about 100 kW.

2.1-2 Estimating the Pipe Diameter

Often we want to be able to visualize the diameter of pipe needed to contain the gas. Figure 2–13 (see p. 2–9) provides this information. To use this graph, we enter it on the ordinate at "the usual velocities" used for transporting gases through a pipe and enter on the abscissa at the desired flowrate. The usual value depends on the molar mass, pressure, and safety.

Example 2–2: What diameter of pipe would be used for the service described in Example 2–1?

An Answer: This problem is deceptively tricky. The data in Figure 2–13 apply to air at 100 kPa. Thus, for the *inlet* to the compressor, the pipe would be, from Figure 2–13, approximately 25 cm in diameter. On the outlet, however, the pressure is 1000 kPa greater than at the inlet and so Figure 2–13 does not apply although we know the diameter will be less than 25 cm. You might take this as an exercise to see what diameter would be needed to convey the same mass at the same velocity as given in Figure 2–13. As an approximation, use the ideal gas law to determine the new density. This is given as a figure in **Data** Part C.

2.1-3 Estimating the Pressure Drop

The pressure drop for fluids flowing in a pipe because of friction is given as:

$$\frac{\Delta p}{\rho g} = 4f\left(\frac{L}{D}\right)\left(\frac{<v>^2}{2g}\right) \qquad (2\text{-}3)$$

where

Δp = pressure drop, kPa

ρ = mass density, kg $\cdot$ m^{-3}

g = local acceleration due to gravity, 9.2 m $\cdot$ s^{-2}

f = friction factor for turbulent flow is about 0.005; this factor has no dimensions

L = equivalent length of piping of internal diameter D

$<v>$ = average velocity of fluid flowing inside the pipe.

Example 2–3: What is the pressure loss per 100 meters of pipeline for a gas of density 1 kg $\cdot$ m^{-3} flowing at 25 m $\cdot$ s^{-1}? The pipe diameter is 10 cm and the flow is turbulent so that f = 0.005.

An Answer: From Equation 2–3, the pressure loss is

$$\frac{\Delta p}{\rho g} = 4f\left(\frac{L}{D}\right)\left(\frac{<v>^2}{2g}\right)$$

$$\Delta p = 4f \cdot \rho \cdot \left(\frac{L}{D}\right)\frac{g<v>^2}{2g}$$

$$= 4 \times 0.005 \times \frac{1\ kg}{m^3} \cdot \left(\frac{100\ m}{0.10\ m}\right)$$

$$\frac{1(25)^2}{2}\frac{m^2}{s^2}$$

$$= 6250\ \frac{kg}{m \cdot s^2}$$

$$= 6.25\ kPa$$

Example 2–4: How many values of L/D are required to have a pressure loss of 1 velocity head?

An Answer: A velocity head is defined as $<v^2>/2g$. From Equation 2–3

$$\frac{\Delta p}{\rho g} = 4f\left(\frac{L}{D}\right)\frac{<v>^2}{2g}$$

$$1 = (4 \times 0.005)\frac{L}{D}$$

Hence,

$$\frac{L}{D} = \frac{1}{4 \times 0.005}$$

$$= 50.$$

This means that every time we encounter a pipe of length 50 times the diameter, the head loss is 1 velocity head.

This result is independent of the fluid and depends only on the friction factor.

The pressure drop in a gaseous system can be generalized by rearranging Equation 2–3 to yield:

$$\frac{\Delta p}{L\rho g} = 4f\frac{<v>^2}{2gD} \qquad (2\text{-}4)$$

Since, from Figure 2–13 the usual range of gas velocities is 10 to 60 m/s, Figure 2–14 (see p. 2–10) shows this velocity as a parameter and the diameter of the duct as the abscissa. The estimates assume f = 0.005 or 4f = 0.02. The results on the ordinate are independent of the gas. If the gas density is known, then the pressure drop per unit length can be estimated. At the left-hand side, the details have been worked out for gas densities of 1.2 kg/m^3, 0.5 km/m^3, and 2 kg/m^3.

Example 2–5: Estimate the pressure drop for a gas of density 1.2 kg/m^3 if the design velocity is 30 m/s and the total flow is 4700 dm^3/s.

An Answer: From Figure 2–13, the diameter is about 40 cm. From Figure 2–14, the pressure drop is 2.7 kPa/100 m.

Example 2–6: Repeat Example 2–5, but assume the gas density is 0.42 kg/m^3.

An Answer: The diameter of the pipe would be the same, 40 cm. From Figure 2–14, the ordinate is 2.3 Δp/L ρg.

Since ρ = 0.42 kg/m^3 and g = 9.8 m/s^2

$$\frac{\Delta p}{L\rho g} = 2.3$$

Hence,

$$\frac{\Delta p}{L} = 2.3 \times 0.42\ \frac{kg}{m^3} \cdot 9.8 \cdot \frac{m}{s^2}$$

$$= 9.47\ \frac{kg\ m}{m^3\ s^2}$$

$$= 9.5\ \frac{Pa}{m}$$

$$= 0.95\ \frac{kPa}{100\ m}$$

Check: This seems reasonable when compared with the scale for a gas density of 0.5 kg/m^3.

For gas, the pressure drop in the ductwork can be expressed as:

$$\text{"velocity heads"}\quad \frac{\Delta p}{\rho g} = \frac{k<v>^2}{2g} \qquad (2\text{-}5)$$

$$\text{"velocity pressure"}\quad \Delta p = k\rho\frac{<v>^2}{2} \qquad (2\text{-}6)$$

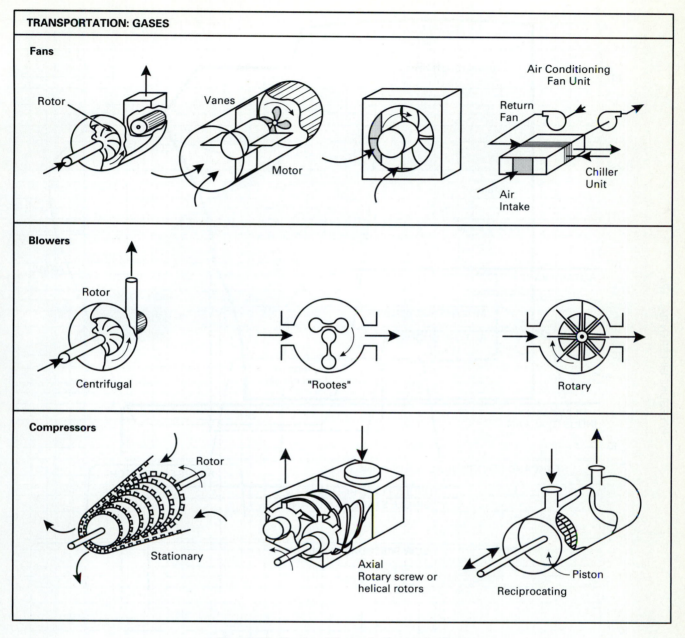

Figure 2–10   Sketches of Devices for Transporting Gases

The "velocity pressure" is

$$\Delta p = k\,0.6\left[\frac{\rho}{1.22}\right]\left[\frac{<v>}{10}\right]^2 \text{ cm of water} \qquad (2\text{-}7)$$

where

ρ = gas density, kg/m^3
v = gas velocity, m/s
k = estimate of the number of velocity heads lost in the ductwork or piping.

The estimates of the pressure drop have been based on flow through a straight pipe. But energy is required to force the fluid through valves, fittings, around bends, and into/out of discharge pipes. We can account for the pressure drop caused by the "fittings" by expressing the loss as:

- the "equivalent length of straight pipe" that would give the same pressure drop as occurs across the fitting; this can then be expressed as the equivalent L/D ratio for the fittings, or
- the number of "velocity heads" (defined as $<v>^2/2\,g$) represented by the symbol "k." From Equation 2–3 we note that "k" equals 4f (L/D), and thus the two methods are interchangeable.

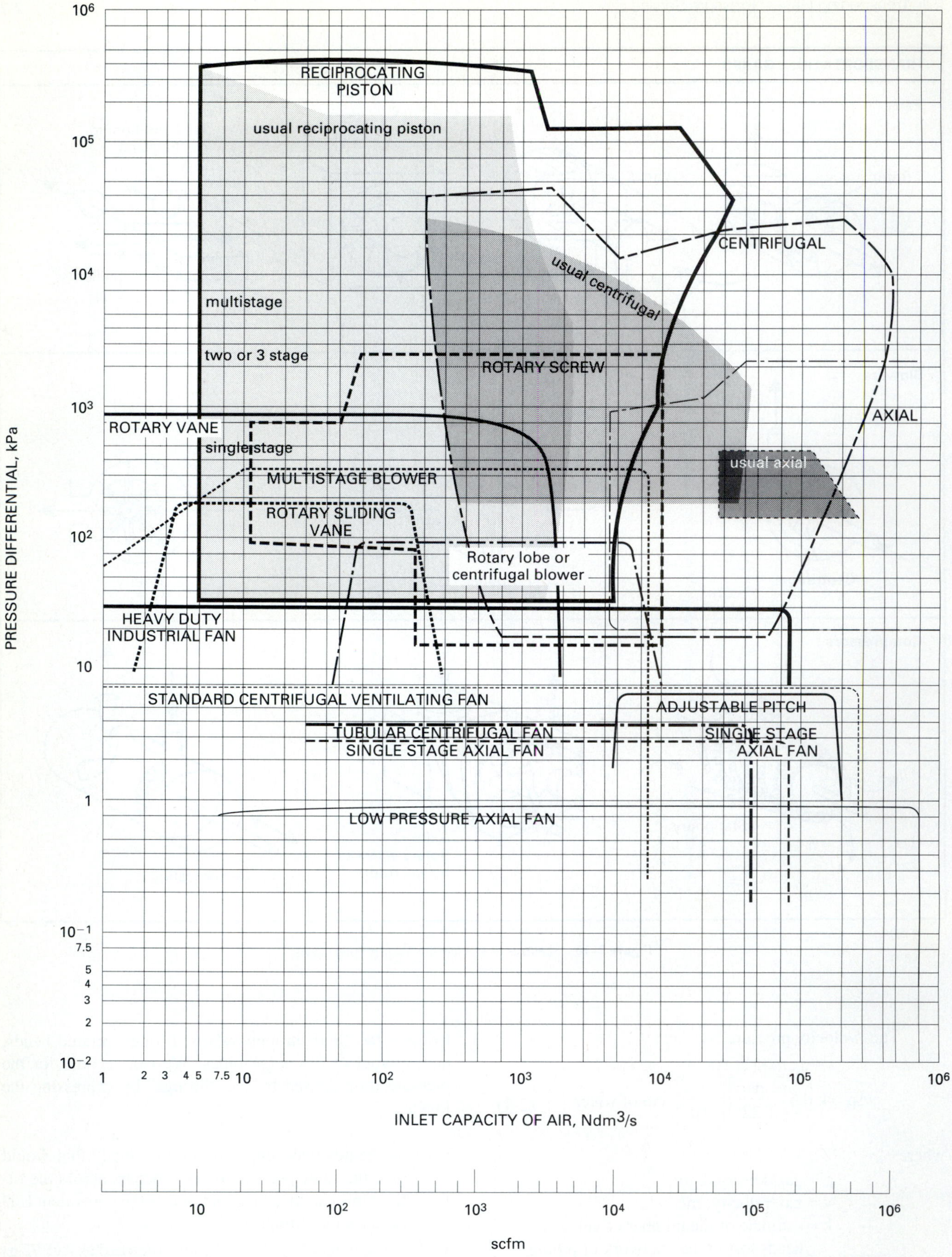

Figure 2–11 General Regions of Applicability for "Pressure" Gas Moving Equipment

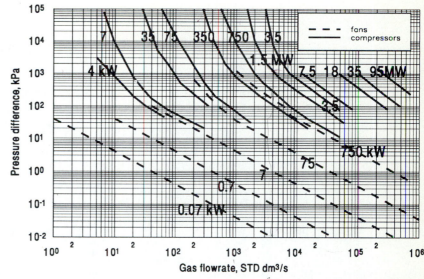

Figure 2–12 Energy Requirements for Gas Moving Equipment Based on Air with Efficiency of 60% for Fans and 80% for Compressors

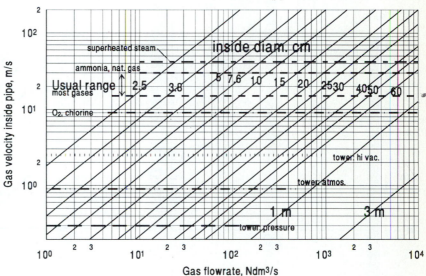

Figure 2–13 Velocity-Capacity Relationships for Gas Flow in a Circular Pipe (for any Gas)

Table 2-1 summarizes approximate values of both of these representations.

Example 2–7: Estimate the static pressure drop for gas flowing through 20 m of 30 cm diameter ductwork at 20 m/s if the system includes a pipe entrance, a heater with 3 rows of tubes, and exits through a stamped grill with 60% free area. The duct is 30 m long. Assume the gas density is 1.4 kg/m^3.

An Answer: From Figure 2–14, from the ordinate of 1.3 the pressure drop is

$$= 1.3 \, L \, \rho g = 1.3 \times 30 \text{ m} \times 1.4 \frac{\text{kg}}{\text{m}^3} \times 9.8 \frac{\text{m}}{\text{s}^2}$$

$$= 0.535 \text{ kPa or } 5.46 \text{ cm water}$$

From Table 2–1, the number of velocity heads, "k," for the various components are:

pipe entrance	$k = 1$
heater (3 rows)	$k = 3 \times 0.4 = 1.2$
stamped grill	$k = 4$
Total	$k = 6.2$

From Equation 2–7, the velocity pressure contribution is:

$$\Delta p = 6.2 \times 0.6 \, \frac{[1.4]}{[1.22]} \left| \frac{<20>}{10} \right|^2 \begin{array}{l} \text{cm of} \\ \text{water} \end{array}$$

$$= 17 \text{ cm of water}$$

Thus the total loss is about 5.5 plus 17 or 22.5 cm of water.

$$\frac{\Delta p}{L\rho g} = \frac{4f}{D} \frac{\langle v \rangle^2}{2g} \quad f = 0.005$$

L = length of equivalent pipe

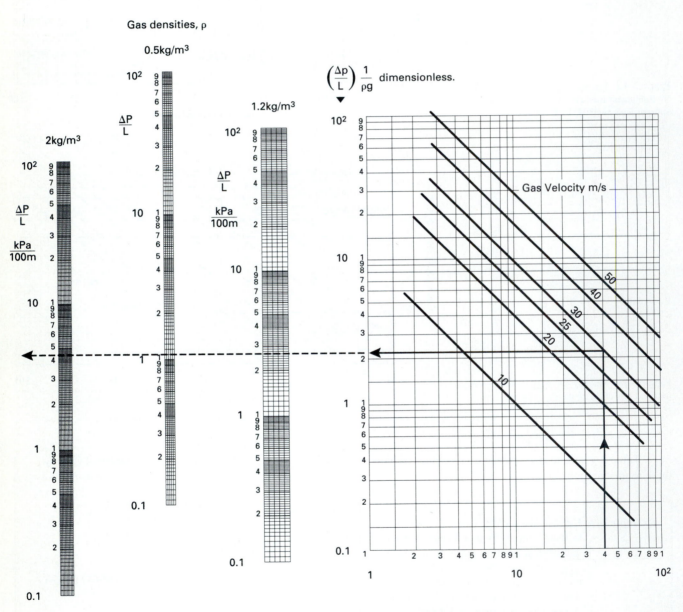

Figure 2–14 Pressure Drops for Gas Flowing Inside Ducts/Pipes

TABLE 2-1. Approximate friction loss through fittings

ESTIMATED FRICTION LOSS

FITTINGS		No. of Velocity heads, k		Equivalent length of pipe, L/D	
		Flanged	Screwed	Flanged	Screwed
	Coupling		0.05		
	45° Elbow		0.35		22
	90° Elbow	0.25	0.70		40
	Tee-run through		0.40		30
	Thru-side		1		70
	180° Reverse	0.4	1.5		75
	Entrance: Into pipe in wall Into pipe Into projecting pipe	0.5 1.0 0.8		40	
	Fumehood	0.5			
	Stamped grill-70% free area 60% 50%	2 3 4			
	Exit: Ex pipe in wall Ex pipe	1 1			
	Stamped grill-70% free area 60% 50%	3 4 5			
	Capped exit 100% free area	3			
	In-line heater with ''n'' rows of tubes in the direction of the flow	0.4n			
	Dust collector	$5+\dfrac{(\rho<v>^2-1)}{2}$			
	Sudden enlargement $D_1/D_2 = 0.25$ 0.50 0.75	0.8 0.55 0.1			
	Sudden contraction $D_2/D_1 = 0.25$ 0.50 0.75	0.35 0.20 0.1			

The usual pressure drop across various pieces of equipment is summarized in Table 2–2. The pressure drop through a bed of packed particles of diameter, D_p, is illustrated in Figure 2–15. The full expressions are complex. Hence, the ordinate is the pressure drop per thickness of bed in the direction of flow multiplied by the cube of the diameter of the particles. The abscissa is the face velocity of the fluid times the diameter of the particles. Figure 2–16 compares friction factor behavior for flow in a smooth pipe, flow past a single sphere, and flow through a packed bed.

TABLE 2–1. (Continued)

ESTIMATED FRICTION LOSS

FITTINGS		No. of Velocity heads, k		Equivalent length of pipe, L/D	
		Flanged	Screwed	Flanged	Screwed
	Butterfly valve	0.7		40	
	Gate valve (open) $^1/_4$ closed $^1/_2$ closed $^3/_4$ closed		0.1 0.8 3		10 200 800
	Angle valve (open)		2.5		160
	Globe valve (open)		7		320
	Swing check valve		1.7		160
	Foot valve		15		
	Ball check valve		65 to 70		
	''Control valve'' (open) = **value for valve x 3.**				

TABLE 2-2. Usual pressure drops across devices		
VESSELS		$\dfrac{\Delta p, \text{kPa}}{20 \text{ kPa/m}}$
Gas through catalyst bed		
Gas through tank		negligible
Liquid through tank		negligible
MIXERS/BLENDERS		
Extruders		
CONTROL SIZE		
L-L homogenizers		
Liquid spray nozzles		
HEAT EXCHANGE		
Gas in furnace tubes		
Shell/tube head exchanger	liquid	70
	gas	5
Plate heat exchanger	liquid	50 kPa/pass
SEPARATORS		
Conventional trays (per theoretical stage)	gas/liquid	0.3 to 0.65
Linde tray (per theoretical stage)		0.25 to 0.4
Packing (per m of packing)		0.2 to 0.75
High vacuum equipment (per theoretical stage)		0.04 to 0.065
		>0.5 MPa
Gas membrane	gas	
Dialysis	liquid	5 to 10 MPa
Reverse Osmosis	liquid	0.1 to 4 MPa
Ultrafiltration	liquid	
PHYSICAL SEPARATORS GAS		
cyclone	gas	0.5 to 1.6 kPa
venturi scrubber	gas	0.5 to 6 kPa
PHYSICAL SEPARATORS LIQUID		
Filter press (plate & frame)	liquid	70 kPa
vacuum drum	liquid	70 kPa

(Pressure drop per unit length (depth) of bed) times the
(bed particle diam)³

$$\frac{\Delta p}{L}(D_p)^3; \quad \frac{kPa}{m}(mm)^3$$

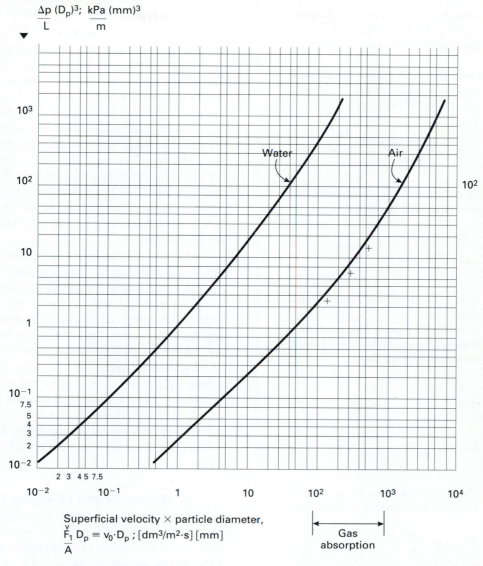

Superficial velocity × particle diameter,

$$\frac{\overset{v}{F_1}}{A}D_p \equiv v_0 \cdot D_p \; ; \; [dm^3/m^2 \cdot s] \, [mm]$$

Gas
absorption

Figure 2–15   Estimating Gas or Liquid Pressure Drop Through Packed Beds (Assume $\varepsilon = 0.40$)

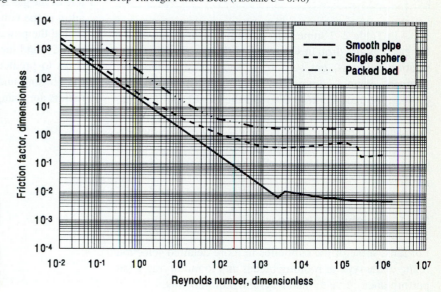

Figure 2–16   Friction Factors for Fluid Flow Inside Pipes, Past Spheres, and Through Packed Beds

Example 2–8: What is the pressure drop for air flowing through a 30 cm bed of 2 mm diameter particles? The bed is in a 10 cm diameter tube and the gas flowrate is 2 dm³/s.

An Answer: To use Figure 2–15 requires that we estimate the abscissa. For a 10 cm diameter tube, the superficial velocity, v_o, is

$$2 \frac{dm^3}{2} \times \frac{1}{\frac{\pi}{4}(0.1 \text{ m})^2} = 255 \frac{dm^3}{s \cdot m^2}$$

Since the particles are 2 mm diameter, the total value of the abscissa is

$$255 \times 2 \text{ mm}$$

$$= 510 \frac{dm^3}{m^2 \cdot s} \text{ mm}$$

From Figure 2–15, the ordinate value is about 15.
To obtain the pressure drop per depth of packing, we need to divide this value by the diameter cubed. Thus

$$\frac{15}{(Dp)^3} = \frac{15}{(2)^3}$$

$$= \frac{15}{8}$$

$$= 1.88 \frac{kPa}{m}$$

The depth of the bed is 30 cm or 0.3 m. Hence the pressure drop would be about

$$0.3 \times 1.88 \frac{kPa}{m}$$

$$= 0.56 \text{ kPa}$$

In Table 2–2 some very general values are given for the gas pressure drop in separators. Separators, such as distillation columns, gas scrubbers, and gas absorbers and strippers, contain commercial trays, packings, or grids to provide intimate contact between the liquid and the gas. Figure 2–17a provides more details about how the pressure drop across a tray varies with the gas flowrate up the column; Figure 2–17b shows the pressure drop for grids and packings per meter of height of packing as a function of the gas flowrate.

2.1-4 Enrichment Ideas

For fans, some of the subtleties in more detailed selection relate to the type of rotating blades and to modifying a fan performance curve to match system requirements.

A. Performance Curves.
Table 2–3 illustrates the different types of blade configurations that are common for fans, the performance curves, and the characteristics of each. The performance curves are expressed in terms of the fan total pressure and the fan static pressure. This highlights some of the care that is needed in selecting fans. Three different pressures can be measured: the static pressure, the impact pressure, and the "velocity pressure." These are illustrated in Table 2–4. Various combinations of these can be measured on the inlet and outlet of a fan as illustrated in Table 2–4. The easiest one to measure is the static pressure difference. The system requirements, however, are the sum of the static and velocity pressure (or the total pressure). The fan performance is given in terms of both the fan static pressure and the fan total pressure. Thus care is needed in reporting and selecting the appropriate pressure difference.

Another precaution is that because we report pressure on the ordinate, the curves depend on the density of the gas being handled. Usually, the performance curves are given assuming that the gas is bone-dry air at 21°C and 101 kPa with a density of 1.20 kg/m³. Thus, we usually have to "correct" the fan performance curve. Fans are inherently constant volume machines. This means that for constant speed, and all other factors equal, the volume of gas delivered will be the same regardless of the density. However, the total head and the power vary directly with the density. Figure 2–18 (see p. 2–18) illustrates this procedure for a case where the gas density is 14% greater than the one upon which the performance curve was based.

B. Similitude and Changing Conditions.
If we have an existing fan, can we change the diameter of the impeller or the rotational speed of the fan to achieve the required set of conditions? For centrifugal fans and axial fans we can use the similarity of the velocity vectors of the gas leaving the tip of the blades to relate the variation in the volumetric flowrate, $\check{F}$, the pressure difference, p, and the power required, P, to the blade diameter, D, the rpm, N, and the gas density, ρ. We can use these relationships to predict new performance curves if the shape of the housing and blade are maintained but the other parameters are changed. The "similitude" relationships for fans are:

$$\frac{\check{F}_1}{\check{F}_2} = \frac{N_1}{N_2} \times \left(\frac{D_1}{D_2}\right)^3 \qquad (2\text{-}8)$$

$$\left(\frac{p_1}{p_2}\right) = \left(\frac{N_1}{N_2}\right)^2 \times \left(\frac{D_1}{D_2}\right)^2 \times \left(\frac{\rho_1}{\rho_2}\right) \qquad (2\text{-}9)$$

$$\left(\frac{P_1}{P_2}\right) = \left(\frac{N_1}{N_2}\right)^3 \times \left(\frac{D_1}{D_2}\right)^5 \times \left(\frac{\rho_1}{\rho_2}\right) \qquad (2\text{-}10)$$

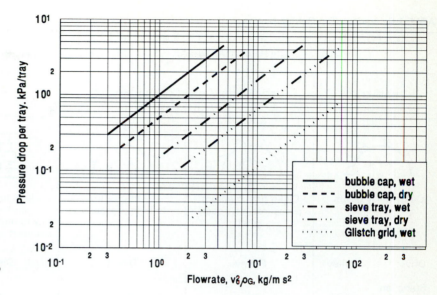

Figure 2–17 Pressure Drop
a) Across Trays

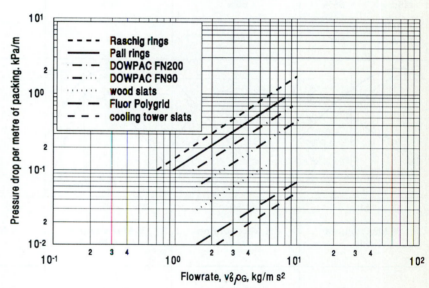

Figure 2–17 Pressure Drop
b) Across Packing

Example 2–9: A 96 cm diameter, 18° aerofoil, single stage axial flow fan runs at 450 rpm and delivers a 2000 dm³/s of gas at a static pressure of 0.75 cm water. The power required is 0.32 kW. To control the operation, we alter the speed to 315 rpm. What will be the new flowrate, pressure, and power required?

An Answer: From the fan laws:

the flowrate is	$2000 \times 315/450 =$ 1400 dm³/s
the pressure is	$0.75 \, (315/450)^2 =$ 0.37 cm water
the power is	0.32 kW $(315/450)^3 =$ 0.11 kW

For more on fans, see Osborne and Turner (1960).

2.2 TRANSPORTING GASES: VACUUM SERVICE

For vacuum service, the equipment is slightly different; the sketches in Figure 2–19 (see p. 2–18) illustrate them. The general regions of application are shown in Figure 2–20 (see p. 2–19). The general relationship between mass flow, inlet or vacuum pressure, and volumetric flowrate are given in Figure 2–21 (see p. 2–20) (for air). Superimposed on the regions of application chart are indications of the energy required for mechanical pumps. Other estimates of energy usage can be made from Figure 2–22 (see p. 2–21). For liquid piston pumps, some of the sealant liquid evaporates. The amount of makeup liquid required is

$$\check{F}, \frac{L}{s} = 0.5 \left[\frac{\text{(g/s of air)}}{\text{(absolute pressure, kPa)}} \right]^{0.9} \quad (2\text{-}11)$$

TABLE 2-3. Various Fan Configurations

CENTRIFUGAL			AXIAL	
RADIAL	FORWARD	BACKWARD		
Dusty, dirty gases	Clean gas, medium pressures, compact, quiet	Clean gas, low to high pressures	Clean to relatively clean gases, low to high pressures, compact, can be noisy	
Drive: Pulley	Pulley	Direct	Direct	
Static fan effic: 65%	60%	70%	75%	
Self limiting power: No	No	Yes	Yes	
% Max volume rate	% Max volume rate	% Max volume rate	% Max volume rate	**Centrifugal fan:** produces radial flow similar to a centrifugal pump
To control volume flowrate: Damper Not bypass	Damper Not bypass	Damper Perhaps bypass	Bypass; speed Not damper	**Axial fan:** produces continuous flow parallel to the shaft of the fan

---

Example 2–10: What device would you use to convey 10 kg · h⁻¹ of air and maintain a vacuum of 1 kPa?

An Answer: From Figure 2–20, we might be able to use a four-stage steam ejector or a two-stage dry reciprocating pump. Both seem to be close to the border. If in doubt, we could use a five-stage steam ejector system or an oil-sealed rotary pump.

Figure 2–23 (see p. 2–22) illustrates the configuration used to hook up "stages" of steam ejectors. Usually we use an open or direct contact condenser between each two steam ejectors so as to condense out and minimize the amount of vapor that has to be handled. An ejector is a high pressure-vacuum differential, low-capacity device; a booster ejector is a small pressure-vacuum differential, high-capacity device. Booster ejectors are large physically, say perhaps 2 to 3 m long; ejectors are small, say perhaps 8 to 10 cm diameter and 1 m long or smaller. Do not use mechanical vacuum pumps for condensable gases; the gas condenses and then flashes on the expansion stroke. Thus, little vacuum is pulled.

Often in vacuum installations, one of the most difficult tasks is to identify the amount of air that needs to be extracted from the system. This is difficult because the air in the system (for steady-state, continuous processing) is the air that leaks in through the cracks and flanges. It is dangerous to generalize, but for many industrial-size units,

TABLE 2-4. Options for Describing Fan Performance

CONCEPT	OPTIONS			
Pressure	static pressure, p_s		impact pressure, p_i	velocity pressure, $p_i - p_s$
Fan performance	**Static pressure difference** $\Delta p = (p_{s2} - p_{s1})$	**Fan static pressure** $fsp = p_{s2} - p_{i1}$ $fsp = ftp - \langle p_v \rangle$	**Fan total pressure** $ftp = p_{i2} - p_{i1}$ $ftp = fsp + \langle p_v \rangle$	**Fan velocity pressure** $\langle pv \rangle_2 = \rho \langle \frac{v}{2} \rangle_2^2$ $\langle pv \rangle_2 = ftp - fsp$
	Easy to measure	Reported on performance curves	Reported on performance curves	
	Relationships	If $Area_1 \gg Area_2$ $fsp = \Delta p$	If $Area_1 = Area_2$ $ftp = \Delta p$	
System needs			Expressed in terms of total pressure	

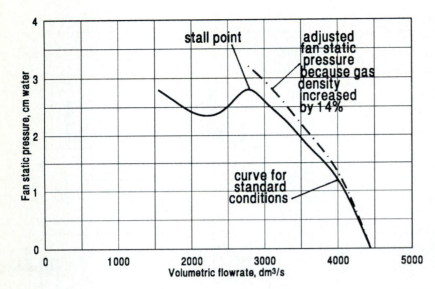

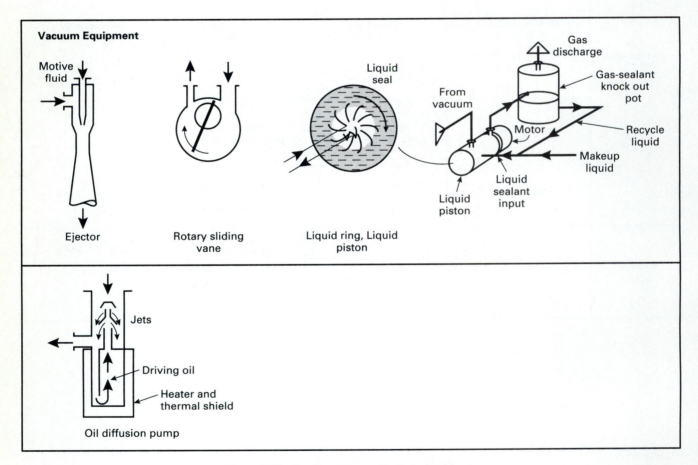

Figure 2–18 Correcting a Fan Performance Curve for Gas Density (Single-Stage Aerofoil Fan, 60-cm Diam., 24°, 1440 RPM)

Figure 2–19 Sketches of Vacuum-Producing Equipment

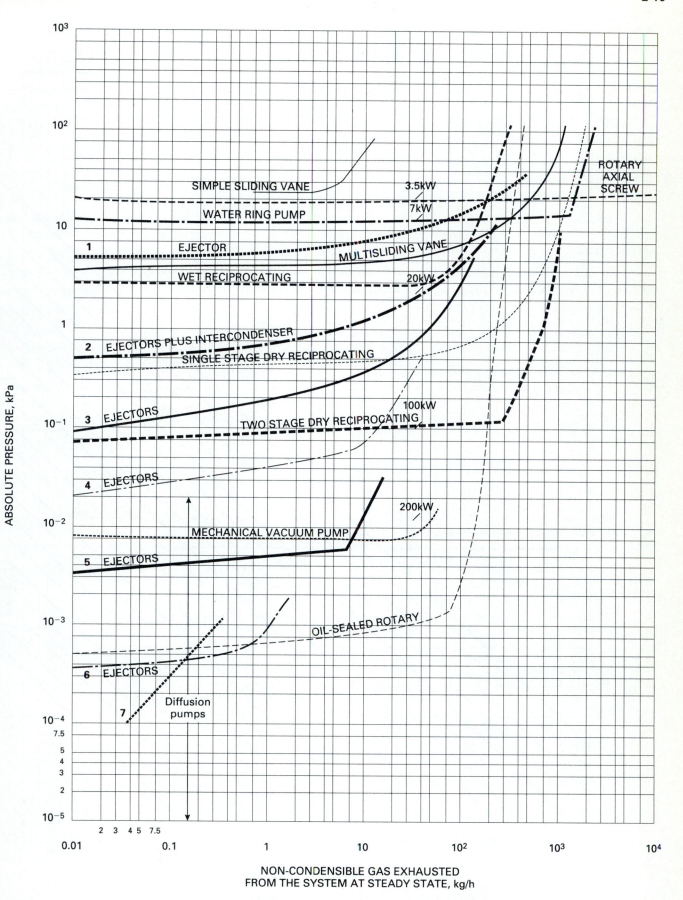

Figure 2–20 General Regions of Applicability for "Vacuum" Gas Moving Equipment

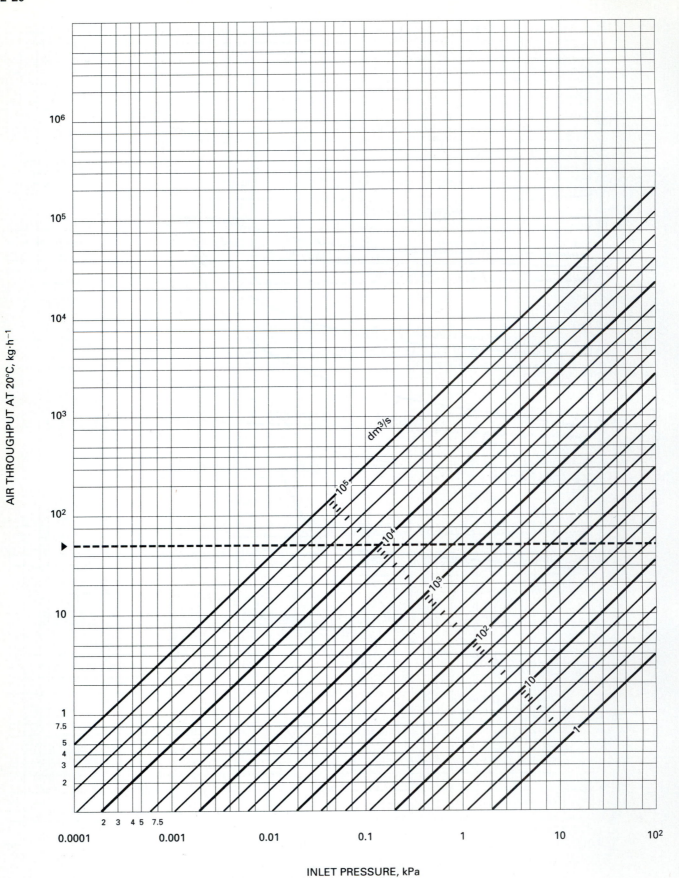

Figure 2–21 Relationship Between Inlet Vacuum Pressure, Mass, and Volume Flowrates (for Air at 20°C)

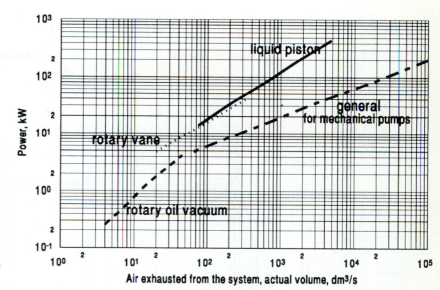

Figure 2–22 Power Requirements for Some Mechanical Vacuum Equipment

this is approximately 50 kg · h^{-1} of air that leaks in and must be exhausted by vacuum device. More details on how to estimate the air leakage are given by Ludwig (1964) and Ryans and Croll (1981).

If an ejector run by steam is used, we often need to estimate how much steam is required to run the ejector for a given duty. Some estimates can be made from Figures 2–24 and 2–25. The fundamental design principles for ejectors are outlined by Hougen, Watson, and Ragatz (1959), pages 715 to 722.

Example 2–11: Estimate the power needed to drive a liquid piston vacuum pump to extract 50 kg · h^{-1} of air at an absolute pressure of 20 kPa.

An Answer: Check first on Figure 2–20 to see if a liquid piston pump is applicable. The term used in this figure is water ring pump. Yes, it seems to apply. Next, to use Figure 2–20 requires that we convert the mass flow into volumetric flow. On Figure 2–21, this corresponds with (at 20 kPa and 50 kg · h^{-1}) 50 dm^3 · s^{-1}. The power needed, from Figure 2–22, is about 10 kW.

Example 2–12: Identify the steam ejector configuration and the steam usage to run a distillation column at an overhead pressure of 5 kPa absolute and extract 10 kg · h^{-1} of air.

An Answer: From Figure 2–20, the tentative application looks like a two-stage ejector plus an intercooler condenser. On Figure 2–25, the steam usage is about 5.5 kg steam per kg of air exhausted. Hence, the steam usage to extract 10 kg · h^{-1} of air is 55 kg · h^{-1} of 10 MPa steam.

2.3 TRANSPORTING LIQUIDS

The types of devices available for transporting liquids are shown in Figure 2–26. Although there are some special types of pumps—such as the Archimedes screw—three basic types are available: the centrifugal, the reciprocating, and the rotary. Table 2–5 (see p. 2–25) classifies the types according to the method used to transmit energy.

2.3-1 Selecting Devices

In selecting a pump, the first consideration is the viscosity; Figure 2–27 (see p. 2–26) illustrates how the viscosity dictates the general class of pump.

Once the general class has been selected, then either Figure 2–27 or 2–28 (see p. 2–27) is used to identify the appropriate subclass of pump. Unlike gas moving equipment, where most of the graphs referred to air (because the behavior depended on the material being pumped), for the liquid moving devices the head and capacity are independent of the fluid for centrifugal pumps.

Example 2–13: Select a pump to handle 20 L · s^{-1} of 100 mPa · s fluid against a 30 m head.

An Answer: From Figure 2–27 at the top for 100 mPa · s fluid, the alternative is to use a centrifugal pump. Since 20 L · s^{-1} = 0.02 m^3 · s^{-1}, from Figure 2–27 a single-stage centrifugal radial flow pump seems appropriate.

2.3-2 Estimating Power

Sometimes we need to estimate the power required. Figure 2–29 (see p. 2–27) shows the power for pumping water. The power depends on the density because to determine the

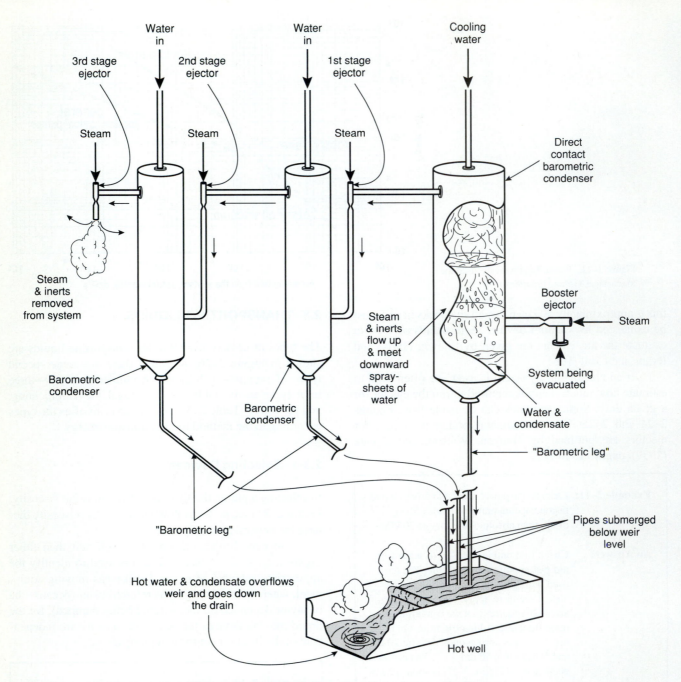

Figure 2–23   Example of a Three-Stage Ejector System with a Booster

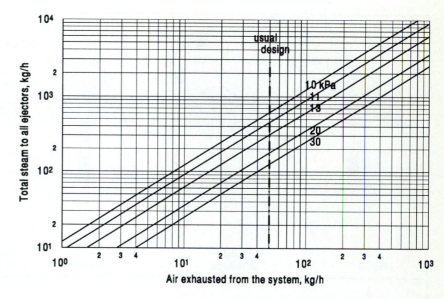

Figure 2–24 Estimates of Steam Requirements for Ejectors for Different Exhaust Rates and Absolute Pressure (for the Industrial Range)

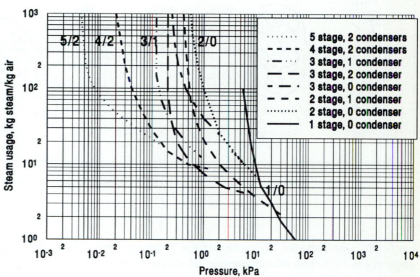

Figure 2–25 Approximate Steam Usage for Steam Ejectors for Different Vacuum Conditions (for Steam at 10 MPa and Cooling Water at 30°C)

power, we multiply the mass of fluid by the head it is being transported against and divide by a reasonable efficiency for a pump of 60%. Thus

$$kW = \frac{\check{F} \times (\rho g) \times head}{0.60} \qquad (2\text{-}12)$$

For example, to pump 30 L · s⁻¹ of water against 100 m of head requires

$$= 1.3 \, L \, \rho g = 1.3 \times 30 \, m \times 1.4 \, \frac{kg}{m^3} \times 9.8 \, \frac{m}{s^2}$$

$$= 50 \, kW$$

Because the data in Figure 2–29 are based on water, the power requirements must be adjusted by multiplying by the density ratio to obtain reasonable value. Methods for estimating the frictional contribution to the power are given in Section 2.3–4.

Example 2–14: Determine the power to pump benzene of density 0.8 Mg · m⁻³ at 100 L · s⁻¹ against a 10 m head.

An Answer: From Figure 2–29, for water the answer is about 20 kW. Hence, for the benzene application, the power needed is corrected by the density ratio and since the density of water is 1 Mg · m⁻³, the answer is

$$(0.8/1.0)20 = 16 \, kW.$$

2.3-3 Estimating Pipe Size

It is very useful to be able to visualize the diameter of the piping needed to convey the liquid. Figure 2–30 (see p. 2–28) shows the relationships together with typical val-

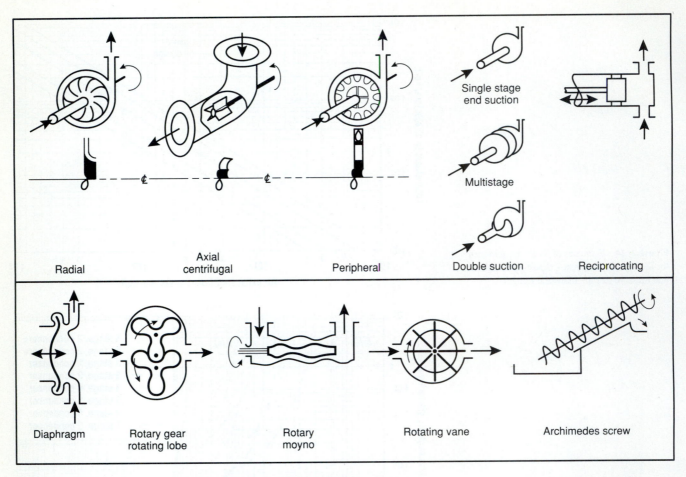

Figure 2–26 Sketches of Pumps

ues for the pipe diameter, flow relationships for pump-pipe networks and for open flow drainpipes for downcomers in a tray for a distillation column.

Example 2–15: The diameter of a pipe on the exit side of a pump is 10 cm. Estimate the flowrate in that pipe.

An Answer: For most pump-pipe systems, the economic diameter of pipe is chosen to minimize the annual operating cost. From Figure 2–30, the economic pump-pipe line intersects the 10 cm line at $9 \text{ L} \cdot \text{s}^{-1}$. Hence, an estimate of the liquid flow is $9 \text{ L} \cdot \text{s}^{-1}$.

Example 2–16: The eavestrough on the house is 5 cm in diameter. Estimate the roof area that this one eavestrough can service if the maximum expected rainfall is $2 \text{ cm} \cdot \text{h}^{-1}$.

An Answer: From Figure 2–30, the gravity flow curve intersects the 5 cm line at $0.25 \text{ L} \cdot \text{s}^{-1}$

flowrate. Since this is the maximum that can be collected from the area, and since the rainfall rate is given, the result is:

$$\text{area, m}^2 = \frac{0.25 \text{ L}}{\text{s}} \times \frac{10^{-3} \text{ m}^3}{\text{L}} \times \frac{\text{h}}{2 \text{cm}} \cdot$$
$$\frac{3600 \text{ s}}{\text{h}} \times \frac{100 \text{ cm}}{\text{m}}$$
$$= 45 \text{ m}^2.$$

Example 2–17: Liquid sodium is being circulated through a heat exchanger system at the rate of $1 \text{ L} \cdot \text{s}^{-1}$. What diameter of pipe seems reasonable?

An Answer: Since this is a pump-heat exchanger system, the recommended economic velocities are higher than the pump-size economic velocity. For $1 \text{ L} \cdot \text{s}^{-1}$, a reasonable diameter would be, from Figure 2–30, 2.5 cm.

TABLE 2-5. Classification of Equipment for Transporting Liquids

Form of Energy Principle	Mechanical		Fluidic
		Reciprocating	
Kinetic Energy Interchange	Rotary Centrifugal Radial Axial Mixed Flow Turbine		Ejector
Volumetric Displacement	Rotary gear — internal gear — lobe — vane — sliding vane — screw — cam-piston — squeegee — moyno Peristaltic	Piston Diaphragm Plunger	Pressure tank Barometric leg

2-25

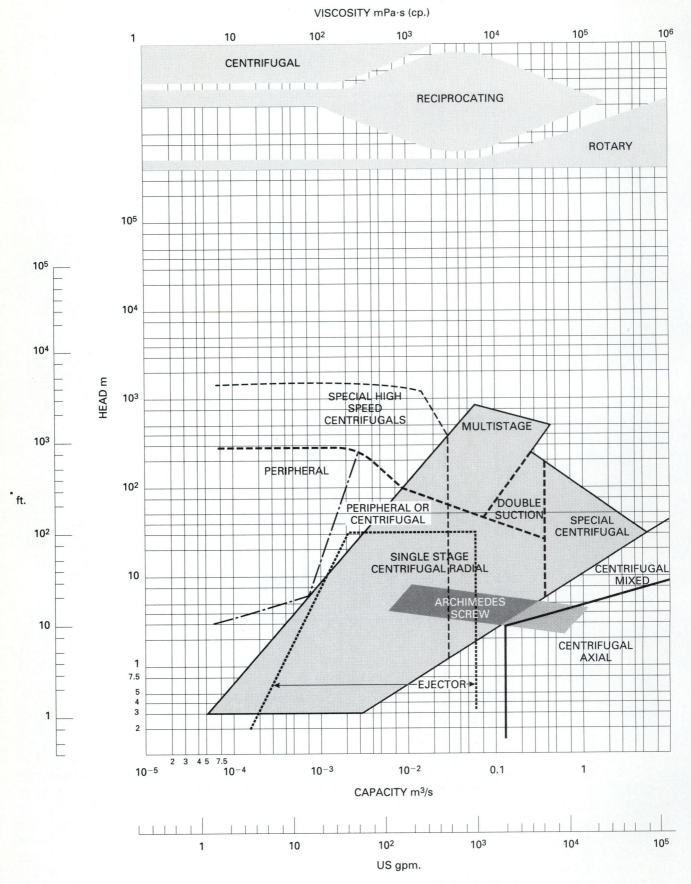

Figure 2–27 General Regions of Applicability for Liquid Pumps: Centrifugal

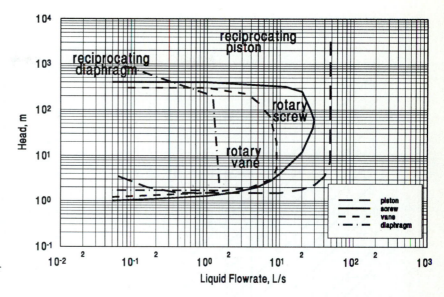

Figure 2–28 General Regions of Applicability for Liquid Pumps: Reciprocating and Rotary

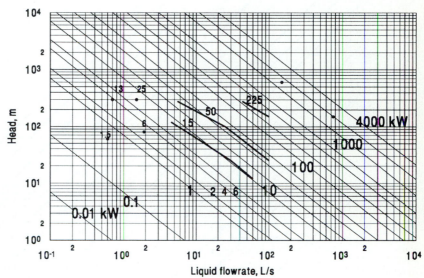

Figure 2–29 Estimating Power for Pumping Water (at 60% Efficiency, with Data Points Added)

2.3-4 Estimating Pressure Loss

The pressure loss because of friction can be estimated from Equation 2–3. The same results apply as for gases:

- the fluid loses one velocity head for every 45 to 50 L/D of equivalent pipe;
- the usual turbulent friction factor is about 0.005.

For water of density 1 Mg · m^{-3} the pressure loss per 100 meters of pipe is given in Figure 2–31. This figure is useful because some prefer to use a rule-of-thumb related to pressure drop (instead of using an economic velocity approach given in Figure 2–30). Thus, Lieberman (1983) uses a guideline of 23 kPa/100 m of pipe; Bell (1985) suggests 10 and 32.5 for water in headers and lateral configurations. A precaution in the use of Figure 2–31 is that the data apply for water; density and viscosity corrections are needed for other fluids and when the friction factor varies with the

Reynolds number. For water flowing at 1 m/s, one velocity head is 0.5 kPa of pressure drop or 0.005 m water head loss.

The friction loss caused by liquids flowing through "fittings" can be estimated using the same approach as with gases and the data in Table 2–1. Usually we express the results as velocity head or equivalent length of pipe.

Sometimes we want to estimate the pressure drop across a control valve. Usually the control valve is put in "one size" smaller than the pipe. Thus, the pressure drop across a fully open valve would be about 3 times the velocity head estimated based on full-pipe velocity. In addition, the valve will likely be about ½ shut so as to effectively fulfill its control function. Different authors make different recommendations about the reasonable allowance; see Smith and Corripio (1985), Lieberman (1983), Oliver (1966). For the situation shown in Figure 2–32, where the control valve is in the system between the pump and the delivery point, then the pressure drop across the control valve at design conditions should be:

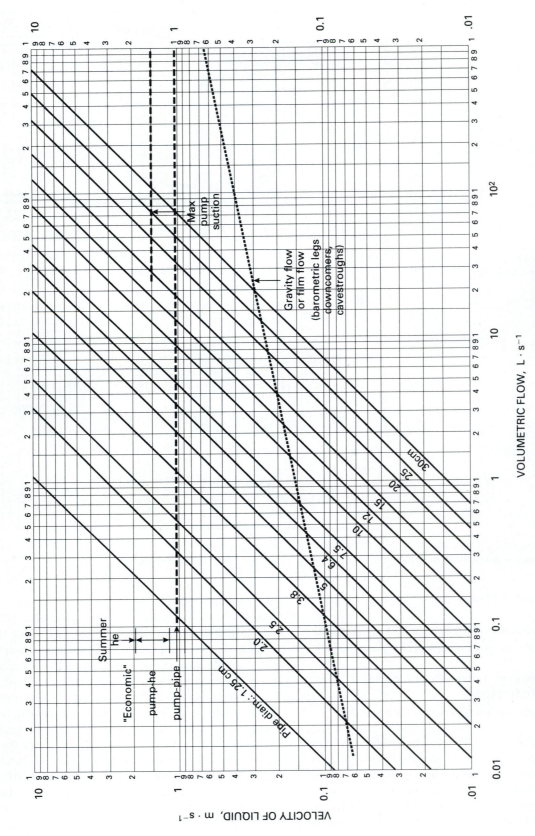

Figure 2-30 Flowrate, Velocity, and Diameter Relationships for Liquids (These Results Are Independent of Density)

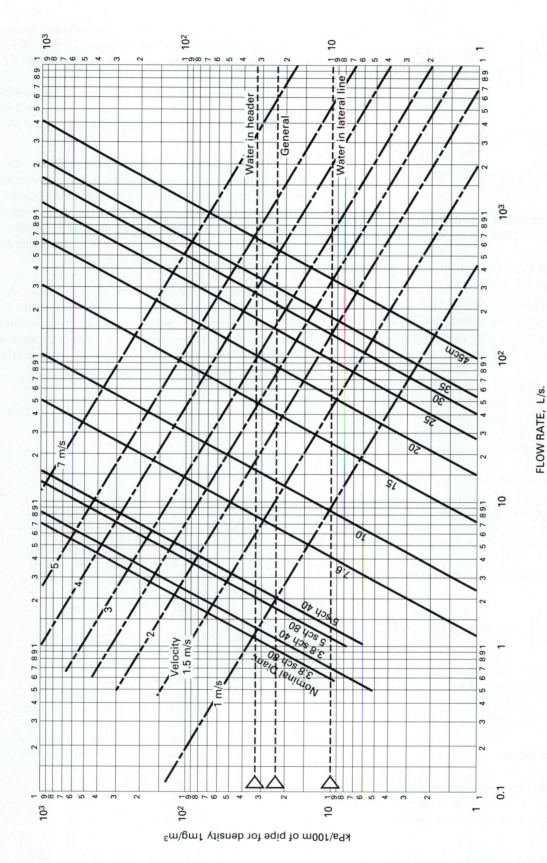

Figure 2-31 Pressure Loss for Turbulent Flow

$$\left(\text{Based on } \frac{\Delta p}{\rho} = 4f\left(\frac{L}{D}\right)\frac{<v>^2}{2g}\right)$$

FLOW RATE, L/s.

kPa/100m of pipe for density 1mg/m³

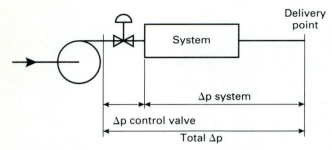

Figure 2–32 Definitions of Pressure Losses
in a Control System

- 20 to 50% of the system total dynamic (representing the sum of the frictional and the $\Delta \langle v \rangle^2/2g$ contributions);
- the larger of 70 kPa (7 m water) versus 25% of the total system dynamic loss; or
- the larger of 140 kPa (or 14 m water) versus 50% of the system frictional loss. A more detailed analysis of estimating the pressure drop across control valves is given by Connell (1987).

The pressure drop for liquid flowing through a packed bed of particles can be estimated from Figure 2–15. Alternatively, Figure 2–33 shows the pressure loss for liquids flowing through various resins typical of ion exchange resins and activated carbon adsorption granules used in separations.

All the ideas presented here refer to Newtonian type materials where, although the viscosity may be large, nevertheless it is constant and a single value. The pumping of ketchup, for example, would not behave this way. Methods for pumping non-Newtonian fluids would have to be consulted.

We can estimate the power needed to *overcome friction* in a pipe to be:

$$\text{Power, Watts} = \left[\frac{L/D}{50} \right] [\check{F}, \text{L/s}][\text{velocity, m/s}]^2 [\rho] \quad (2\text{-}13)$$

where

  L/D = length to diameter ratio
  $\check{F}$ = volumetric flowrate, L/s
  v = average velocity, m/s
  ρ = fluid density, Mg/m^3

2.3–5 Enrichment Ideas

The preceding information is sufficient to select, rough-size, and cost a pump. However, some additional information is needed to improve our selection of pumps.

A. Performance Curves. Pump performance curves are expressed in terms of the "head of liquid" supplied versus the volumetric capacity. Such performance is independent of the density of the fluid. Thus, whether the pump is pumping mercury or hexane, the performance curve is the same; the pressure inside the pipes is very different but the head supplied by the pump is the same. Provided the liquid is Newtonian and has a kinematic viscosity less than 1 cm^2/s, no corrections to the head-capacity curve are needed. For Newtonian fluids with higher kinematic viscosities, the Standards of the Hydraulics Institute provide correction charts.

B. Similitude and Changing Conditions. If we have an existing pump, can we change the diameter of the impeller or the rotational speed of the pump to achieve the required set of conditions? For centrifugal pumps we can use the similarity of the velocity vectors of the liquid leaving the tip of the impeller to relate the variation in the volumetric flowrate, $\check{F}$, the pressure difference, h, and the power required, P, to the impeller diameter, D, and the rpm, N. The similarity is illustrated in Figure 2–34 between impellers of different diameters. Figure 2–34 also illustrates the relationship between the curve of the vanes on the impeller and the direction of rotation. We can use these similitude relationships to predict new performance curves if the shape of the housing and blade are maintained but the other parameters are changed. The "similitude" relationships for pumps are:

$$\frac{F_1}{F_2} = \left(\frac{N_1}{N_2} \right) \times \left(\frac{D_1}{D_2} \right) \quad (2\text{-}14)$$

$$\left(\frac{h_1}{h_2} \right) = \left(\frac{N_1}{N_2} \right)^2 \times \left(\frac{D_1}{D_2} \right)^2 \quad (2\text{-}15)$$

$$\left(\frac{P_1}{P_2} \right) = \left(\frac{N_1}{N_2} \right)^3 \times \left(\frac{D_1}{D_2} \right)^3 \quad (2\text{-}16)$$

Example 2–18: The performance curve for an existing pump is given in Figure 2–8 for 1800 rpm. If we altered the pump speed to 3600 rpm, could this pump be used to provide 15 L/s and 30 m head?

An Answer: Although we could consult the pump manufacturer to obtain a new head capacity curve for this new condition, we can use the similitude laws to estimate the curve for the new conditions. The approach is to arbitrarily select one point on the curve in Figure 2–8 (say 5 L/s and 10 m head) and apply the adjustments to the flowrate and head in turn. Thus, the flowrate under the new conditions would be, from Equation 2–14:

$$F_{\text{new}} = 5 \text{ L/s} \left| \frac{3550}{1750} \right| = 10 \text{ L/s}$$

and from Equation 2–15, for the head:

$$h_{\text{new}} = 10 \text{ m} \left[\frac{3550}{1750} \right]^2 = 40 \text{ m}$$

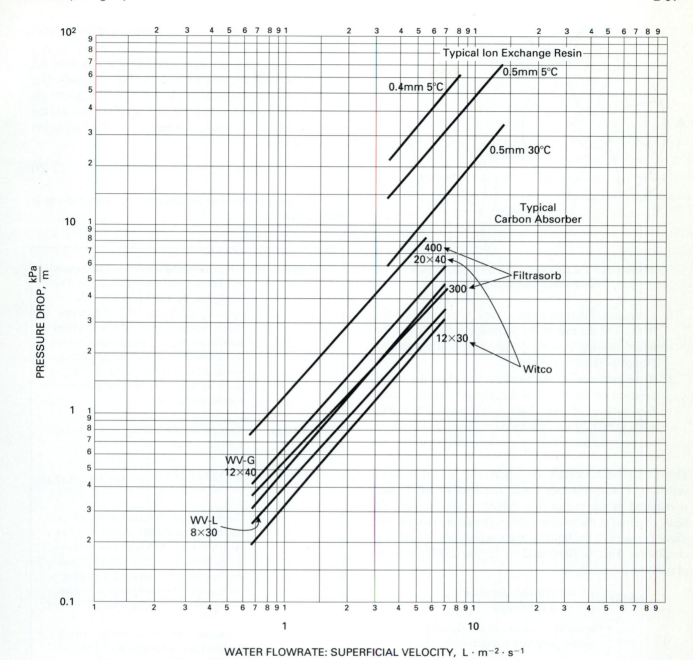

Figure 2–33 Pressure Drop for Liquid Water Flowing Through a Typical
Ion Exchange or Activated Carbon Bed

Thus, the point on the 3600 rpm curve is 10 L/s and 40 m head.

Similarly, other points can be estimated. Figure 2–35 illustrates the curve supplied by the manufacturer for this pump; the estimated data are also shown. The agreement is very good.

C. Special Considerations for the Suction Side of the Pump.   When considering the pumping of

liquid from a sump, or when we are pumping liquid that is close to its boiling point, special considerations are needed for the suction side of the pump. Indeed can the pump be expected to lift the liquid to the pump? Is sufficient pressure kept on the liquid to prevent it from boiling or flashing in the pump or pipe? These two conditions are illustrated in Figure 2–36. For a pump to work well, homogeneous liquid (that is, free of bubbles) must be transported into the rotating impeller inside the pump. We note that the lowest pressure in the system is at the eye of the impeller inside the pump. This leads to two fundamental ideas:

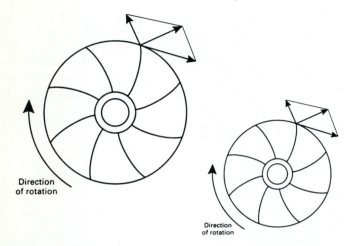

Figure 2–34   The Velocity Vectors Are Similar for Both Impellers for a Centrifugal Device

1. a pressure difference must exist between the inlet flange and the eye of the rotating impeller to get the liquid flowing into the pump. This is illustrated in Figure 2–37a.
2. if the pressure on the liquid falls below the vapor pressure of the liquid at the pumping temperature, then bubbles of vapor form in the liquid. The liquid starts to boil. This is illustrated in Figure 2–37a where the pressure has fallen below the vapor pressure.

These two fundamentals are combined in the concept called Net Positive Suction Head "required" or $NPSH_r$ that is required by the pump. The $NPSH_r$ is the pressure, or head, required at the inlet or suction flange in excess of the vapor pressure of the liquid at the temperature of pumping conditions. This is illustrated in Figure 2–37b and expressed by the equation:

$$NPSH_r = \begin{pmatrix} \text{Total pressure supplied} \\ \text{at the flange} \end{pmatrix} - \\ \begin{pmatrix} \text{The vapor pressure of the liquid} \\ \text{at the pumping temperature} \end{pmatrix} \quad (2\text{-}17)$$

This value depends on the design of the pump, is expressed in meters of water, and must be supplied by the pump manufacturer. An example is shown in Figure 2–38 for the pump characterized by the performance given in Figure 2–8. For estimation purposes, for optimum pump design the $NPSH_r$ may be estimated to be:

$$NPSH_r = \left| \frac{N\check{F}}{5400} \right|^{4/3} \text{meters} \quad (2\text{-}18)$$

where
   N = rpm
   $\check{F}$ = flowrate in L/s

If the $NPSH_r$ required by the pump is not supplied, the pump "cavitates." A loud chattering noise is heard; the

pump performance drops; unwanted erosion of the impeller occurs.

Thus, we need to supply a $NPSH_s$ that is at least 0.5 m more than that required, $NPSH_r$. The $NPSH_s$ supplied by the system can be calculated from Equation 2–1, where the first reference condition is at the free liquid surface and the second reference condition is at the centerline of the pump suction flange:

$$NPSH_s = p_1 - p_2 + z_{2-1} - \text{friction loss}_{2-1} \quad (2\text{-}19)$$

Vapor pressure data for different liquids are given in **Data** Part C.

Example 2–19:   We are selecting a pump for the suction lift situation given in Figure 2–36a. The details are: the vapor pressure at the operating temperature is 5 kPa; atmospheric pressure at p_1 is 90 kPa; the fluid density is 1 Mg/m^3. The pump is located 5 m above the liquid level; the friction loss in the suction line is 1.2 m. The pump $NPSH_r = 4$ m. Will the pump work without cavitating?

An Answer: Thus the supplied $NPSH_s$ is $p_1 - p_{vp}/\rho g + z_2 - z_1 -$ friction or

$$\left(\frac{90 \text{ kPa} - 5 \text{ kPa}}{1.0 \times 9.8} \right) + (0 - 5) - 1.2 = 2.47 \text{ m}$$

For the pump to operate without cavitating, the supply $NPSH_s$ should be 0.5 m greater than the $NPSH_r$ or greater than 4.5 m. Since the $NPSH_s$ is 2.47 m, and thus less than that required, the pump will cavitate. Another pump or another configuration should be selected.

Sometimes the term *suction lift* is used. Suction lift is defined as the NPSH – static height of water at 21°C that would be supported by a perfect vacuum (or 10.06 m). Thus, if a manufacturer says that the pump has a maximum suction lift of 6 m, then the $NPSH_r$ is 10.06 – 6 or 4.06 m.

Example 2–20:   The bottoms in a distillation column are to be pumped to heat exchange and storage. The configuration is illustrated in Figure 2–36b. The liquid level in the column is 3.3 m above the pump; the friction loss in the line is 0.5 m. The pump $NPSH_r$ is 2 m. Will the pump work without cavitating?

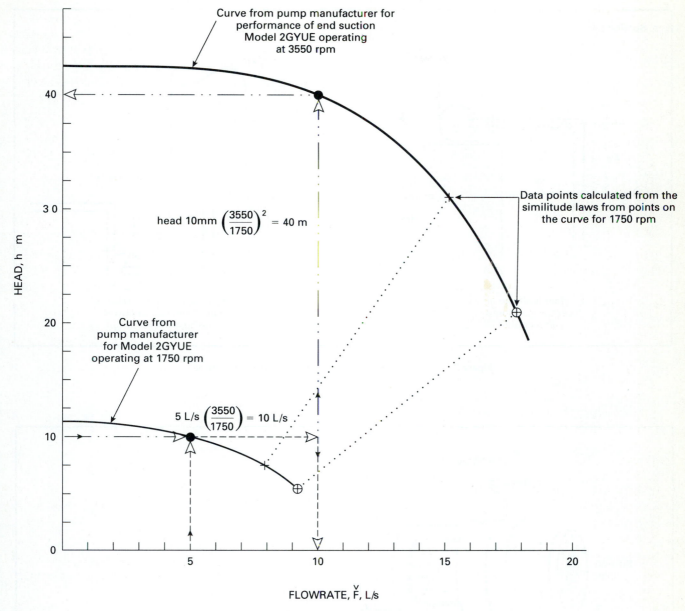

Figure 2–35   Use the Similitude Laws To Predict the Performance
Curve Under New Conditions

An Answer: Because the liquid is boiling, the $p_1 - p_{vp}$ term is zero. Thus the $NPSH_s$ is $z_2 - z_1$ – friction or

$$(3.30 - 0) - 0.5 = 2.8 \text{ m}$$

The $NPSH_r$ is 2 m. Thus, $NPSH_s$ is 0.8 m greater than $NPSH_r$. The pump should work without cavitating.

in the suction line contains no bubbles. If bubbles were entrained, then the density of the fluid would decrease, and the static head contribution to the $NPSH_s$ would decrease. Often we install vortex breakers to prevent the entrainment of vapor in the liquid. Example designs of vortex breakers are given in Figure 2–39. Also shown is the degree of submergence that would be needed to prevent vapor entrainment.

This analysis has interesting implications for the installations. First, to maximize the $NPSH_s$ we should try to minimize the pressure drop in the suction line. Hence, normally, we install larger diameter pipes on the suction side. Second, the static head contribution assumes that the liquid

Third, if a pump in an existing facility begins to cavitate, a temporary troubleshooting activity might be to pour cold water on the pump housing to try to decrease the vapor pressure of the liquid. Fourth, usually the practical height that we can lift liquids out of a well or sump on the suction side of a pump is about 5 m; the maximum possible is 10.06 m.

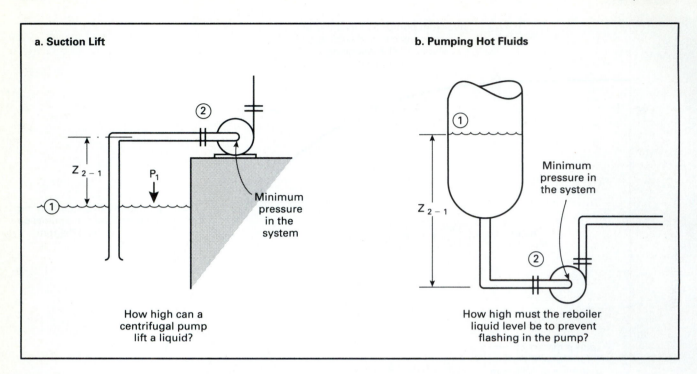

Figure 2–36   Two Operating Conditions Where Suction Is Important

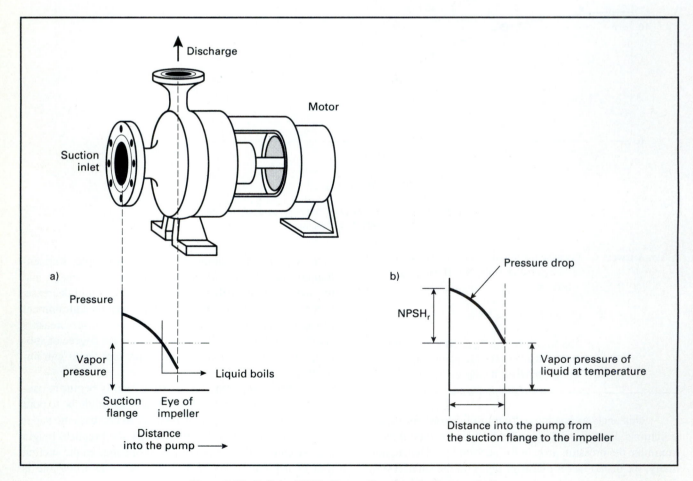

Figure 2–37   Defining NPSH$_r$: Pressure Drop from the Flange to the Eye

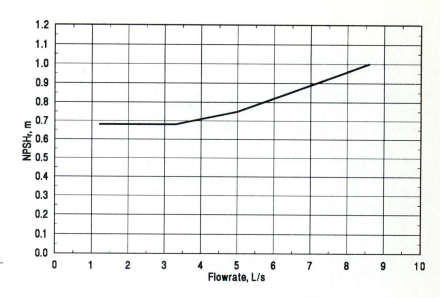

Figure 2–38 Manufacturer's NPSH$_r$ Data for an End-Suction Pump Operating at 1750 RPM

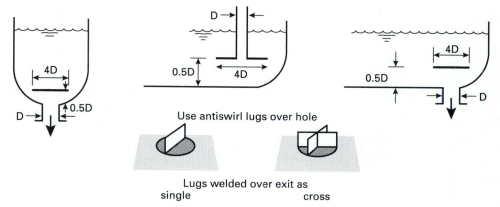

Use antiswirl lugs over hole

Lugs welded over exit as
single   cross

a) Some Vortex Breaker Designs (based on Kerns, 1960 and reprinted with permission from "New Graphs Speed Drum Sizing," Petroleum Refiner, July, 1960, copyright Gulf Publishing Co. 1960, all rights reserved)

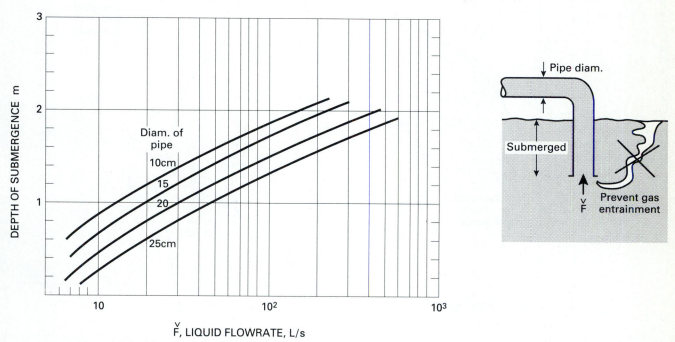

b) Use Submergence (based on Yedidiah 1977 and reproduced with permission from Chemical engineering, copyright 1977)

Figure 2–39 Controlling Gas Entrainment

2.4 PUMPING SLURRIES

In Section 2.3, we considered pumping clean liquids. What if the liquids contain some solids? As the solids concentration increases, we then have to consider the size of the particles. Figure 2–40 illustrates the general applicability of centrifugal pumps for pumping slurries. More details are given on page 2–57, hydraulic conveying. Indeed, one of the ways to transport solids is to mix them with water, or "repulp" to say about 30% w/w solids. Then the slurry is pumped. Figure 2–40 helps us to realize when different considerations and precautions should be taken.

2.5 TRANSPORTATION OF SOLIDS

A wide variety of devices are available to convey solids; some are illustrated in Figure 2–41. Some are mechanical, some pneumatic, and some are short-distance feeders. The choice depends often on the solid characteristics, the space

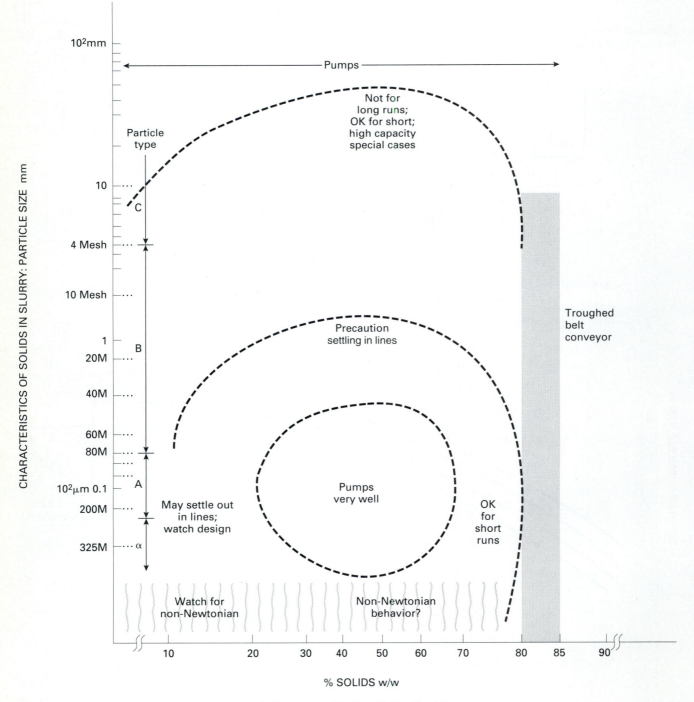

Figure 2–40 Pumping Slurries with Centrifugal Pumps

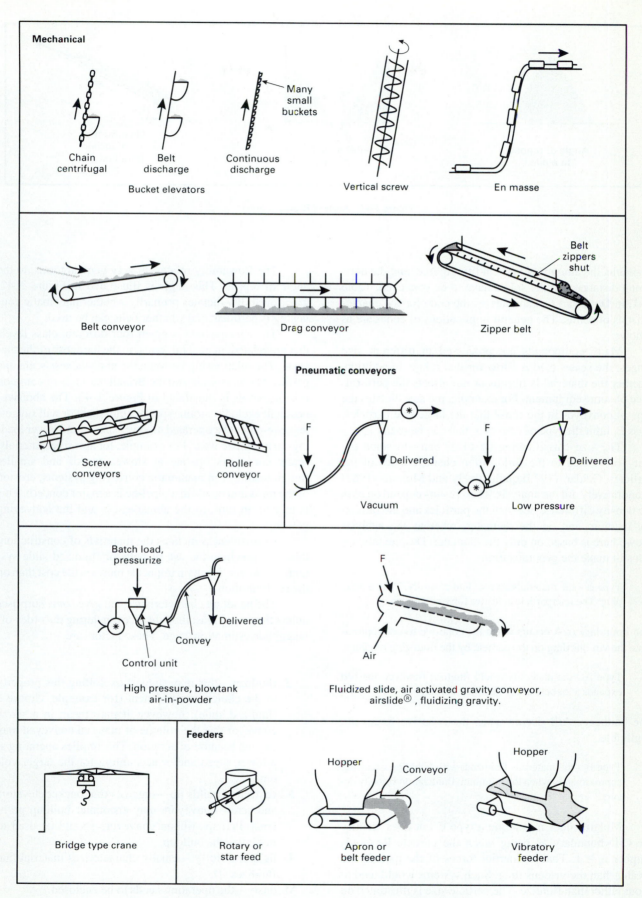

Figure 2–41 Sketches of Solids Transportation Equipment

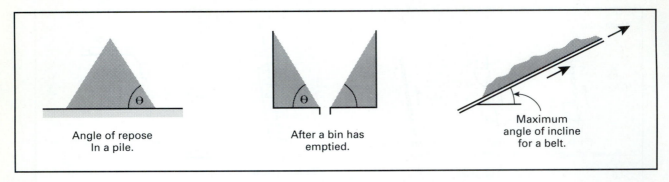

Figure 2–42   Angle of Repose of Solids

available, the amount of vertical lift required, and the horizontal distance that the solids have to be conveyed. Table C–1, in **Data** Part C, summarizes the code for the characteristics of solids. The general implications of codes are as follows:

Many compounds are processed in different size ranges: the scales α to Δ allow for that range. In addition, whether the material is fibrous or not affects the performance of some equipment. For example, the size dictates the type of conveyor in the sense that the large size particles, class E, limit the type of conveyor that can be used.

The size classification (α to Δ) can—to some extent—be related to the fluidization characteristics of the particles. Geldart (1973) qualitatively and Molerus (1982) quantitatively did the analysis. Their results depend on *both* the density difference between the particles and the fluidizing medium and on the diameter, whereas the analysis shown here is based on only the diameter. Despite this, we can still make the generalization:

> Type α—the materials do not fluidize easily. They tend to plug. The interparticle cohesive forces dominate.

The boundary α/A occurs when the cohesive forces approximate the viscous drag on the particle by the fluidizing medium.

> Type A—the materials would fluidize; however, the bed expands a lot before any bubbling occurs.

The boundary A/B occurs when the cohesive forces are negligible.

> Type B—the materials fluidized, with bubbling occurring approximately when the minimum fluidization velocity occurs.

Molerus does not define a type C but does identify the C/D boundary as being when the particle Reynolds number is ≥ 1. That is, inertial forces of the particle are greater than the viscous drag. Such systems would tend to spout rather than fluidize. The particle size is illustrated on the ordinate of Figure 2–40.

The flowability relates, in very general terms, to the angle of repose. This angle is illustrated in Figure 2–42. The flowability dictates primarily whether pneumatic conveying is pertinent and whether belts can be used.

The abrasiveness is given four different class levels that are related, in general terms, to the hardness of the material. The relationship between the abrasiveness scale, the particle hardness scale and the Brinell hardness of common pipe materials is illustrated in Figure 2–43. The abrasiveness will exclude certain types of equipment or will suggest that equipment be operated at lower speeds and lower loadings to minimize wear. For example, for abrasive materials, screw conveyors operate at slower speeds and smaller trough loadings; in pneumatic conveying systems, erosion of the turns and bends in a pipeline is a major concern. This is related, in turn, to the abrasiveness and the conveying velocity.

Corrosiveness dictates the materials of construction; this may preclude the use of belts or fluidized slide systems. However, the main impact is more on the cost than on the configuration.

The handling characteristics can give some surprises unless they are at least thought of, even during the order-of-magnitude estimation stage. These factors are:

J fluidizes—this may suggest exploiting this property in the choice of equipment (for example, choose a fluidized slide); however, it may mean in a screw conveyor that the volume of material conveyed may expand because of aeration. This implies operating at a lower speed and/or accounting for the larger volume.

K cakes and builds up—suggests that bucket elevators and screw conveyors may encounter buildup problems. Perhaps ribbon conveyors should be used to minimize the buildup.

L light and fluffy—usually characterizes material that fluidizes (J).

M dusty—the operating needs to be enclosed.

N sticky—like K in its implications. Usually in bucket

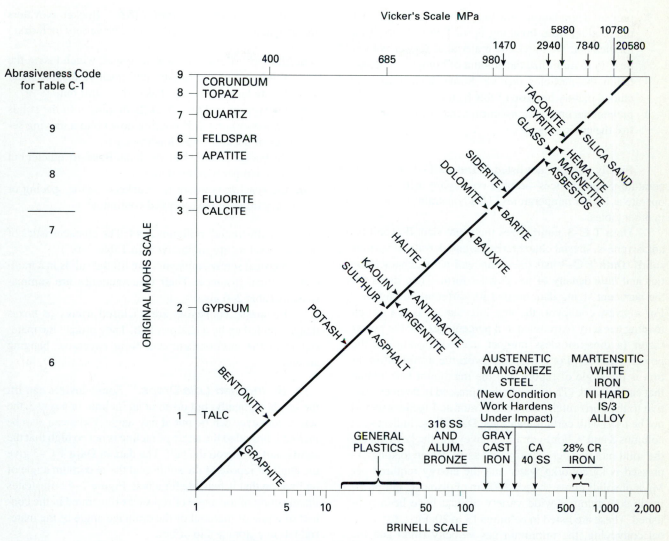

Approximate Comparison of Hardness Values of Various Common Ore Minerals and Metals

Figure 2–43 Abrasiveness and Hardness of Solids
(from Wasp et al. 1977 and reproduced with permission)

elevators, a positive displacement discharge (as op-posed to a centrifugal discharge) is needed.

O hygroscopic—tightly sealed units may be used, or the resulting density change should be anticipated when estimating the power requirement.

P is similar to N.

Q packs under pressure—suggests that care be taken in designing the feeder. Indeed a feeder that aerates the material might be used for applications that normally do not require one. For example, a screw conveyor normally does not need an upstream feeder; however, for this application, it might be needed.

R similar to Q.

The safety hazard ratings also have implications.

S many powders generate static electricity, especially when they are pneumatically conveyed. This usually means that the moisture content of the gas should be increased and this, in turn, affects the sizing proce-dures outlined here.

T toxic fumes or toxic materials—usually we try to op-erate under a vacuum; the devices need to be tightly sealed, and care is needed in designing the exhaust and separation device.

U flammable or explosive—the system should be
and tightly sealed and perhaps explosive suppressors
V added.

In general, the codes in this category affect the extent and types of enclosures needed.

The product considerations include:

W "prevent contamination"—affects the choice of bear-ings and the ease with which the device can be cleaned frequently.

X prevent degradation—means that we want to prevent the product from breaking up or falling apart. This implies that we move the material at slower velocities, try to minimize the amount of bumping and agitation the particles receive. Some devices are excluded mainly because of this factor.

Y prevent decomposition—means that we need to control the environment.

Finally, a factor not listed in **Data** T C–1 is the temperature. Some devices—rubber conveying belts—cannot operate at higher temperatures. This constraint is included in other tables.

Data T C–3 summarizes the usual size, flowability, abrasiveness, special characteristics, and density of various solids. **Data** T C–3 lists the compound name, characteristics and *bulk* density of the powder form. Then listed are the pertinent sizing data needed for different alternatives. For a screw conveyor, the key information is the trough loading capacity (expressed as a percentage) and the power factor (a dimensionless number needed to estimate the power required). For a belt conveyor, the useful information is the angle of repose and the maximum belt incline that can be used. (These are both expressed in degrees relative to the horizontal.) Bucket elevator and feeders should not be used with certain materials. Data are so indicated in columns 8 and 9. For hydraulic conveying, the key data are the solid (not powder) density, the usual design loading expressed is kg solid/kg water and the energy required per Mg/h of solids conveyed per 1000 m of distance. For pneumatic conveying, a wide variety of data have been published. These are listed in columns 13 to 20. First, for vertical conveying, the minimum gas velocity (m/s) and the maximum solids loading (kg solid/kg gas) are given. Then, the analogous data for horizontal conveying are given. Usually the horizontal conditions are the constraining conditions. In columns 17 to 20, the usual design gas velocity (m/s) and the usual solids loading (kg solid/kg gas) are given for vacuum and for pressure conveying respectively operating at a distance of 120 m equivalent pipe length.

The appropriateness of the various transportation alternatives can then be assessed initially based on the material characteristics and is illustrated in Table 2–6. The methods available are classed as mechanical and fluidic.

2.5-1 Mechanical Devices

Mechanical devices are designed for primarily vertical lifts in small spaces (bucket elevators and vertical screw conveyors); combination of vertical and horizontal via inclined lifts (primarily belt conveyors); primarily horizontal conveying (belt, drag, and screw conveyors); primarily feeders (star valves, apron and vibratory feeders); and special designs (en masse and zipper).

A. Primarily Vertical Lifts. Bucket elevators are designed to lift solids vertically. The variety includes:

a. the type of discharge—centrifugal, which tosses the solids out of the bucket, and positive displacement for stickier solids that shakes the solids out.

b. the type of pickup—buckets dipping into the solids (in a boot) or solids pouring down into a moving series of buckets (via a loading leg).

c. the bucket support-belt or chain. Belts are quieter but have temperature limitations.

d. the spacing between the buckets—some spacing or very little for the so-called continuous.

These are illustrated in Figure 2–44. The characteristics of bucket elevators are summarized in Table 2–6.

Vertical screw conveyors can lift the solids in a moderate vertical distance. Their characteristics are summarized in Table 2–6 also.

En masse conveyors are enclosed trunks or boxes that are pulled up by a chain or belt. They protect the material from the environment and from excessive banging around.

B. Inclined Lifts/Drops. Some devices can lift the solids vertically, but do so at an incline. In a sense, the screw conveyor can be put at any angle. Belts can also be inclined, provided the angle of incline is not so high that the solids start to slip on the belt. The data in **Data** T C–3 give the angle of repose of the solids and the maximum angle of inclination that is feasible for a belt. Figure 2–42 illustrates the meaning of the angle of repose as illustrated in the context of a pile of material or the drainage angle of the material left in a storage silo or bin.

Mechanical rollers or chutes can be used to lower boxes and cartons.

C. Primarily Horizontal Movements. The belt and screw conveyors are the workhorses for horizontal movement. Many variations on the belt conveyor have been developed:

1. flat belt.

2. troughed belt with idlers at the side to angle the belt. An angle of 20° is often used. This is illustrated in Figure 2–45.

3. a drag conveyor with flat paddles that drag the solids along.

A belt needs to have a feeder to provide a *steady* flow of solids to it. Other characteristics of belts and screw conveyors are given in Table 2–6. In selecting belt width, belt speed, and solids handling, Figures 2–46 and 2–47 are useful. Figure 2–46 (see p. 2–44) relates the maximum speed to the width for different solids. Figure 2–47 (see p. 2–45) relates *capacity* to the speed/width product and the thickness of the solids layer.

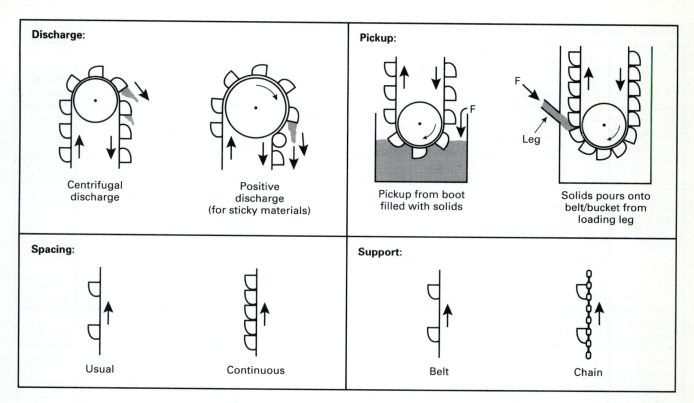

Figure 2–44 Sketches of Bucket Elevator Configurations

D. Feeders. Feeders convert a variable feedrate into a controlled one. The usual feeders are:

1. Star valves.
2. Apron feeders—these are like short belt conveyors.
3. Vibratory feeders—these are like troughs that shake the material into the desired location.
4. Screw feeders.

These are illustrated in Figure 2–41.

E. Special Devices. To minimize contamination, to prevent breakage, and to handle toxic materials, a wide variety of specialty conveyors have been devised. These include the zipper and en masse conveyors.

Figure 2–48 (see p. 2–46) shows a very general relationship between tonnage capacity, distance, and power for various mechanical devices used for primarily vertical lifts; Figure 2–49 (see p. 2–47) shows the same relationship for primarily horizontal movement.

The power required, as shown in Figure 2–49, depends very much on the type of material transported. For example, for belt conveyors, the capacity depends on the density of the material, the belt speed, the center-to-center distance, and details about idlers and trippers. Nevertheless, the power data are typical of material with density 1.6 Mg/m³ and belt speed of 1 m/s. For screw conveyors, the drive power depends on the type of bearing used to support

the screw, how deep the screw is submerged in the particles, and the material factor (as given in **Data** T C–3). The power values given in Figure 2–49 are for densities of 0.8 Mg/m³, a material factor of 2.0 and a screw submergence of 30%. Table 2–7 (see p. 2–48) illustrates how the submergence is dictated by the type of material and how this affects the capacity. Usually, single screw conveyors are not built longer than 90 m because of the limitations in the torque that can be applied to shafts of this length. Figure 2–50 (see p. 2–48) shows how the degree of submergence affects the diameter, rotational speed and the capacity.

Example 2–21: What is a reasonable rpm and diameter for a screw conveyor at 15% trough loading to handle 1 dm³/s of solids?

An Answer: From Table 2–7, the usual rpm is 9 to 60 with the maximum reducing to 35 rpm at the large diameter. From Figure 2–50, corresponding to 1 dm³/s, the ordinate is 0.3 rpm (m)³·⁵. To estimate reasonable rotational speeds, we visually check where 1 dm³/s intersects the full length of the 15% line. It intersects about midway. Hence, the maximum rpm would be about midway between 60 and 35 rpm or about 45 rpm. (If the capacity had been around 8 dm³/s, then this is near the maximum diameter of screw so that the maximum rpm chosen would be about 35

TABLE 2-6. Selection criteria for solids conveyors

| | VERTICAL | | | PRIMARILY HORIZONTAL | | | | | | | | |
	Usual Bucket	Cont. Bucket	Screw	Belt	Screw	Zipper	Drag	En Masse	Apron	Vibratory	Pneumatic (Air)	Hydraulic (water)
Size of particle A,B,C,D,H	BC	ABCD	BC	ABCD H	ABC			H	BCD	BCD	A	
Flowability	1,2	1,2	2	1,2,3	1,2			1	2	1,2	1,2	
Abrasiveness	7,8	7,8,9	7,8	7,8,9	7,8				7,8	7,8,9		
OK	N PRS	N RSTW XY Z	RS WY	LN RSWX Z	N RSW Y	TY		K	N RS	NKRS WZ		
NOT	KL TWX YZ	KL	KL TXZ	K TY	KL TXZ	X			KLW XYZ	L		
Horizontal Distance, m	←	NA	→	not limited	<75				<75			
Vertical Distance, m	>25	>25	usual 14	not limited	<10		<10	>25	<20			
Angle Incline,°				<10	<10		<10		<18	<10		
Temp., °C				<80	<150				<150	<150		
Handle Uncontrolled Feedrate?	No	No		No	Yes		Yes	Yes	Yes	Yes		
Space	Need clear vertical height.			Need relatively straight horizontal line							Flexible	
Speed m/s	1 to 1.5	0.6		1 to 2.5			<0.1					
Usual Capacity, Mg/h	14 to 150	30 to 120					<20					

rpm.) Thus, the range of rpm would be between 9 and 45 rpm. We arbitrarily select about 35 rpm. Now, from the ordinate and the right-hand scale of Figure 2–50, we can estimate the screw diameter.

$$0.3 \text{ rpm (m)}^{3.5} = 35 \text{ rpm (D)}^{3.5}$$
$$D^{3.5} = \frac{0.3}{35}$$
$$= 0.009$$
$$D = 28 \text{ cm.}$$

Hence, a reasonable diameter and rpm would be 28 cm diameter screw rotating at 35 rpm.

Example 2–22: Select a device to transport oats at 50 Mg/h a distance of 10 m horizontally.

An Answer: From **Data** T C–3, oats is C27 and so could be handled by most of the devices.

From Figure 2–49, since this is simply a horizontal transportation, a screw conveyor could be used. The approximate power would be about 4 to 7.5 kW.

Comment: The power will depend on the degree of submergence and type of bearing. However, this gives an order-of-magnitude estimate.

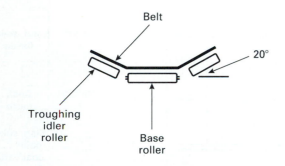

Figure 2–45   Sketches of Belt Conveyors

Example 2–23: What precautions should we consider in conveying wheat?

An Answer: From **Data** T C–3, the dust may be explosive, code V.

Example 2–24: We wish to transport 15.5 dm³/s of wheat horizontally a distance of 15.2 m. What might we use?

An Answer: From **Data** T C–3, the coding for wheat is C27. From the size of the particle and Table 2–6, any type of mechanism would be appropriate.

Since this is horizontal, we would try either a belt or a screw conveyor. From Table 2–6 (and Table 2–7), the loading would be 45% and the power factor is 0.4. The density of wheat is 0.72 to 0.77 Mg/m³ from **Data** T C–3.

The mass conveyed is:

$$15.5 \times 10^{-3}\,\frac{m^3}{s} \times 0.75\,\frac{Mg}{m^3} \times \frac{3600s}{h}$$
$$= 42\ Mg/h$$

The data in Figure 2–49 for 42 Mg/h and 15.2 m suggests about 5 kW drive needed for a screw conveyor. However, Figure 2–49 was based on a 30% trough loading (and we can take more capacity at 45% loading) and a power factor of 2 whereas wheat has a power factor of 0.4. If the power was used only to push the wheat and the bearing load was negligible, then the adjusted power would be 5 kW ×

(0.4/2) = 1 kW. The trough loading factor affects the diam/rpm combination more. For a 45% loading, the ordinate from Figure 2–50 is about 2. If the rpm is 100 rpm (from Table 2–7), the screw diameter would be about

$$2 = 100\ \text{rpm}\ (D)^{3.5}$$
$$0.02 = (D)^{3.5}$$
$$D = 0.32\ m$$

Thus, we might estimate a 32 cm diameter screw at 45% trough loading rotating at 100 rpm with about 1 to 5 kW power.

Comment: This example was worked out by Link-Belt Manual (1969) that suggests that a 30 cm diameter screw operating at 104 rpm and requiring 2.7 kW would be appropriate.

Example 2–25: Select a device to transport 8.5 dm³/s of raw gypsum a distance of 5.2 m horizontally. The maximum lump size is 2.5 cm and this size makes up about 15% of the volume.

An Answer: From **Data** T C–3, the raw gypsum is D27 coding handled in a 30% trough with a power factor of 2.0. The density is 1.12 to 1.28 Mg/m³.

From Table 2–6, a screw conveyor might be difficult to use because of the large size of particle. Table 2–6 suggests that usually ABC size particles are transported by screw conveyor. Probably a belt conveyor would be appropriate.

The capacity is

$$8.5 \times \frac{dm^3}{s} \times 1.15\,\frac{Mg}{m^3} \times \frac{3.600}{1000}\,\frac{h}{s}\,\frac{m^3}{dm^3}$$
$$= 3.5\ Mg/h$$

From Figure 2–49, usually a belt conveyor is not used for such a short distance. If the diameter of the screw is kept large, sometimes a screw conveyor would be appropriate.

From Figure 2–49, the power usage would be about 2 kW and this is for 30% trough loading and a power factor of 2.0.

From Table 2–7 and Figure 2–50, the ordinate value is about 1.8 and we want

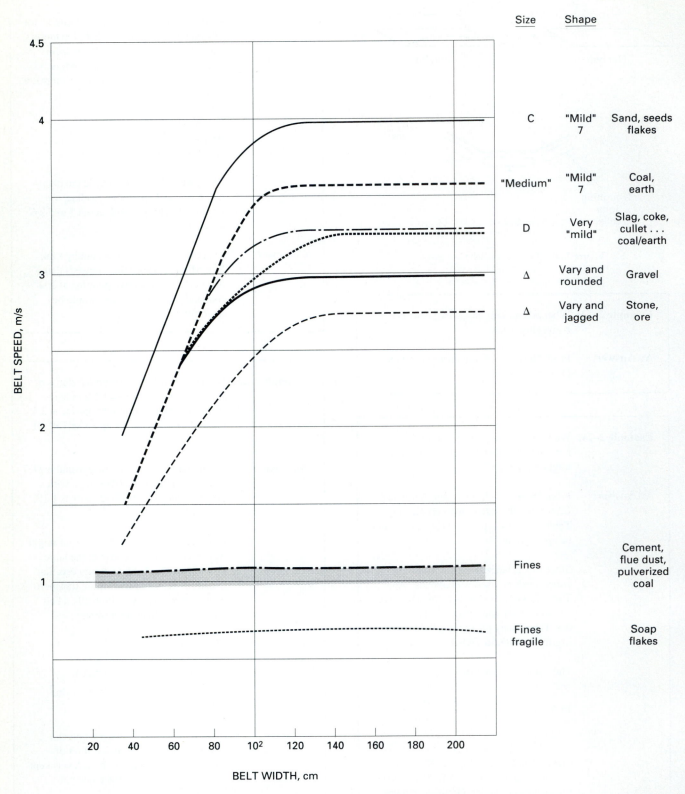

Figure 2–46 Characteristics of Belt Conveyors: Particle Characteristics, Capacity, Speed, and Width

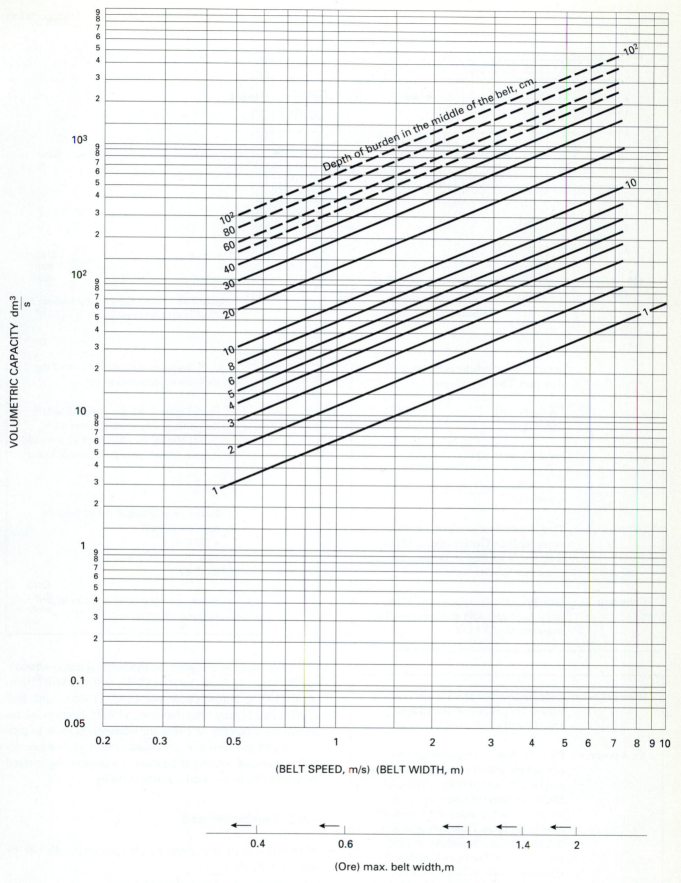

Figure 2–47 Characteristics of Belt Conveyors: Volumetric Capacity, Speed, Width and Burden Thickness (Depth (Cm) = 0.17 (Volumetric Capacity, Dm^3/s/) (Belt Speed, m/s) (Belt Width, m)

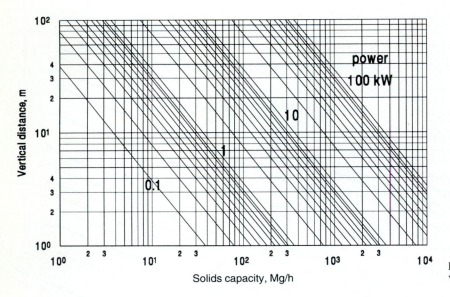

Figure 2–48 Power, Capacity, and Distance for Vertical Transportation of Solids

to use a large diameter and a relatively slow rpm. Thus, try 30 rpm

$$1.8 = 30 \, (D)^{3.5}$$
$$0.06 = (D)^{3.5}$$

From Figure 2–50

$$D = 0.45 \text{ m}$$
$$= 45 \text{ cm}$$

Thus, we might check with a screw conveyor supplier. Our estimate is 45 cm diam. screw rotating at 30 rpm requiring about 2kW power.

Comment: Link-Belt (1969) suggests that 40 cm screw at 35 rpm with an estimate power requirement of 3.1 kW.

Example 2–26: Estimate how 100 Mg/h of dry, crushed bauxite ore can be transported horizontally 150 m and vertically 10 m.

An Answer: First, we should summarize the characteristics of the solid. From **Data** T C–3, the solid is D38, with density 1.2 to 1.360 Mg/m^3, an angle of repose of 31°, and a maximum incline angle of 17° for belt conveying. From Table 2–6, probably a belt plus elevator or perhaps an inclined belt is feasible. For an inclined belt the maximum angle needed is 10 m in 150 m or an angle of 3.8°. Hence, provided the space was available, the belt could travel horizontally and then go up an incline of about 10° (from Table 2–6). From **Data** T

C–3, the solids will not slip even if the incline was increased to 16°.

From Figure 2–49, the power to satisfy the horizontal function is about 4 kW. To lift 100 Mg/h, a height of 10 m would require, from Figure 2–48, an additional

$$= 2.7 \text{ kW}$$

Hence, an estimate of the total power required is 4 kW
+ 2.7 kW.

$$= 6.7 \text{ kW}$$

A belt conveyor needs a feeder to provide a constant flow.

For more information on mechanical transportation, see Buffington (1969), Strube (1954), Sullivan Mill Equipment (1972), Screw Conveyor Corp. (1966), Link-Belt (1969), and Jeffrey Manufacturing (1967). Feeders are required for a number of conveyor alternatives so as to produce a uniform feed rate. Johanson (1969) gives more details. Overhead cranes can be used to transport larger-sized products. For more, see La Pushin (1969).

2.5-2 Fluidic Devices

Materials can be conveyed by air (pneumatically) or by water (hydraulically).

A. Pneumatic Conveying. Free-flowing—classes 1 and 2—and relatively small diameter, class A, materials can be blown along by an air stream.

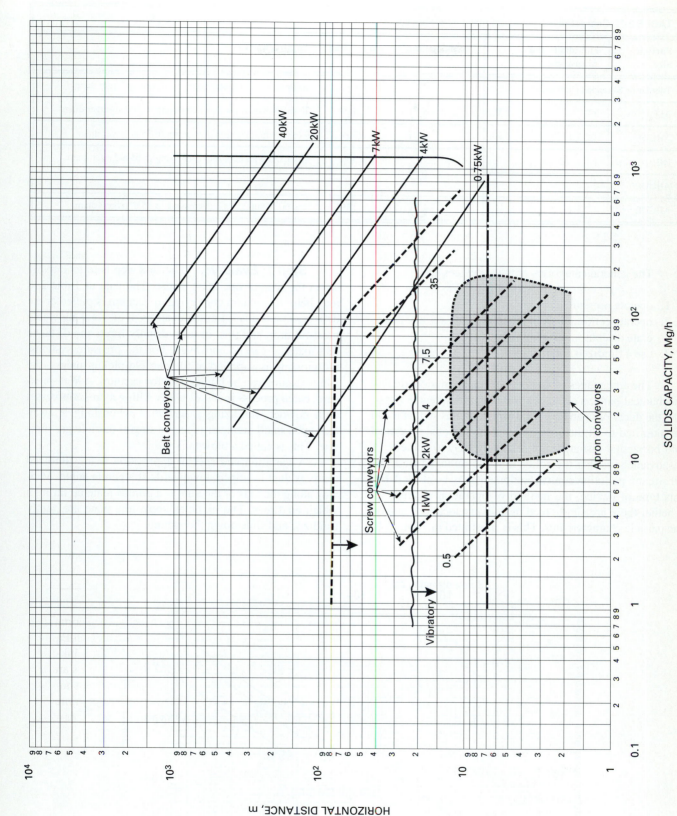

Figure 2-49 Power, Capacity, and Distance for Horizontal Transportation of Solids

SOLIDS CAPACITY, Mg/h

HORIZONTAL DISTANCE, m

TABLE 2-7. Characteristics of horizontal screw conveyors

Particle Size	Flowability and Abrasiveness		Recommended Trough Loading		Approx. RPM
Tubular or Shrouded Conveyor				95%	10 to 80 with rpm decreasing to 15 rpm at larger diam.
ABC	1,2	7		45%	20 to 180 with max rpm decreasing to 100 at larger diam.
ABCDE	1,2,3,4	8		30%A	9 to 60 with max rpm decreasing to 35 at larger diam.
DE	1,2	7		30%	15 to 120 with max rpm decreasing to 65 at larger diam.
ABCDE	3,4	7			
ABCDE	1,2,3,4	9		15%	9 to 60 with max rpm decreasing to 35 at largest diam.

The general categories of these devices are

1. operate pressure or vacuum;
2. operate continuously at relatively low pressure or operate discontinuously at high pressure;
3. use a fluidized slide.

The advantages of this type of transportation are few mechanical parts, a clean controlled environment is possible, the ductwork can be placed around bends so that it can be modified to fit the space, and combinations of vertical and horizontal distances are relatively easy to achieve. A disadvantage is that the solids must be removed from the air stream; pollution control equipment may be needed. For this type of device, the method of operation depends on whether there are single feed or delivery locations and on the particle characteristics. This is illustrated in Table 2–8.

Step 1: Evaluate. The first step is to tentatively select the type of operation to use. The data in Table 2–8 may help. In vacuum conveying, a vacuum is pulled at the delivery point; this is limited in terms of the total allowable pressure drop and this, in turn, usually limits the distance that material can be conveyed. For the pressure systems, the pressure is supplied at the particle pickup point—the solids loading can be higher than for vacuum. A high-pressure, batch process involves partly filling a pressure vessel with powder, fluidizing it and then whooshing it away through a small diameter pipe as a slug of particle/gas mixture. This is called a blow tank.

A variation and special application is the fluidized ramp. A duct is sloped downwards at a gentle angle. Midway along the axis of the duct is a porous plate up through which air flows. The particles above the plate are gently fluidized and slide down the incline.

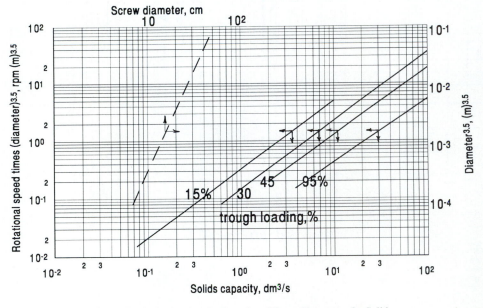

Figure 2–50 Estimating the Capacity of Screw Conveyors for Solids

TABLE 2-8. Selecting pneumatic conveying options

	Vacuum	Pressure Low	Medium	High Blow Tank	Fluidized Slide
Delivery point	one ——→ several		possible		one
Collection point	several possible	one ——→			one
Solid Characteristics from Table C-1					
Size	ABC lumpy ok	ABC lumpy ok	A	ABC	A
Abrasiveness	6,7	6,7	6,7,8,9	6,7,8,9	6,7,8,9
OK other codes	T,O,U,V,X	O,U,V,X	U,V	friable	U,V
Not OK for	T	T	T,O,X	T,V,O,X,U	T,O
Usual transportation distances, m	<450		<600	2300	
Max. allowable Δp, kPa	50	70	70	700	
Usual line velocity, m/s	30	25	25	3 to 5	
Solids loading, kg solid/kg gas	3.5 to 11	3.5 to 12		30 to 60	75 to 200
Power kW.h/Mg	3 to 5	2 to 3		0.8 to 1.5	
Comments	←—— Bends are critical because they erode . . . minimize bends.		Challenge is to introduce solids against pressure and to allow enough air leakage. Usually at least 30% leaks.	Bends not very critical.	Cannot elevate or lift material. Bends not critical.

Example 2–27: What might we use to convey 35 Mg/h of lumpy alum a distance of 25 m?

An Answer: From **Data** T C–3, lumpy alum would be B27. From Table 2–8, this can be handled by vacuum, low-pressure systems, or blow tank. The lumps could not be handled by a fluidized slide. If it was powdered, we could.

Step 2: Estimate the Solids Loading.

Once a type of device has been selected tentatively, the solids loading in the gas needs to be estimated.

In solids conveying gas, three different methods are used to relate the solids/gas concentration.

1. The amount of gas needed to convey a unit mass of solids N dm^3 gas/kg solids (with the solids expressed at its bulk density). This method is useful for sizing the pipe diameter because the gas velocity is a key design variable. Stoess (1970) calls this the "saturation." Usually, the amount of gas under standard conditions is used.
2. The mass of solids conveyed per mass of gas. This expression is needed to estimate the pressure drop. The data are commonly tabulated in this form. To convert to dm^3 gas/kg solids, we require only the density of the gas. If we are conveying using gas from a drying operation, then the gas may be saturated with moisture and at a higher temperature. The density may be 0.8 kg/m^3. For ordinary air, the density is about 1.2 kg/m^3. Figure 2–51 relates these two forms of representing the data.
3. The final density of *mixture* is sometimes quoted. The conversion combines both of the above ideas.

$$\rho_{mix} = \rho_{air}\left(1 + \frac{\text{mass solids}}{\text{mass gas}}\right) \qquad (2\text{-}20)$$

This relationship is illustrated in Figure 2–52 and is often used to delineate when different types of conveying are used.

Example 2–28: A reasonable design value for conveying rubber pellets is 4.6 kg/kg. If room air is used in the conveying, estimate both the density of the mixture and the volume of the gas/unit mass.

An Answer: For room air, assume the density is 1.2 kg/m^3. Hence, from Figure 2–51, about 180 dm^3/kg are needed. From Figure 2–52, the resultant density of the mixture is about 6 kg/m^3.

Figure 2–53 summarizes some data for the solids loading. The data show the maximum safe loading before plugging occurs. A lower load of solids can be handled in horizontal lines so this (shown as squares) is the limiting condition. Shown in the 3 to 10 loading range are some data for a wide range of materials and for both vacuum and low pressure conveying. Although there seems to be an overlap between the regions, an analysis of the individual data points shows that for a given product, the actual design loading is about ½ of the maximum. Also shown in this graph is the loading given by Kraus (1968) who describes the data as "somewhat obsolete but are still useful and safe for conveying some types of solids." The data and recommendations given in the mid-1950s and 60s seem to work with low-pressure loadings of 0.1 to 1, whereas more recently, both vacuum and low-pressure operations use loadings in the range 1 to 10 kg/solids/kg gas. The loading value depends on:

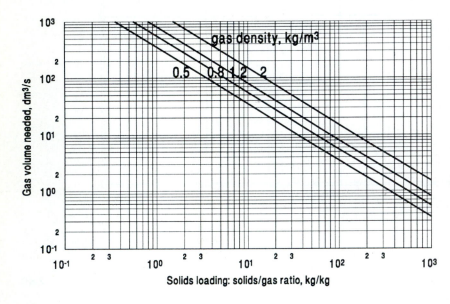

Figure 2–51 Relating the Solids Loading to the Volume of Gas Needed for Pneumatic Conveying

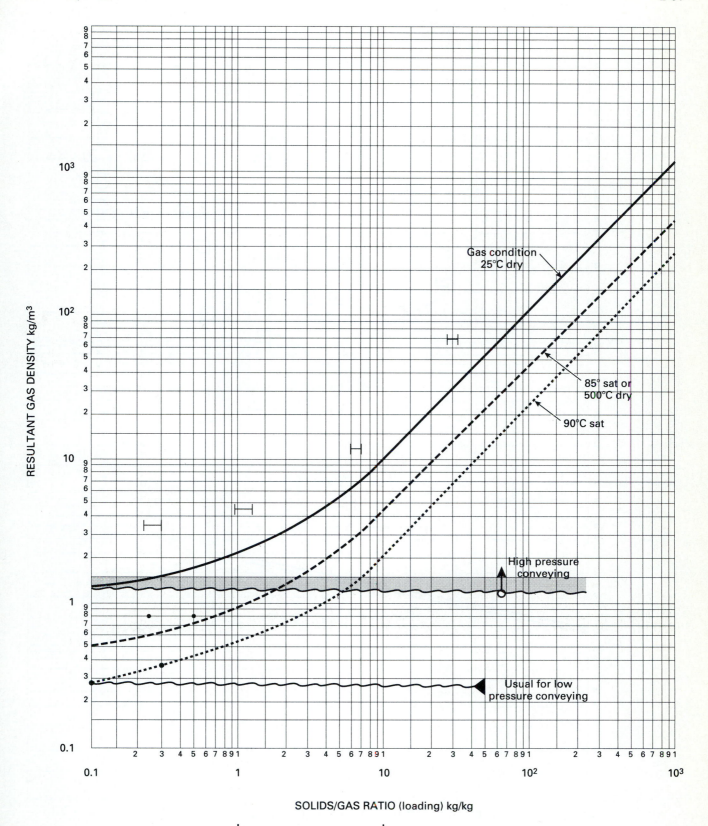

Figure 2–52 Estimating the Resultant "Gas Density" from the Solids to Gas
Mass Loading Ratio and the Conditions of the Gas

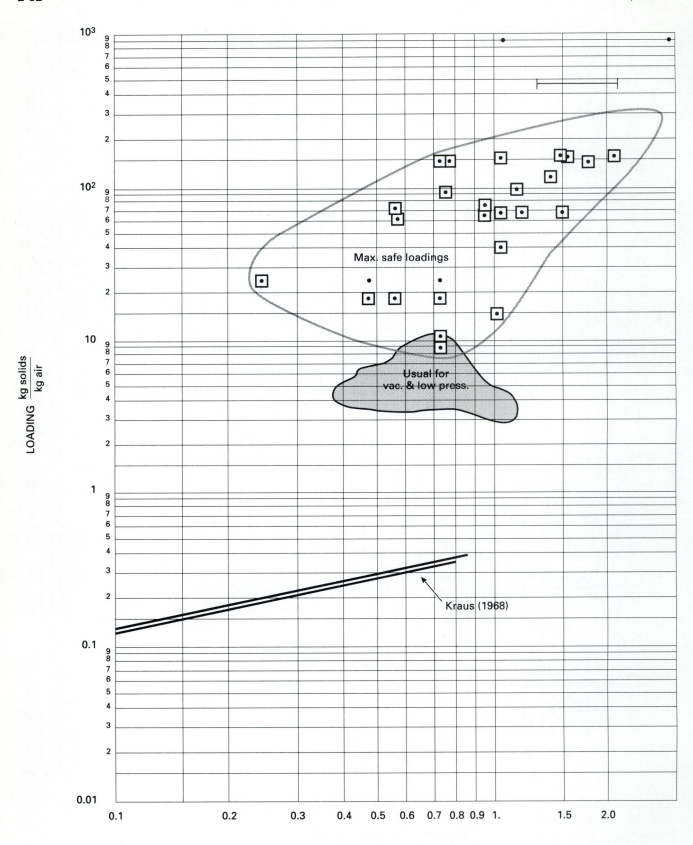

Figure 2–53 Solids Loadings for Pneumatic Conveying as a Function
of the Bulk Particle Density

1. The type of solid. For alum, it could be 2.8 kg/kg and for analogous operating condition wheat flour is 5.3 kg/kg.
2. The conveying velocity. Although, as an approximation, we will standardize on 25 m/s as being a reasonable gas conveying velocity, for low-pressure conveying systems, we may design on 15 to 20 m/s, while for vacuum, we may use 30 m/s. For these two conditions, for conveying alumina powder, the loadings are 7 kg/kg and 3.4 kg/kg respectively.
3. The distance the material is conveyed. If, for example, we standardize on design loading ratio for conveying 120 m distance, then Figure 2–54 shows how that loading ratio should be adjusted to account for distance conveyed. The longer the distance, the lower the loading or the more gas that is needed to keep the solids from dropping out.
4. The solids loading may be dictated by the device. For example, the ratio may depend on the amount of heating or cooling expected to occur during the conveying. The ratio may be determined by the configuration of the pertinent equipment—for example, a pulverizer determines the solids/gas ratio.

To sum up, the solids loading depends on many design factors; for order-of-magnitude estimations and for vacuum or low-pressure conveying values between 1 and 10 kg solids/kg gas are usually reasonable. Higher values, but not in excess of the maximum safe value are used for high-pressure conveying. The ratio is important because it dictates the gas flow needed, it affects the pressure drop dramatically, and that, in turn, may mean that a vacuum system is not applicable.

Example 2–29: Estimate a reasonable solids loading for vacuum conveying of polyethylene pellets.

An Answer: From **Data** T C–3, the maximum loading horizontally is 19 kg/kg air. For 120 m, also from **Data** T C–3, loadings of 12 kg/kg for low pressure and 7.2 kg/kg for vacuum are given. A design value around 10 kg/kg for most applications would be conservative provided the distance is less than 120 m of equivalent pipe.

Step 3: Estimate Gas Velocity and Duct Size.
Minimum gas velocities are reported in **Data** T C–3. Design values about 50% greater than the horizontal value can be used or an average design value from **Data** T C–3 may be available. In general, 25 m/s gas velocity is a reasonable design value.

The gas volume, for standard conditions, can be estimated from the solids loading and the density of the conveying gas.

$$\text{Gas rate} = \frac{1}{\text{solids loading}} \times \frac{1}{\text{gas density}} \times \text{design solids rate} \quad (2\text{-}21)$$

Example 2–30: If the loading is 10 kg solids/kg air and we wish to convey 10 Mg/h, what is the gas volume?

An Answer:
$$\text{Gas volume} = \frac{\text{kg air}}{10 \text{ kg solids}} \times 10{,}000 \, \frac{\text{kg}}{\text{h}}$$
$$\times \frac{\text{m}^3}{1.2 \text{ kg}} \times \frac{1}{3.6} \frac{\text{dm}^3 \text{ h}}{\text{m}^3 \text{ s}}$$
$$= 230 \text{ Ndm}^3/\text{s}$$

To estimate the pipe diameter, we use the design value of 25 m/s and the actual gas flowrate.

Example 2–31: What diameter pipe would we use for a pressure operation operating at 80 kPa and the conditions in Example 2–30?

An Answer: The standard volumetric flow is converted to actual volumetric flow and then the diameter estimated

$$\text{dm}^3/\text{s} = \frac{230 \text{ Ndm}^3}{\text{s}} \times \frac{101 \text{ kPa}}{181 \text{ kPa}}$$
$$= 128 \text{ Ndm}^3/\text{s}$$

Thus, the pipe cross-sectional area needed is:

$$128 \, \frac{\text{dm}^3}{\text{s}} \times \frac{10^{-3} \text{ m}^3}{\text{dm}^3} \times \frac{1}{25} \frac{\text{s}}{\text{m}}$$
$$= 5.1 \times 10^{-3} \text{m}^2$$
Diameter = 8 cm

Step 4: Estimate the Pressure Drop and Confirm Our Configuration.
The pressure drop for gas only is predicted from Figure 2–14 based on the gas velocity, the density of the gas and the diameter of the pipe. The pressure drop when solids are conveyed is approximated by

$$\frac{\Delta p \text{ solids}}{\Delta p \text{ gas}} = \left(1 + a \left(\frac{\text{kg solid}}{\text{kg gas}} \right) \right) \quad (2\text{-}22)$$

where a = constant between 0.3 and 0.5 depending on the solid.

This relationship is shown in Figure 2–55. To this should usually be added the pressure drop across the gas-solids separation device at the discharge end of the system.

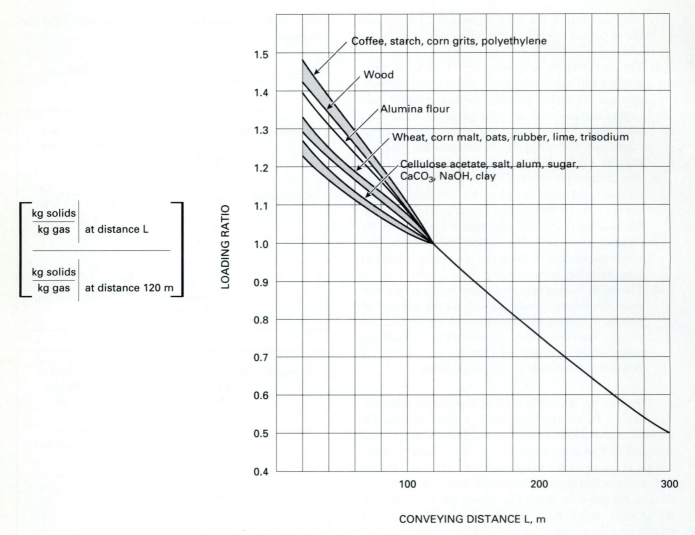

$$\left[\dfrac{\dfrac{\text{kg solids}}{\text{kg gas}}\bigg|_{\text{at distance L}}}{\dfrac{\text{kg solids}}{\text{kg gas}}\bigg|_{\text{at distance 120 m}}}\right]$$

Figure 2–54   How Pneumatic Conveying Distance Affects the Loading
for Vacuum/Low Pressure Pneumatic Conveying of Solids

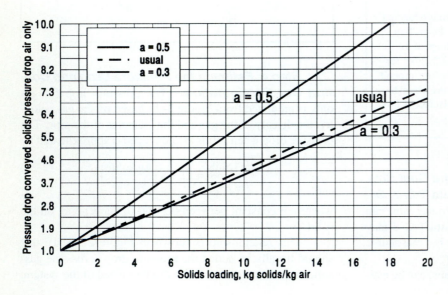

Figure 2–55   Estimating Pressure Drop for Pneumatic Conveying

Example 2–32: What is the pressure drop for conveying at a loading of 10 kg solid/kg gas in an 8 cm diameter pipe at 25 m/s if the gas density is 1.2 kg/m^3?

An Answer: From Figure 2–14, the pressure drop for gas only is about 9 kPa/100 m.

From Figure 2–55, at a 10/1 loading, the conveying system pressure drop would be 4 to 6 times the gas only drop.

That is:

Δp = 36 to 54 kPa/100 m.

Thus, if we were trying to operate this as a vacuum system, and we had 300 m of equivalent pipe, the total pressure drop would be between 108 and 160 kPa. Since we cannot usually tolerate more than 50 kPa for a vacuum system, we would rethink the design and probably go to a high-pressure discontinuous system or decrease the solids loading.

Trying to visualize the tradeoffs is difficult. Figure 2–56 shows how diameter, loading, and pressure drop are interrelated for a standard set of conditions (1, 3, and 10 kg solids loadings; gas density of 1.2 kg/m^3; and conveying velocity of 25 m/s). Thus, to convey 12 Mg/h at 3 kg solid/kg air, we would use a 20 cm diameter pipe and have an effective Δp of 7 kPa/100 m of equivalent pipe. This plot is interesting. If we go to low loadings, then the diameter of pipe increases, but the pressure drop decreases. Higher loadings give smaller diameter pipes and higher pressure drops. In general, Stoess (1970) provides a rule of thumb design curve for vacuum conveying systems. On the pressure drop scale, the regions indicated as total vacuum, total low pressure, etc., are a little misleading. The scale is Δp per *100 m* equivalent length. The regions called "total vacuum" refer to the *total* pressure drop for whatever length is used in that system. The total allowable pressure drop information is useful to include on Figure 2–56, yet caution is needed in appreciating the difference in meaning.

Example 2–33: Pebble lime is to be loaded from a storage bin to hopper railway cars at a rate of 25 Mg/h. The distance is 54 m in a straight line. Rough-size a conveying system.

An Answer: *Step 1: Evaluate*
From **Data** T C–3, the characteristics of pebble lime are C27 OY. Thus, from

Table 2–8, vacuum or low pressure would do. The hygroscopicity limits the applicability of the other devices. The net equivalent distance we might estimate to be 75 m to allow for changes in elevation and fittings.

Step 2: Estimate Solids Loading
From **Data** T C–3, the solids loading at 120 m equivalent length of pipe is about 3.4 kg solids/kg. Our conveying distance is less than 120 m and so, from Figure 2–54, we could increase the loadings for shorter distances.

Step 3: Estimate Velocity and Diameter
We will use 25 m/s (although we have specific data in **Data** T C–3). Hence, from Figure 2–56, we enter the figure on the lower abscissa with the mass rate of 25 Mg/h. A vertical line intersects the loading line of about 3.4 kg/kg at a pipe diameter of 29 cm. If we used Stoess's generalized line for vacuum operation, the value would be 20 cm. Both conditions are within the vacuum conveying region and so seem acceptable.

Step 4: Estimate the Pressure Drop
To estimate the pressure drop, move horizontally on Figure 2–56 to the appropriate solids loading line and read off 5 kPa/100 m and 175 kPa/100 m, respectively. The latter corresponds to 20 cm diameter and a loading of 10 kg/kg because that is the general loading value that corresponds with Stoess's line. For an equivalent length of pipe of 75 m, the result would be between 3.75 and 12.75 kPa. Assume 10 kPa for the line only. To this, we would have to add the pressure drop across the separation device.

Comment: Stoess (1970), p. 52, cites a 20 cm diameter line with a pressure drop of 34 kPa.

Step 5: Estimate the Power Requirements.

From the pressure drop in the pipe, the pressure drop across the gas-solid separator; from the conveying gas needed plus about a 30% allowance for leakage, the blower, fan, or vacuum device can be sized. As an approximation, however, we can use values of the power required per Mg/h of material conveyed. For vacuum systems with 120 m distance, a reasonable value is 4 to 5 kWh/Mg; for low-pressure systems, the value is 2 to 4 kWh/Mg for 120 m distance.

So far, the emphasis has been on the vacuum or low-

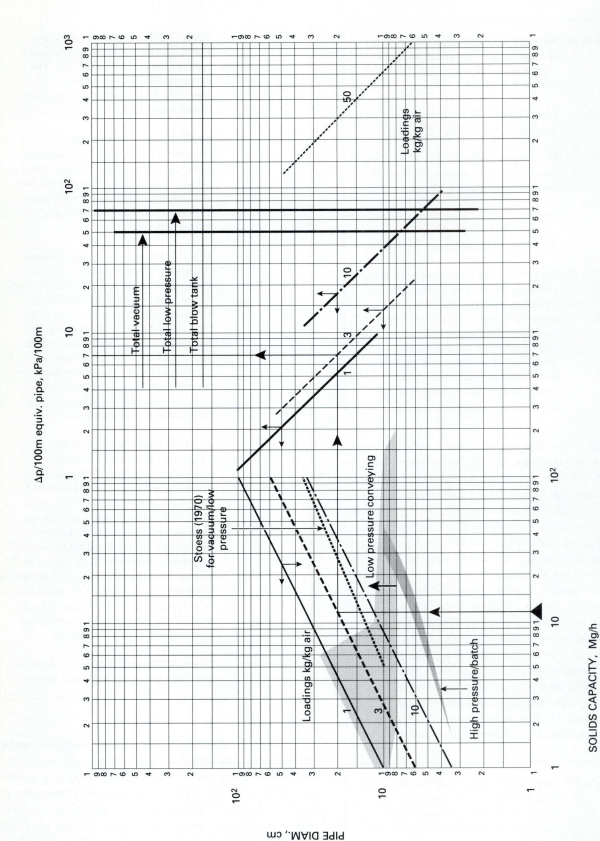

Figure 2–56 Sizing Pneumatic Conveying Systems

pressure options, although the same general principles apply for a high-pressure batch process, sometimes called fluidized transfer, blow tank systems, or air-in-powder system. In this system, a "fluidizer" is loaded with powder and pressurized in about 10 to 30 s. The powder is then conveyed as a fluidized slug; the process repeats. For sand conveyed a distance of 60 m, the loading is about 48 kg solids/kg air. The power usage ranges from 0.7 to 1.3 kWh/Mg. The general range of data are loadings 30 to 200 kg solid/1 kg gas; conveying velocities of 3 to 25 m/s. Some data are shown in Figure 2–56 for high-pressure batch operation.

Example 2–34: Portland cement is to be conveyed at a rate of 12.6 kg/s up to a distance of 20 m. What might we use?

An Answer: *Step 1: Evaluate*
From **Data** T C–3, cement, portland is A28 J. From Table 2–8, the options are medium pressure, blow tank, or fluidized slide because of the abrasiveness.

Since this is a vertical lift, the fluidized slide is not a possibility. Hence, consider a blow tank.

Steps 2 and 3:
The solids loading varies throughout the conveying cycle and the velocity is usually much slower than 25 m/s. Hence, we will just use the general curve on Figure 2–56 which corresponds to about 50 kg/kg equivalent at 25 m/s conveying velocity on the average. Thus for

$$12.6 \text{ kg/s} = 12.6 \times 3.6$$
$$= 45 \text{ Mg/h}.$$

From Figure 2–56, this corresponds to a diameter of 10 cm and, moving to the right, a pressure drop of about 500 kPa/100 m.

For 20 m, this would be 100 kPa excluding the pressure loss across the gas-solid separator. The general power requirement would be about 0.7 to 1.3 kWh/Mg and hence would be

$$0.7 \text{ to } 1.3 \times 45 \frac{\text{Mg}}{\text{h}}$$
$$= 30 \text{ to } 58 \text{ kW}$$

Comment: For this plant, 12.6 kg/s was conveyed 20 m in a 10 cm diameter pipe, with a total pressure drop of 160 kPa and a power requirement of 55 kW. The agreement is very good.

A fluidized slide can convey material downhill. The relationship between conveyor width and solids conveyed capacity is shown in Figure 2–57. The data are similar for both cotton fabric and porous plates. One air inlet supplies enough air to sustain a 45 m length of slide. The usual angle of inclination is 2° to 6°. The air requirements are in the range 10 to 150 dm³/s · m².

For more: see Stoess (1970), EEUA (1963).

B. Hydraulic Conveying.
Water can be used as the conveying medium. The distance conveyed could be short—say, several meters as in conveying activated carbon to and from regeneration—or very long, say 50 to 100 km as in coal, limestone and mineral conveying. The solids loading tend to be 35% to 40% by volume. This can be ex-

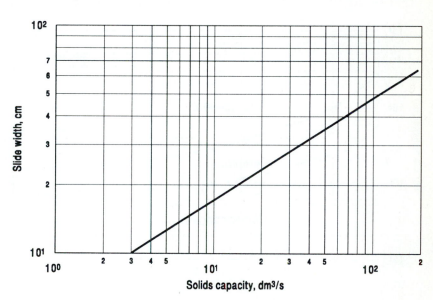

Figure 2–57 Estimating the Width of a Fluidized Slide to Convey Solids

pressed as a solids loading, kg solid/kg water, or as wt%. The interrelationships are given in Figure B–2 in **Data** Part B; here, the density is the solid—not bulk—density. This is summarized in **Data** T C–3. In section 2–4, the focus was on the selection of a centrifugal pump to handle slurries. Here, the focus is on transporting the solids (with water as a possible medium).

In *pneumatic* conveying, the loading was restricted primarily by the loading that would clog the pipe. Tradeoffs were made between pressure drop and loadings, although this primarily affected the choice of operating conditions. In *hydraulic* conveying, only one general method of operation applies and so the tradeoff between pressure drop and solids loading is the prime criterion. The pressure drop is complex because the slurry may behave as a non-Newtonian material and in different flow regimes. To obtain some idea of the parameters that play a role in this decision, Figure 2–58 illustrates how two factors affect the flow regime. Usually if we can get the slurry into the heterogeneous turbulent flow regime, the friction loss will be minimized. Hence, the choice of loading is complex and we depend on experimental data to help select the conditions. Table 2–9 summarizes some practical data. Figure 2–59 (see p. 2–62) shows these data on a graph similar to Figure 2–40, where the emphasis was on pumps.

The minimum liquid velocity can be estimated from the Durand equation

$$v_{min} = 1.34[2g\,D\,(p_{II} - 1)]^{0.5} \qquad (2\text{-}23)$$

where v_{min} = minimum liquid velocity, m/s

g = acceleration due to gravity, 9.8 m/s²
D = inside pipe diameter, m
p_{II} = density of the pure solid (as opposed to the average density of the bulk powder), Mg/m³

In this approximation, the velocity is independent of the particle size. In general, v_{min} is 0.6 to 1.2 m/s. The power requirement can be expressed as kW · h/Mg · km and is in the general range of 0.05 to 0.5.

Example 2–35: Estimate the conditions needed to pump limestone 27 km by hydraulic conveying. The limestone is B size and the amount to be conveyed is 225 Mg/h.

An Answer: From **Data** T C–3, the general mass loading is 1 to 2.4 kg solids/kg water. Hence, for 2 kg solids/kg water, the mass of water would be

$$225\,\frac{Mg}{h}\,\text{solids} \times \frac{1\text{ kg water}}{2\text{ kg solid}} \times \frac{m^3}{1\text{ Mg}}$$
$$\times \frac{1}{3.6}\frac{h}{s}\frac{L}{m^3}$$
$$= 31\text{ L/s}$$

From Figure 2–30 and a velocity of 1 m/s, a pipe diameter of about 18 cm would be needed.

The power is between 0.09 and 0.4 kW · h/Mg · km. Assume it is 0.2 kW · h/Mg · km, then the power required would be

$$0.2\,\frac{kW \cdot h}{Mg \cdot km} \times 225\,\frac{Mg}{h} \times 27\text{ km}$$
$$= 1215\text{ kW}$$

Comment: The Calaveras pipeline is 17 cm diameter with 1043 kW installed power and delivers 225 Mg/h a distance of 27 km.'

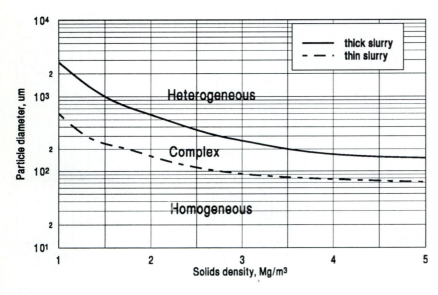

Figure 2–58 Characteristics of Two-Phase Flow: Liquid-Solid (reproduced with permission from Chemical Engineering, June 28, 1971, McGraw Hill Inc.)

TABLE 2-9. Hydraulic conveying data

Material	Mat. class code Table C-1	Solid Density Mg/m³	Conc. wt %	Conc. vol %	Pipe Diam cm	Pipe Length km	Flow Characteristics	Solids Cap. Tg/a	Power kW-h/Mg-km	Velocity m/s	Comments
Borax	A	2.6	26	12	12.7	1	Comp	0.5		3.7	US Borax
Borax tails	A	2.28	30		15		Hetero				
Carbon, Active											
Coal	B	1.4	50	42	25	172	Hetero/Comp	1.3	0.12	1.52	Consolid Coal Ohio
Coal	B	1.4	50	42	45	436	Compl.	4.8	0.06	1.52	Black Mesa.
Coal refuse	A	1.4	4	3	10	1	Homo	0.023		3.4	
	B	1.79	22		15		Hetero				
Copper Concentrate	A	4.2	26	8	15	Vert	Comp/Homo	0.3		1.52	Kennecott
	B		58		15	27		1	0.34		Bougainville
	A		60-65		10	110		0.3	0.12		West Iran
	A		45		12.7	60		1	0.11		KBI, Turkey
	A		55		10	17.6		0.4	0.64		Pinto Valley

Copper Tailings	α		18		30	70		0.6	0.23		Japan
Copper sulfide flotation conc	α	5.0	29		15		Homo				
Copper sulfide tailings	α	2.73	45				Homo				
	α	2.73	50		15		Homo				
Gilsonite	B	1.04	46	45	15	115	Comp	0.4	0.13	1.2	American Gilsonite
Kaolin	α	2.6	33	16	20	25	Homo	0.8		1.7	
Limestone	B	2.8	70	46	17	27	Hetero/Comp	2	0.17	1.92	Calaveras
	α	2.68	38-69	67							
	B		50-60		25	91		1.7	0.09		
	B		60		20	9.6		0.6	0.41		
	α	2.7	44-71								
	α	2.68	38-69		15		Homo				
	B	2.7	52-64		15		Hetero				
	α		44-71		15		Homo				
Limestone-clay	α	2.67	45-66		15		Homo				
Limestone-shale	α	2.67	35-66		20		Homo				
	B	2.67	39-67		20		Hetero				

TABLE 2-9. (Continued)

Material	Mat. Class Code Table C-1	Solid Density Mg/m³	Conc. Wt %	Conc. Vol %	Pipe Diam cm	Pipe Length km	Flow Characteristics	Solids Cap. Tg/a	Power kW-h Mg-km	Velocity m/s	Comments
Magnetic conc.	A	5.2	60	22	23	85	Compl.	2.5	0.07	1.7	Savage River
	B		45		20	8		1	0.61		Waipipi, NZ
	A		55-65		20	48		1.8			Pena Color, Mexico
	A		55-65		20	32		2.1	0.17		Sierra Grande
	A		55-65		20	27		1.5	0.34		Las Truchas
	A		60-70		71	400		25	0.023		Samarco
Magnetite	A		55-60		22.9	85		2.3	0.08		Tasmanic
Concenc.	α	4.7	40-70		15		Homo				
Nickel refinery tailings	B		44-60		10	6.9		0.1	0.57		Western Mining
Phosphate concentrate	α	2.8	55-65		22.9	104		2	0.07		Valep
	A	2.87	37-64				Hetero				
	A	2.87	22-34		15		Hetero				
Phosphate semi-concentrate	A	2.49	43-58		15						
	A	2.60	29-40				Hetero				
Red mud tailings	A	2.73	22		15		Hetero				
			18-34		15		Hetero				
Silica sand tailings	A	2.63	27-46		15		Hetero				
Taconite tailings	B	2.89	41-67		15		Hetero				

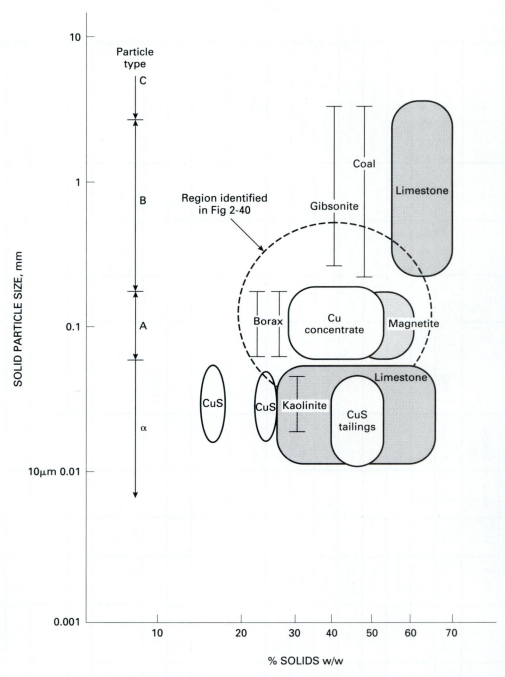

Figure 2–59 Example Hydraulic Conveying Conditions for Solids

2.6 SUMMARY

Transporting material from one location to another is often a key operation. The general considerations are to match rate of flow with the pressure drop encountered to minimize the power required. This balancing results in rules of thumb about velocities in the pipelines. For gases and pressure operation, gases and vacuum operation, and for liquid conveying, estimation procedures were given to determine the line size, the types of fans, blowers, or compressors, the power requirement, and the pressure drop.

For solid conveying, the characteristics of the solids play a major role. Tables of data were given that could be used to select the most appropriate device from among bucket elevators, belt conveyors, screw conveyors, pneumatic conveyors, and hydraulic conveyors.

Figure 2–60 summarizes this chapter.

2.7 REFERENCES

BELL, N. J. 1985. "Shortcut Methods for Determining Optimum Line Sizing in SI Units." *Chem Eng* (Oct 14): 120–122.

BUFFINGTON, M. A. 1969. "Mechanical Conveyors and Elevators." *Chem Eng* **76,** 22: 33

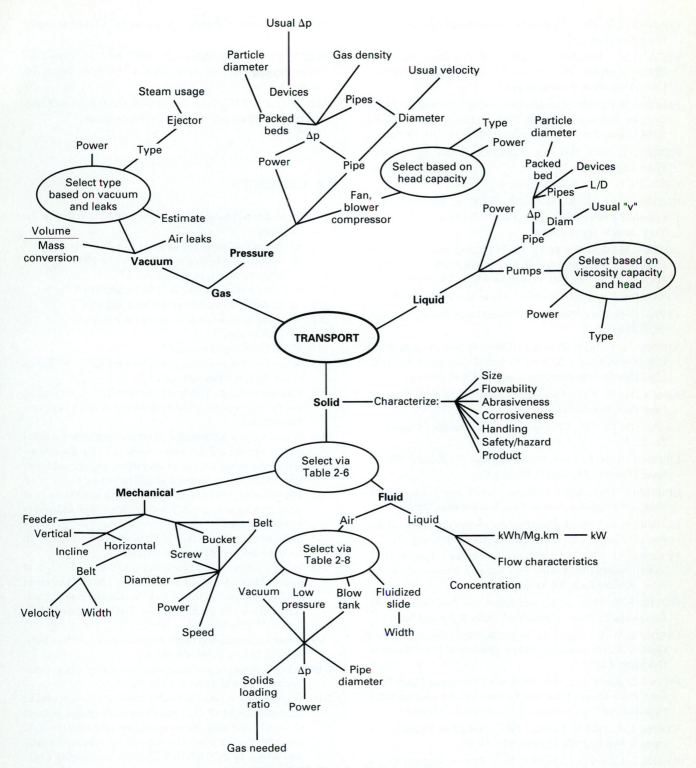

Figure 2–60   Overview for Transportation

CONNELL, J. R. 1987. "Realistic Control-valve Pressure Drops." *Chem Eng* (Sept. 28): 123–127.

EEUA. 1963. *Pneumatic Handling of Powdered Materials.* EEUA Handbook 15. London: The Engineering Equipment Users Association, Constable and Co.

FADDICK, R. R. 1976. Hydraulic Design Considerations for Slurry Systems. Proceedings of the First Technical Conference on Solid-Liquid Slurry Transportation, Columbus, OH, February.

FISCHER, J. 1958. "Practical Pneumatic Conveying Design." *Chem Eng* (June 2): 114.

GELDART, D. 1973. *Powder Technology* **7:** 285.

HOUGEN, O. A., K. M. WATSON, and R. A. RAGATZ. 1959. *Chemical Process Principles, Part II Thermodynamics.* New York: John Wiley and Sons.

JEFFREY MANUFACTURING. 1967. Catalog and Engineering Manual for Screw Conveyors, Manual No. 5000. Jeffrey Manufacturing Co., P.O. Box 1259, Fort Worth, TX 76101.

JOHANSON, J. R. 1969. "Feeding." *Chem Eng* **76,** 22: 75–83.

KERNS, G. D. 1960. "New Graphs Speed Drum Sizing," Pet. Ref. 39, 7: 168–170.

KIEFNER, J. F. 1976. Review of Slurry System Projects in the U.S. Proceedings of the First Technical Conference on Solid-Liquid Slurry Transportation, Columbus, OH, February.

KRAUS, M. N. 1968. *Pneumatic Conveying of Bulk Material.* New York: Ronald Press.

LA PUSHIN, G. 1969. "Transportation and Storage." *Chem Eng.* **76,** 22: 19.

LIEBERMAN, N. P. 1983. *Process Design for Reliable Operations.* Houston, TX: Gulf Publishing Co.

LINK-BELT. 1969. *Screw Conveyors and Screw Feeders,* Book 3089, 1960 Eglinton Avenue East, Scarborough, Ontario.

LUDWIG, E. E. 1964. *Applied Process Design for Chemical and Process Plants,* vol. 1. Houston, TX: Gulf Publishing Co.

MOLERUS, O. 1982. "Interpretation of Geldart's Type A, B, C, and D Powders by Taking into Account Interparticle Cohesion." *Powder Tech,* **33:** 81–87.

OLIVER, E. D. 1966. *Diffusional Separation Processes: Theory, Design and Evaluation.* New York: John Wylie and Sons.

OSBORNE, W. C., and C. G. TURNER. 1960. *Woods Practical Guide to Fan Engineering,* 2nd ed. Colchester, U.K.: Woods of Colchester Ltd.

PITTS, J. D., and T. C. ANDE. 1976. Iron Concentrate Slurry Pipelines. Proceedings of the First Technical Conference on Solid-Liquid Slurry Transportation, Columbus, OH, February.

RYANS, J. L., and S. CROLL. 1981. "Selecting Vacuum Systems." *Chem Eng,* **88,** 25 (Dec. 14): 74–89.

RYANS, J. L., and D. L. ROPER. 1984. *Process Vacuum System Design and Operation.* New York: McGraw-Hill.

SCREW CONVEYOR CORP. 1966. "Screw Conveyor Catalog and Engineering Manual," Cat. No. 166, 700 Hoffman St., Hammond, Indiana 46320.

SMITH, CARLOS A., and A. B. CORRIPIO. 1985. *Principles and Practice of Automatic Process Control.* New York: John Wiley and Sons.

STOESS, H. A., JR. 1970. *Pneumatic Conveying.* New York: Wiley-Interscience.

STRUBE, H. L. 1954. "Conveyors and Elevators." *Chem Eng* (April): 195.

SULLIVAN MILL EQUIPMENT LTD. 1972. *Sullivan Screw Conveyor Engineering Catalogue 7300,* 130 Milvan Drive, Toronto, Ontario.

WASP, E. J., et al. 1977. *Solid-Liquid Flow Slurry Pipeline Transportation.* Switzerland: Trans. Tech. Publications.

2.8 EXERCISES

2-1 What size of pipe/duct would be needed to transport the following:
  a) 30 dm^3/s of air at 101 kPa and 30°C?
  b) 300 dm^3/s of hydrogen at 150 kPa and 100°C?
  c) 3000 dm^3/s of superheated steam?
  d) 150 dm^3/s of chlorine at 500 kPa and 70°C?
  e) 10^3 dm^3/s of natural gas at 8 MPa and 25°C?

2-2 What size of pipe would be needed to transport:
  a) 1 L/s of water?
  b) 20 L/s of liquid benzene?

2-3 What size of downcomer would be needed in a distillation tray if the liquid flowrate is 6 L/s?

2-4 If a plant produces 3 million L per annum of liquid product and the plant operates 330 days/annum, is this a large production rate?

2-5 What device or combination of vacuum equipment would I need to operate a distillation column at 1 kPa absolute assuming the vacuum is pulled after the overhead condenser? How much power, or steam, would be needed?

2-6 The blowring system on a bag filter requires 1000 dm^3/s of air at 55 kPa pressure differential. What type and power would be needed?

2-7 Estimate the pressure drop in 60 m of pipe of diameter 10 cm if the density of the gas is 5 kg/m^3.

2-8 Select a conveyor and rough-size it to convey 27 Mg/h of hydrated lime an equivalent distance of 120 m. The system has four storage bins from which the material is to be transported to a single loading hopper. The bulk density of the lime is 0.35 Mg/m^3.

2-9 Select a conveyor and rough-size to convey salt cake a distance of 45 m at a rate of 6 Mg/h.

2-10 Select a conveyor and rough-size to convey wheat a distance of 120 m at a rate of 20 Mg/h.

2-11 We have a blow tank pneumatic conveying system with an equivalent length of 21 m and a maximum charge pressure of 600 kPa including the pressure drop across the gas-solid separator. For air alone sent through the system, the pressure drop is 50 to 70 kPa. If baby chick feed, having a bulk density of 0.77 Mg/m^3, was charged to the system, estimate how many Mg/h might the system convey. The blower power is 22 kW. What complications might you encounter? Estimate the diameter of the line.

2-12 Estimate what we would need to convey 20 Mg/h of mineral tailings 3 km.

2-13 Estimate what we would need to hydraulically convey 4.8 × 10^6 million Mg/a of class B size coal 436 km.

2-14 A Process and Instrument Diagram is given in PID–2A. Estimate an overall mass balance about this column.

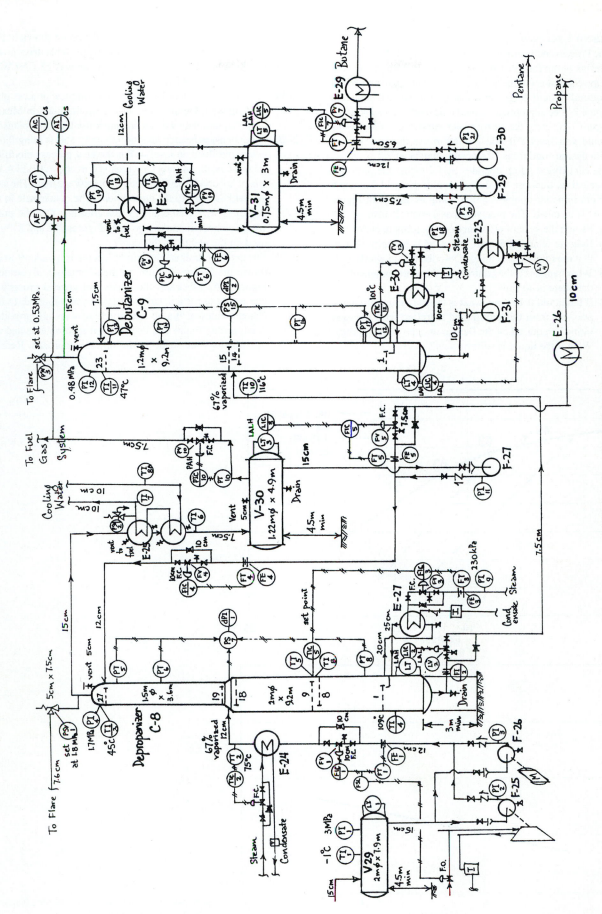

PID-2A Depropanizer/Debutanizer

Enrichment Exercises:

2-15 The Process and Instrument Diagram given in PID–2A shows a reflux pump for a distillation column. If the plate spacing in the column is 0.6 m, estimate the type of pump and power required. Would the pump whose performance is shown in Figure 2–8 be appropriate? Could it be modified in any way?

2-16 Figure 2–61 gives the performance curve for a pump. What would you specify if you wanted to use this in a system with design conditions 10 L/s and 20 m head?

2-17 The centrifugal pump operates by following the performance curve. Suppose the design conditions is 11 L/s and a 25 cm impeller double suction pump as shown in Figure 2–61 is installed. The pipework has a control valve.

   a) Sketch the system resistance or requirement curve.

   b) Indicate what happens if the valve is partially shut so that the total flow is 8 L/s. Describe this in words.

   c) What happens if the control valve is shut completely so that there is no flow?

   d) If the liquid is hexane, what is the maximum discharge pressure from this pump (and therefore what pressure rating is required on the flanges and the outlet piping)? What if the liquid is sulfuric acid?

2–18 Sketch the head capacity curve for the pump shown in Figure 2–61 if it were operated at 3600 rpm. What drive power would be needed if the design flowrate was 25 L/s? What happens to the NPSH$_r$?

2–19 We have completed the design of a complete process plant for Cornwall, Ontario. We need a similar plant in Mexico City. Explain why we can or cannot use the same plans.

2–20 A condensate header runs horizontally and is hung 0.6 m below the floor on level 4. For the new process modifications we need, on Level 5 which is 6 m above Level 4, 6.3 L/s of hot condensate pumped into a storage tank. The storage tank is open to the atmosphere. The condensate in the header is 96.7°C and at 115 kPa absolute. Can the pump whose characteristics are given in Figures 2–8 and 2–38 be used on Level 4 to do this duty?

2–21 The detailed design/selection of a pump is a tradeoff to optimize the financial return on the company's investment. Figure 2–62 illustrates the issues and general procedures one would use if a detailed analysis were required. In this chapter, rules of thumb and quick answers are provided. Analyze this figure and decide what assumptions and approximations are made in the quickie approach. Redraw

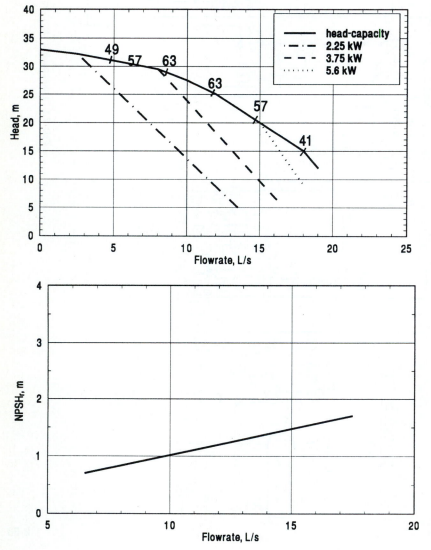

Figure 2–61 a) Head Capacity Curve for a Pump
b) NPSH Required by Pump

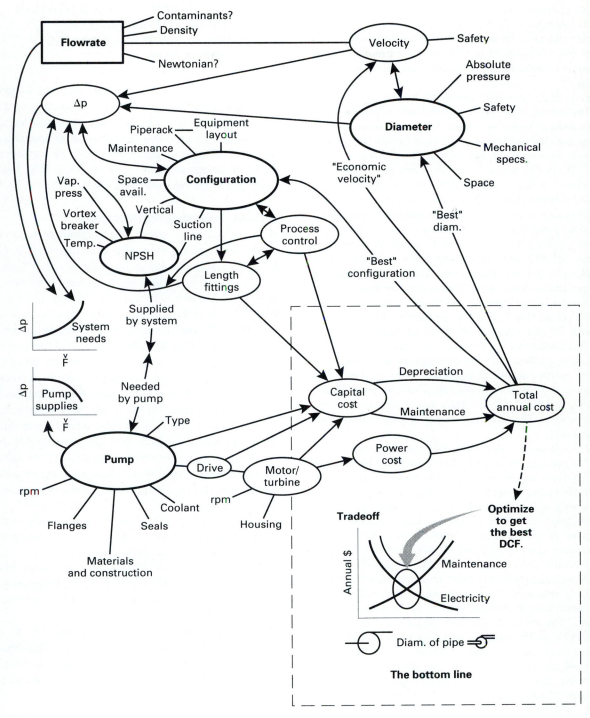

Figure 2–62 Issues To Consider in Selecting Pumps

this diagram to represent where and what approximations are made.

2–22 *Terry Sleuth and the Case of the Recirculating Blower*
Harold whipped through the last calculations of the power required for the gas recirculating blower on the catalytic reactor. "Let's see, that's finished—calculate the density of the gas via the ideal gas law based on the temperature, pressure, and composition of the recirculating gas, multiply the

actual volumetric flow for the recirculation conditions to get the ideal power required. A reasonable value . . . oh, hi, Terry . . . I'm just finishing up the design calculations for the new reactor system."

Terry Sleuth looked over Harold's shoulder. "That's the recirculating system operating on predominantly hydrogen being recirculated at about 2 psi and 150°C?" "Right!" replied Harold. "I was just saying that I adjusted my ideal power calculation by dividing it by a reasonable value of the motor-blower efficiency. I used 60%." Terry Sleuth perused the calculations done so far. To Harold's surprise, the Sleuth cautioned, "I don't think we'd better install the motor based on those calculations." Do you see what Terry sees?

2–23 *Terry Sleuth and the Case of the S12 Pump*
The murmur of the argument grew louder and louder. Terry meandered over to Bill's desk. Terry didn't want to butt in, but both Bill and Marge were Terry's friends. Both of them were showing all the signs of losing their tempers.

"It can't be that serious," said Terry Sleuth, hoping to ward off a nasty situation. Bill, relieved to have a sympathetic listener, launched into his story. "I'm working on the S9 project to install a circulating pump to handle 10 L/s of acid and supply 30 m of head. I know I can save the project a bundle by using the pump we scrapped on the old S12 unit. That pump handled the same acid. However, the pump/motor combination operated at 1800 rpm and can only produce 6 L/s and 8 m head. My idea is to buy a 3600 rpm motor. Thus, by doubling the speed, I'll get the flow I need. Since I have a 1 kW motor on it now, I'll just order a 1 kW motor that turns at 3600 rpm. That's where we get the argument. Marge says I need to order a 2 kW motor." "Yes," Marge cut in, "we are working on the principle of similitude. Since we doubled the speed, we should also double the power; we need at least 2 kW. Tell Bill I'm right." Terry looked puzzled. Terry had spotted the key—but how best to respond? What would you say?

2–24 *Terry Sleuth and the Case of the Cavitating Pump*
The rain beat down on the Operating Room door so hard that Terry had to struggle hard to force it open. The red lights by the pump transfer station on the control panel were lit up like a Christmas tree. "Boy, are we ever glad to see you!" the process engineer, Jim, exclaimed. "That pump on the hot water circulation system has started to cavitate again. We've had to shut it down." Terry was pleased by Jim's warm welcome. "Don't take off your coat. Please come out to the transfer area and look at that baby!" Terry responded, "Let's ask some questions first. Is the suction from an overhead drum?" "No," replied the operator. "The hot water goes into an open pit and the pump sucks it up. The pump operates on a float level control that just keeps the sump from overflowing . . . " "Whoa, I've heard enough," said Terry. "Take off your coat and rest awhile, Jim." "But shouldn't we go out and see what's the matter?" asked Jim. "Not today," replied Terry. What had Terry spotted?

2–25 "**PID–2A** might be used to treat the bottoms of the deethanizer, **T102,** in the ethylene process shown in **PID–1A.**" What would be the conditions on the ethylene plant to make it appropriate to add a depropanizer and a debutanizer to treat the bottoms? For the conditions that you specified, what would have to be modified on **PID–2A** for it to handle the situation? For example, are the flowrates compatible? Are the pressures and temperatures of the bottoms of **T102** consistent with those for the feed to the depropanizer? Consider the vapor pressures given in **Data Part C** and infer from the temperatures and pressures something about the compositions of the streams.

2.9: PROCESS AND INSTRUMENTATION DIAGRAM FOR STABILIZER DEPROPANIZER ILLUSTRATES APPLICATIONS

The processes chosen to enrich this chapter illustrate a variety of methods to transport liquids, slurries, and solids. They provide an opportunity to use your knowledge about fluid and particle mechanics to do order-of-magnitude tests to see if the size of pipes, the width of conveyors, and choice of pumps and conveyors are reasonable. Enjoy! Indeed, you might uncover some errors.

PID–2A: Depropanizer/Debutanizer
Here our goal is to "stabilize" an organic by removing the more volatile components. These include C1, C2, C3, and C4. Some example applications include the "light ends" section of a petrochemical plant, to treat the bottoms of column **T102,** the deethanizer, in the ethylene plant shown in **PID–1A** and in the gasoline stabilizing section of a refinery. This PID and the enrichment exercises focus on the transportation of liquids, although some of the lines contain gases. All of the pumps are centrifugal pumps (F-25 through F-31). Because distillation is involved, a consideration of the NPSH is a critical part of assessing this PID.

This particular PID provides more details about the instrumentation than is given in the other PIDs in this book. Things to reflect on while you are perusing this diagram are: how does the instrumentation account for safety? for startup and shutdown? Note which instruments are shown on the control room (indicated as ⊖) and suggest why was this choice made.

For gasoline stabilization, the sequence of the columns is shifted so that the debutanizer is upstream of the depropanizer. What are the implications?

For more, see ASEE Design Case Study #11 "Bid Proposal for the Star Oil Limited: Nevod Processing Plant" by E. C. Roche, Jr., Newark College of Engineering and F. Isaacson, Foster Wheeler.

For more on gasoline stabilization see C. K. Rayner and L. H. Appleby "The Performance of a Gasoline Stabilizing Unit," International Symposium on Distillation (1960) Institution of Chemical Engineers, London, p. 245.

2.10: PROCESS AND INSTRUMENTATION DIAGRAM FOR BRANNERITE ORE MILLING ILLUSTRATES APPLICATIONS

Whereas the transportation issues in **PID–2A** are for fluids, consider now a process where the transportation issues are for bulk solids and slurries.

PID–2B: Brannerite Ore Milling
Uranium milling is the process by which uranium naturally present in an ore, in the tetravalent or hexavalent state, is

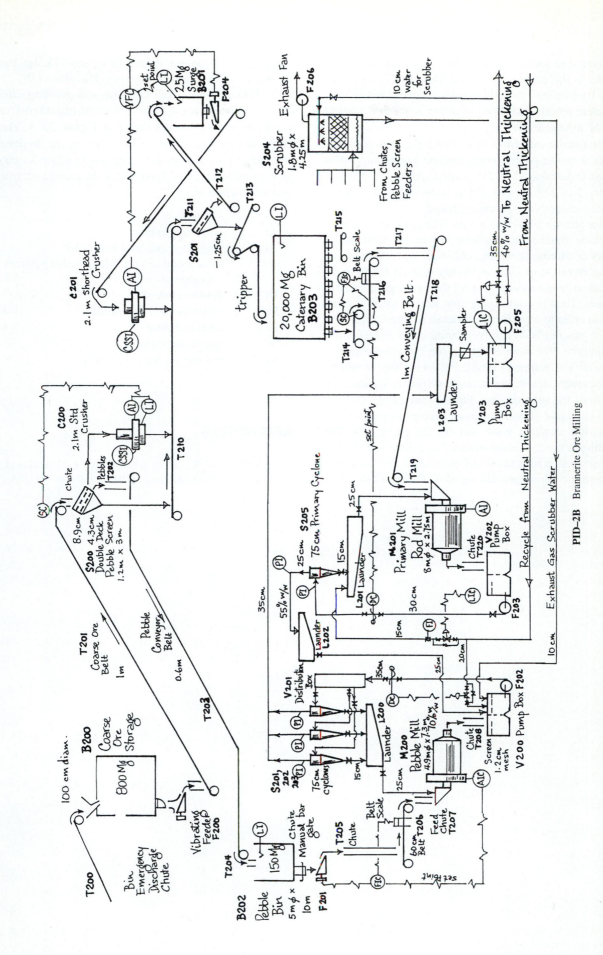

PID–2B Brannerite Ore Milling

concentrated to yield "yellow-cake." An example of yellow-cake is the magnesium form, MgU_2O_7. The yellow-cake is later refined to yield uranium metal, U_2O or UF_6. For nuclear power, each MW_e of power requires 4.3 Mg U (based on a 30-year life).

Brannerite ore, from the Elliot Lake region in Ontario, is a quartz-pebble conglomerate containing 40 to 60% quartz pebbles, 15 to 25% matrix quartz, 2 to 8% sulfides, and 5 to 20% sericite. Pyrite is the main sulfide mineral. The radioactive minerals are found in association with the pyrite and are within the matrix. The radioactive minerals include brannerite, uraninite, monazite with minor amounts of zircon, thucolite, coffinite, and uranothorite. The concentration of U_3O_8 in the ore is about 0.12% w/w. The brannerite grains are < 1 μm; the pyrite grains 150 to 1,000 μm. The "liberation size" (or size we need to grind the ore in order to separate the minerals of interest from the waste or gangue) is 40 μm for uraninite and around 1 μm for brannerite. For more about the ore characteristics, see Honeywell and Kaiman (1966). Following this would be thicken-densify, leach, neutralize/precipitate, separate the solids from the uranium-rich liquid, concentrate the uranium in the liquid via ion exchange, precipitate the iron and the uranium, dewater, wash and dry the precipitate. The solid tailings need to be neutralized and disposed of. The overall process is illustrated in Figure 2B–1. For more details about the whole process see B. O'Reilly, et al., "De-velopment of Uranium Milling Processes," Design Project Report 1982, McMaster University, Hamilton, ON.

PID–2B shows the crushing and grinding circuits. This accounts for about 25 to 35% of the capital cost for the complete milling process shown in Figure 2B–1. The capacity of this plant is about 5000 Mg/d of ore. In developing such a process, novices often fail to allow for enough storage and surge bins. Symbols for the process control and instrumentation are:

AI	amperage indicator
AIC	amperage indicator controller
CSSI	closed side setting indicator (for gyratory crushers)
DC	density controller
FIC	flow indicator controller
LI	level indicator
SC	speed controller
VFC	volumetric flow controller

REFERENCES

HONEYWELL, W. R., and S. KAIMAN. 1966. "Flotation of Uranium from Elliot Lake Ores." *CIM Bulletin* **59**:347–355.

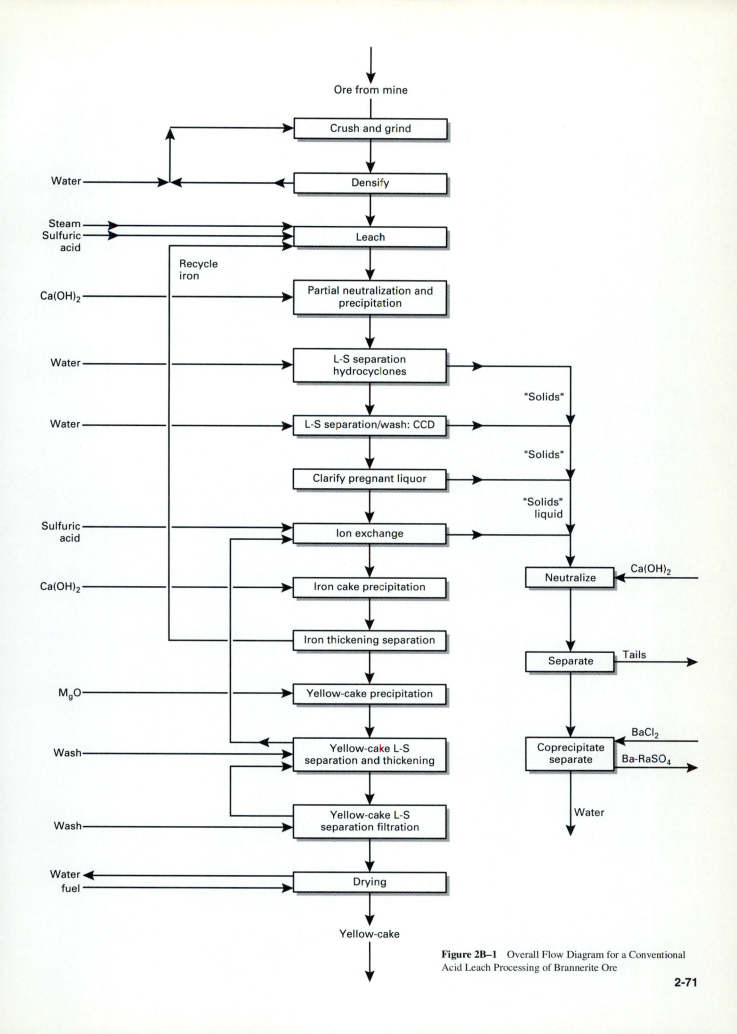

Figure 2B–1 Overall Flow Diagram for a Conventional Acid Leach Processing of Brannerite Ore

Chapter 3

Energy Exchange: Mechanical— Electrical—Thermal

For reactions or for other processing needs energy must often be added or removed from the system. This can be in the form of mechanical, electrical, or heat energy. Light and noise are other forms of energy, but they are not discussed here.

The three major forms of energy can be obtained directly or indirectly from:

1. materials that react or burn,
2. moving fluids such as water and air,
3. the sun or center of the earth.

The relationships among these and the types of equipment are shown in Figure 3–1. Starting with the most common source, materials, we note that they can react or be burned. Reactions produce heat or an electrical current, as in a battery. Combustion yields hot gases that can either drive a piston or turn blades in a turbine to create mechanical energy. In most of these conversions the thermal energy may be transferred to steam or an intermediate fluid that can then serve all these functions. The energy from falling or moving fluids can be collected as mechanical energy while the thermal/light energy from the sun or the center of the earth can be converted to steam, thermal, or electrical energy, as shown in Figure 3–1.

Each form of energy has special characteristics. Thermal devices that transfer sensible heat are called exchangers; those that transfer latent heat are called boilers or condensers. Mechanical energy can be transmitted from one device to another by direct drive or connection or via gears, v-belts and pulleys, or chain and sprockets. Electrical energy can be converted from AC to DC forms by rectifiers.

Moving fluids may interchange energy through ejectors. These are discussed under Transportation, Chapter 2.

In this section we discuss equipment for manipulating mechanical and electrical forms of energy. Later, thermal energy is considered.

3.1 ENERGY EXCHANGE: MECHANICAL—ELECTRICAL

Mechanical energy is usually in the form of rotating or oscillating equipment or pressure. Pressures are usually supplied via gas compressors and are discussed in Chapter 2.

Motors, turbines, and engines are the usual types. Some pictures of these are shown in Figure 3–2. The range of application depending on the drive speed and the power required is given in Figure 3–3. The usual motor drive speeds are 1200, 1800, and 3600 rpm.

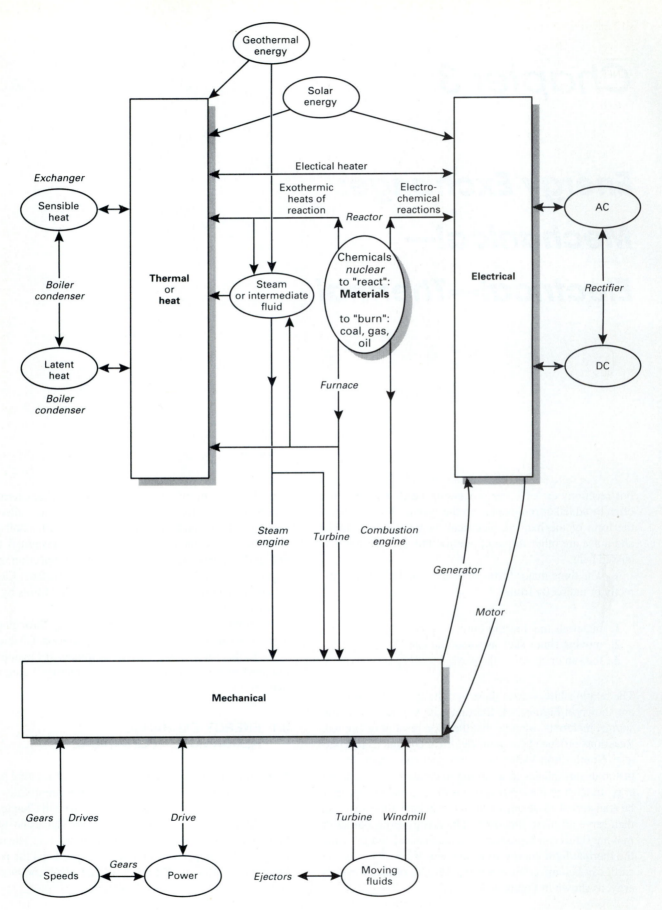

Figure 3–1 Sources and Forms of Energy

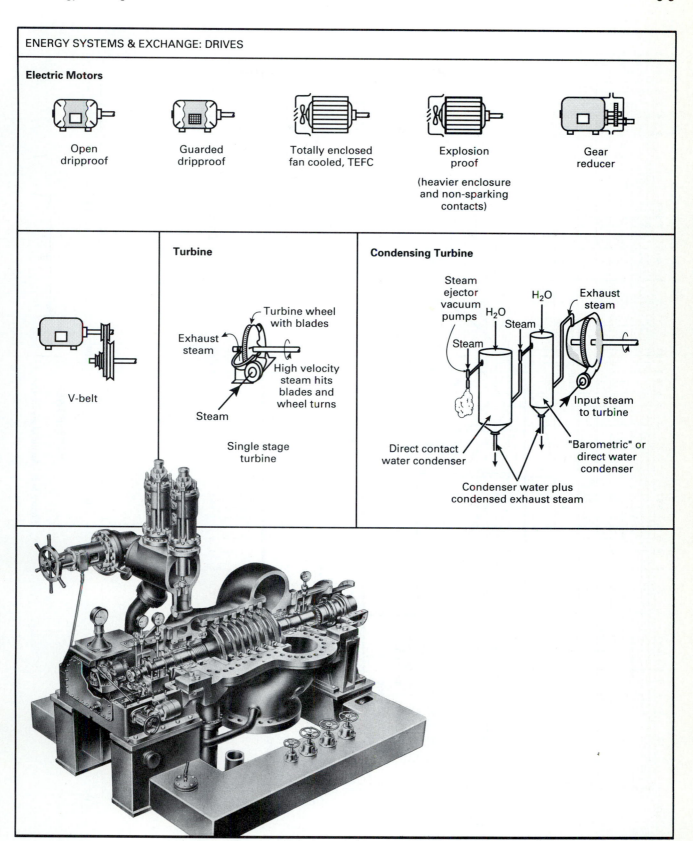

Figure 3–2 Sketches of Energy Exchange Systems: Drives and Steam Backpressure Turbine (reprinted courtesy of GEC Alstrom Turbine Generators Ltd.—Medium Turbo Machines Group)

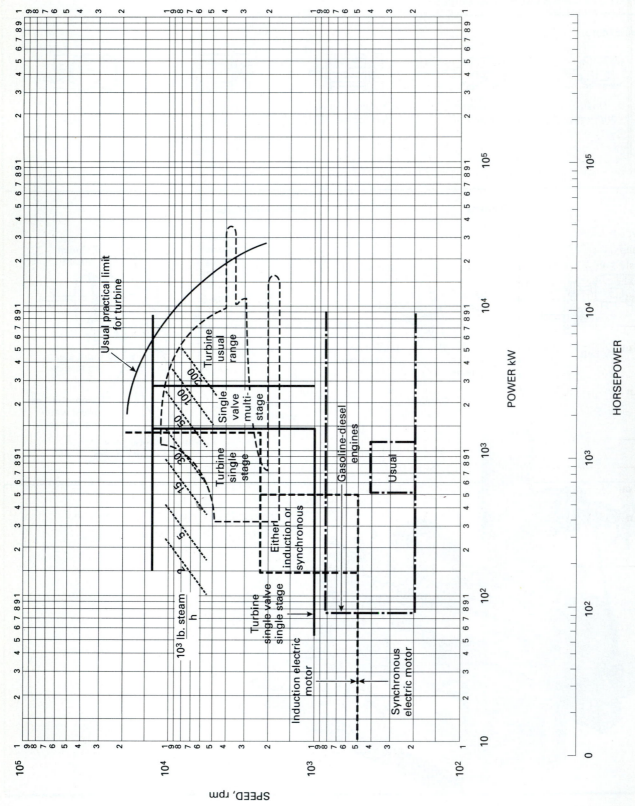

Figure 3–3 Applications for Drives

Equipment	Comments
TABLE 3-1. Some efficiency for mechanical energy exchange	
Electric Motor	90%
Gear Units	
Turbines	2 kg/h of steam /kW for energy generated from a condensing turbine. Need 1.8m^2 of condenser surface /kg/h steam for a condensing turbine system.
Pumps	40 to 60%
Compressors	60 to 80%

The energy required, or the efficiency for energy conversions for different types of mechanical/electrical exchange equipment, is summarized in Table 3–1. Figure 3–4 shows the approximate steam usage for different inlets to exhaust conditions to produce different amounts of power.

Example 3–1: Select a drive for 100 kW power at 300 rpm.

An Answer: From Figure 3–3, a synchronous electric motor or a gasoline-diesel engine might be appropriate.

Example 3–2: Select a drive for 2000 kW and 1800 rpm.

An Answer: From Figure 3–3, an induction motor or a multistage steam-driven, single-valve turbine could be used.

Example 3–3: Estimate the steam usage for a 600 kW steam-driven turbine if the steam is at 1.8 MPa, 230°C.

An Answer: We cannot answer this question unless we identify the exhaust conditions for the steam. If the steam exhausts at 170 kPa, then the amount is about 11.2 Mg/h from Figure 3–4.

3.2 ENERGY EXCHANGE: THERMAL

The first question to consider in selecting a method of exchanging thermal energy is whether the process material (that is wanted) can be contacted directly with the source of heat or whether a wall or barrier must separate them. This is illustrated in Table 3–2. If contact is allowed, then direct contact exchangers should be considered. These are discussed later. If no direct contact can be allowed, then the next question concerns the temperature required.

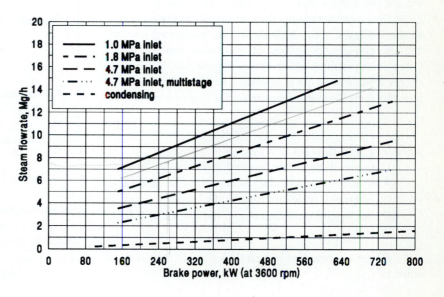

Figure 3–4 Approximate Steam Usage for Turbines

TABLE 3-2. General selection route for thermal energy exchange				
● Can heating/cooling medium contact the process fluid?	NO			YES
● What is the general temperature range?	Use Figure 3-5 to select			Consider direct contact devices Section 3.2-2
	Furnace	Usual exchanger	Cryogenic	
● Consider key criteria other than total duty and heat flux allowed. These are:	Temperature control	Temp. driving forces, configurations, approach temperature, and maintenance	pressure	

3.2-1 No Direct Contact Devices

The general classification of heat exchange equipment is furnaces, exchangers, boilers and condensers, and cryogenic units. The general ranges of application are given in Figure 3–5a for heating and cooling. For boiling and condensing, the general ranges of application are given in Figure 3–5b. Furnace configurations are shown in Figures 3–6 and 3–7. Other types are shown later in Figure 3–13.

Example 3–4: Estimate the conditions needed for a CO shift reactor.

An Answer: From Figure 3–5, a furnace would handle the usual temperatures of 400 to 500°C.

A. Furnaces. In furnaces, the heat is transferred (a) by radiation or in simplest terms from the place where we see the flames and (b) by convection where the hot gases flow over or around tubes. In simplistic terms, this is shown in Figure 3–6, which shows the "radiation" section and the convection section of a furnace configuration. Other configurations are shown in Figure 3–7. Since heat energy flows from a hot temperature to the colder temperature, the overall equation governing the rate of heat flow is

$$q = UA \, \Delta T \qquad (3-1)$$

where q = heat flow per unit time
  U = overall heat transfer coefficient
  A = area perpendicular to the flow of heat and through which the heat flows
  ΔT = temperature difference in the direction of heat flow

Hence, the heat flux or heat flow per unit area, q/A, is equal to

$$\frac{q}{A} = U \, \Delta T \quad \frac{Watts}{m^2} \qquad (3-2)$$

The furnace features that are important are:

1. the heat flux allowed in the radiation section is restrained by tube materials of construction and the surface temperature.
2. the ability to control the tube surface temperature throughout the furnace.

If we have just a single row of tubes surrounded by flames on both sides, the maximum heat flow at some location in the tubes is very close to the average heat flux that any fluid receives anywhere in the tubes. If, however, there are banks of tubes, then larger heat flux (and hence surface temperature) conditions occur. In the latter, the control will not be as good as the former. The ratio of the maximum to the average heat flux for various tube configurations is given in Figure 3–8.

Example 3–5: A reaction is to be carried out at 300°C and the unwanted side reactions occur if the temperature in the tube is not controlled to within ±10°C. What does this suggest?

An Answer: From Figure 3–8, we should try for a single-row configuration. From Figure 3–7, an A-frame (No. 4), or units 12 to 16 might be used.

Figure 3–7 shows different furnace-boiler configurations for these two factors: amount of radiant heat flux and variation in heat flux.

An order-of-magnitude design of furnace requires a choice of the configuration and the specification of the heat duty it must supply. The general ranges are shown in Figure 3–9, which gives the radiant section heat flux, the *total*

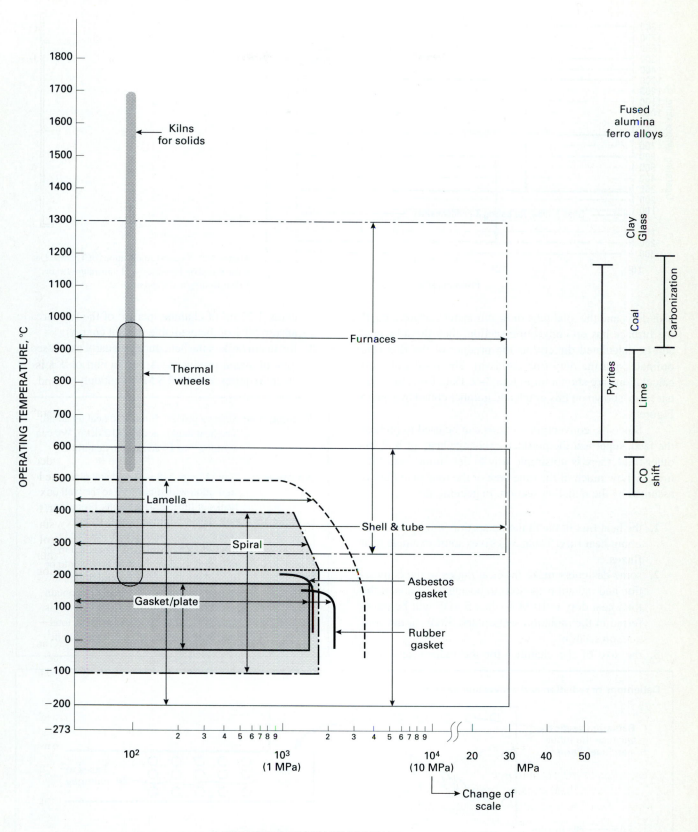

Figure 3–5 General Applications of Thermal Devices Based on Pressure
and Temperature Levels a) For Heating and Cooling Only

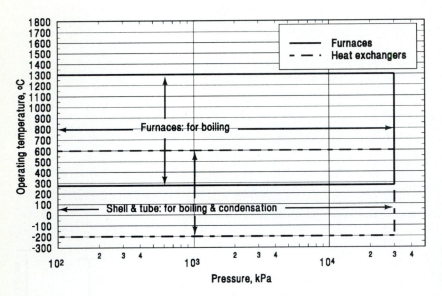

Figure 3–5 General Applications of Thermal Devices Based on Pressure and Temperature Levels. b) For Boiling and Condensation

heat duty, and the total tube or heat transfer surface area. If the furnace has *no* convection section, then the tube area can be calculated directly as the product of the flux (the ordinate) and the duty (the abscissa). The results of such calculations are shown on Figure 3–9. Data for actual furnaces are shown on this graph and number coded by *type* to Figure 3–7.

Since the convection sections can be used to preheat the feed, superheat the product, exchange heat, or boil another fluid, there is no simple way to determine from Figure 3–9 how much of the tube area or the total heat duty is assumed by the radiation section. In general, the

1. the heat flux in the radiation section will be the maximum heat flux; Table 3–3 gives some example heat fluxes.
2. some designers make the heat duties in the convection and radiation sections to be equal. That is, if the total heat duty is 10 MW, then 5 MW will be transferred in the radiation section and 5MW in the convection section.
3. the size of the chamber for the radiant section is

about 1.22 m^3 of chamber per m^2 of the tube area to absorb 50% of the available radiant energy.
4. for steam boilers the heat duty is usually expressed as kg/s of steam produced. A production of 1 kg/s of steam requires about a 2.8 MW heat duty.

Example 3–6: Select a process furnace to heat up an oil. The heat load is about 2×10^7 kJ/h. We can accept a 3 to 1 variation in heat flux.

An Answer: From Figure 3–8 we see that a piping configuration with tubes next to the wall or banks of tubes is acceptable. From Figure 3–9 standard furnace types ④, an A frame, or ⑧, a vertical furnace, can be used. If all the heat load is handled in the radiation section, the tube area would be about 200 m^2, for the general design level of heat fluxes that we usually can obtain with this configuration. However, this is unlikely so that we might expect a total area of more than 200 m^2 or say 300 m^2.

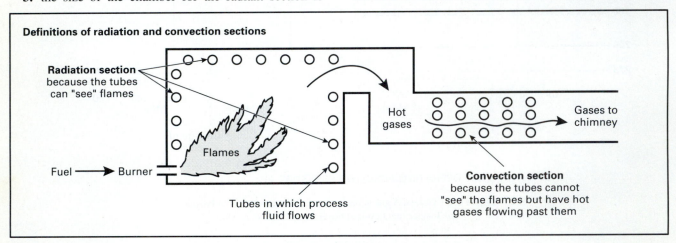

Figure 3–6 A Furnace Configuration Illustrating the Radiation and Convection Sections

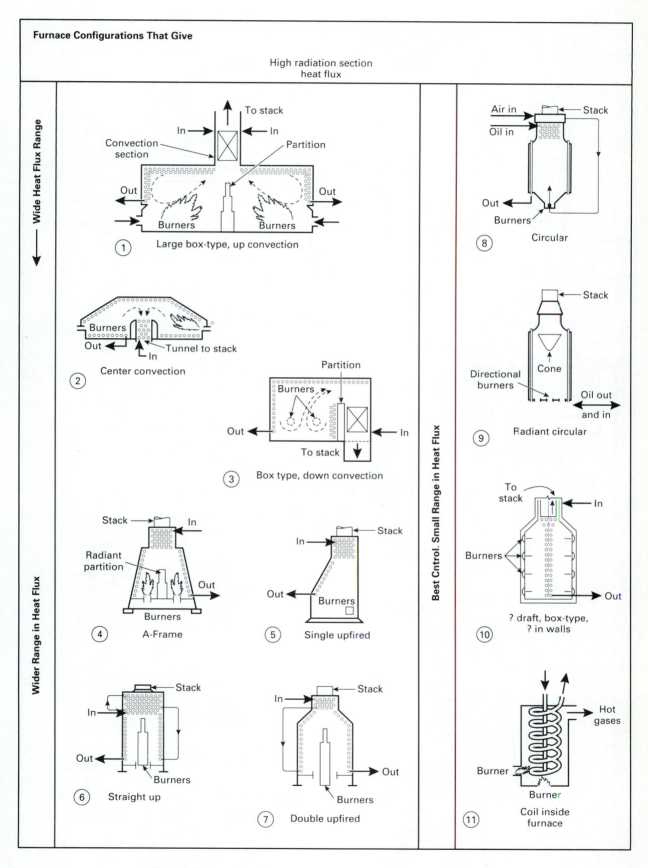

Figure 3–7 Some Furnace Configurations (reprinted courtesy of Nelson [1949], McGraw-Hill Book Co. and Ellwood and Donatos (1966) Buffington (1975) Chemical Engineering, McGraw-Hill Publishers)

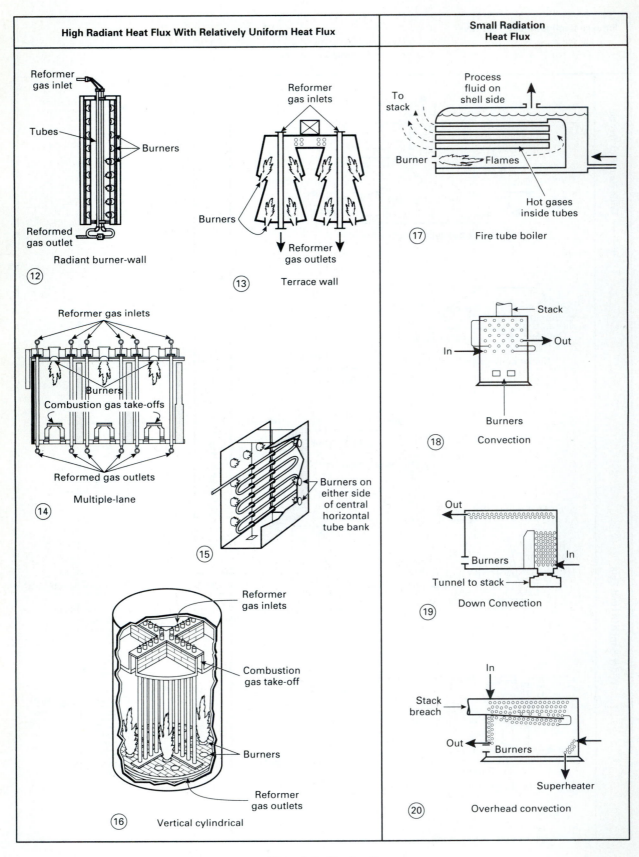

High Radiant Heat Flux With Relatively Uniform Heat Flux	Small Radiation Heat Flux

Figure 3–7 (Continued)

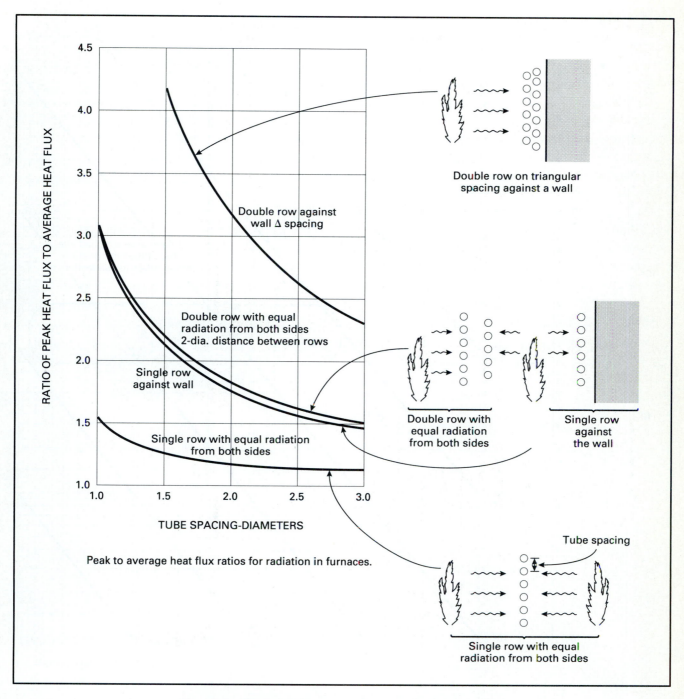

Figure 3–8 How Tube Configuration Affects the Uniformity of Heat Flux
and Hence the Tube Wall Temperature (Gunder [1969]) reprinted by permission from
"How to Specify Process Heaters and Evaluate Bids," Hydrocarbon Processing,
Oct. 1969, copyright Gulf Publishing Co. 1969, all rights reserved.)

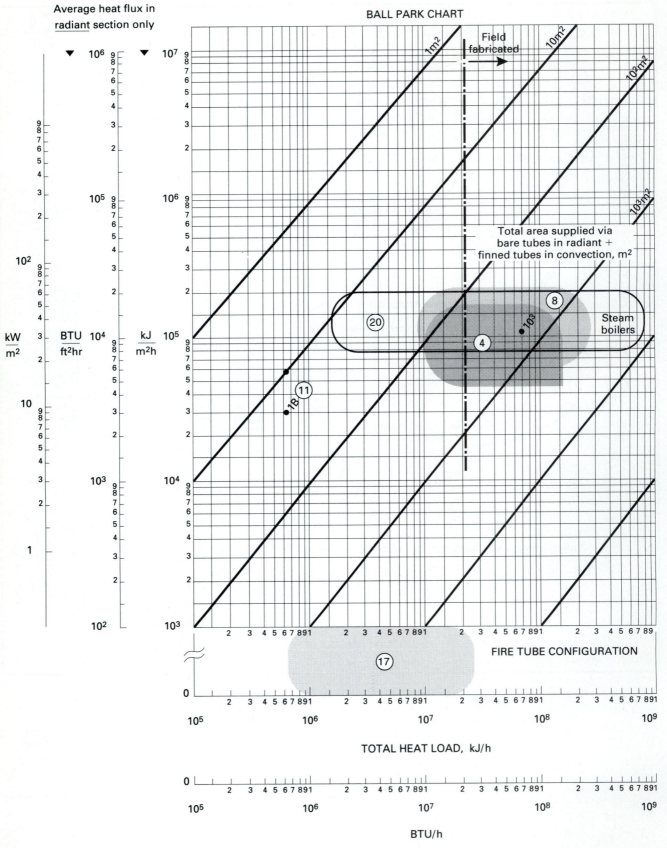

Figure 3–9   Rough Sizing of Furnaces

TABLE 3-3. Example heat fluxes for furnace applications

	kW/m² based on outside tube area*	Fluid velocity (at 15°C) m/s
Heating		
General	50	2 to 3
Feed for cat cracker	30	0.3 to 0.6
for cracking still/soaker	25 to 50	1.5 to 2.1
Boiling/Reboiler		
Low temp. & high vaporization	60	
Vacuum distillation	25 to 30	0.15 to 1.2
Delayed coking	30	2
Visbreaking	30 to 60	0.6 to 1.8
Vac distillation lube oil	25	1.2
heavy lube	17	
Reactors		
Reformers/polymerizers gas oil	40 to 50	1.5 to 2.5
light oil	25 to 40	1.4 to 2.3
heavy oil	25 to 35	1.7 to 2.1
Cracking to produce ethylene ex		
ethane (50 to 60% conversion)	23 to 28[†]	
propane (70 to 85%)	14 to 17[†]	
butane	11 to 15[†]	
naphtha	11 to 15[†]	

*the flux is three times larger if based on projected area.

[†]based on exit conditions. At the inlet, the value is double.

B. Exchangers, Boilers, Condensers, Heaters, Coolers. Consider now energy exchange in devices other than furnaces: exchangers, boilers, and condensers. The configuration and size of exchangers, boilers, and condensers are chosen based on the amount of heat that can be transferred and the mechanical energy required to push the fluids through the exchanger. Usually if the energy exchange is high, the mechanical energy demand is high. A high-energy exchange means the heat exchange surface area required can be small and hence the purchase expenditure is lower but the pumping or mechanical energy cost will be high. These trends are shown in Figure 3–10.

In design we need to calculate both the heat and mechanical energy requirements. For estimations, we do not calculate mechanical energy or pressure drop requirements; rather we assume heat transfer conditions corresponding to the *usual*, compromise pressure drop allowances and estimate the exchange *area* required.

For estimation purposes, the six steps required to size and select a heat exchanger are:

1. determine the heat load or how much heat is to be exchanged, q,
2. select the heating/cooling medium if both are not given in the problem,
3. make a preliminary overall selection based on absolute temperature ranges,
4. decide on inlet and exit temperatures of all streams in the system and estimate the effective temperature driving force difference, ΔT,

5. select the heat transfer coefficient values, U,
6. select-size the exchanger from the equation

$$\text{Area} = q/U\Delta T \tag{3-3}$$

Step 1: Determine the Heat Load. If sensible heat is exchanged, the load is

$$q_s = \hat{F}\hat{c}_p(T_1 - T_2) \tag{3-4}$$

where q = heat exchanged per unit time
$\hat{F}$ = mass flowrate of material being heated up or cooled down
$\hat{c}_p$ = heat capacity of the material
T_1 = inlet temperature of the material
T_2 = exit temperature of the material.

If latent heat is exchanged, the load is

$$q_c = \hat{F}\lambda \tag{3-5}$$

where λ = latent heat of evaporation, fusion (condensation, melting)
If a reaction is going on, the load is

$$q_r = \hat{F}\Delta H_{react} \tag{3-6}$$

ΔH = heat of reaction at the temperature and pressure of the reaction.

If all three are going on, then the total heat is the sum of the three contributions. **Warning:** the units used for F and c_p, λ and ΔH_{react} must be consistent and F refers to only that amount of material that is undergoing the heating, the condensing and the reacting.

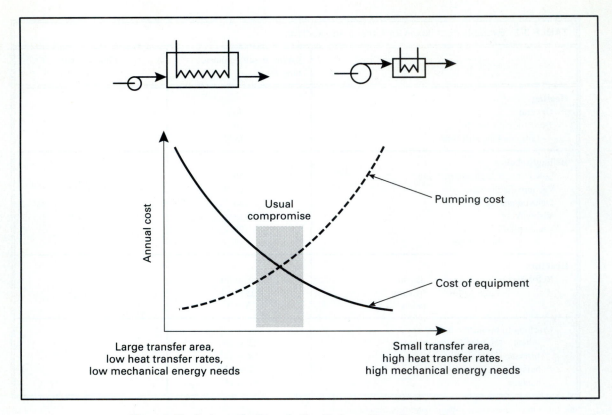

Figure 3–10 Optimum Conditions for Heat Exchange Versus Pumping Power

Example 3–7: A stream of 100 kmol/h of material is heated up from 18°C to 100°C. This stream is a mixture of a number of components. During this process 10 kmol/h evaporate at 30°C to create a vapor whose heat capacity is 30 kJ/mol°K and 10 kmol/h of liquid components react at 70°C. This reaction is endothermic and at the reaction conditions *requires* 200 MJ/kmol of A reacted. The reaction is

1 mol A + 1 mol B = 1 mol C + 1 mol D

and the stream initially contains 10 mol/h of A and 10 kmol/h of B. Calculate the heat load.

An Answer: In this example it is easiest to keep each part of the stream separate:

- The sensible heat material only:
  the flow: 100 kmol/h − 10 kmol − 20
  = 70 kmol/h.
  the heat capacity: assume 75 kJ/kmol. K. Hence,

$$q_s = 70 \frac{kmol}{h} \; 75 \frac{kJ}{kmol°K} \; (100 - 18)$$
$$= 431 \; MJ/h$$

- The evaporation process:
  the flow 10 kmol/h
  the latent heat for a substance with an assumed molar mass of 50 is

500 kJ/kg or 2.5 MJ/kmol

In this process there are three steps:

1. heat up liquid:

$$q_s = 10 \frac{kmol}{h} \; 75 \frac{kJ}{kmol°K} \; (30 - 18)$$
$$= 9 \; mJ/h$$

2. evaporate:

$$q_c = 10 \frac{kmol}{h} \; 2.5 \frac{MJ}{kmol} = 25 \; MJ/h$$

3. heat up product gas:

$$q_s = 10 \frac{kmol}{h} \cdot 30 \frac{kJ}{kmol.K} \; (100 - 30)$$
$$= 21 \; MJ/h$$
Total = 55 MJ/h.

- The reaction. In this process there are three steps.

1. heat up reactants:

$$q_s = 20 \frac{kmol}{h} \; 75 \frac{kJ}{kmol.K} \; (70 - 18)$$
$$= 78 \; MJ/h$$

2. react. Here we are reacting 10 kmol of A with 10 kmol of B. Since the heat of reaction is defined in terms of the amount of A that reacts the result is:

$$q_r = 10 \frac{kmol}{h} \cdot 200 \frac{MJ}{kmol} \text{ of A.}$$
$$= 2 \text{ GJ/h}$$

3. heat up products. We assume products are liquids with a specific heat capacity of 75 kJ/kmol.K. Note here that for every mol of reactants we have the *same* number of mols of product. Hence, there are 20 kmol/h of product. The result is:

$$q_s = 20 \frac{kmol}{h} \cdot 75 \frac{kJ}{kmol.K} (100 - 70)$$
$$= 45 \text{ MJ/h}$$

The total for the reaction step is 2.123 GJ/h.

The total heat load is

sensible	431 MJ/h
evaporation	55 MJ/h
react	2123 MJh
	2609 MJ/h

Step 2: Select Medium. The next decision to make is to identify the two streams that are going to exchange heat. Usually these are called the process fluid (namely the one we are interested in) and the medium fluid (which is the fluid that exchanges the heat). Sometimes we exchange heat between two process fluids.

The basic principles are that:

1. heat flows from a high temperature to a lower temperature; the difference between these two is called the temperature driving force difference.
2. when a fluid condenses, or solidifies, heat is released by that fluid and the temperature is constant if only one pure component is present and the pressure is constant.

3. when a fluid boils or a solid melts, heat is required by that fluid and the temperature is constant if only one pure component is present and the pressure is constant.

Some common media are listed in Table 3–4. These can serve as either heating or cooling media depending on the temperature. Often the medium is recirculated and reused. Some example circulation systems are shown in Figure 3–11.

Wherever possible we would like to exchange the heat via condensation, evaporation, freezing, or sublimation processes. Such a procedure assists in the control of the temperature driving force causing the heat to transfer, and this makes it easy for us to use the media over and over again. From the phase rule we know that the phase transition temperature is set if we specify the pressure for a one-component system. Figure 3–12 shows the phase change temperature for different high temperature and refrigeration media. Also shown on this figure are the usual operating pressures for steam.

Example 3–8: We wish to condense pure butane coming off the top of a distillation column operating at 0.5 atmos. What do we use as the condensing medium?

An Answer: Since this is *condensation* we must use a medium whose temperature is less than the condensation temperature of butane at 0.5 atmos. From vapor pressures data (or in this case these data are given in Figure 3–12) the condensation temperature of butane is about −15°C. Hence the medium must be colder than −15°C. From Figure 3–12 the medium could be refrigerants R-13, 12, 22 operating at a medium pressure of 1 atmos., ammonia or propane operating at pressures of less than 2 atmos., since these conditions give media condensation temperatures less than −15°C.

TABLE 3-4. Some common heat exchange media

High Temp.	liquid sodium	liquid sodium combustion gases thermal fluids hot mineral oils
Usual Temp.	boiling water air water—river or well city cooling tower or recirculated	condensing steam electricity ammonia
Very Low Temp.	ammonia butane ethylene	ethylene

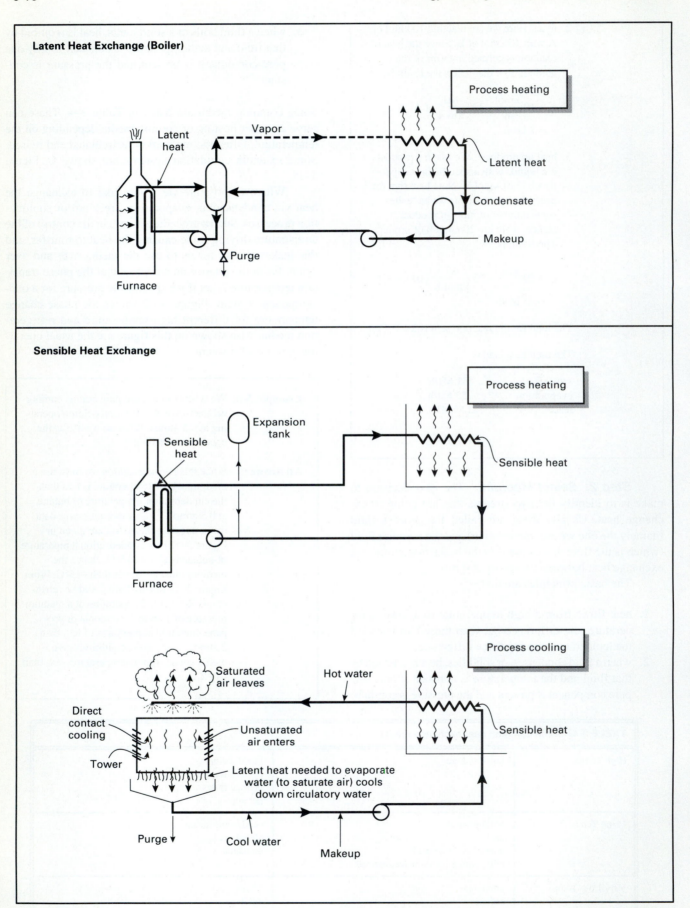

Figure 3–11   Some Example Circulation Systems for Heat Exchange

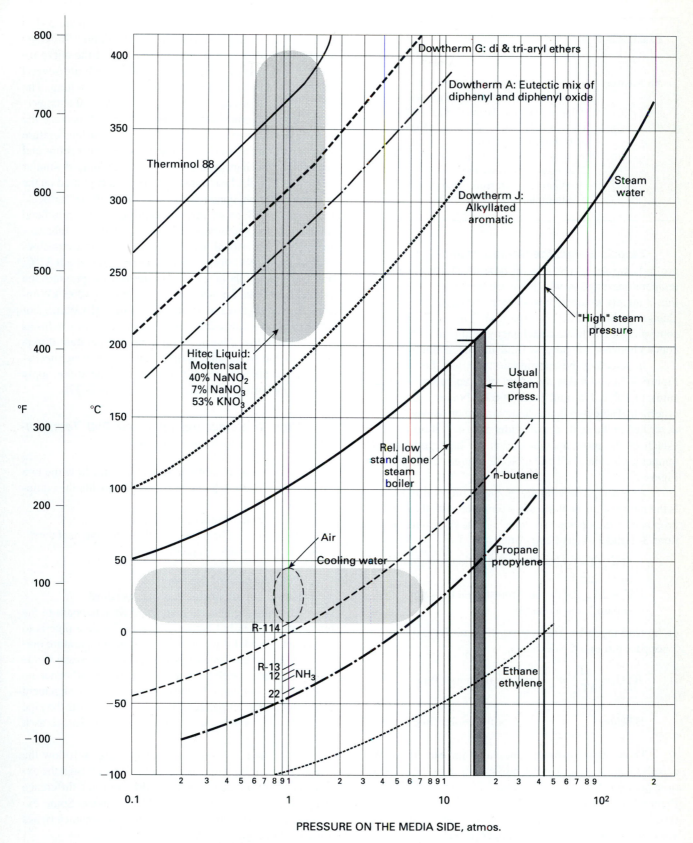

Figure 3–12 Choice of Medium Based on Temperature and Pressure (based on Fried (1973) Chemical Engineering, a McGraw-Hill publication)

Example 3–9: We wish to heat a system to 200°C. What might we use?

An Answer: Since this is *heating* we must select a medium whose operating or condensation temperature is greater than 200°C. This, from Figure 3–12, can be achieved with Dowtherm J at pressures greater than 2 atmospheres, steam at about 20 atmospheres, Hitec molten salt, Dowtherm A at pressures greater than 0.2 atmospheres.

Step 3: Select the General Range. From Figure 3–5, we select the general type of equipment for the temperature range in which we are working. At this stage this is merely to identify whether a cryogenic or usual exchanger configuration is preferred. Figure 3–13 shows some pictures of the heat exchange equipment with indirect contact between the process fluid and the medium.

In selecting the appropriate type, use the criteria in Figure 3–5a and b (for the effect of temperature and pressure), in Table 3–5 (and Table 3–6 for boilers). One of the criteria in Table 3–5 is the "temperature crossover"; this is described in the next section under Step 4. A more detailed sketch of the most popular option for high pressure and temperature—the shell and tube configuration—is given in Figure 3–14.

Another useful guideline for selecting configurations is the rating of the heat transfer compared with the pressure drop across the device. To "rate" a heat exchanger, Equations 3–1 and 3–4 can be rewritten as:

$$q = \rho \check{F}_1 \hat{c}_{p1} (T_1 - T_2)_1 \text{ for the sensible heat acquired}$$
$$\text{by stream 1}$$
$$= U A \Delta T_{lm} \text{ for the heat transferred between}$$
$$\text{stream 1 and stream 0}$$

or, by rearrangement, the heat transfer unit HTU for stream 1 per pass through the device is:

$$\mathrm{HTU}_1 = \frac{\Delta T_1}{\Delta T_{1m}} = \frac{UA}{\check{F}_1 \rho c_{p1}} \quad \text{for the hotter fluid} \quad (3\text{-}7)$$

$$\mathrm{HTU}_2 = \frac{\Delta T_0}{\Delta T_{1m}} = \frac{UA}{\check{F}_2 \rho c_{p2}} \quad \text{for the cooler fluid} \quad (3\text{-}8)$$

Thus we can estimate one value for each fluid. We can also estimate the specific pressure drop as the pressure drop per unit HTU (with units kPa_1/HTU_1 and kPa_0/HTU_0). Figure 3–15 illustrates typical values of these for different types of exchangers. Figure 3–16 shows the effect of fluid viscosity on the choice of exchanger. If we are selecting a boiler from the criteria given in Table 3–6, we need to consider the different boiling mechanisms as shown in Figure 3–17. Figure 3–17a shows the boiling characteristics of "pool" boiling where a "quiet pool" of liquid is boiling be-

cause a hot surface is immersed in the pool. Figure 3–17b shows the boiling characteristics of liquid "flowing" through a heated tube. Here the first part of the curve increases with increasing liquid velocity. There are several features to note about Figure 3–17a for pool boiling. The heat flux increases up to "the critical point" as the temperature difference increases. During this initial increase the boiling mechanism is "nucleate" boiling. For temperature differences greater than the critical, the heat flux drops and the boiling mechanism becomes "film" boiling. A similar curve results if the heat transfer coefficient is plotted on the ordinate. One challenge is that the value of the "critical" temperature difference varies depending on the liquid and on the type and smoothness of the heat exchanger tube surface. Thus, the critical heat flux for water and acetone occurs at about 25°C, for ethanol and i propanol about 33°C, and for butanol about 44°C. At atmospheric pressure the critical heat flux for water is about 360 to 1260 kW/m^2 whereas for organics it is about 120 to 400 kW/m^2. For quick estimates, however, this type of information helps us to appreciate that there is an optimum temperature difference and that we usually want to keep the boiling mechanism in the nucleate regime. Similar considerations apply for flowing boiling systems shown in Figure 3–17b.

Step 4: Select the Inlet and Exit Temperatures. These temperatures depend on:

1. the equipment configuration and hence how the two streams flow relative to each other inside the equipment,
2. the approach temperature allowed,
3. the amount of heat transferred, called the heat duty,
4. whether changes in phase occur.

Consider each of these in turn.

How configuration affects our decisions:

Consider the heat exchange and what happens to the temperature when hot benzene flowing inside a pipe is to be cooled. To cool it we need to have the temperature outside the pipe to be *everywhere* less than the benzene inside the tube. Otherwise there is no driving force and hence no heat transfer. Figure 3–18 shows the physical arrangement and a plot of the temperature of the benzene inside the pipe as it is cooled down and the temperature of the ice as it melts.

The features to note are that we need to follow the temperatures of both the fluids as they flow through the exchanger, and it is the temperature driving force difference at any location that dictates the heat transfer. Some exchanger configurations and the temperature characteristics are shown in Figure 3–19.

Normally, we use countercurrent configurations.

If water is the cooling medium, then assume the water is available about 18°C; to prevent erosion and corrosion of the pipes carrying the water, we rarely let the water

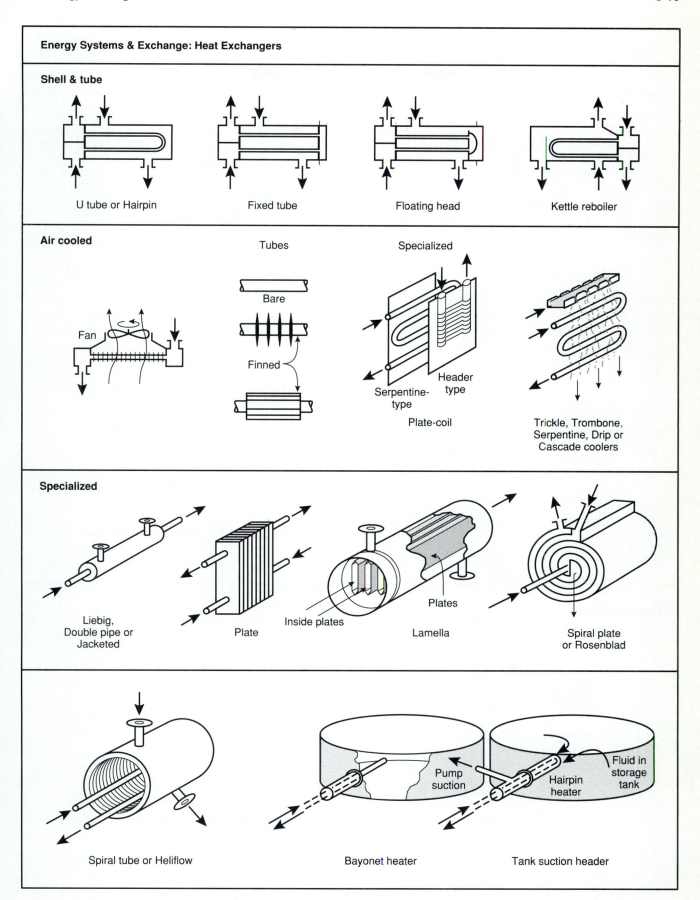

Figure 3–13 Pictures of Heat Exchange Equipment

Energy Systems & Exchange: Heat Exchangers

Specialized

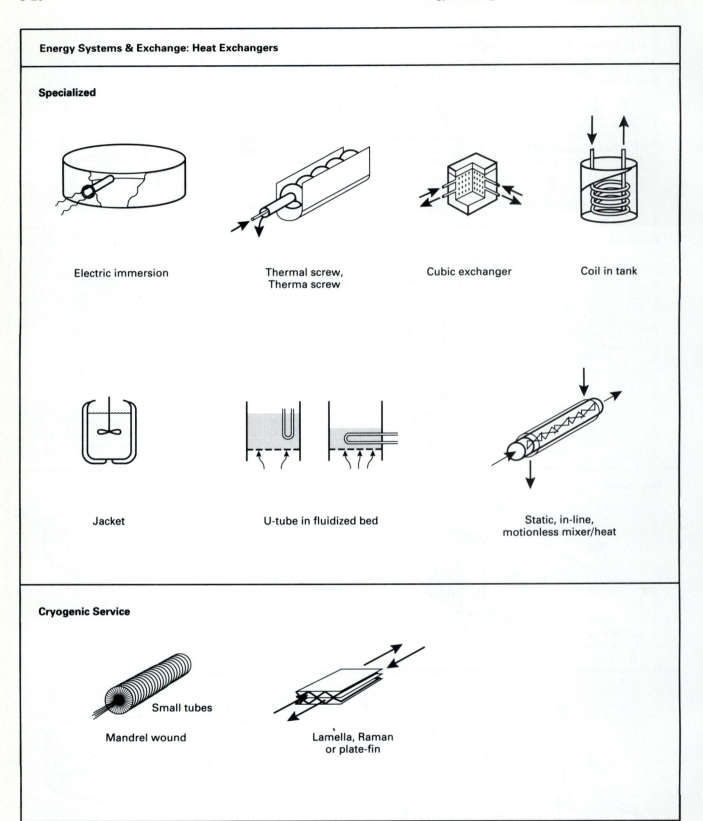

Electric immersion

Thermal screw,
Therma screw

Cubic exchanger

Coil in tank

Jacket

U-tube in fluidized bed

Static, in-line,
motionless mixer/heat

Cryogenic Service

Small tubes

Mandrel wound

Lamella, Raman
or plate-fin

Figure 3–13 (Continued)

TABLE 3-5. Comparison criteria for selecting the type of exchanger

Type	Feed Conditions									Handle		Hold up L/m²	Area m²	Flow rate ratio F_1/F_0	Max flow L/s	Strengths	Weaknesses
	Gas	Scale Foul	Crystal Form	Sus. Solids	Fibers	Heat Sensitive	Viscosity mPa.s	Exit* Temp Cross	Corrosive	Boil?	Condense?						
Shell Tube general	OK	rel. clean NO	NO	NO	NO	NO ?	<10	Not much		YES	YES		10 to 10^3			mainly high temp/pressures flexible	fluctuating temp. or extremes cause fatigue/stress; vibration
U Tube																cheap OK for thermal expansion	no dirty fluids on shell
Fixed																cheap	not for thermal expansion
Floating																OK for thermal expansion	
Air Cooled	OK	NO rel. clean	NO	NO	NO					NO	OK					air is cheap	high initial cap investment usually need trim cooler
Plate Coil	NO	OK	YES	YES	YES		OK		OK	NO	NO					easy to remove & clean	high initial cost
Trombone	OK									NO						cheap; preferred when cooling high temp. streams where fouling on outside can be excessive	
Double Pipe	OK	NO rel. clean					<10^2	OK		OK	OK		0.25 to 1200			modular, low flowrates small area. Usually used for <10m² OK high temp/ press	

TABLE 3-5 (Continued)

Type	Gas	Scale Foul	Crystal Form	Sus. Solids	Fibers	Heat Sensitive	Viscosity mPa.s	Exit* Temp Cross	Corrosive	Boil?	Condense?	Hold up L/m²	Area m²	Flow rate ratio F_1/F_0	Max flow L/s	Strengths	Weaknesses
Plate/Gasket	NO	Some OK		<70 ppm		Ok	OK $<3\times10^4$	OK some		NO	NO	Low 1.5	1 to 10^3	0.7 to 1.3	550	usually most economical especially if alloys used. Negligible vibration stress fatigue	temp/pressure limitations; not for two phase or gases. HTU>0.2 Not when $\Delta T_1 >> \Delta T_0$
Plate/ Welded		NO	OK					Some		OK	OK	Low	$\propto 10^3$			High temps pressure	
Lamella		OK		OK						OK	OK	Low	<800	1:1.15 to 1:8 Good	1000	Passage size 3 to 5X size of largest particle Handle large flow ratios	
Spiral		OK	NO	OK	OK <0.5 %	OK	OK for high	OK	OK	OK	OK		<300	Not for large diff. <3.5	110	Low maintenance self cleaning, true counter current	Not for deposits. Difficult to repair in the field. HTU>0.2 not when $DT_1 >> \Delta T_0$
Spiral Tube Heliflow																	
Bayonet Tank Suction	NO	NO	NO	NO	NO	NO										Small area	
Thermal Screw	NO															For solids	
Cubic	OK								YES	NO	NO					Fabricate out of carbon/low strength materials	
Coil Jacket																	

TABLE 3-5 (Continued)

Type	Feed Conditions									Handle		Hold up L/m²	Area m²	Flow rate ratio F_1/F_0	Max liquid flow L/s	Strengths	Weaknesses
	Gas	Scale Foul	Crystal Form	Sus. Solids	Fibers	Heat Sensitive	Viscosity mPa.s	Exit Temp Cross	Corrosive	Boil?	Condense?						
Mandrel wound.	YES									NO	NO					For cryogenic gases	Made out of Al. Limited to noncorrosive
Raman, Plate/Fin	YES	NO					GAS	OK					<10³			High thermal efficiency	
Fluidized bed	YES			YES						NO	NO						

TABLE 3-6. Criteria for Selecting Type of Boiler (Based on Shah (1979) American Institute of Chemical Engineers).

Condition	Choice of Reboiler		Boiling Mechanism	Amount of Vaporization	Design is Sensitive to
Relatively low pressure	**Kettle**	at higher pressures, V-L separation is difficult	Pool boiling. Keep Δt <50°C to prevent film boiling.	75 to 100%	Fouling
	Vertical/ horizontal thermosiphon		Nucleate	5 to 25%	Liquid level
Vacuum operation or viscous/ fouling liquids	**Forced circulation**	suppress fouling by placing a valve on exit to cause flashing	Sensible heating only		Setting position of valvestem on exit line
>250°C	**Fired heater**		Nucleate		

Vertical thermosiphon   Horizontal thermosiphon   Kettle

exit temperature exceed 50°C. For air as the cooling medium, we expect that 50°C is the upper limit for the air leaving the exchanger, based on experience for air-cooled heat exchanger design.

From Figure 3–19 comes the concept of temperature crossover—a concept used as one of the selection criteria in Table 3–5. This is illustrated in Figure 3–20.

The approach temperature is defined as the temperature difference between the entering and leaving fluids. In Figure 3–19, the approach temperature is as follows:

- the temperature of the "leaving" fluid is 30°C
- the temperature of the "entering" fluid is 0°C
- the approach temperature, or difference, is 30°C

This usual value of the approach temperature is an order-of-magnitude rule of thumb based on good general practice.

For usual heat exchange/condensation, use 5 to 8°C.

For usual boiling operations, use 25°C. (We note from Figure 3–17 that this is slightly below the critical heat flux and keeps the boiling mechanism in the nucleate regime.)

For porous heat exchangers, use 2°C.

For cryogenic or brazed aluminum exchangers, use 2°C.

Sometimes, at this stage, we should also consider which fluid should be on the shell side and which on the tube side. Table 3–7 summarizes some of the considerations.

Tubeside for Pull-Through Floating Head

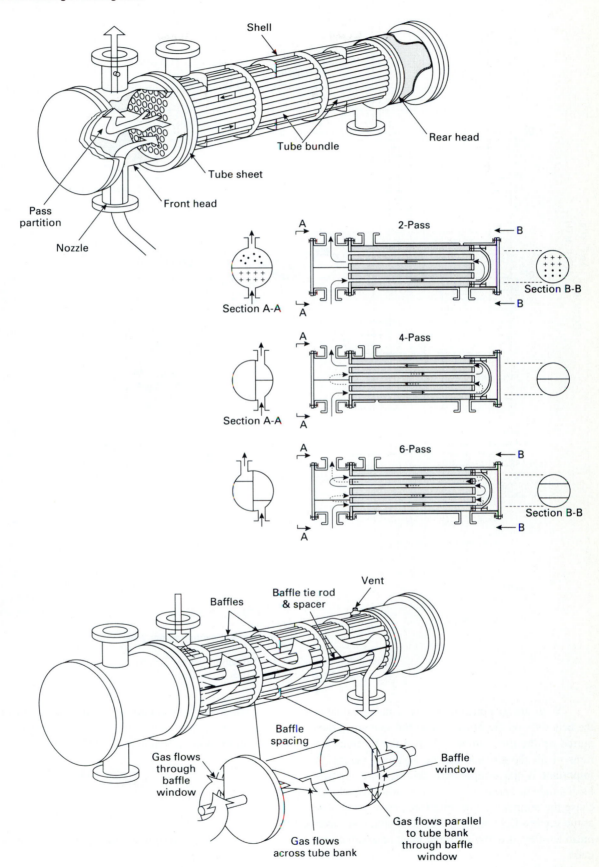

Figure 3–14   Details of a Shell-and-Tube Heat Exchanger

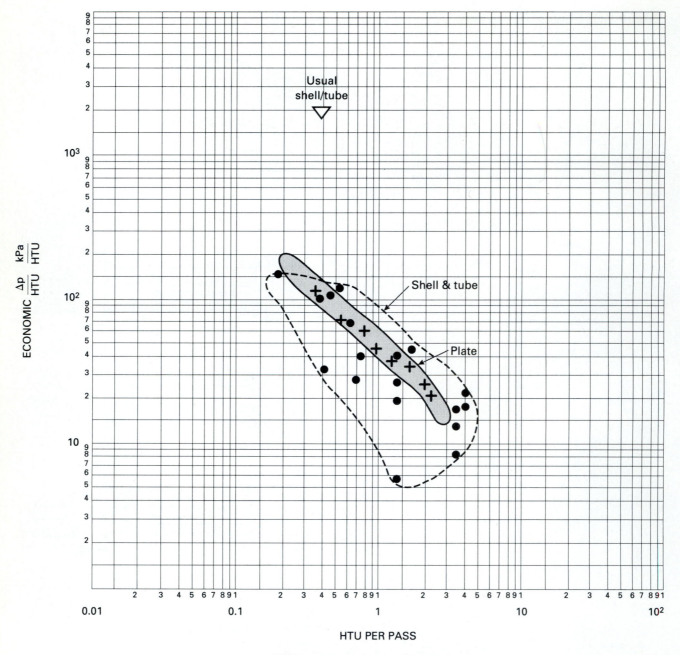

Figure 3–15   Selection of Heat Exchangers: Energy Transfer
and Usage for Different Heat Exchangers

With these constraints, we can calculate usually the heat duty on the process side; the same heat duty is required on the medium side. Hence the terminal temperatures of all the streams can be assigned values. What is important is the temperature driving force between the local temperatures at all locations inside the exchanger. Because the temperature driving force changes continuously as the streams flow through the exchanger, some average or mean temperature driving force must be assumed or calculated.

The log mean temperature difference for true coun-

tercurrent, true cocurrent, or flow through a condenser for one pure gas component is a good average. The log mean temperature difference, or LMTD, is

$$\text{LMTD} = \frac{\Delta T_1 - \Delta T_2}{\ell n \dfrac{\Delta T_1}{\Delta T_2}} \qquad (3\text{-}9)$$

where ΔT_1 is the temperature driving force at one end of the exchanger

ΔT_2 is the temperature driving force at the other.

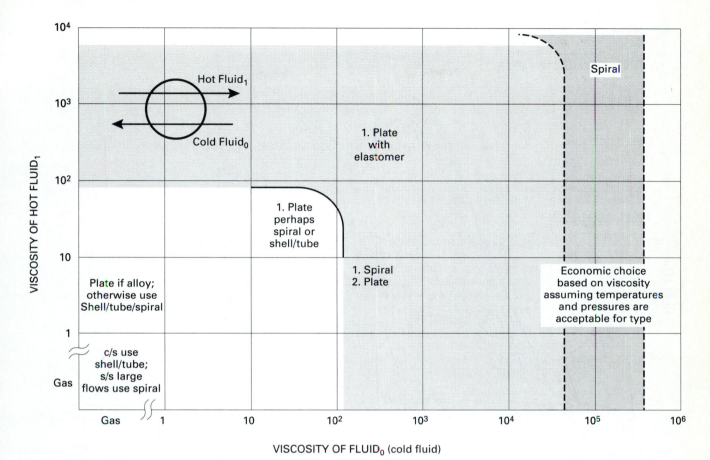

Figure 3-16 Selection of Heat Exchangers: the Effect of Fluid Viscosity

That is,

$$\Delta T_1 = T_1 - T_{20} \qquad T_2 - T_{10}$$

for countercurrent.

$$T_1 - T_{10} \qquad T_2 - T_{20}$$

for cocurrent.

Example 3-10: Estimate the LMTD for the following:
hot fluid cools from 80°C → 20°C
cold fluid heats to   15°C ← 5°C
                      65°C   15°C

An Answer: log mean $= \dfrac{65 - 15}{\ell n \dfrac{65}{15}} = 34°C$

The log mean is approximately equal to the arithmetic mean if the largest terminal temperature difference $\gg 2$ (smallest temp. difference). Figure 3-21 (p. 3-34) is a nomograph to assist in the estimation of the LMTD.

However, for most heat exchanger designs a truly countercurrent heat exchanger is specified. For many shell and tube exchangers (and for many other cases) the condi-

tions are NOT satisfied for which the log mean is a good approximation to the average. To handle this case, correction factors, f, are applied to give an empirically correct mean temperature difference from the log mean temperature difference, i.e.

$$MTD = f(LMTD) \qquad (3\text{-}10)$$

Values of f can be found in most texts on heat exchange. However, for a good design the f value should never be below 0.75 so as to prevent temperature crossover internally. For a conservative quick estimate:

1. Calculate LMTD for the system
2. Use a corrected MTD of 0.75 (LMTD).

This completes the estimate of the temperature driving force.

Step 5: Estimate the Transfer Coefficient.
Values for the overall heat transfer coefficient, U, must be estimated. The overall coefficient is related to the individual coefficient for the fluids on either side of the tubing by the equation:

$$\frac{1}{U} = \frac{1}{h_i} + r_i + \frac{1}{h_o}\frac{A_i}{A_o} + r_o + r_m \qquad (3\text{-}11)$$

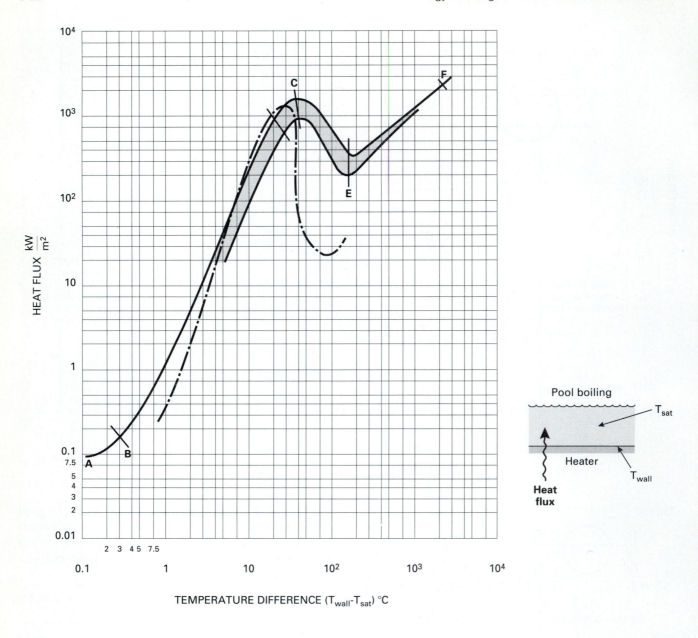

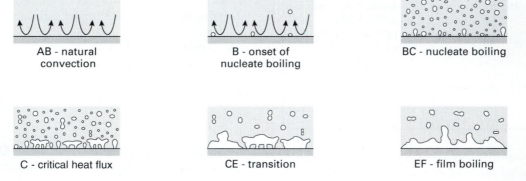

Figure 3–17   Boiling Phenomena: a) Pool Boiling

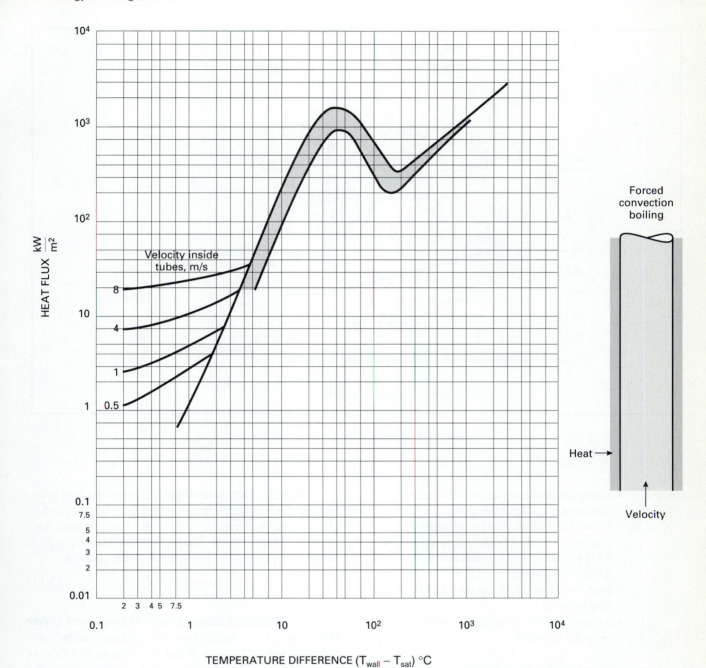

Figure 3–17 Boiling Phenomena b) Convective Boiling
for Liquid Flowing in a Hot Tube

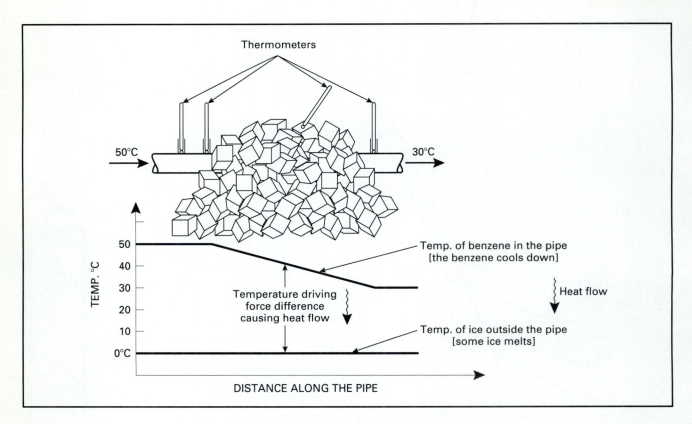

Figure 3–18 Concept of Temperature Driving Force

where h_i = calculated heat transfer coefficient for the fluid on the inside of the tubes.

r_i, r_o = resistances to heat transfer because of scaling and fouling of the exchange surface, for the inside conditions and outside conditions respectively.

$A_{i, o}$ = areas for the inside or outside of the heat transfer areas.

r_m = resistance of the material separating the exchanging fluids.

As an approximation, the resistances can be added to each individual coefficient and the result is:

$$\frac{1}{U} = \frac{1}{h_{id}} + \frac{1}{h_{od}} \qquad (3\text{-}12)$$

Figure 3–22 shows some order to magnitude values for the individual coefficients and a nomograph to calculate the overall transfer coefficient.

For shell and tube exchangers, Figure 3–23 gives some values. Table 3–8 gives some values for the individual coefficients for practical operating conditions for shell and tube exchangers. Values for the reasonable dirt factors or fouling resistances are given by Ludwig (1965) Vol. 3, pp. 57 and 58, metal resistances p. 59. Practical values for overall heat transfer coefficients are given in Table 3–9 for different types of exchangers.

Step 6: Estimate Area. The general type of exchanger can be selected from the heat flux (i.e., the product of the overall heat transfer coefficient U and the mean temperature difference), the total heat load. The selection is shown in Figure 3–24. This is based on the equation:

$$\text{Area} = q/U\Delta T \qquad (3\text{-}13)$$

C. Fluidized Beds, Static Mixers, and Other Devices. The individual and overall heat transfer coefficients given in Tables 3–8 and 3–9 assume that there is reasonable forced convection or that good mixing and a reasonable velocity for the fluids occur on either side of the barrier. Indeed, the heat transfer coefficients are usually correlated by an equation of the form:

$$Nu = a\, Re^n Pr^m \qquad (3\text{-}14)$$

where Nu = Nusselt number, $h\, D/k$, dimensionless

Re = Reynolds number, $\rho\, v\, D/\mu$, dimensionless

Pr = Prandtl number, $c_p\, \mu/k$, dimensionless

h = heat transfer coefficient, $W/m^2{}^\circ C$

D = characteristic diameter or length, m

v = velocity of the fluid, m/s

ρ = density of the fluid, kg/m^3

$\hat{c}_p$ = heat capacity of the fluid, kJ/kg.K

μ = viscosity of the fluid, Pa.s or kg/m.s

k = thermal conductivity of the fluid, W/m.K

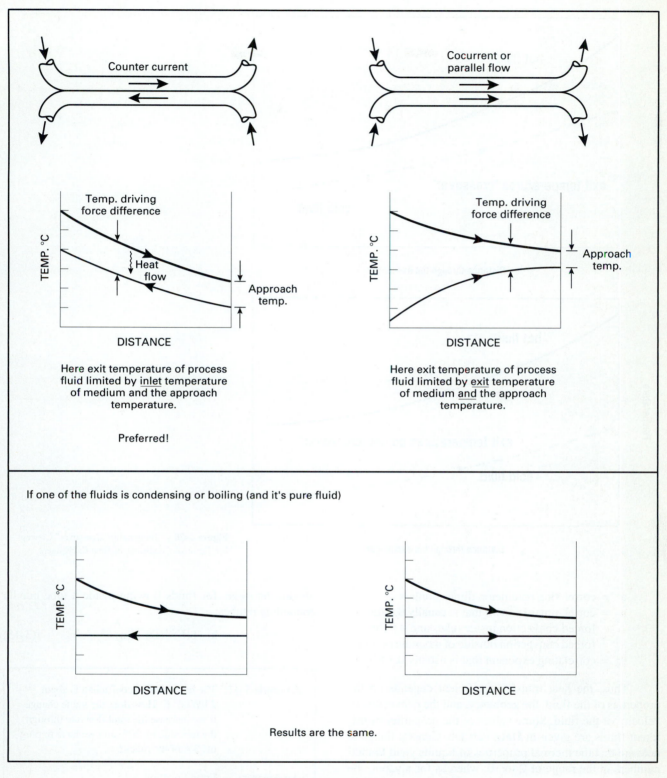

Figure 3–19 Effect of Configuration on Temperature Paths

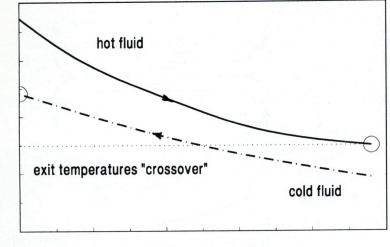

Distance through the exchanger

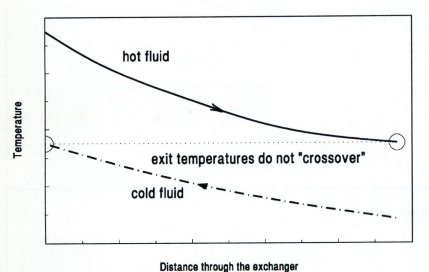

Distance through the exchanger

Figure 3–20 "Temperature Crossover" Concept for Terminal Conditions of Heat Exchangers

a = correlating parameter, dimensionless.

n = correlating exponent that is usually 0.8 for forced convection inside tubes and 0.6 for forced convection outside of a bundle of tubes.

m = correlating exponent that is usually 0.3 to 0.4.

Thus, the heat transfer coefficient depends on the properties of the fluid, the geometry, and the convection or velocity of the fluid. Some values of the properties of different fluids are given in **Data** Part Db. General things to note are that the thermal properties of liquids yield Prandtl numbers in the range of 2 to 10; whereas for a vapor, the Prandtl number is about 0.7. The thermal conductivity of a liquid is about 0.6 W/m.K; of a gas, 0.02 W/m.K. Water has unique properties; hydrogen has a thermal conductivity about five times greater than that of most gases.

Equation 3–14 is very useful to see trends, to suggest corrections to the values given in Tables 3–8 and 3–9, and for trouble-shooting. For example, rearrangement of Equation 3–14 yields, when the thermal properties of the fluids

remain the same, for fluids flowing inside a tube bundle containing n tubes:

$$h = b(\hat{F})^{0.8}(D)^{-1.8}(n)^{-0.8} \qquad (3-15)$$

Example 3–11: The heat transfer coefficient is about 2 kW/m^2.K. How does the value change if we increase the total flowrate through the tubeside by 20% and we have to plug off 3% of the tubes?

An Answer: From Equation 3–15,

$$h_2 = h_1(\hat{F}_2/\hat{F}_1)^{0.8}(n_2/n_1)^{-0.8}$$
$$= 2 \text{ kW/m}^2.\text{K}(1.2)^{0.8}(0.97)^{-0.8}$$
$$= 2.37 \text{ kW/m}^2.\text{K}$$

This is a 19% increase in the heat transfer coefficient.

TABLE 3-7. Guidelines for deciding on fluids for shell and tube exchangers

Tube Side	Shell Side
More corrosive fluid (to minimize cost of alloy shell). Higher pressure fluid (to minimize cost of shell). Dirtier fluid (because tubes are easier to clean). Hotter material (so as to minimize heat loss through the shell). Viscous materials (so as to minimize the dead pockets).	Cleaner fluid. If fluid is dirty, then use square pitch on tube placement. Stream where pressure drop is to be minimized.
	Vaporization & condensation cooling water.

Example 3–12: What happens to the shell side coefficient if 3% of the tubes are blocked off and the baffle spacing and the baffle area are both increased by 30%? The total flowrate on the shell side remains unchanged.

An Answer: We must be careful in responding and return to Equation 3–14 because this is the shell side coefficient where n = 0.6. Since blocking off the inside of the tubes does not affect the outside coefficient, rearrangement shows that:

$$h \text{ prop to } (v)^{0.6}$$

Increasing the baffle spacing and window area, while keeping the total flowrate constant, will decrease the velocity by 30%. Thus, the heat transfer coefficient on the shell side will be $(1-0.30)^{0.6}$ or 81% of its value before the baffle spacing was changed.

Example 3–13: In designing a gas cooler we assumed that the overall gas side coefficient controlled the heat transfer and that the gas contained about 30 mol % hydrogen and the rest is a light hydrocarbon mixture like methane. On the plant, we have now removed the hydrogen from this stream before it enters the cooler. What effect does this have on the performance of the cooler?

An Answer: We are not given a temperature range, so we assume that the properties of the gas

are similar to those at 25°C. At this temperature the heat capacity and thermal conductivities of hydrogen to methane are 14.4/2.24 and 183/80, from Appendix Db. For the design of the unit the properties of the mixture would have been used. Since the Prandtl numbers of most gases are about 0.7, the main effect of properties will be through the thermal conductivity. An estimate of the thermal conductivity of a gas mixture is based on a mol fraction weighting: the mixture value would be about $0.7 \times 80 + 0.3 \times 183$ = 110 mW/m.K. Removing the hydrogen has two effects. It reduces the volume of gas by 30% and thus the velocity by 30% and reduces the thermal conductivity from 110 to 80 mW/m.K. Substitution into Equation 3–14 yields:

$$\frac{h_{now}}{h_{before}} = \frac{k_{now}}{k_{before}} \left(\frac{v_{now}}{v_{before}} \right)^{0.8}$$

or

$$h_{now} = h_{before} \frac{80}{110} \left(\frac{0.7}{1.0} \right)^{0.8}$$
$$= 0.55 \, h_{before}$$

Thus, the heat transfer coefficient will be about 1/2 of the value used for the design because the hydrogen has been removed from the stream. Part is because of the property change and part because of the reduction in gas velocity. Hence, we need about double the size of heat exchange to do the required heat exchange duty.

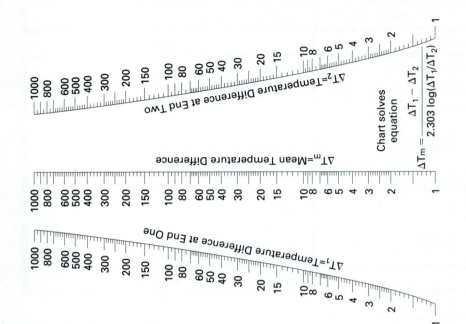

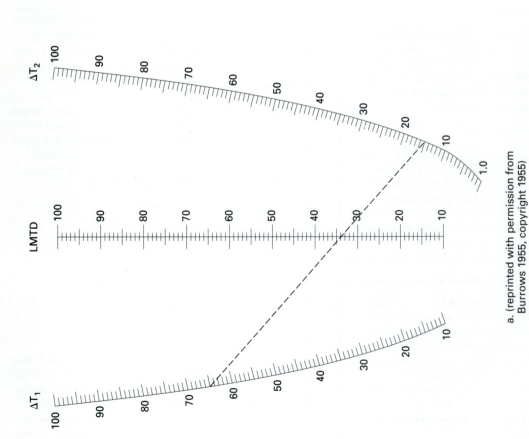

Figure 3–21 Estimation of the Log-Mean Temperature Difference (LMTD)

a. (reprinted with permission from Burrows 1955, copyright 1955)

(reprinted with permission from Burrows, 1955, copyright 1955; American Chemical Society; Clarke, L. "Manual for Process Engineering Calculations" (1962) McGraw Hill Book Co., reproduced with permission of McGraw-Hill)

b. (Clarke, L. "Manual for Process Engineering Calculations" (1962) McGraw Hill Book Co., reproduced with permission of McGraw Hill)

$$\Delta T_m = \frac{\Delta T_1 - \Delta T_2}{2.303 \log(\Delta T_1/\Delta T_2)}$$

Chart solves equation

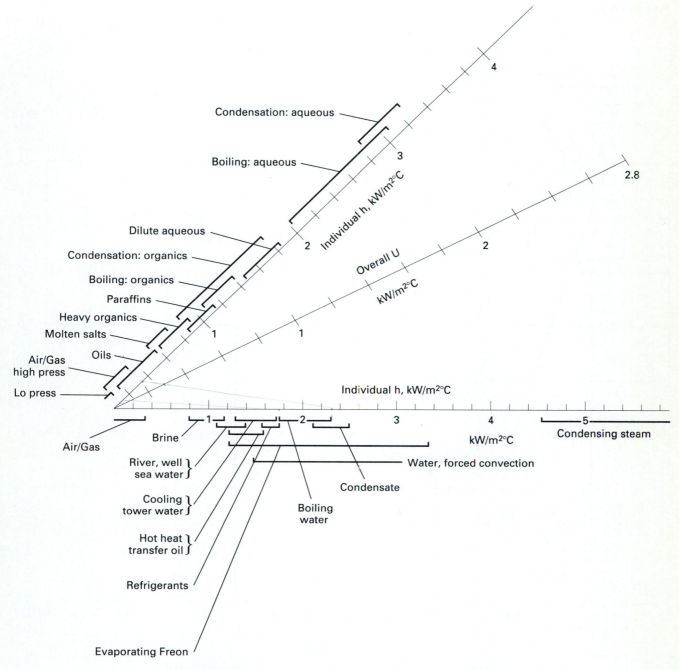

Figure 3–22 Overall Heat Transfer Coefficients for Shell-and-Tube Exchangers
(based on Frank (1974) and reprinted courtesy of Chemical Engineering,
McGraw-Hill Publishing Co.)

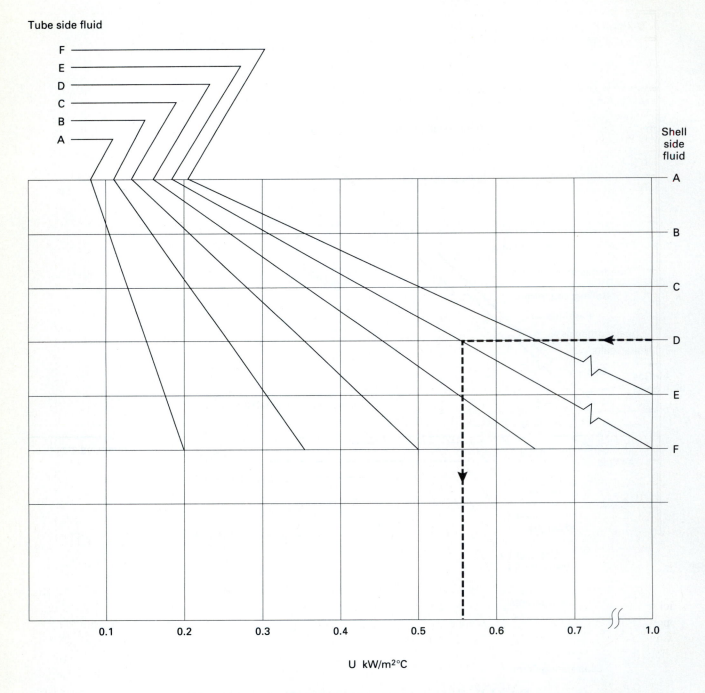

U kW/m²°C

	No Phase Change	Phase Change
Heavy hydrocarbon residue	A	
Middle distillate h/c	B	
Gasoline & light h/c	C	D
Water (steam)	E	F

(from Chauvel et al. (1981) and reprinted courtesy of Societé des Editions Technip, copyright 1981)

Figure 3–23   Estimation of the Overall Transfer Coefficients for Shell-and-Tube
Exchangers (from Chauvel et al. (1981) and reprinted courtesy
of Societé des Editions Technip, copyright 1981)

TABLE 3-8. Individual coefficients (kW/m² °C) for shell and tube exchangers

	Liquid		Gas		Boiling	Condensing
	Inside	Outside	Inside	Outside		
Water	4.8 1.7 to 11	4.8			2.8 4.5 to 11	8.5 5.7 to 17
Pentanes & Lighter					1.4	1.1
Gasoline					0.85	0.5 to 0.85
Petroleum 500 mPa.s		0.1				
100		0.15				
10		0.3				
1		0.6				
0.5		0.9				
0.2		1.1				
Organic Solvents		0.4 to 2.8			0.6 to 1.7	0.9 to 3.
Light Oils		0.06 to 0.7			0.9 to 1.7	1.1 to 2.2
Heavy Oils (vacuum)					0.06 to 0.3	0.12 to 0.3
Ammonia					1.1 to 2.2	2.9 to 5.7
Gas Oil						
Kerosene		0.25				0.3
Freon					1.2 to 3.4	
Gases		0.02 to 0.3				

Thus, we note the sensitivity of the heat transfer coefficient to the mixing, the fluid velocity, and physical properties of the media. Two devices that provide good mixing are fluidized beds for gases and static in-line mixers for liquids. Consider these in turn.

1. Fluidized Bed Heat Exchange. Fluidized beds provide excellent mixing within the bed and the presence of the solids in the gas increases the heat transfer coefficient to 5 to 25 times that of the gas alone. The net result is that the temperature within the bed is relatively uniform and does not vary by more than ±6°C. These characteristics also improve the heat transfer coefficient between the "bed" and the outside walls or a tube bundle immersed in the bed. The overall heat transfer coefficient is given in Table 3–10.

To take advantage of these increased heat transfer coefficients we must be able to create acceptable fluidized bed operation. In Chapter 2 we saw how we could transport solids by pneumatic conveying. To do this, we used a high velocity gas to pick up the solids. To understand how a flu-idized bed works, Figure 3–25 (p. 3–45) illustrates what happens when a gas is blown up through a bed of particles. As the gas velocity increases, so does the pressure drop until the flow past the particles is sufficient to begin to lift the particles so that they can wiggle around slightly in the bed. The particles move around until, although the particles still are touching each other, they are oriented so that the pressure drop through the bed is at minimum. This is shown at position "B," the point of minimum fluidization or the "onset" of fluidization. This incipient fluidization velocity is about 70 times smaller than the "settling" velocity of a single particle in an infinite fluid. Figure 3–26 shows the "settling" velocity of a single particle of density 2 Mg/m³ in air at room temperature. Also shown is the incipient fluidization velocity of a group of particles.

For higher gas velocities, the particles move about freely; the volume of the "bed" has increased. The bed is fluidized; the pressure drop across the expanded bed is relatively independent of gas velocity. However, the gas velocity is not high enough so that it would carry the particles along with it (as we did in pneumatic conveying). Thus, the

TABLE 3-9. Overall heat transfer coefficients, kW/m² °C

Shell & Tube Liquids	Cooling				Heating			Boiling			Condensing				
	Water	Brine	Light Oil	Organic Solvents	Steam	Dow Therm	Heavy Oil	Steam	Dow Therm	Hot Oil	Water	Brine	Oil	Propylene	Air
Water	0.85 to 1.7	0.6 to 2.			2 to 7.			2 to 4.3							
Ammonia	2.5 to 5.				2 to 7.			2 to 4.3			0.8 to 0.9				
Organic Solvents	0.3 to 0.85	0.17, 0.51	0.12, 0.40	0.115, 0.34				0.6 to 1.1							
Light Oils	0.35 to 0.9														
Naphtha	0.06 to 0.28							0.09 to 0.12							
Heavy Oil							0.045, 0.28								
Gas Oil					0.07										
Butadiene														0.4 to 0.45	
C$_4$ Olefins								0.5 to 0.6						0.35 to 0.40	
Light Olefinic hydrocarbons														0.3 to 0.35	
Chloroethanes & Lightends														0.09 to 0.15	
Chlorinated Olefinic h/c											0.5 to 0.7				
Chlorinated h/c											0.12 to 0.17				
Ethylene								0.87 to 1.2						0.35 to 0.52	
Ethanol Amine					1 to 2										
Light h/c	0.75 to 1.5	0.15 to 1.0									0.15 to 0.90				
Refrigerant											0.23 to 0.50				
Dichloroethane								0.4 to 0.52							
Heavy Solvent								0.46 to 0.63							
Water/organic acid	2.5 to 5.				2 to 7			0.35 to 0.6							
Water/amine					1 to 5			0.7 to 0.8					0.4 to 0.6		

TABLE 3-9. (Continued)

Shell & Tube, Gases	Cooling				Heating			Boiling			Condensing				
	Water	Brine	Gas	Steam	Steam	Dow Therm	Flue Gas	Steam	Dow Therm	Hot Oil	Water	Brine	Oil	Propylene	Air
Propylene												0.75 to 0.90			
Medium organics	0.5 to 1.25				0.5 to 0.6	0.5 to 1.									
HCl gas	0.04 to 0.09														
Steam atmos											2 to 4.3				
vac											1.7 to 3.4				
Ammonia															
Organic Solvents atm.															0.57 to 1.14
Vac. plus small non-C															0.3 to 0.7
Atm. plus high non-C															0.1 to 0.45
Vac. plus high non-C											0.08 to 0.14				0.06 to 0.3
Low boil hydrocarbon											0.5 to 1.1				
High boil hydrocarbon											0.06 to 0.18				
Air & chlorine	0.06 to 0.1														
Air & steam	0.12 to 0.20										0.05 to 0.09				
Gases	0.017, 0.28	0.02, 0.28													
Benzene/toluene vapor.							0.03 to 0.05				0.75				
Stabilizer overheads											0.5				
Cooler condenser benzene											0.3				
Gas	0.02 to 0.5		0.05		0.05 to 0.5										

TABLE 3-9. (Continued)

Air/Cooled; Plate-Coil; Trombone; Double Pipe		Cooling			Heating		Boiling				Condensing			
		Water	Brine	Air	Steam	Dow Therm	Steam	Dow Therm	Hot Oil	Water	Brine	Oil	Propylene	Air
AIR	Water/Steam			0.38 to 0.41										0.40 to 0.41
	Ammonia													0.35 to 0.39
	LPG			0.34 to 0.375										0.32 to 0.35
	Light Hydrocarb.			0.32 to 0.35										0.30 to 0.32
	Light Naphtha													0.27 to 0.30
	Heavy Naphtha													0.25 to 0.27
	Gasoline													0.30 to 0.33
	Gas-Oil													0.25 to 0.27
	Medium h/c (1 mPa.s)			0.27 to 0.30										
	(5 mPa.s)			0.17 to 0.19										
	(10 mPa.s)			0.13 to 0.14										
PLATE/COIL														
TROMBONE	Milk	0.9												
	HCl (karbate)	1.7												
	50% sugar (glass)	0.28 to 0.34												
DOUBLE PIPE	Water (karbate)	1.7 to 2.8												
	Water (glass)	0.45 to 0.58								0.58 to 0.70				
	Water		0.8 to 1.0											

TABLE 3-9. (Continued)

Electric Immersion, Thermal Screw; Cubic; Coil in tank; Jacket

	Cooling			Heating		Boiling			Condensing				
	Water	Brine	Vaporize Liquid	Steam	Dow Therm	Steam	Dow Therm	Hot Oil	Water	Brine	Oil	Propylene	Air
COIL IN TANK: AGITATED													
Boiling liquid reactants			0.7 to 1.4										
Viscous liquid reaction			0.06 to 0.17										
Liquid reaction			0.5 to 1.2										
Water	0.85 to 1.7	0.55		0.40		1.2 to 3.5							
Amino acids		0.55				0.23 to 0.35							
Oleum	0.11					0.7 to 2.3							
Sodium hydroxide	0.88					2.8							
Sugar solution	0.28 to 0.34												
Milk	1.7												
COIL IN TANK: NON-AGITATED													
Sugar/molasses solution.				0.28 to 1.4									
Fatty acid				0.55									
Water	0.12												

TABLE 3-9. (Continued)

Jacket; Mandrel wound; Lamella (plate/fin); Cryogenic	Cooling				Heating		Boiling			Condensing				
	Water	Brine	Vaporize Liquid	React	Steam	Dow Therm	Steam	Dow Therm	Hot Oil	Water	Brine	Oil	Propylene	Air
JACKET: AGITATED Liquid reaction			0.17 to 0.35				0.46 to 1.4							
Boiling liquid reaction			0.35 to 0.58				0.7 to 1.75							
Viscous liquid reaction			0.06 to 0.14				0.17 to 0.3							
Reacting PVC				0.4 to 0.74										
Reacting butadiene sulfone				0.5										
Reacting polystyrene				0.03 to 0.12										
LAMELLA Water	2 to 3													

Light   viscosity <0.5 mPa.s: benzene, toluene, acetone, ethanol, naphtha
Medium 0.5 to 1. mPa.s: straw oil, hot gas oil, hot absorber oil
Heavy   >1 mPa.s: cold gas oil, fuel oil, lube oil, reduced crude oil, tars

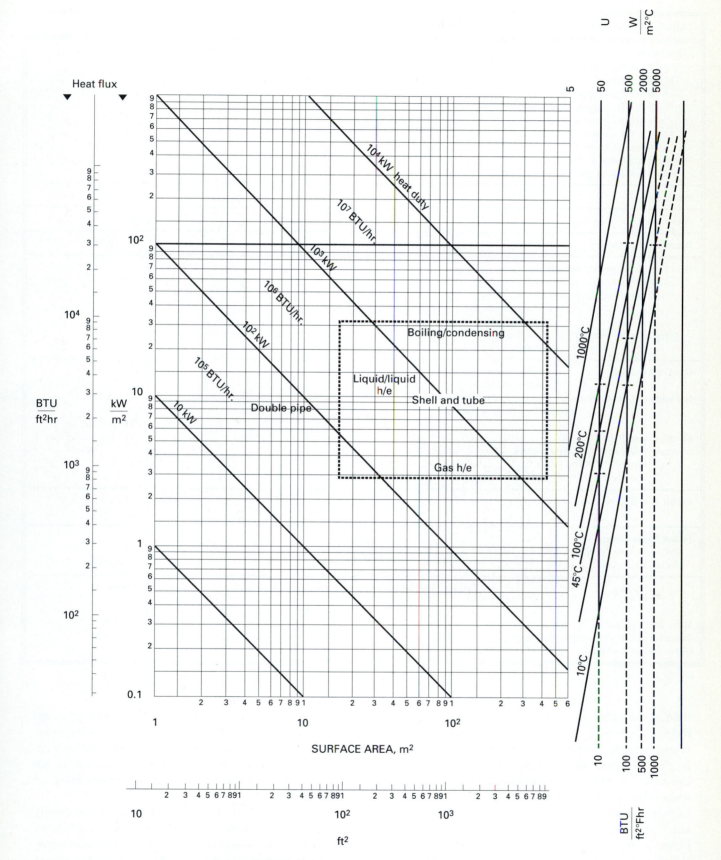

Figure 3–24 Estimation of the Heat Transfer Area

TABLE 3-10. General Overall Heat Transfer Coefficients

	$kW/m^2\ °C$
Evaporators	
Wiped film evaporators	1 to 2
Thin film evaporators:	
—dehydration organic solutions	0.8 to 1.1
—distillation of organic solutions	0.8 to 1.1
—stripping organic solutions	0.45 to 0.7
—reboiler service	1. to 1.1
—solvent recovery	0.8 to 0.93
Forced circulation evaporators	0.75 to 3.
Natural circulation (short tube) evaporators	1 to 1.5
Water Bath	0.56 to 1.7
Pneumatic Conveying h_{inside}	0.2
Drum and Pan Dryers	0.02 to 0.5
Heat Loss: from vessels into still air.	
—distillation columns, uninsulated	0.01
—water storage tank, uninsulated	0.005 to 0.2
(increases with ΔT)	
insulated	0.003
Affect of wind velocity: 10 km/h mult.by 2	
20 km/h mult.by 3	
Gas Fluidized Beds to immersed tube bundle	0.2 to 0.4
solids to gas in bed	0.017 to 0.055
In Fluidized Coal Combustor to immersed tube bundle	0.2 to 0.5
(value increased as particle size decreases)	
to walls	0.15 to 0.28
Fluidized Sand Bath	0.45 to 1.1
(thermal conductivity of fluidized bed.	
vertical: 35 to 50 kW/m.°C	
horizontal: 1.7 kW/m.°C)	
Pebble Bed Regenerators (1 cm pebbles) air	0.022
hydrogen	0.125
In-line Static Mixers:	
Polymerization reactions	10
Jacket heating/cooling of melts	0.15 to 0.25
Volumetric Coefficients	$kW/m^3\ °C$
Spray dryers (based on T_{gas} exit - Tadiabatic saturation)	30 to 50
Direct contact condensers/humidification by spray evaporation	100 to 150
Direct contact gas-liquid	0.5 to 15

fluidization velocity is somewhere between the minimum fluidization velocity and the terminal or pneumatic conveying velocity. Some recommend that the design value for fluidized bed operation be 2 to 3 times (and some indeed 10 to 60 times) the minimum fluidizing velocity while others recommend that it be 0.1 of the terminal velocity of the particles for particles of diameter less than 2 mm and 0.2 of the terminal velocity for larger particles. Figure 3–26 gives the gas velocity versus the particle diameter. Sometimes we identify different classes of particles; these are shown on the abscissa of Figure 3–26. Also shown is a qualitative description of the type of fluidization

that occurs. For example, very small particles tend to agglomerate and the bed tends to spout, as do very large particles of class D. Slugging and bubbling fluidized beds describe the condition where the bed composition is not homogeneous. Rather, void spaces or bubbles move up through the bed.

Figure 3–26 illustrates the behavior for one density of particles in air at room temperature. Figure 3–27 shows the fluidization conditions pertinent to coal combustion. Figure 3–28 shows the behavior in terms of dimensionless variables: the particle Froude number, $v^2/D_p\ g$, and the particle Reynolds number, $v\ D_p\ \rho/\mu$. The various terms can be

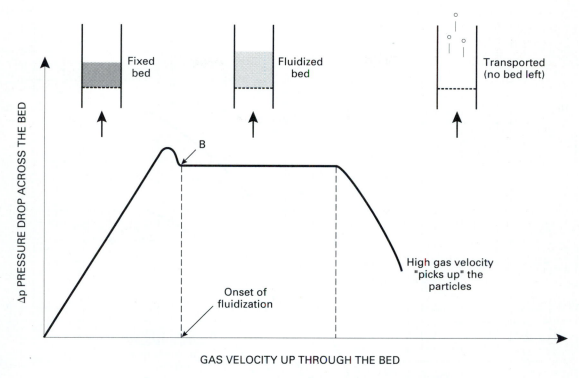

Figure 3–25   Particle Behavior As Upward Gas Flow Increases

combined to yield a dimensionless velocity number, v^+, defined as:

$$v^+ = \left[\frac{v^3 \rho_g}{g\mu}\right]\left\{\frac{\rho_g}{(\rho_s - \rho_g)}\right\}, \qquad (3-16)$$

The ratio of the Re^2/Fr is called the dimensionless Galileo number, Ga. The Archimedes number, Ar, is the Galileo number times the density ratio:

$$Ar = \left[\frac{\rho^2 D_p^3 g}{\mu^2}\right]\left\{\frac{(\rho_s - \rho_g)}{\rho_g}\right\}, \qquad (3-17)$$

Figure 3–28 also shows the particle settling velocity, the minimum fluidization curves, and the location of different types of practical application of equipment: fluidized bed, fixed bed (moving bed, traveling grate, shaft or blast furnace), circulating fluidized bed and venturi fluidized beds, and transported beds (or pneumatic conveying).

The minimum fluidization velocity can be estimated from Figure 3–28 or from a plot of the Archimedes number and the particle Reynolds number for minimum fluidization. Figure 3–29 shows such a plot. The "fluidization velocity" determines the diameter of the fluidized bed. The usual range of design values for the superficial gas velocity are between 0.5 cm/s to 3m/s (depending on the size of the particle and the gas properties). Usually the density of the gas can be estimated using the ideal gas law:

$$\text{density} = p\,\frac{M}{\mathbb{R}T} \qquad (3-18)$$

Data Part C-6 gives a graphical method for quickly estimating the gas density.

In general, fluidized bed conditions are used as reactors (for fast, highly exothermic catalytic reactions when temperature control is important), as boilers for steam generation from the combustion of coal and from the regeneration of catalyst by burning off the coke. In all of these applications the key is the removal of the heat. Some applications are summarized in Table 3–11. Other applications have been suggested for the production of ethylene oxide from ethylene, pyridine from acrolein and ammonia, vinyl acetate, vinyl chloride, and the gas phase polymerization of polyvinyl chloride, polyethylene and polypropylene but either the process has not been implemented industrially or else details are not available (Albright [1974]). The general sizing procedure for a fluidized bed heat exchange system is to:

1. select the size of the particles or catalyst. The general range for most fluidization is 50 to 300 μm although the preferred range is 60 to 80 μm with the ratio of the maximum diameter to the minimum diameter of about 11 to 25 (Bergognou et al. [1986]). Usually there is a range of particle size, in which case the Sauter mean, or volume to surface average diameter, is the best way to represent the distribution for estimation purposes. High conversions in a reacting system occur when the particles contain 20 to 30% "fines" or particles of size less than 40 μm. But such fines are difficult to keep in the bed. The smaller

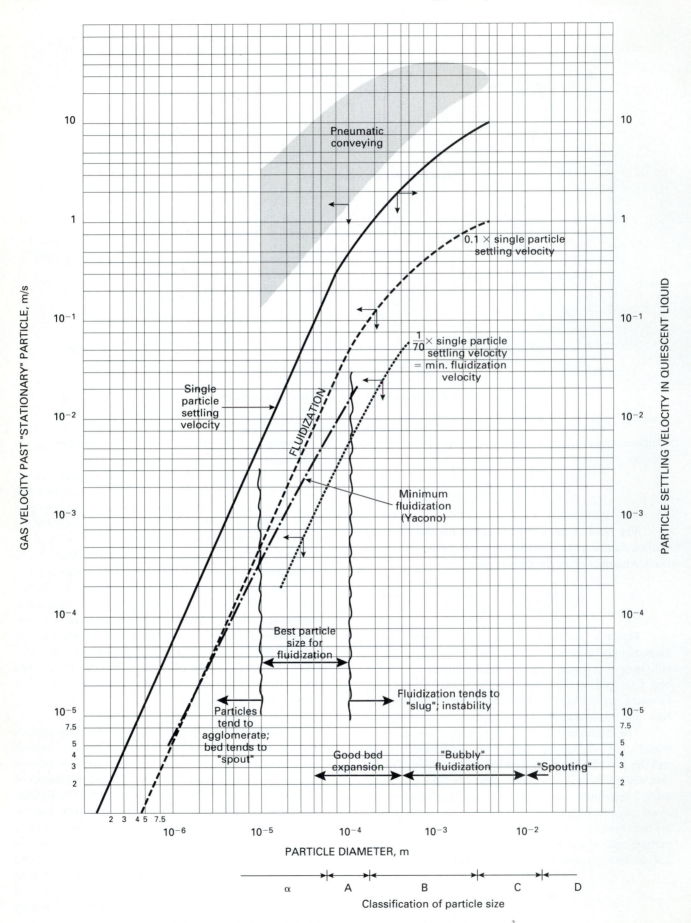

Figure 3–26 Behavior of Particles in "Air" at 25°C; Particle Density 2 Mg/m³

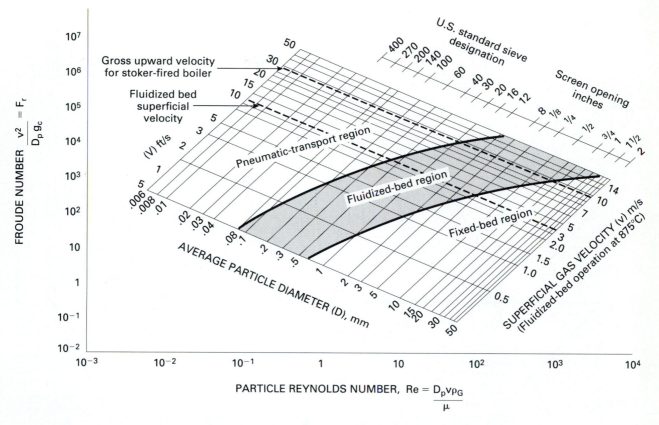

Figure 3–27 Behavior of Coal Particles at 875°C (from Daman [1979]; reprinted from Heat Engineering, courtesy of Foster Wheeler Corporation)

the particles, the lower the gas flowrate should be and hence the bigger the fluidized bed diameter for the same gas throughput. On the other hand, if the particles are larger, then the fluidization velocity can be larger, and the bed diameter smaller for the same gas throughput. However, the larger the particles, the more the bed tends to operate in the "bubbling regime." In such operations, the gas in the bubble tends to pass through the bed unreacted (and thus the overall apparent reaction rate goes down). The values in Table 3–11 represent the compromise that is usually used for different applications.

2. estimate the operating fluidization velocity based on the particle size, the operating gas conditions, the minimum fluidization velocity and the design choice relative to the minimum. From this design value and the inlet gas flowrate, the diameter of the bed can be estimated.

3. determine the height of the bed based on four criteria:

   • usually the height to diameter ratio is 1:1 or greater;

   • the bed must be deep enough to allow for enough residence time for the reaction to occur; as the size of the catalyst increases, the gas velocity tends to

increase, the operation moves into the bubbling regime and deeper beds are needed. (For more detailed estimates, the depth is correlated in terms of the gas exchange characteristics between the bubbles and the "emulsion" phase, see Matsen [1985].)

   • the bed must be deep enough to cover the heat exchanger; for very exothermic reactions, and/or for high pressure systems (where the heat generated per unit volume is large) this criteria tends to dominate;

   • check with the values given in Table 3–11; usually the bed depth is in the range 0.3 to 15 m.

   • for cat crackers the bed is usually not greater than 3 m regardless of the diameter of the bed.

4. estimate the bed expansion and the amount of disengagement height above the bed. Figure 3–30 illustrates how the disengaging height varies with the bed diameter and the superficial gas velocity. This now gives the total height of the bed.

5. estimate the amount of heat exchange surface area required. In general, the tubes are put in vertical in the bed to minimize corrosion and minimize tube sag, unless the bed is very shallow having a bed

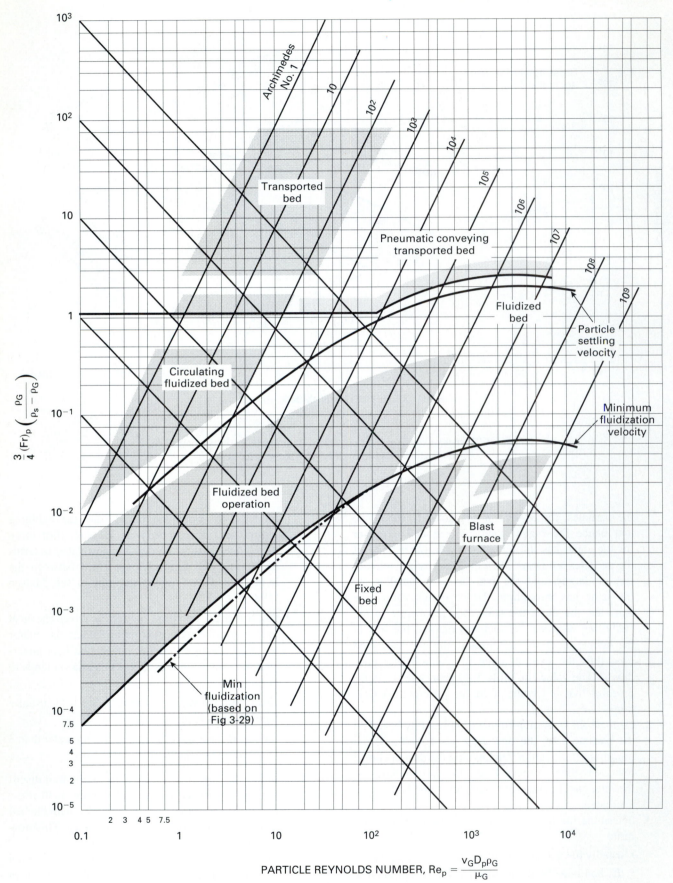

Figure 3–28   Generalized Behavior of Particles Expressed in Terms of Dimensionless
Numbers: Galileo, Particle Reynolds, and Archimedes Numbers

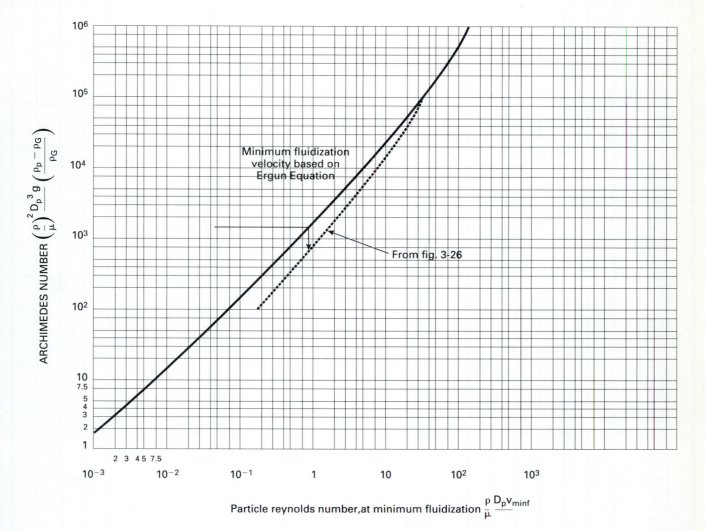

Figure 3–29 Estimate of the Minimum Fluidization Velocity
(Based on an Extrapolation of the Ergun Equation)

depth of 1 to 2 m and the superficial velocity is very high. Vertical tubes are preferred at temperatures above 1000°C because at high temperatures the tubes must support their own weight (Bergognou et al. [1986]). To estimate the required area, estimate the temperature difference and the heat load from the heat of reaction or of combustion, and use the heat transfer coefficients from Table 3–10. Rossi (1984) suggests that about 5 to 30 m² of heat transfer area per m² of bed cross-sectional area can be used before the exchanger starts interfering with the fluidization.

6. to complete the equipment sizing requires an estimate of the blower for the fluidizing gas. The pressure drop across the bed itself is the weight of the bed and is independent of the gas velocity (as illustrated in Figure 3–25). To this add the pressure drop across the inlet distributor, which is sometimes taken to be about 30% of the pressure drop across the bed or a maximum of 2.5 kPa.

Example 3–14: Pyrite ore, 43% FeS, 40% FeS$_2$, 17% silicates, is to be roasted. The feed is 2.76 kg/s of wet ore (7% water). The gas produced should be about 14% SO$_2$ so that we can produce sulfuric acid; to achieve this, the excess air should be 5%. For this low amount of excess air and an operating temperature of the bed of about 900°C, Fe$_2$O$_3$ is not formed, so that the major reactions are:

$$FeS + \frac{5}{3}O_2 \rightarrow \frac{1}{3}Fe_3O_4 + SO_2$$

and

$$FeS_2 + \frac{8}{3}O_2 \rightarrow \frac{1}{3}Fe_3O_4 + SO_2$$

and

$$H_2O_{liquid} \rightarrow H_2O_{vapor}$$

From a mass balance and for 5% excess air, the mass of air required is 6.29 kg/s.

TABLE 3-11. Example Applications of Fluidized Beds as Reactors

Product	ΔH MJ/kmol	Catalyst Size μm	Superficial Velocity m/s	Bed ht m	Disengag ht m	Temp °C	Pressure MPa	Space Velocity	Comments
Acrolein ex propylene	-840					300-400	0.2 to 0.3		
Acrylonitrile ex propylene	-515	51	0.5	9					
Allylchloride ex propylene		150	0.18			510-532		2.6 s⁻¹	
Catalytic cracking		40	0.8	3	10	500			
		20-150	0.5		7				
			0.3 - 0.7	5.5					
Catalyst regeneration by decoking	-393					600			
Chlorine: ex HCl via Shell	-219		0.15		0.1 to 0.3	350		WhW 10h⁻¹	
Chlorosilane: ex methyl chloride									
Coal combustion	27 MJ/kg	<32,000	>0.3			800			Limestone reacts with sulfur
incl. limestone		<3,000	0.6						No slag because temp < ash melt
Coke: ex resid. via fluid coking		150-200				500-600			
Ethylene oxide ex ethylene	-104								Control temp to prevent runaway; not commercial yet
Ethylene: ex cat crack		200 to 3000				760-790		0.3 to 0.5 s	
Hydroforming									
Phthalic anhydride ex -naphthalene -orthoxylene	-1284		0.3-0.6					10 to 20 s	careful temp control vital to minimize side reactions
Polyvinyl chloride		polymer	0.03 to 0.12			38-74	0.4 to 1.2		
Pyridine ex acrolein/ammonia						400	0.15	5 s	
Roasting			2	0.5	0.9	940			
Vinyl acetate	-117								
Vinyl chloride									

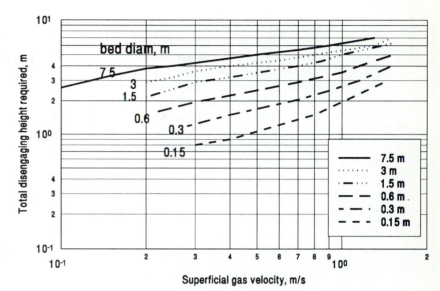

Figure 3–30   Estimate of the Disengagement Height Required at the Top of a Fluidized Bed (based on Matsen et al. (1985) in Scaleup of Chemical Processes, ed. A. Bisio and R. L. Kabel, copyright © 1985, J. Wiley and Sons. Reprinted by permission of John Wiley & Sons, Inc.)

Assume that 10% of the total heat released is lost due to radiation. Thus, an energy balance results in a net heat to be removed of 1.625 MJ/kg of dry feed. The ground ore range from +4 Mesh to −200M with a volume to surface average of about 0.6 mm. Assume that the density of the ore is 5 Mg/m³. Size the heat exchange system to recover the heat as steam at 3.2 MPa.

An Answer: 1. Since we are given the size of the ore, we will use 0.6 mm as the starting diameter to estimate the fluidization conditions. Details of the different mesh sizes used to measure the diameter of solid particles are given in **Data** Part C-2 and the abscissa of Figure 3–26.

2. To estimate the superficial velocity, from **Data** Part C, at 900°C the density of air is about 0.3 kg/m³, the viscosity for air can be estimated from **Data** Part Db, is about 54 μPa.s. Hence, the ordinate of Figure 3–29 is:

Archimedes Number =

$$\frac{\left[\dfrac{0.3\ \text{kg}}{\text{m}^3}\dfrac{\text{m.s.}}{54\times10^{-6}\ \text{kg}}\right]^2 (0.6\times10^{-3}\text{m})^3}{9.8\ \dfrac{m}{s^2}\left\{\dfrac{5000-0.3}{0.3}\right\}}$$

$= 1{,}090$

Note that at this temperature, the physical properties are likely to be different from the linearized approximations given in **Data** Part Db but for order-of-magnitude estimating they should suffice.

$$0.60 = \frac{0.3\ \text{kg}}{\text{m}^3}\frac{\text{m.s}}{54\times10^{-6}\ \text{kg}}\,0.6\times10^{-3}\ \text{m.}$$

$$v_{\text{min fluid}}$$

or

$$v_{\text{min fluid}} = 0.18\ \text{m/s}$$

If we assume that we operate about 2 times the minimum, then the superficial velocity should be about 0.36 m/s.

As a check, from Figure 3–27 for the fluidization of coal by air at about the same temperature, the value is about 0.3 m/s.

The total gas flowrate is 6.29 kg/s with a gas density of 0.3 kg/m³ so that the cross-sectional area needed is

6.29/0.3 × 0.36 = 58 m²

The corresponding diameter of the fluidized bed is 8.6 m.

3. Estimate the bed depth:
   • at 1:1 height to diameter ratio, the height would be 8.6 m.
   • this is a very fast reaction, hence we would not need a very deep bed. From Table 3–11 we might expect the depth to be around 0.5 m.
   • since we have not estimated the heat exchanger area, it is difficult to know if we have enough depth. Since the temperature is less than 1000°C we might elect to put in a horizontal tube, especially if the bed is only 0.5 m deep.

No general trend emerges; try 1 m.

4. Estimate the freeboard height. From Figure 3–30 we estimate it to be about 5 m (corresponding to a bed diameter of 8 m and a superficial velocity of 0.3 m/s).

5. Estimate the heat transfer area. The net heat of reaction to be removed to keep the temperature in the bed constant is:

$$\frac{1.625 \text{ MJ}}{\text{kg dry feed}} \times 2.76 \; \frac{\text{kg wet feed}}{\text{s}} \times 0.93$$

$$\frac{\text{dry solids}}{\text{wet feed}}$$

$$= 4.17 \text{ MW}$$

Assuming that the sensible heat for the boiler water is negligible and that the latent heat of water boiling at 3.2 MPa is the main duty, the temperature inside the tubes is 225°C. The bed temperature is 900°C. Hence, the LMTD is 900 − 225 = 675°C. From Table 3–10, assume the overall heat transfer coefficient is 0.4 kW/m². °C. Since q = U A LMTD, rearrangement and substitution yields:

$$A = 4170 \text{ kJ}/0.4 \times 675$$
$$= 15 \text{ m}^2$$

We can check that the heat flux is not excessive. It is 4170 kW/15 m² = 278 kW/m², which is reasonable since the critical heat flux for water is 360 to 1260 kW/m². A check on the heat transfer area per fluidization cross-sectional area gives 15 m²/58 m², which is well below the guideline suggested by Rossi.

6. The pressure drop across the bed is the mass of the bed. Assume that the expanded bed has a porosity of 0.5 (or that 50% of the volume is gas of density 0.3 kg/m³ and 50% is solids of density 5000 kg/m³). Thus, the force, Newtons, per square meter of cross-section is:

[effective bed density] × g × bed volume per square metre area

$$= \frac{2500}{\text{m}^3} \times 0.9 \; \frac{\text{m}}{\text{s}^2} \times 1\text{m} \times 1\text{m}^2$$
$$= 24.5 \text{ kPa}$$

The pressure drop across the distributor is 0.3 × 24.5 or 2.5 kPa, whichever is less. Thus the total pressure drop is about 24.5 + 2.5 = 27 kPa.

Comment: This situation is similar to that installed at Sulfuric Acid Ltd. in South Australia. Uhlherr and Vogl (1968) report that the

fluidization velocity = 0.3 m/s
reactor diameter = 9 m
bed depth = 1.52 m
reactor freeboard = 3 m
pressure drop = 30 kPa
heat exchange area = 13.1 m² in the bed
heat transfer coefft. = 0.397 kW/m²°C

The agreement is very good.

For more on fluidization and heat transfer see Kunii and Levenspiel (1969); Froment and Bischoff (1979); Zenz and Othmer (1960); Botterill (1975); and Matsen (1985).

2. Static In-line Mixers and Heat Transfer. Static mixers, shown in Figure 3–13, are inserts placed inside a pipe to divide and bring together sequentially different parts of the fluid flowing in the pipe. Although static mixers have been used extensively for mixing liquids and blending solids, they are very effective for promoting heat transfer especially for the production of nylon, silicone, styrene, and polypropylene. The overall heat transfer coefficients are given in Table 3–10. For more, see Mutsakis et al. (1986).

3. Other Heat Exchange Configurations. Table 3–10 gives heat transfer coefficients for other processing equipment. Details about evaporation are given in Chapter 4; about drying, in Chapter 5.

3.2-2 Direct Contact Devices

Heat can be exchanged directly without tube walls separating the phases exchanging heat. This is the usual configuration for the high temperature processing of solids and for "quenching" high temperature gases. However, this approach is also very applicable for some low temperature processing where intermixing of the two streams is acceptable. Sketches of the devices are given in Figure 3–31; the devices are classified according to the major phase being handled: solids, liquids, or gases.

A. Solids. For solids the major devices are direct contact furnaces or kilns, fluidized beds and "flash" conveyers, and multiple-hearth furnaces.

1. Direct contact furnaces and kilns are illustrated in Figure 3–31 and the general region of applicability is illustrated in Figure 3–32. Figure 3–32 is based on a volumetric solids loading of 3 to 12% within the kiln. The heat supplied is 25 to 60 kW/m³ of kiln volume. The approximate diameters of the kilns are also shown in Figure 3–32.

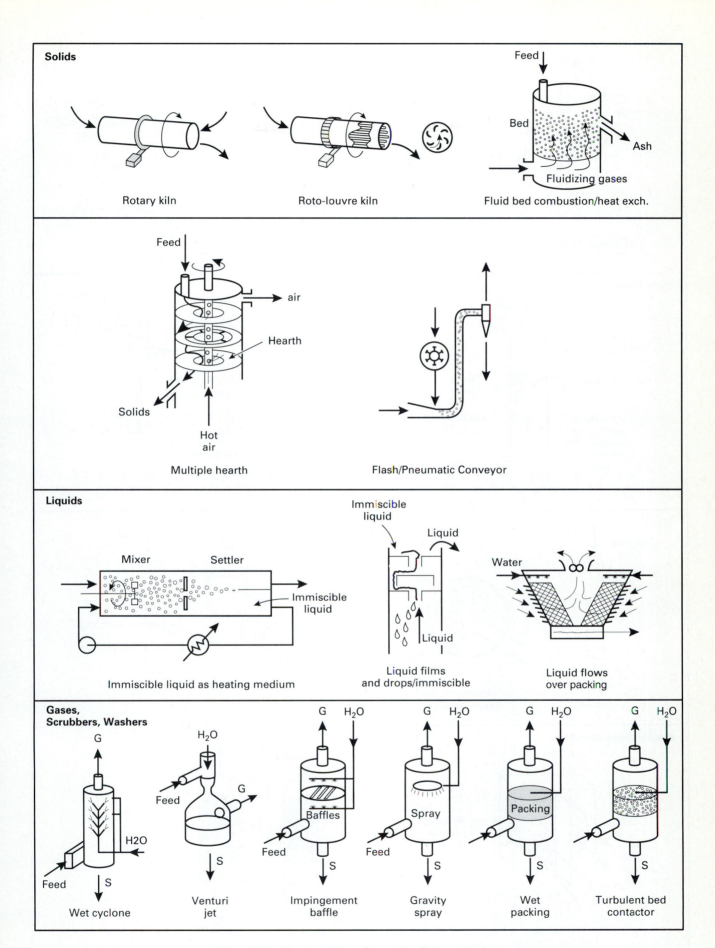

Figure 3–31 Sketches of Direct Contact Heat Exchange Devices for Solids, Liquids and Gases

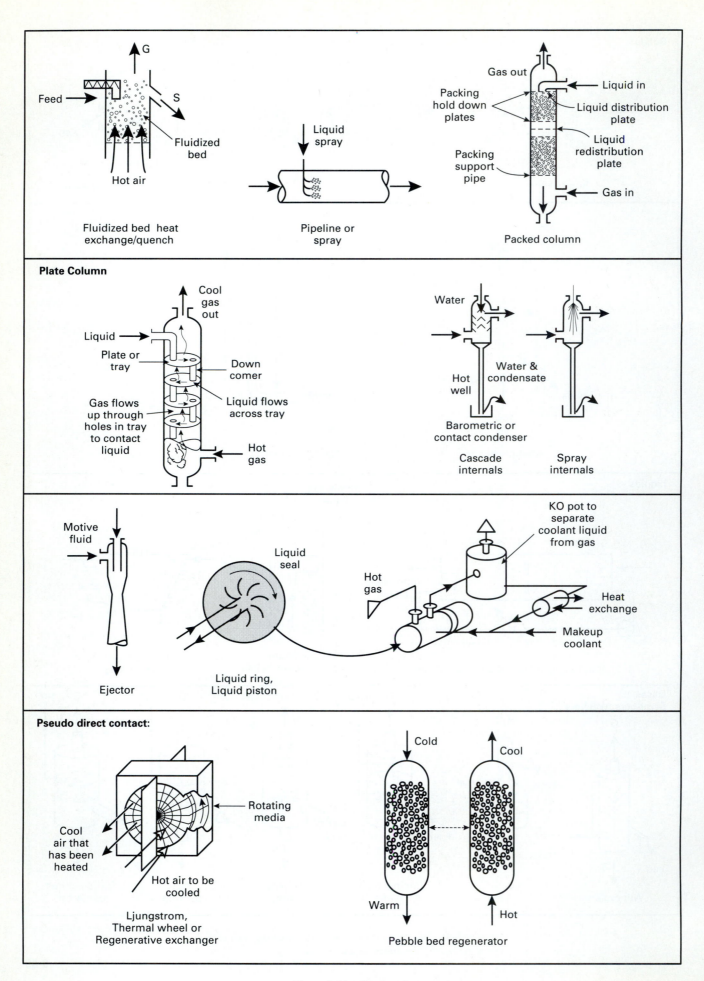

Figure 3–31 (Continued)

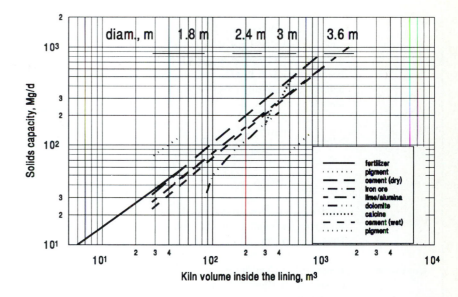

Figure 3–32   Rough Sizing of Rotary Kilns

Example 3–15:   We wish to make 50 tonnes/day of cement. What should be the size of the kiln?

An Answer: From Figure 3–32, we note that the total volume needed is about 50 m^3 and that the kiln will be about 2 m in diameter and approximately 16 m long. The heat load is about 2000 kW.

---

   2. Fluidized bed systems are direct contact heat exchangers that can be used as dryers, as reactors, and to change the size of the solid particles. Here the focus is on the heat transfer characteristics and on processing the solids. (Fluidized beds can be used to quench gas reactions; this is explored under topic C.) The principles of fluidization and sizing the unit from the fluidization point of view are given in the previous section 3.2–1 (c). For the heat transfer characteristics, as an approximation the particles and the gas tend to leave the bed at the same exit temperature (Kunii and Levenspiel [1969] and Rossi [1984]), the particle-gas heat transfer coefficients are given in Table 3–10, and the surface area per unit volume of fluidized bed is in the range 20,000 to 100,000 m^2/m^3 of bed for particles in the size range 40 to 100 μm. The area decreases as the particle size increases.

Example 3–16:   We wish to heat a fine, fragile solid from 10°C to 260° to burn off trace impurities. The design is for 25 kg/s of particles of less than 75 μm diameter. The heat capacity of the solid is 0.85 kJ/kg °C. Assume that combustion gas is the fluidizing gas and the maximum inlet gas temperature is 650°C. Estimate the configuration.

An Answer: First we need to do a heat balance to estimate how much fluidizing gas is required to supply the heat. Assume that the fluidized bed is well mixed as far as the solids are concerned and that the gas flows up through the bed in "plug flow." Assume that the exit gas temperature is 260°C. (For more detailed calculations see Kunii and Levenspiel [1969] p. 455.) For these assumptions from **Data** C and Db for a bed inlet temperature of 650°C and an exit temperature of 260°C a reasonable average density of the gas is 0.63 kg/m^3 with a corresponding gas heat capacity of 1.7 kJ/kg°C. The solids heat load is 25 kg/s × 0.84 kJ/kg°C × (260–10)°C = 5.25 MW = gas heat supplied = $\hat{F}_1$ × 1.7 /kJ/kg°C × (650–260)°C

Hence, $\hat{F}_1$ = 7.92 kg gas/s

The next step is to check out the fluidization conditions. Based on the particle size, and the properties of the system, the principles outlined in Section 3.2–1 (c) can be used to estimate the fluidization velocity to be 0.5 m/s. Thus, the required cross-sectional area is:

$$\frac{7.92 \text{ kg}}{\text{s}} \quad \frac{\text{m}^3}{0.63 \text{ kg}} \quad \frac{1\text{s}}{0.5 \text{ m}} = 25 \text{ m}^2$$

To estimate the depth we consider the heat transfer coefficient between particles and gas, the amount of area supplied per cubic meter of fluidized bed and the temperature driving force. From Table 3–10, the heat transfer coefficient is about 0.017 kW/m^2.K. Assume an area per volume of 30,000 m²/m³. The LMTD for solids at 260°C and gas going from 650 to 260 is 195°C. Hence, the height of bed needed is q = U A LMTD and for a bed height of "h" per unit cross-sectional area:

$$5250\,\text{kW} = \frac{0.017\,\text{kW}}{\text{m}^2.\text{K}} \times 30000\,\frac{\text{m}^2}{\text{m}^3} \times \text{h},$$
$$\text{m}^3 \times 195°\text{C}$$
$$= 0.05\,\text{m}$$

From the section on fluidized beds, the minimum bed depth is about 0.3 m. Hence, use 0.3 m.

Comment: Rossi (1984) gives this example. The cross-sectional area is 25 m², and the bed depth is 0.61 m.

Flash conveyors and dryers are sized on the principles given in Chapter 2 based on pneumatic conveying. The heat transfer between the wall and the transported solid/gas mixture is about 0.1 kW/m²°C. More details about flash drying are given in Chapter 5.

3. For multiple hearth furnaces, the general loading is 1.25 to 2.5 g/s.m². Thus, for a solids throughput we can estimate the size of the hearth needed. Table 3–12 lists the loadings in general. These are expressed as a solids loading per single hearth area and a solids loading per total area in the hearth (or single hearth area × no. of hearths).

B. For Liquids. Liquids can exchange heat with immiscible liquids. For approximate sizing, a *volumetric* heat transfer coefficient can be used to estimate the total volume of the contacting device. The transfer coefficient depends on the type of internals in the vessel. The log mean temperature difference between the entering and leaving

Example 3–17: We wish to regenerate activated carbon by means of a multiple hearth furnace. If the total flowrate is 125 kg/h, what size of furnace is required?

An Answer: From Table 3–12, the approximate loading is 2 g/s.m². The carbon flowrate is

$$\frac{125\,\text{kg}}{\text{h}} \times \frac{10^3\text{g}}{\text{kg}} \times \frac{\text{h}}{3600\text{s}} = 34.7\,\text{g/s}$$

Thus, the nominal area needed would be:

34.7/2 = 17.4 m²

If a six-hearth furnace is used, then for each hearth the area would be

17.4 m²/6 = 2.87 m²

The diameter would be 1.92 cm.

Comment: The South Tahoe waste water treatment facility regenerates 125 kg/h of carbon in a 6-hearth furnace that is 1.37 m in diameter. Our estimate was a 6-hearth furnace with 1.92 m diameter. (Incidently, for activated carbon regeneration the downtime for the regenerator is estimated to be 60% of the time. Thus, it is on-stream 40% of the time.) EPA (1973)

streams is used similar to the approach taken for shell and tube type exchangers. Values are given in Table 3–13. For more see Sideman (1966) and Hoffman (1980).

Example 3–18 concerns liquid-liquid direct contact heat exchange. Liquids can also be heated or cooled by contacting them with gases or solids. For example, in cooling towers, liquid can cool itself through humidification. That is, unsaturated air is blown past the hot water. Because

TABLE 3-12. Solids Loading for Multiple Hearth Furnaces

Material	Loading per single hearth area, g/s.m²	Loading per total hearth area in furnace, g/s.m²
Activated carbon, regeneration	2	2.7 to 5.4
Bauxite adsorbent, regeneration		8 to 16
Bone char adsorbent, regeneration		20 to 35
Calcination of kaolin for pigment		4 to 6.5
Charcoal from wood		2.5 to 5.2
Foundary sand reclamation		20 to 35
Lime sludge recalcination		2 to 3.5
Pyrites, roasting	1.25 to 2.5	
Sludge incineration	1.8	9 to 16

TABLE 3-13. Volumetric Heat Transfer Coefficients for Liquid-Liquid Heat Exchange

Contacting Configuration	U_v,kW/m³°C	Ref.
Wetted wall	20	Hoffman
Spray tower	2 to 70 65 to 100	Fair Hoffman
Sieve tray	160	Hoffman
RTD	260	Hoffman
Baffles (Donut shape)	80	Hoffman
Pipeline	110 to 3800 with 200 to 1000 being the usual 275 to 930 110 to 1110	Fair Fair
Turbulent Re 7x10⁶	6600	Hoffman

Example 3–18: Fat, such as bonegrease, can be hydro-
lyzed into fatty acid and glycerine. This
splitting can be done continuously at
5MPa, 255°C by dispersing the fat as
drops that rise through a vertical column
of downward flowing hot water. The
water leaves the bottom of the column as
16 to 20% glycerol. Consider the direct
contact heat exchanger at the top of the
column where we wish to heat up the
0.35 kg/s of incoming water at 60°C and
cool down the 0.80 kg/s of product fatty
acid from 255°C to 140°C. The heat
capacity of fatty acid is 2.8 kJ/kg°C. The
diameter of the column, 0.9 m, is chosen
based on the hydrolysis reaction time and
the rise velocity of the drops. Estimate
the height of heat exchange region
needed if we use film-flow type plates.
Figure 3–33 illustrates the system.

An Answer: First, do an overall energy balance to
check out the system and determine temperature and heat load. The cooling heat
load for the fatty acid is:

$$q = \frac{0.80 \text{ kg}}{s} \times \frac{2.8 \text{ kJ}}{\text{kg·°C}} \times (255 - 140)°C$$
$$= 257 \text{ kW}$$

Assuming negligible heat loss, this can be
transferred to the incoming water:

$$257 \text{ kW} = 0.35 \text{ kg} \times 4.2 \times (T - 60)°C$$

Solving for T yields 235°C. This is
reasonable since one might expect the
temperature to be as high as 257–5°C
= 252°C.

Estimate the LMTD 255 → 140

$$\frac{235}{20} \leftarrow \frac{60}{80}$$

From Figure 3–21, LMTD = 43°C.

From Table 3–13, this configuration
might be described as a spray tower or a
wetted wall since plates and downcomers
are installed. Hence, try a value of
20kW/m³°C.

Substitution gives:

$$q = U_v V \text{ LMTD}$$
$$257 \text{ kW} = \frac{20 \text{ kW} \times V \text{ m}^3}{\text{m}^3°C} \times 43°C$$

$V = 0.3 \text{ m}^3$; since the diameter is 0.9 m
the height or the heat exchange region
would be about 0.5 m.

Comment: Potts (1978) provides a sketch of a fat
splitter. The overhead fatty acid flowrate
and the exit temperatures were estimated
from experience. The height of the heat
exchange region, based on the sketch, is
about 0.9 m to 1.2 m. Other information
is provided by Swern (1964) p. 939.

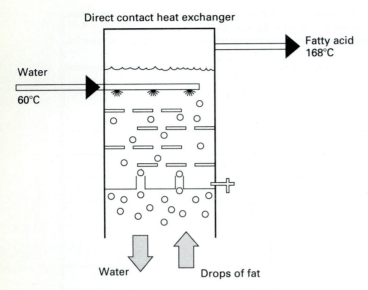

Figure 3–33   Sketch of the Fat Splitter Direct Contact Heat Exchanger

of the concentration driving force between the liquid and the gas, some of the water evaporates. The heat of vaporization comes from the sensible heat of the water. Thus, by evaporating about 4% of the water, the temperature of the water is cooled down about 15°C. Because understanding this process requires an appreciation of humidification, it is discussed in the next section.

C. For Gases.   Direct contact between a gas and a liquid is commonly used for quench cooling, direct contact condensing, barometric condensers, humidification, dehumidification, gas scrubbers, and cooling towers.

In this section we discuss a wide variety of functional operations. Because many of them involve the evaporation or condensation of vapor we start with a general discussion about "humidification" and "dehumidification" and then

consider separately the different functions that might exploit this simultaneous mass and heat transfer.

1. Background Ideas about Simultaneous Heat and Mass Transfer.   Direct gas-liquid contacting operations often involve the simultaneous transfer of both heat and mass (because some liquid is either condensing or evaporating). Such simultaneous exchange means:

1. that the latent heat of vaporization will cool down the gas if the liquid evaporates as in humidification; or the latent heat of vaporization must be supplied from the liquid as an additional coolant load if liquid is being condensed from the hot gas as it cools down. Thus, care is needed to identify the effect of the heat of vaporization related to the mass transfer; does it cool down the gas or heat up the liquid?

2. that the rate of heat transfer may be "enhanced" or "reduced" from what we would expect if heat were being transferred without the mass transfer. Thus, if the gas is cooling down and liquid is being condensed from the gas, the heat and mass transfer are both in the same direction (from the gas to the liquid) and the heat transfer coefficient might be 1.5 to 3 times higher than we would expect for heat transfer alone. On the other hand, if the gas is cooling down and liquid is evaporating into the gas (humidification), then the heat transfer coefficient might be 0.3 to 0.8 times what we would expect from heat transfer alone. The amount of correction depends on the direction of the transfer and the relative amounts. This is illustrated in Figure 3–34. For more, see Bird, Stewart and Lightfoot (1960) p 664.

3. Table 3–14 illustrates some of the options that can occur in direct gas-liquid heat and mass transfer. It is a challenge to keep them straight in that we wish to focus on the direction of heat transfer, the direction

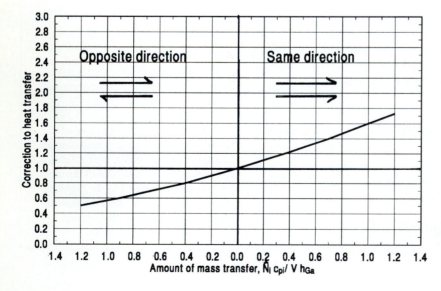

Figure 3–34   Corrections for Simultaneous Heat and Mass Transfer (Reprinted by permission of John Wiley & Sons, Inc., from "Transport Phenomena" by R.B. Bird, W.E. Stewart and E.N. Lightfoot, copyright © 1960, John Wiley & Sons, Inc.)

of mass transfer, and where the latent heat associated with the evaporation (or mass transfer) is accounted for. Consider first gas cooling. The heat transfer is from the hot gas to a cold liquid. In spray cooling, the liquid evaporates; the mass transfer is from the liquid into the gas. As an approximation the latent heat of evaporation comes from the gas and results in sensible heat loss in the gas. The gas cools.

In hot gas cooling with humidification, only some liquid evaporates. Thus, now the latent heat of evaporation *plus* the sensible heat in heating up the cold liquid represents the sensible heat of the gas cooling down.

For hot gas cooling with dehumidification, some of the vapor in the gas condenses (dehumidification) and thus the mass transfer is from the gas to the liquid. So is the heat transfer. Now the latent heat of condensation and the sensible gas cooling heat load must both come from the sensible heating of the liquid.

For hot gas cooling and complete condensation, this is similar to hot gas cooling with dehumidification except that now the major heat power load is the condensation that must be supplied by sensible heating of the liquid.

On the other hand, with cold gas heating and humidification, the heat transfer and the mass trans-

fer are in the same direction. The heat load from the sensible cooling of the liquid accounts for heating up the gas and supplying the latent heat of vaporization.

If we assume that the heat transfer coefficient for vaporization/condensation is very high, then the expressions for heat transfer coefficient should scale the appropriate heat transfer coefficient according to the sensible heat load. Fair (1972) has summarized these expressions in terms of the volumetric heat transfer coefficient U_v. These are summarized in Table 3–14. In these expressions, α is the correction for the enhancement or reduction in heat transfer because of the direction of the mass transfer, h_{Ga} is the volumetric heat transfer coefficient for the gas film; h_{La} is the volumetric heat transfer coefficient for the liquid film, q_T is the total heat load and λ is the latent heat load. Thus, if the total heat load is sensible heat, the ratio in brackets becomes 1.

4. that we have to estimate "how much" liquid might transfer and the temperatures for these conditions. Gases can only hold or contain a limited amount of liquid as vapor. If we try to add more than the saturation amount, the liquid condenses. Figure 3–35 shows the humidity chart for the air water system. The absolute humidity is defined as the mass of liquid vapor per mass of bone-dry or vapor-free air, H.

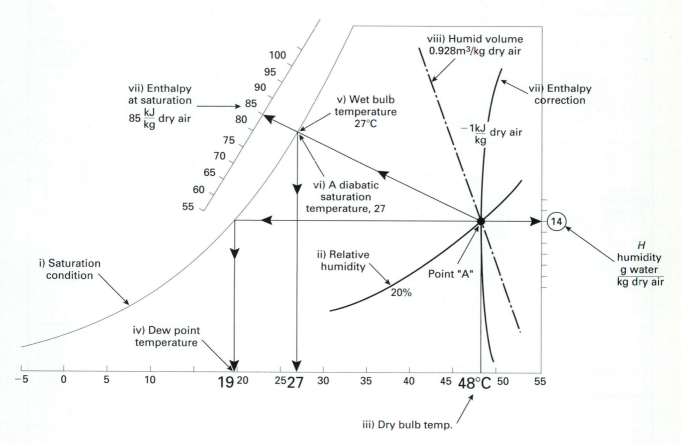

Figure 3–35 Defining Terms on an Air-Water Humidity Chart

TABLE 3-14: Options for Direct Gas-liquid Heat Transfer

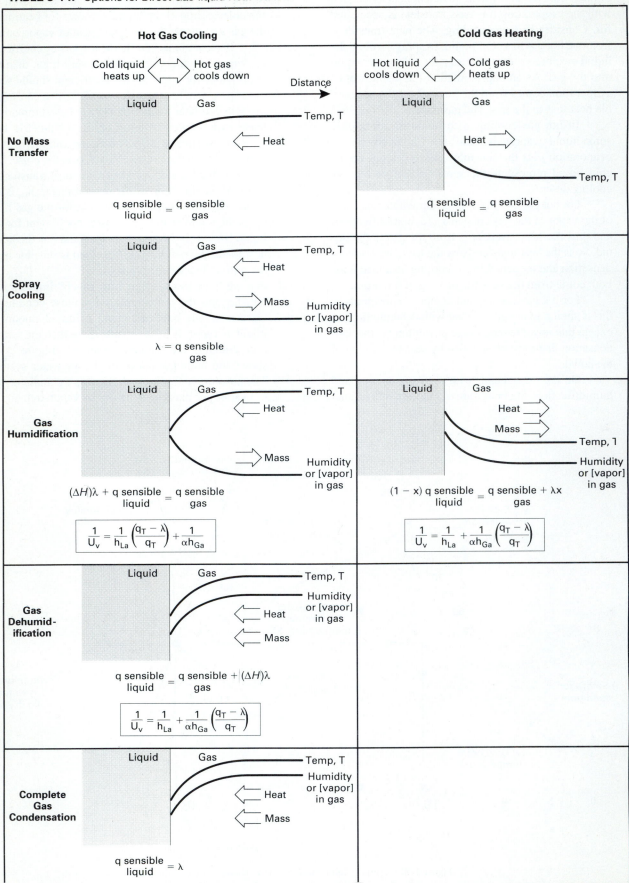

For the water-air system the liquid vapor is "water" vapor. Different humidity charts would be needed for different pressures and for different liquid-air combinations. The humidity chart for sea level pressure and for water-air expresses:

(i) the saturation condition: the absolute humidity for saturated conditions as a function of temperature. For example, at 19°C, the saturation humidity is about 14 g/kg. Any condition below the saturation line represents an "unsaturated" condition. The humidity at a temperature T for saturated conditions is (for ideal gases where mol % equals the partial pressure %)

$$H = p \frac{M_v}{(\pi - p) M_g} \qquad (3\text{-}19)$$

where p   = vapor pressure at temperature T
    π = total pressure
    M_v = molar mass of the vapor
    M_g = molar mass of the vapor free gas

If more than one condensible species is present as vapor then, assuming Raoult's law applies, we obtain the more general equation:

$$H = \tilde{x} p° \frac{M_v}{(\pi - \Sigma\, p) M_g} \qquad (3\text{-}20)$$

where $\tilde{x}$ = mol fraction of the vapor component in the liquid.
    $p°$ = pure component vapor pressure. For data see Appendix C.
    $\tilde{x} p° = p$ = partial pressure of the vapor components.

(ii) the degree of unsaturation. This can be defined by specifying two conditions from among the relative humidity, the dew point temperature, or the wet bulb temperature (or the adiabatic saturation temperature). Consider each of these terms in turn.

(ii) the relative humidity. The relative humidity or percent relative humidity is the ratio of the partial pressure of the vapor to the partial pressure of liquid at the dry bulb temperature. This is usually called just the "relative" humidity. This quantity can be defined at temperatures greater than 100°C. Care is needed to distinguish the relative humidity from the percent absolute humidity. This latter cannot be defined above 100°C because by definition it is the ratio of the absolute humidity at the dry bulb temperature to the saturation humidity at the same temperature and this ratio cannot exist when the vapor pressure becomes greater than atmospheric pressure. Shown on Figure 3–35 for position "A" is a condition represented by 20% relative humidity.

(iii) the dry bulb temperature, or the temperature that an ordinary thermometer would read (this is the abscissa on Figure 3–35), and for point "A" is 48°C;

(iv) the dew point temperature, or the temperature when liquid would just start to condense out of the gas; the gas is saturated. Another way of expressing this is that the partial pressure of the vapor in the gas equals the vapor pressure of the liquid at that temperature. For point "A," the dew point is 19°C.

(v) the wet bulb temperature, or the temperature of a thin film of water surrounding a thermometer if the liquid film was in dynamic equilibrium as the water evaporates to try to saturate the surrounding air. At this condition, the rate of heat transfer to the liquid surface by convection equals the rate of mass transfer away from the surface because of the water evaporating. However, the gas never becomes saturated. The key idea here is the small amount of water and the equality of the two rates of heat and mass transfer. On Figure 3–35 the wet bulb lines slant upwards to the left and have the following meaning. If we had a "dry" thermometer at 48°C with 20% relative humidity, then the temperature that a "wet film" thermometer would read would be found by following the wet bulb line upwards until it intersects the saturation line, in this case at 27°C. The slope of the wet bulb line is the convective heat transfer coefficient/(gas phase mass transfer coefficient times the latent heat) all evaluated at the wet bulb temperature. The wet bulb line is very important. While water evaporates from a spray into a gas, the process follows the wet bulb line. In drying, the true surface temperature of the liquid on the evaporating surface, during the constant rate period and if convection is the mechanism of heat transfer, is the wet bulb temperature.

(vi) the adiabatic saturation temperature is the temperature the gas would achieve if, in a closed cycle, the gas contacts a large volume of liquid and there is no heat exchange with the environment so that all the heat lost by the gas is converted quantitatively into latent heat. There is a redistribution of the liquid so that enough is now in the gas to saturate it. In this process, the gas becomes saturated. The slope of the adiabatic saturation line is the ratio of the gaseous molar heat capacity/latent heat.

We can use the Chilton-Colburn j-factor analogies between heat and mass transfer to help contrast wet bulb conditions with adiabatic saturation conditions. Because the fundamental equations and boundary conditions for many heat and mass transfer problems are analogous, Chilton-Colburn introduced the dimensionless j-factors to be:

for heat transfer: $j_H = \dfrac{h\, Pr^{2/3}}{\rho \hat{c}_p y}$ (3-21) for mass transfer: $j_D = \dfrac{k_x}{\tilde{c}v} Sc^{2/3}$ (3-22)

where: $\tilde{c}$ is the mol concentration

k_x is the local mass transfer coefficient.

Since these can be equated and since $\rho\hat{c}_p = \tilde{c}\tilde{c}_p$ we obtain the Lewis expression:

$$\frac{h}{k_x} = \tilde{c}_p Le^{2/3} \qquad (3-23)$$

and Le = Lewis number = Sc/Pr.

Note that the term on the LHS of Equation 3–21 is the slope of the wet bulb line multiplied by the latent heat and the first term on the RHS is the slope of the adiabatic saturation line multiplied by the latent heat. Thus, if the Lewis number is 1, the slopes of the wet bulb and the adiabatic saturation lines would be the same. For the air-water system, the Lewis number = 1. Thus, in this example, point "A" has an adiabatic saturation temperature of 27°C. For systems where the Lewis number is not unity, the slopes of the adiabatic saturation lines and the wet bulb lines are not equal. This is illustrated in Figure 3–36 for a variety of systems.

(vii) the total enthalpy along the adiabatic saturation line should be constant. Thus, it is convenient to show on the humidity chart, the enthalpy of the mixture along the adiabatic saturation line. The enthalpy can be calculated above any base condition. For these charts the base condition is 0 enthalpy at 0°C. On the humidity charts, the adiabatic saturation lines are not straight, whereas the enthalpy lines are. Hence, we correct this by showing only the adiabatic saturation lines and the enthalpy at saturation conditions. Then corrections are given along the adiabatic saturation lines. At point "A," for example, the enthalpy at saturation condition is about 85.kJ/kg dry air. The correction is −1 kJ/kg and so the enthalpy at A would be 85.−1 = 84.kJ/kg.

(viii) the humid volume is the volume of the mixture per unit mass of dry air. Again, this information is often shown on the chart for convenience. At A this is about 0.928 m^3/kg dry air. Alternatively, this could be shown as the reciprocal, namely, the kg dry air per m^3 of mixture.

Detailed humidity charts for the air water system are given in **Data** Part C.

With this background, consider now the various applications of simultaneous mass and heat transfer in the context of the different functions we want to achieve. Table 3–15 lists four different prime functions and illustrates how

humidification does or does not play a role. Details of each of these are given in the next sections.

2. Function: Quenching of a Hot Gas. Quenching can be used to cool down a gas rapidly and is particularly useful when we must stop possible unwanted side reactions. Some examples include quenching the gas from the pyrolysis reactor in a plant to produce ethylene as illustrated in Figure 1–1 or the exit gas from an ethylene dichloride cracker for the production of vinyl chloride monomer.

For quenching, Figure 3–37a shows five common options of cold walls (not a direct heat transfer option); fluidized bed, gas mixing, and spray and gas-liquid contacting.

Example 3–19: We want to set out a process for the production of ethylene from ethane or propane feedstock. Outline the tasks, and create a process flow diagram to represent one option. In particular, note the role and placement of any quenching task.

An Answer: Figure 1–1 outlines the tasks. We note that in this proposal we alternate between react and separate and continually consider energy and heat integration. In the first reaction step, the hydrocarbon feedstock is to be cracked to produce the major component, ethylene. The operating temperature in the pyrolysis furnace depends on the feedstock but is in the general region of 850°C with the key design parameter being the length of time the reactants remain at that high temperature. For ethane and propane, the residence time in the pyrolysis furnace is of the order of 0.5 to 1 s. Keeping the reactant mixture at this high temperature for longer times promotes coke formation. This not only decreases the yield of ethylene but it fouls the furnace tubes. Hence, as soon as the prescribed residence time has been reached we want to "stop the reaction" by reducing the temperature rapidly. Mol and Westenbrink (1974) suggest that the temperature should be quenched from 850 to 400°C in less than 150 ms. Further quenching should reduce the temperature to about 40°C. Such is the function of the quench that is proposed after the pyrolysis furnace.

Comment: Also shown in Figure 1–1 is a photograph of an ethylene plant that relates the flow diagram to the physical hardware. For more see Zdonik et al. (1974), Zdonik and Haywood (1975), and Kneil et al. (1983).

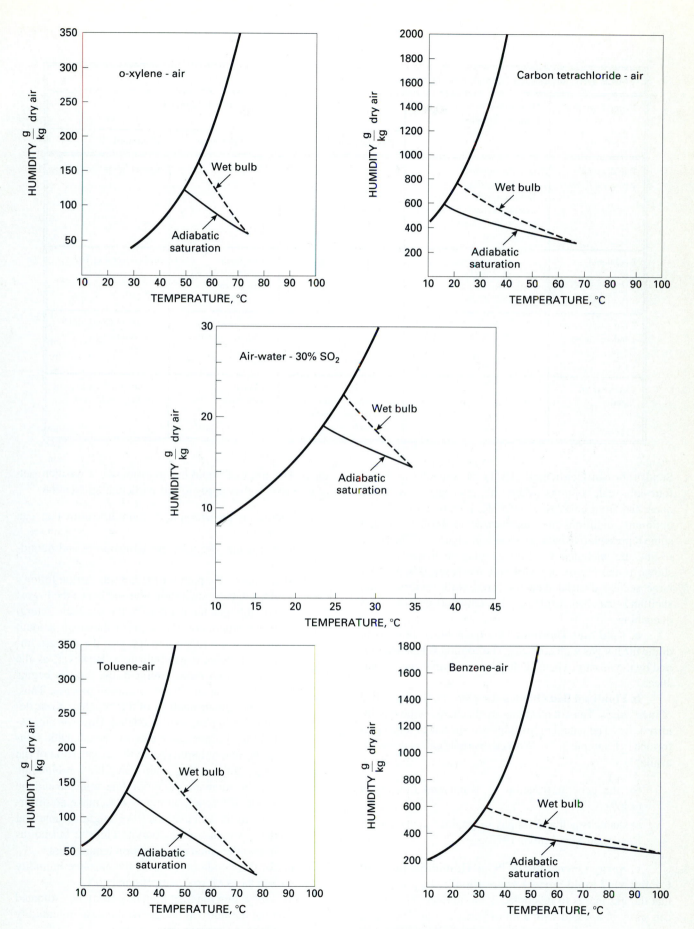

Figure 3–36 How the Adiabatic Saturation Lines Differ from the Wet Bulb Lines
for Different Binary and Ternary Systems: o-xylene/air; Carbon tetrachloride/air;
Toluene/air; Benzene/air and SO₂/Water/air

TABLE 3-15. Alternative Functions for Direct Contact

Prime Function	Description of Non-Liquid Contact Options	Direct Gas-Liquid Contact		
		Gas Behavior	**Liquid Behavior**	**Comment**
Rapidly cool gas/quench	-cold gas dilution: straight blending -fluidized bed: exploit thermal properties of inert solids in bed	*(HUMIDITY vs TEMP. plot)*	All liquid evaporates	Called Spray Quenching
Cool liquid		*(HUMIDITY vs TEMP. plot)*	Some liquid evaporates; liquid cools down	Called Cooling Tower
Cool gas: perhaps condense some liquid		*(HUMIDITY vs TEMP. plot)*	Some liquid may condense; liquid temperature increases	Occurs in Gas-Liquid Contactors; Gas Scrubbers
Condense all liquid		Usually not a two component system so a humidity chart is not pertinent	All liquid condenses	Barometric or Direct-contact condensers

Sundstrom and DeMichiell (1971) theoretically analyzed four lab scale options when the inlet gas was to be quenched from 2800°K to 1000°K. For the lab scale the optimum conditions are summarized in Table 3–16 and some representative data are shown in Figure 3–37b. These results are included to illustrate the phenomena. Also shown in the Figure are Mol and Westenbrink's data for large scale industrial data for "cold wall," indirect, and shell-and-tube heat exchange for ethylene pyrolysis gas quenching.

a. Cold Gas Dilution: One quenching option is to dilute the hot gas with cold gas. The cooling is fast and limited by the mixing. The calculations are simple energy balances.

b. Fluidized Bed Quenchers: Here we fluidize a bed of inert solids. The hot reactants are discharged into the fluidized bed. The calculations done are similar to those performed in Example 3–16. Worked example calculations are available

- for the general situation by Kunii and Levenspiel (1969) p. 455.
- to quench methane flames to produce acetylene by Russo and Massimilla (1972).

c. Spray Quenching or "Gas Humidification and Cooling via Liquid Spray Evaporation": The principle is that the gas that is to be cooled is unsaturated. Hence, we can spray in a liquid with a lot of surface area. The liquid evaporates completely and in so doing humidifies and cools the gas. This cannot work if the gas is already satu-

rated. The spray of liquid has to evaporate. Consider each of the steps as they apply to this particular application.

1. Decide on the amount of humidification that can occur by
   a) estimating the inlet gas temperature and humidity.
   b) assuming the path for the humidification follows the adiabatic saturation line and hence the lowest possible gas temperature is the adiabatic saturation temperature. (Because of the small amount of the spray relative to the gas, the process initially follows a wet-bulb line. However, as the gas becomes more saturated, the process begins to shift to an adiabatic saturation process. Thus, for an accurate analysis of this problem, these details would have to be modeled. However, for estimating purposes, we can use the adiabatic saturation line to approximate the evaporation of the spray. If the system is air-water, the wet-bulb line and the adiabatic saturation line coincide and the subtle differences in the process make no difference in the calculations.) We might assume that the gas temperature cools to close to but not as low as the adiabatic saturation temperature.
   c) for this condition read off the absolute humidity and the "exit gas" temperature.
   d) subtracting the inlet humidity from the estimated final humidity to obtain the increase in humidity per unit mass of vapor-free gas.
2. Estimate the amount of liquid spray required from

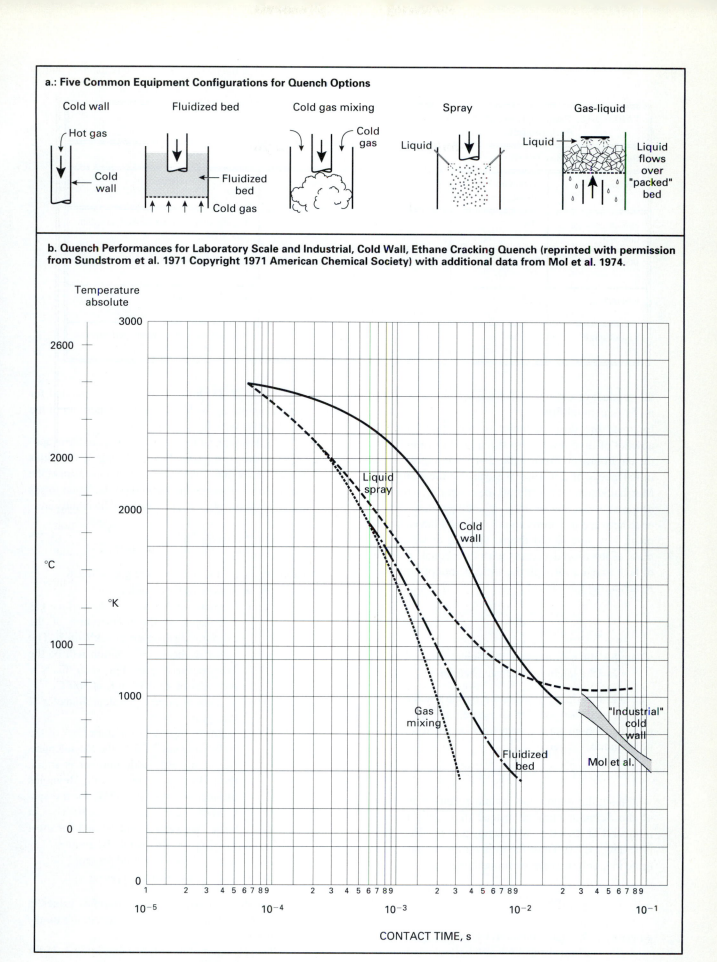

Figure 3–37 Quench Operations a) Five Common Equipment Configurations for
Quench Options b) Quench Performance for Laboratory Scale and Industrial, Cold Wall,
Ethane Cracking Quench (reprinted with permission from Sundstrom et al., 1971
copyright 1971 American Chemical Society, with additional data from Mol et al. 1974)

TABLE 3-16. Example Laboratory Scale Comparison of Optimum Quenching Performance

Hot Gas: 9 kg/m²·s at 2800°K

Option	Inlet Gas Pipe diam., cm	Optimum Conditions	Comment
Non-direct cold tube wall	2.5	Side wall temp. of 37°C	Results very sensitive to feed tube diam. rate: 0.06×10^6 °K/s
Fluidized bed	2.5	0.3 m/s fluidizing velocity with 0.4 mm diam. particles	rate: 0.17×10^6 °K/s
Gas mixing	2.5	Coolant gas velocity=2.0 hot inlet gas velocity	rate 0.7 $\times 10^6$ °K/s
Liquid spray	—	250 μm diam. spray with $\hat{F}_G/\hat{F}_L = 1.0$	rate of vaporization limits performance rate: 0.006×10^6 °K/s

the product of the mass of dry gas flowrate and the increase in humidity.

3. Estimate the drop diameter of the spray and the amount of time it takes for the spray to evaporate in the environment. From Table 3–16 one might use an initial guess of 250 μm. We can use the Ranz and Marshall equation to estimate the evaporation time of drops as a function of drop size, temperature difference between the drops and the gas and the environment. For pure liquid drops, the rate at which the mass of the drop decreases with time equals the rate of heat transfer through the surface area of the drop. Rearranging this yields:

$$\frac{2\langle \Delta T_{\ell m}\rangle}{\rho_L \lambda}\, dt = \frac{dD_p}{h} \qquad (3\text{-}24)$$

where h = the heat transfer coefficient
λ = the latent heat of vaporization

The heat transfer is represented by the usual type of Nusselt equation similar to Equation 3–14 plus a term to account for the heat transfer when the Reynolds number is zero:

$$Nu = \left(1 + 0.30\,(Pr)^{1/3}\,(Re)^{1/2}\right) \qquad (3\text{-}25)$$

with a similar expression for mass transfer in terms of the Schmidt number instead of Pr, the Prandtl number. For Re = 0 or no flowrate past the drop, then Nu = Nu$_{AB}$ = 2.0. Thus, the time for the drop to evaporate in quiet, stagnant surroundings is:

$$t = \frac{\lambda\, \rho_L\, D_p^2}{8\, k_g \Delta T_{\ell m}} \qquad (3\text{-}26)$$

where k$_g$ = thermal conductivity of the gas (often about 40 to 260 mW/mK depending on the gas temperature)
D$_p$= the original drop diameter.

Figure 3–38 shows Equation 3–26 plotted for water evaporating with a thermal conductivity of the gas of 260 mW/m.K.

Laskowski and Ranz (1970) analyzed the evaporation of drops in a gas stream moving at velocities greater than 60 m/s. Their results are independent of the temperature difference but depend on the liquid to gas ratio and on the physical properties. The results are:

$$t_q = 1.7 \times 10^{-13}\left[\frac{c_{pg}\, \rho_L}{k_s}\right]\frac{s}{m^2}\frac{D_p^2}{\alpha\,(F_L/F_G)} \qquad (3\text{-}27)$$

where $\hat{c}_{pg}$ = heat capacity of the gas, kJ/kg.K
ρ$_L$ = density of the liquid, kg/m³
k$_s$ = thermal conductivity of the surface gas and vapor at the temperature of the evaporating drop surface, kW/m.K. The vapor in the surface region will be the evaporating species. For water this is in the order of 42 mW/m.K at 300°C.
D$_p$ = average volume equivalent diameter of the spray, μm
α = factor to represent the net effect of the RHS Equation 3–25 on the Nusselt number. If the Reynolds number is small, the RHS approaches 2; if the Reynolds number is large, the RHS, for the systems of interest, approaches 10. Thus, α varies between 2 and 10 and for most flowing systems would be about 6.
$\hat{F}_L$ = average mass of liquid present
$\hat{F}_G$ = average mass of gas present.

The dimensional property group has values in the range 6×10^6 to 80×10^6 s/m² depending on the liquid used.

This equation is also plotted on Figure 3–38 for different values of the property ratio. Laskowski and Ranz (1970) suggest further that the temperature pro-

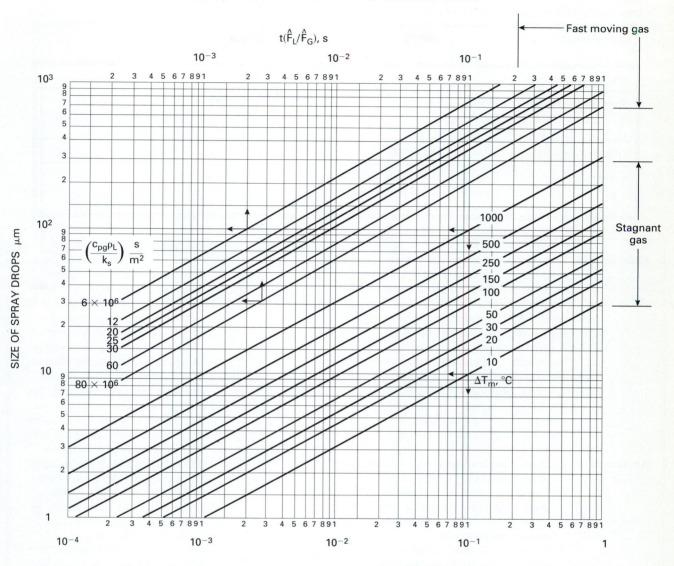

Figure 3–38 Evaporation of Drops: in Quiescent Conditions Using the Ranz-Marshall Predictions and in Flowing Conditions Using the Laskowski-Ranz Predictions

file in the pipeline as the spray evaporates can be estimated from:

$$\frac{(T_{G2} - T_L)}{(T_{G1} - T_L)} = \exp[-t/t_q] \qquad (3\text{-}28)$$

where: T_{G1} = initial gas temperature
T_L = boiling temperature of the liquid
T_{G2} = gas temperature at time t.

The considerations outlined in Equation 3–28 become important when the exit gas temperature ap-

proaches the boiling temperature of the liquid in the spray.

4. Estimate the length of the pipeline required to provide the evaporation time.

As an alternative short-cut procedure when selecting a spray-type chamber as a quencher instead of a pipeline spray, Schifftner and Hesketh (1983) suggest that the gas velocity be less than 15 m/s and that the gas contact time in the scrubber depends on the type of spray used and the inlet gas temperature as summarized in Table 3–17.

TABLE 3-17. Contact Times for A Spray Quencher (Reprinted with permission from Schifftner and Hesketh, 1983, Copyright CRC Press, Inc., Boca Raton, FL)

Type of Spray	Inlet Gas Temperature °C	Residence time, s
Pressure atomized	850 to 1100	1.5 to 2
Spray, approx 200 μm	550 to 850	1 to 2
	250 to 550	0.75 to 1.5
Twin fluid atomization	850 to 1100	1. to 2
approx 90 μm	550 to 850	0.75 to 1.5
	250 to 550	0.5 to 1.5

Example 3–20: In the production of acetic anhydride from acetone, acetone is cracked to produce ketene at 760°C. Because ketene rapidly decomposes at temperatures above 500°C we wish to quench this reactor output. An attractive quench material is an equal mass mixture of liquid acetic acid and acetic anhydride. The data are as follows: reactor flowrate 2.46 kg/s; mean molar mass of this stream is 45.2 kg/kmol. The heat required for the cooling is 1.436 MW. Rough-size a spray cooling system.

An Answer:

1. Our first task is to check on the amount of liquid that can evaporate. In this example, this is not a constraint. Indeed, we will absorb ketene in acetic acid to produce acetic anhydride in a downstream processing unit.

2. The amount of spray needed will be such as to provide the heat load basically through the latent heat of evaporation. Neglecting the sensible heat to heat the liquid to its boiling temperature, and with the latent heat for acetic acid of 405 kJ/kg and for acetic anhydride of 385 kJ/kg (from **Data** Part Db), the amount of spray needed is:

$$q = \hat{\lambda}\hat{F}$$
$$1436\text{kW} = 400 \text{ kJ/kg} \times \hat{F}$$
$$\hat{F} = 3.59 \text{ kg/s}$$

3. Assume the drop size in the spray is 100 μm. From **Data** Part Db the physical property group for Equation 3–25 at about 600°C is as follows:

	AcOH	Ac$_2$O
$\hat{c}_{pg}$	2.30	2.05 kJ/kg.K
ρ_L	1.04	1.075 Mg/m^3
k_{gs}	50.	54. mW/m.K

The group is about 50×10^6 s/m^2. From Figure 3–38 corresponding to a drop size of 100 μm, the abscissa is about 1.4×10^{-2} for a flowing system. Since the ratio of liquid to gas is 3.59 kg/s/2.46 kg/s, the time required is $t = 1.4 \times 10^{-2}\text{s} \times 2.46/3.59 = 0.95 \times 10^{-2}\text{s}$. To estimate the length of pipeline needed, we calculate the volumetric flowrate and the cross-sectional area of the pipe to get a velocity. The average gas flowrate in the pipeline at about 600°C is 2.46 kg/s of reactor effluent (with a mass density of 101 kPa $\times$ 45.2/8.314 $\times$ (273 + 600) = 0.63 kg/m^3) or 3.88 m^3/s; plus 3.59/2 kg/s acetic acid (gas density of 101 kPa $\times$ 60.1/8.314 $\times$ [273 + 600]) or 2.15 m^3/s; plus 3.59/2 kg/s acetic anhydride (gas density of 101 kPa $\times$ 102/8.314 $\times$ (273 + 600)) or 1.24 m^3/s. The total gas flowrate is 7.27 m^3/s. From Figure 2–30 for a velocity of 40 m/s a 50 cm diameter pipe would be used. At a velocity of 40 m/s, the length of pipe needed to supply 1/100th of a second is 0.4 m. As a check on the temperature we can use Equation 3–28. Since the boiling point of the liquid is about 100°C:

$$[(510 - 100)/(760 - 100)] = \exp - t/t_q$$

Hence, $t = 0.62 t_q$ or even less than 0.4 m.

Comment: This is an interesting problem because it emphasizes the limitations of approximations and the need for a sound understanding of physical properties. First, Jeffreys (1961) sized this quench situation. Instead of assuming that the latent heat alone provided the coolant, he included the sensible heat to heat up the liquid to the boiling temperature and to heat up the vapor to the exit temperature. His results show a requirement of 1.34 kg/s of liquid spray (instead of the 3.59 kg/s estimated here). The second observation is that acetic acid vapor often partially dimerizes so that a molar mass of about 100 might be a better choice in estimating the density of the gas. The correlation used to estimate the evaporation time assumes that the gas in the pipeline is traveling at 60 m/s. This would reduce the diameter to 40 cm and increase the length to 0.6 m.

Alternative: To illustrate the spray chamber approximations outlined in Table 3–17, the recommended gas velocity should be less than 15 m/s. For this situation of 2.46 kg/s of gas of 0.63 kg/m^3 density, the cross-sectional area for the vessel would be 2.46/(0.63 × 15) = 0.26 m^2 or 58 cm diameter. From Table 3–17, for an inlet temperature of 760°C and pressure atomization the residence time should be 1 to 2 s. Hence, the length of the vessel should be 15 m/s × 2 s or 30 m assuming plug flow. This result is much larger than the more detailed analysis done above with 50 cm diameter pipe that is 0.4 m long. Another option is to consider the drops as evaporating in stagnant gas. The temperature difference is 760 − 100°C at the start and about 510 − 100°C at the end or a mean temperature difference of about 500°C. For 100 μm diameter drop the evaporation time, assuming the drops behave like water, is, from Figure 3–38, about 0.2 s. For this situation the length of pipe would have to be (at 40 m/s gas velocity) 8 m.

Example 3–21:   Hot exhaust gas from a cold-blast type cupola melting 14.4 Mg/h of grey iron is to be treated. The mass flowrate is 10.7 kg/s of "air" at 1150°C. The initial humidity of the gas is 50 g/kg dry air. This corresponds with an adiabatic saturation temperature of 83°C. We wish to cool the gas to 260°C. Assume the heat capacity of the gas is 1.25 kJ/kg.K. Estimate the configuration.

An Answer: 1. To estimate the amount of liquid that evaporates we can (i) consider the humidity chart, follow an adiabatic saturation line up to the target temperature, estimate the amount of water the gas acquires, and multiply by the mass of gas being processed or (ii) we can do a heat balance on the gas and assume all the heat comes from the latent

heat of evaporation. By method (i) with reference to the high temperature humidity chart in **Data** Part C, the adiabatic saturation line of 83°C intersects the 260°C dry bulb temperature line at about 550 g/kg dry air. Thus, 550 g-50 g of spray water must be supplied/kg dry air. The flowrate is 10.7 kg/s and thus we would need about 0.5 × 10.7 or 5.35 kg/s of spray. This is illustrated in Figure 3–39.

   By method (ii) the heat required to cool the gas equals the latent heat provided because the spray evaporates. An estimate of the sensible heat required to cool the hot, "dry" gas is 10.7 kg/s × 1.25 kJ/kg. K × (1150 − 260)°C = 11.9 MW. The latent heat of vaporization is about 2000 kJ/kg. Hence, the mass of spray required is 5.95 kg/s. Both of these methods are approximations but are in reasonable agreement.

2. To estimate the time required for the spray to evaporate, assume the spray is 200 μm. The physical property group for water at about 600°C is, from **Data** Part Db, 4.2 × 1000/80 mW/m.K = 52 × 10^6. From Figure 3–38 the abscissa is 4 × 10^{-2}s. Since the water to gas mass ratio is 5.95 to 10.7, the time required is estimated to be 0.04 × 10.7/5.95 = 0.07 s. If the gas flows at 30 m/s then this requires about 2 m of pipe length.

Comment: W. W. Sly Manufacturing Co., Cleveland, reported this situation in their Data Study 64-7. They inserted cooling sprays in the inlet duct to an Impinjet scrubber. The total water flowrate to both the sprays and the scrubber is 7.37 kg/s. An analysis of the split between the spray and the scrubber suggests that about 4.8 kg/s goes to the spray and 2.5 goes to the scrubber. Hence, the estimates by the approximate methods outlined here are in reasonable agreement with the plant installation.

Schifftner and Hesketh (1983) p. 74 give some background information about this type of application.

d. Gas Quenching via Liquid or Gas Cooling by Gas-Liquid Contact and Either Humidification or Dehumidification.

Gas-liquid contacting can provide a variety of functions other than gas quenching. The function might include humidification or dehumidification. In gas-liquid contacting, a wide range of options are available as outlined in Table 3–14. The options vary in terms of the direction of heat and mass transfer, the location in the en-

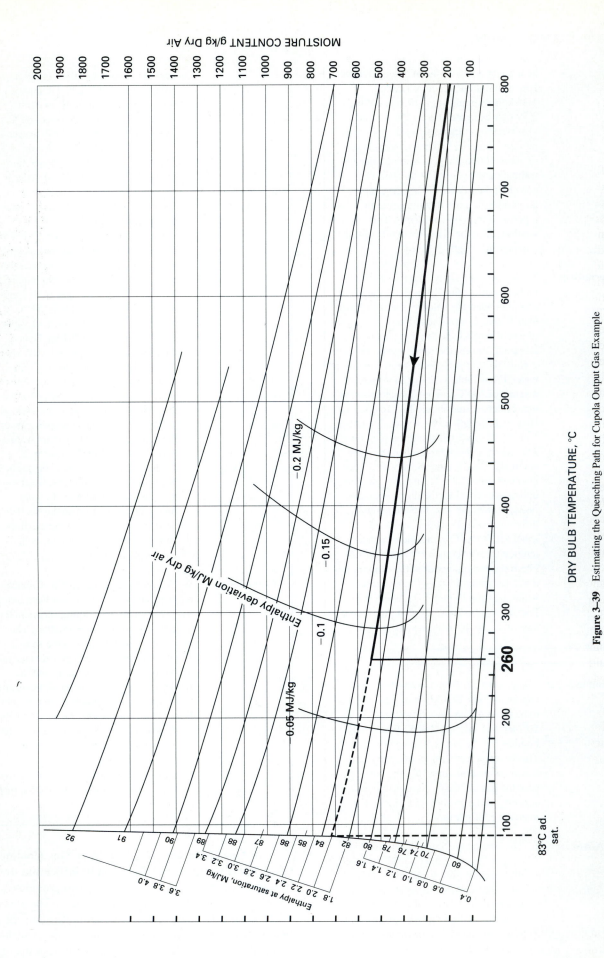

MOISTURE CONTENT g/kg Dry Air

DRY BULB TEMPERATURE, °C

Figure 3–39 Estimating the Quenching Path for Cupola Output Gas Example

ergy balance where the latent heat is accounted for, and the form of the equations used to represent the overall volumetric heat transfer coefficient. We will refer to Table 3–14 continually throughout the discussion in this section. The general sizing/selection approach is as follows:

1. find out what is going on. This will be best done by drawing the given conditions on the humidity chart and relating the operating conditions to the adiabatic saturation line. From this, we can decide whether humidification or dehumidification is going on. Sometimes, in part of the column humidification occurs followed by dehumidification. Do not consider such fine details, just look at the overall terminal conditions: does the humidity increase or decrease? Use Table 3–14 to identify the appropriate process, energy balance equations, and overall heat transfer coefficient equations.

2. estimate the temperature conditions, the amount of liquid needed to achieve the task, the total heat load, and the latent heat load. Which ones we have to calculate will depend on what information is given to us in the design situation. Estimate the LMTD. Usually the heat load can be estimated directly from the humidity chart from the enthalpy of the streams. Care is needed, however, because different charts use different bases. The charts in this text use 0°C as the basis. Many use 0°F. The calculations will be correct if a consistent set of charts and bases is used.

3. sizing the device will be a tradeoff between the heat transfer efficiency and the power required to overcome the pressure drop through the unit. Figure 3–40 and Table 3–18 provide typical application conditions for some of the more common internals. Figure 3–40 shows two types of data: data points from calculations where the emphasis is on heat exchange, humidification, or dehumidification; and general liquid-to-gas operating conditions for gas scrubbers where the prime function is to remove particulates and absorb soluble fumes. Thus, a baffle configuration is shown to operate at large liquid-to-gas ratios for heavy condensation loads but very low ratios for scrubber application. From Table 3–18, trays tend to provide the best heat and mass transfer but they also require more pressure drop in the general regions of applicability. More details about pressure drop are given in Figure 2–17. Also shown is the general range of liquid-to-gas mass flowrates. Thus, poorer heat transfer leads to larger equipment but less pressure drop. We start by choosing an internal based on the general guidelines given in Table 3–18. This table provides the criteria for choosing the cross-sectional area. Usually we select the diameter based on the gas flowrate of 1 m³/m².s. From this we can estimate the cross-sectional area for the device and thus the superficial mass flowrates of the liquid and the gas.

4. from the known gas and liquid flowrates we can then estimate the overall volumetric heat transfer coefficient and the pressure drop from the correlations given in Table 3–19.

5. estimate the volume of the device needed knowing the total heat load, the LMTD and the overall volumetric heat transfer coefficient. Volume = q/LMTD × U_v. From the cross-sectional area we can estimate the height.

6. check that this is reasonable.

Example 3-22: Hot flue gas leaves a waste heat boiler at 149°C and a dew point of 37.7°C. We want to cool it to 32°C. The flowrate is 2.52 kg/s. Cooling water is available at 26.8°C but it should not be heated hotter than 49°C. Assume that the flue gas behaves like air. We can accept a reasonable pressure drop.

An Answer:

1. Find out what is going on. Figure 3–41 illustrates the gas conditions. The gas starts at the inlet conditions of 149°C dry bulb, 37.7°C dew point (with a corresponding humidity of about 42 g/kg d.a. and a corrected enthalpy of 276 less about 10 kJ/kg or 266 kJ/kg d.a. and an adiabatic saturation temperature of about 50°C). It moves up the adiabatic saturation line and may go up to saturation before some condenses or it may shortcut. After the gas reaches about 50°C it condenses and follows the saturation line down to the exit temperature of 32°C. This corresponds with an exit humidity of about 31 g/kg d.a. and an enthalpy of about 116 kJ/kg d.a. Thus, the overall process is one of *dehumidification* and the appropriate equations and viewpoint shown in Table 3–14 should be taken.

2. Estimate the total heat load, the latent heat load and the amount of liquid and the temperature conditions. The total heat load is the difference in enthalpies or 2.52 kg/s (266 − 116 kJ/kg) = 378 kW. The latent heat load is the amount of water that has to be condensed or (42 − 31) g/kg × 2.52 kg/s × 2000 kJ/kg = 55 kW. Thus, the sensible heat load is 378 − 55/378 = 85%. From Table 3–14, the sensible heat load for the liquid must equal the total heat load for the gas = 378 kW. The allowable temperature increase for the water is (49 − 26.8)°C and with a liquid heat capacity of 4.2 kJ/kg.K the water requirement is 378 kW/(4.2 × 22.2) or 4.05 kg/s of water. The temperature profiles are:

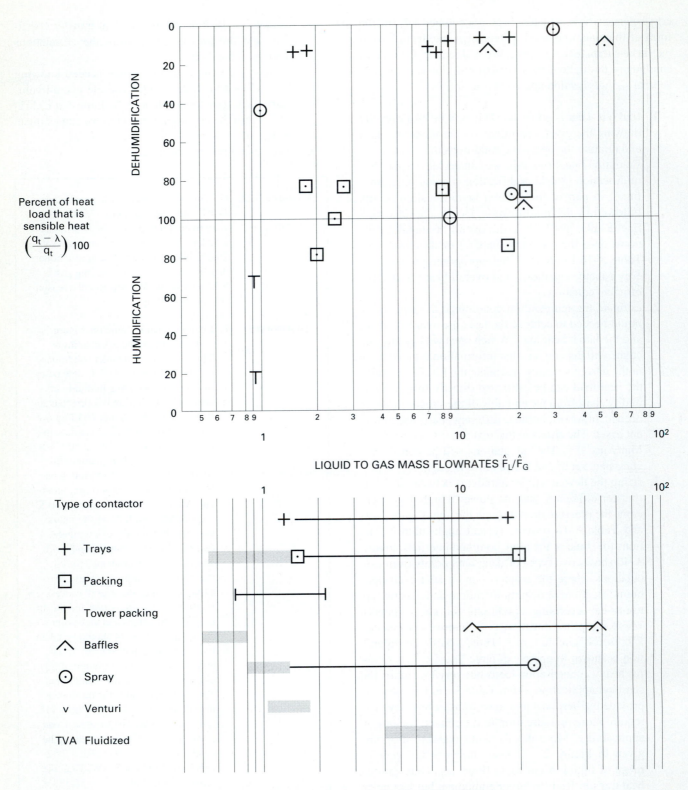

Figure 3–40   General Applications for Gas-Liquid Direct-Contact
Heat Exchangers (Data points show heat exchanger/quench applications;
shading shows gas scrubber applications.)

TABLE 3-18. Typical Application Conditions for Different Internals for Direct Gas-Liquid Heat Exchange

	Volumetric heat transfer coefficient kW/m³°C	Pressure drop, Δp kPa	Select cross-sectional area based on . . .	Usual loadings for $\hat{F}_L/\hat{F}_G$	
				Heat exchange	Scrub
Plates or Trays	3 or 7 to 20 kW/m² tray *area* °C	0.9 to 1.7 kPa per tray	Gas 1 to 1.6 m³/m².s	10 to 20	
Packing	3	0.35 kPa/m	Gas 1 m³/m².s	1.5 to 3	0.7 to 1.5
Cooling tower packing			Gas 1 m³/m².s Liquid 1 L/m².s	1	NA
Baffles	0.5		Gas 0.7 to 1m³/m².s	15 to 60	0.3 to 0.7
Sprays	1.8 to 5		Gas 1 m³/m².s	1 to 50	0.9 to 1.2
Peabody Impingement jet					0.2 to 0.7
TVA fluidized					4 to 8
Venturi					1.3 to 1.6

TABLE 3-19. Heat Transfer and Pressure Drop Correlations

	Heat transfer, kW/m³.°C	Pressure drop
Plates or trays	$U = h_{GV} = 3(\hat{F}_G)^{1.0}$ $h_G = 7{-}20 \, kW/m^2.tray \, area. \, °C$	see Figure 2-17
Packing	$h_G = 6(\hat{F}_G)^{1.0}(\hat{F}_L)^{0.2}$ $h_L = 42(\hat{F}_G)^{0.7}(\hat{F}_L)^{0.5}$	see Figure 2-17
Baffles	$U_a = h_G = (0.5{-}3.0)(\hat{F}_G)^{0.7}(\hat{F}_L)^{0.4}$	$= (1.2{-}2) <v>^2/2g$
Spray	$U = 1.2 \, (\hat{F}_G)^{0.8}(\hat{F}_L)^{0.4}/h^{0.5}$ $= 1.8{-}5.6$	negligible

Symbols:

h	= height of tower, m
$\hat{F}_G$	= gas mass flowrate, kg/m². s
$\hat{F}_L$	= liquid mass flowrate, kg/m². ss

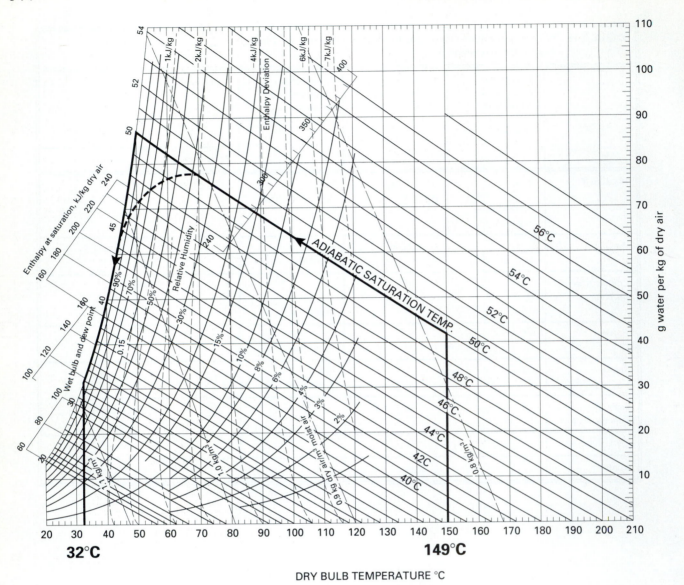

Figure 3–41   Estimating the Quenching Path for Flue Gas Example

hot cools from $149 \rightarrow 32$
cold heats from $\dfrac{49}{100} \leftarrow \dfrac{26.8}{5.2}$
LMTD from Figure 3–21 is 32°C.

3. Select a configuration and estimate the cross-section area. For this system the liquid-to-gas flowrate ratio is 4.05 kg/s/2.52 or 1.61. From Table 3–18, for this flow ratio and accepting a reasonable pressure drop, we can select a packed tower. To rough-size these we use a gas velocity of about 1 m/s. From Figure 3–41 the gas density varies from about 0.8 kg/m^3 at the inlet to about 1.1 at the outlet. Assume an average of about

1kg/m^3 and hence, the cross-sectional area needed is 2.52 kg/s/[1 kg/m^3 × 1 m/s] or 2.52 m^2. The corresponding diameter is 1.79 m.

4. From this choice and this cross-sectional area, the gas and liquid mass loadings on the column are:

gas: 2.52 kg/s/2.52 m^2 = 1 kg/s. m^2
liquid: 4.05 kg/s/2.52 m^2 = 1.61 kg/s. m^2.

From Table 3–19, the overall volumetric heat transfer coefficient is made up of a gas and a liquid phase coefficient. The gas phase coefficient is

$h_G = 6(1 \text{ kg/s. m}^2) (1.61 \text{ kg/s.m}^2)^{0.2}$
$\quad = 6.6 \text{ kW/m}^3.°C$

The liquid phase coefficient is:

$$h_L = 42 \, (1 \text{kg/s. m}^2)^{0.7} \, (1.61 \text{ kg/s.m}^2)^{0.5}$$
$$= 53 \text{ kW/m}^3. \, °C$$

From Table 3–14, since the heat and mass transfer are going in the same direction, from Figure 3–34 α will be greater than 1. Hence, if we leave it unity our sizing will be conservative. The overall heat transfer coefficient, from Table 3–14 is:

$$\frac{1}{U_v} = \frac{1 \times 0.86}{1 \times 6.6} + \frac{1}{53}$$
$$U_v = 6.7 \text{ kW/m}^3. \, °C$$

5. Estimate the volume, V, and then the height. The total heat load is:

$$q_T = U_v \times V \times \text{LMTD}$$
Hence, V = 378 kW/6.7 kW/m^3. °C × 32°C
$$= 1.76 \text{ m}^3$$

For a cross-sectional area of 2.52 m^2 the height will be 0.7 m.
The pressure drop would be, from Table 3–18, about 0.7 m × 0.35 kPa or 0.2 kPa.

6. Most of the results sound reasonable although the height to diameter ratio diameter is smaller than I would expect at 0.7/1.79. I would expect a ratio of 1 or greater.

Comments: Sawistowski and Smith (1963) worked out this example p. 456. Their humidity and enthalpy data were calculated so that they obtained a total heat load of 405 kW (instead of the 378 kW obtained here). The calculated volumetric heat transfer coefficient was 2.4 kW/m^3. °C with a resulting column configuration of 1.83 m diameter and a height of 1.98 m. The agreement is reasonable.

Example 3–23: Hot pyrolysis from the reactor in the ethylene plant given in Figure 1–1 comes out of an initial steam generation quench at 115°C. The gas flowrate we estimate to be 55 kg/s of gas. In the first section of a direct quench we want to keep the pressure drop to a minimum and reduce the temperature to 75.4°C. We plan to use 1722 kg/s of water coming in at 60°C. Picciotti (1977) suggests that the heat load for this duty is 110 MW with about 4% of it being sensible heat cooling of the gas. The remainder of the heat load is condensation to dehumidify the gas. Rough-size the unit.

An Answer:

1. Figure out what is going on. This information is given; the process is dehumidification with 4% sensible heat load.
2. Estimate the heat load (given as 110 MW), the latent heat load (given 96%), the water flowrate and the LMTD. From Table 3–14, the liquid must absorb all the heat. Hence, the liquid exit temperature can be calculated from the heat balance:

$$q_T = \hat{F} \times \hat{c}_p \times (T - T_{in})$$
$$110000 \text{ kW} = 1722 \text{ kg/s} \times 4.2 \text{ kJ/kg. °C} \times (T - 60°C)$$
$$T_{out} = 75.2°C$$

Now we can estimate the LMTD

hot gas cools from 115 → 75.4
cool liquid warms 75.2 ← 60
$$\frac{39.8 \quad 15.4}$$
LMTD, from Figure 3–21, = 25.5°C

3. Select the configuration and estimate the cross-sectional area. The liquid-to-gas mass ratio is 1722 kg/s/ 55 or 31. From Table 3–18, baffles, spray, or perhaps plates are the options. However, to keep the pressure drop to a minimum, a plate is not a reasonable choice. The gas density we estimate to be about 0.88 kg/m^3 for ethylene at these temperatures. Hence, choosing baffles and a gas superficial velocity of 1 m/s, the cross-sectional area is:

$$\text{Area} = 55 \text{ kg/s/}(0.88 \text{ kg/m}^3 \times 1 \text{ m/s})$$
$$= 62 \text{ m}^2 \text{ or } 8.9 \text{ m diameter.}$$

4. Estimate the volumetric heat transfer coefficient. The mass loadings for the cross-section chosen in step 3 are:

gas 55 kg/s/62 m^2 = 0.89 kg/s. m^2
liquid 1722 kg/s/62 m^2 = 27 kg/s. m^2.

From Table 3–19, the overall heat transfer coefficient for sensible heating is

$$h_G = 0.5 \, (0.89)^{0.7}(27)^{0.4}$$
$$= 1.72 \text{ kW/m}^3. \, °C$$

From Table 3–14, for gas dehumidification with only the gas phase coefficient being important:

$$U_v = \alpha \, h_G/\text{fraction sensible heat load}$$

For this case, α will be greater than 1; assume 1 to be conservative.

Hence, $U_v = 1 \times 1.72/0.04$
$\qquad\quad = 43 \text{ kW/m}^3.°C$

5. Estimate the volume required from:

$$q_T = U_v \times V \times LMTD$$
$$V = 110\,000/43 \text{ kW/m}^3.°C \times 25.5°C$$
$$= 100 \text{ m}^3 \text{ or a height of } 100/62 \text{ or}$$
$$1.62 \text{ m}$$

6. Check. The pressure drop through baffles should be a minimum. We might want to use a reduced gas superficial velocity of 0.7 m/s. This would require larger height in the baffles section.

Comments: Sauter and Younts (1986) and Picciotti (1977) describe the bottom, baffle section of the quench cooler systems on ethylene plants. Unfortunately, few actual dimensions are given. From the series of articles by Picciotti and from the photographs given in the articles, for the conditions described in this situation I estimate the diameter to be 11 m, the full height of the section to be 5 m with 3 m of active section. Thus, the overall superficial gas velocity is 0.66 m/s and the estimated overall heat transfer coefficient is about 15 kW/m³.°C (compared with the estimate of 43 kW/m³.°C). The procedures outlined here give reasonable order-of-magnitude values.

3. Function: Complete Condensation via Direct Contact Condensers.

Steam is often condensed by direct contact condensers. This is particularly applicable for interstage condensation for steam ejectors. The arrangement was illustrated in Figure 2–23. The device is illustrated in Figure 3–31 and is usually a vertical cylinder containing chimneys, and a distributor plate into which water spray gushes. These can also be used for condensing vapors other than water and when the gas phase contains some non-condensables. The general procedure for sizing selecting these condensers is as follows.

1. find out what is going on. Identify the pressure, composition and temperature constraints. The condensa-

tion temperature depends on the gas phase composition and the absolute pressure. Figure 3–42 shows the vapor pressure-temperature relationship for water. Other vapor pressure data are available for pure liquids in **Data** Part C. When the gas phase contains non-condensables, the relationship between the composition, the total pressure, and the temperature can be estimated from a combination of Dalton's and Raoult's laws as given in Equations 3–19 and 3–20.

Figure 3–42 shows the "dew point temperature" for the air water system for different pressures and ratios of water to air. This type of information is vital to understanding what is going on. If, for example, we were planning to condense pure water at 3 kPa and we had contacting water available at 30°C, we can see from Figure 3–42 that no condensation occurs because the condensation temperature is 25°C. Thus, the first step is to understand what is going on, and in particular, to identify the dew point temperature. (The data in Figure 3–42 can also be used to predict the amount of water that saturates the non-condensable gas stream leaving a contact condenser.)

2. the second step is to sort out the temperature-profiles around the device. There are three constraints to be satisfied:

a) the cooling water must be colder than the dew point or condensation temperature. This dew point has been discussed in section 1.

b) the *amount* colder than the dew point depends on the heat transfer temperature driving force or the "approach temperature." For heat transfer to occur we expect that a reasonable temperature driving force exists between the hot gas stream and the cold liquid stream. In general, for condensation of pure steam (containing no non-condensables) an approach temperature difference at both ends of the device of 2.75°C is expected. (Ryans and Roper (1986) suggest that the approach temperature varies from 5.5°C to 1.7°C as the diameter of the condenser increases from 0.25 m to greater than 0.5 m. The value suggested here for quick estimates is midway between these recommendations.) Figure 3–43 shows this information for the cooling water temperature required for condensing pure steam and for condensing saturated water from air. For gas cooling, the temperature driving force should be 16 to 18°C.

c) only a limited temperature increase should occur for the cooling water. While the cooling water absorbs the heat of condensation, its temperature increases. Part of the limit occurs because of constraint "b" above. These include: Ryans and Roper (1986) recommend for operating pressures less than 6 kPa, the maximum temperature increase is 8.3°C; for 6 to 25 kPa, 17°C and above 25 kPa, 25°. Furthermore, the maximum outlet

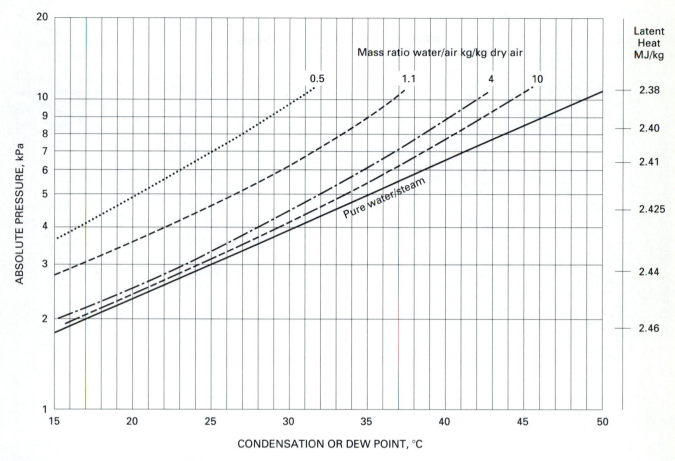

Figure 3–42 Vapor Pressure-Temperature Data for Air-Water System: Showing the Dew Point or Condensation Temperature for Different Ratios of Air/Water and the Latent Heat of Condensation of Water As a Function of Pressure

cooling water temperature should not exceed 50°C to minimize corrosion in the water lines.

Figures 3–43 and 3–44 include most of these constraints. Thus, from a series of trial and error considerations, all temperature constraints will be satisfied: dew point, temperature driving difference between the hot and cold streams, and the temperature increase in the cold stream.

3. From a heat balance, from Table 3–14 where we assume the sensible gas cooling heat load is negligible, estimate the amount of cooling water needed. For steam condensed by water, the results are given in Figure 3–44. The "usual" value is 30 kg water per kg of steam.

4. Estimate the diameter of the condenser. The cross-sectional area of the condenser is selected based on liquid loading. (The liquid loading is the sum of the coolant water plus the amount condensed. However, for approximation purposes use only the coolant water loading.) The maximum loading values vary from 42 to 85 L/m².s depending on the diameter. Figure 3–45 gives the maximum and usual relationships between liquid flowrate, cross-sectional area, and diameter.

5. Estimate the height of the condenser. Ryans and Roper (1986) suggest the following height to diameter ratios:

	for spray	for cascade
height in condensing zone	3.5	3.0
total overall height	5.5	4.5

As an alternative to the last step we can estimate the overall volumetric heat transfer coefficient and, from the heat load and the LMTD, estimate the volume required. Data in Table 3–19 can be used.

Example 3–24: Steam is to be condensed at a rate of 1.07 kg/s. The steam contains 0.01 kg/s of air and is at about 11 kPa pressure. Cooling water is available at 29.3°C. Rough-size the direct contact condenser.

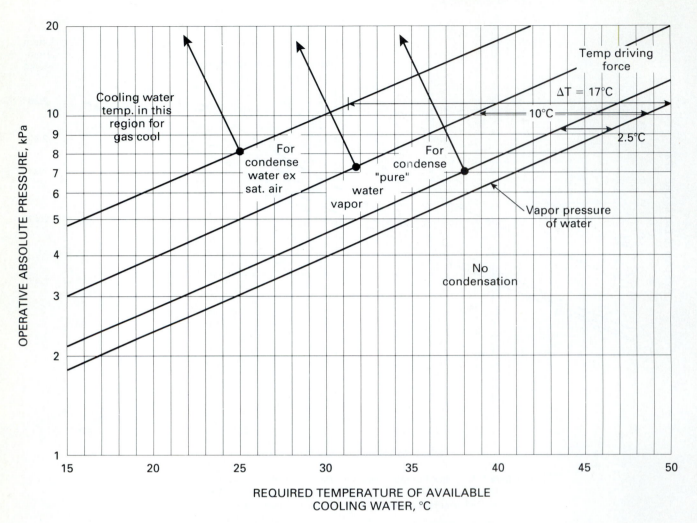

Figure 3–43 Operating Pressure-Cooling Water Temperature for Direct Contact "Barometric Condensers" Showing Temperature Driving Force Requirements

An Answer:

1. Find out what is going on. Here, from Figure 3–42 the gas has a negligible amount of air (the water to air ratio is 107:1). At 11 kPa pressure, the condensing temperature is about 49°C. Hence, since the water is available at 29.3°C we have plenty of potential for making this system work.

2. Estimate the temperatures.
   a) The temperature of the cooling water is colder than the dew point.
   b) For the approach temperatures, we start by drawing a diagram, as in Figure 3–46 and show on it the approach temperatures of 2.75°C at both ends.
   c) The cooling water temperature increase is 46.25 − 29.3 or 16.95°C. From Figure 3–44 for a pressure above 6 kPa, the temperature

   crease can be up to 20°C. Hence, this is acceptable. Thus, the temperatures shown are reasonable.

3. Estimate the water flowrate. From Figure 3–44 with an inlet temperature of 29.3 and an exit temperature of 46.25 the water flowrate is about 32 kg water/kg steam. For a steam flowrate of 1.07 kg/s the water flowrate becomes 34 kg/s or 34 L/s.

4. Estimate the cross-sectional area. From Figure 3–45, the cross-sectional area of about 0.6 m² or a diameter of about 0.85 m.

5. Estimate the height. Assume we install a cascade condenser, then the contacting height is 3 × 0.85 = 2.55 m and the overall height is 4.5 × 0.85 = 3.83 m.

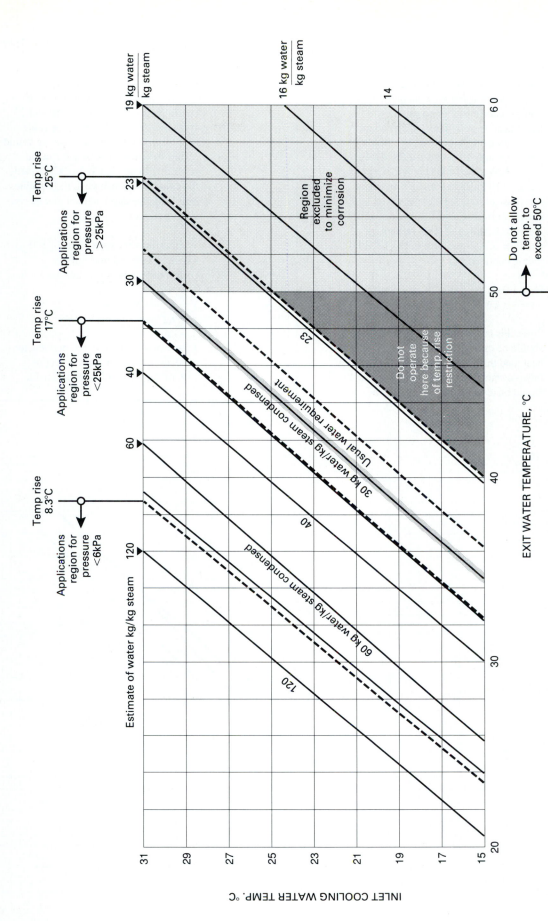

Figure 3-44 Water Requirements for Condensing Steam As a Function of Inlet and Exit Cooling Water Temperatures with Constraints for the Operating Pressure

Comments:

Ludwig (1964) works this example:

As a check using the volumetric heat transfer coefficient approach, the latent heat, from Figure 3–42, is about 2.38 MJ/kg and hence the heat load is approximately 1.07 kg/s × 2.38 MJ/kg = 2.55 MW. The LMTD is 2.75°C. From Table 3–19 the heat transfer coefficient for a cascade condenser is based on a baffle configuration. The mass velocities are $F_G = 1.07/0.6$ m^2 or 1.78 kg/s. m^2 and $F_L = 32/0.6$ or 53 kg/s.m^2. The volumetric coefficient is:

$$U_v = 1.0 \, (1.78)^{0.7} \, (53)^{0.4}$$
$$= 7.3 \text{ kW/m}^3.°C$$

Since this is condensation, this value should be divided by the fraction of the heat load that is sensible heat load. If this is complete condensation, the load would be 0% and the volumetric co-

efficient would become infinity. In practice, the volumetric heat transfer coefficients range between 250 and 1000 kW/m^3.°C. Thus if the sensible heat load is between 1 and 7% we would obtain volumetric coefficients in the order of 100 to 730 kW/m^3.°C. Because the value is so sensitive to the sensible heat load, this approach will not be considered further.

To illustrate this, for a value of 100 kW/m^3.°C the volume and the corresponding height would be:

$$V = 2550/100 \times 2.75$$
$$= 9.27 \text{ m}^3$$

This would yield, for a 0.8 m diameter condenser, a height requirement of 11.6 m. This answer is unreasonable.

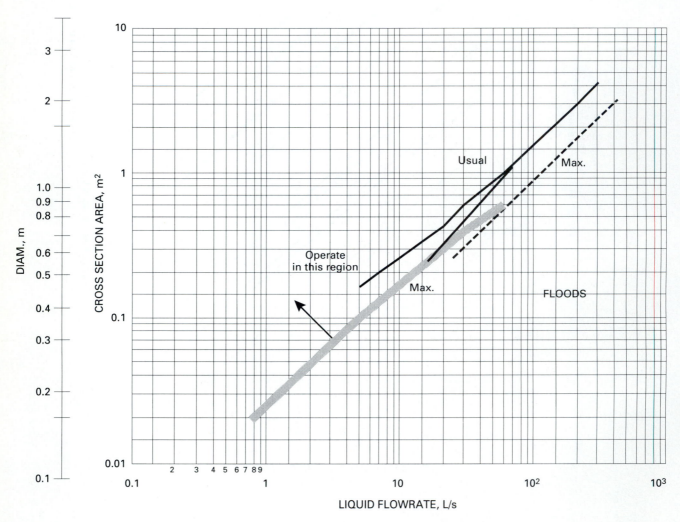

Figure 3–45 Estimating the Cross-Sectional Area for Barometric Condensers

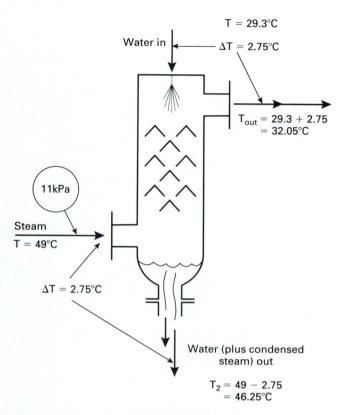

Figure 3–46 Temperature Conditions in a Barometric Condenser: Example 3–24

4. Function: Cooling of Liquids, Cooling Towers. Often the focus is on heat transfer to and from the gas. One important application of direct contact heat exchange is where the focus is on cooling down the liquid. For a cool-

ing tower, hot liquid contacts a cool gas that is unsaturated in the liquid vapor. Hence, some liquid evaporates to saturate the gas. The incoming gas can indeed start off saturated. However, the gas heats up and becomes unsaturated such that the evaporation occurs. Figure 3–47 illustrates what happens to the gas. On the psychrometric chart, the gas enters at a given wet bulb temperature and gradually gets heated up and increases in humidity. The liquid, on the other hand, enters the top and cascades down over slats or packing inside the tower. The incoming temperature, the anticipated decrease in liquid temperature (called the "cooling range") and the approach temperature between the incoming gas and the exit liquid are all key parameters. Some of the equipment options are given in Figure 3–48. Four different variations on equipment design are:

- fully packed, counterflow induced draft; here the inside of the vessel is filled with slats or packing over which the liquid cascades and the fan to draw the gas through the packing is at the top exit of the tower;
- fully packed, cross + counterflow cross-draft forced circulation; now a fan is on the inlet side blowing the gas into the tower.
- fully packed, counterflow natural circulation; here the vessel looks like a chimney extending about 30 m into the air. Thus, the gas circulation occurs because of density differences between the hot exit gas and the surrounding environmental air. The packing fills a thin layer at the base of the chimney;
- periphery-packed, cross-draft induced circulation; here the central part of the tower is open, the sides are

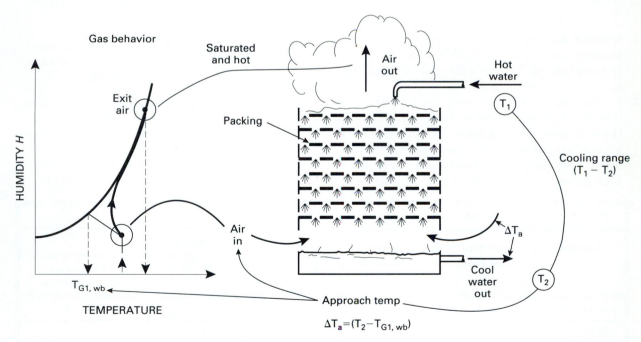

Figure 3–47 Estimating the Gas Behavior in a Cooling Tower

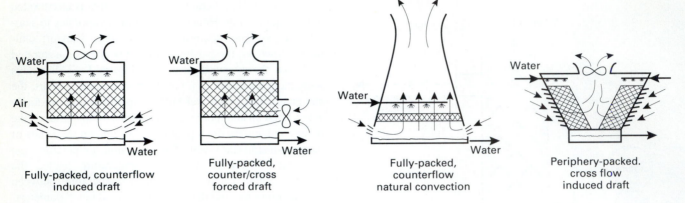

Figure 3–48 Sketches of Cooling Towers

open, and the gas flows across the packing in toward the center. For induced draft the fan is at the top or gas exit;

The key occurrence is the evaporation of the liquid caused by a difference in moisture content in the vapor. Although a humidity difference represents the driving force for the mass transfer to occur, for detailed calculations we replace this with a gas enthalpy driving force. For details see Sawistowski and Smith (1963). Because the enthalpy of a gas is constant along a line of constant wet bulb, the key design inlet gas condition is the wet bulb temperature of the air that is not exceeded for more than 5% of the time during the "summer months of June to Sept. inclusive." Some typical conditions are given in Table 3–20.

Consider now two alternatives for rough-sizing a cooling tower. The first approach draws solely on experience factors.

1. Estimate the required tower cross-sectional area based on liquid loading; 1.7 L/m².s for a fully packed cross-section and 2.4 L/m².s for cross-flow configuration with an open central core.
2. Estimate the air flowrate based on a liquid mass flowrate to gas mass flowrate ratio of 1.0. Ludwig (1964) suggests that, in general, this ratio varies from 0.9 to 2.7 but 1 is reasonable approximation. Alternatively, the volumetric gas flowrate should be about 1800 dm³/m² of cross-sectional area. (The power to

run the fans can be estimated from the guideline that 1 kW supplies 500 dm³/s of damp air.)
3. Check on the target exit liquid temperature to ensure that it is consistent with the wet bulb temperature for the proposed location. Table 3–20 provides example data. The exit liquid temperature is higher than the inlet gas wet bulb temperature by the "approach temperature." Thus, from a specified wet bulb temperature and approach temperature check that the cooling range is reasonable. If not adjust these.
4. From the cooling range and the approach temperature, use the data in Table 3–21 to estimate the height of packing in the tower.

Example 3–25: A cooling tower is to be selected to cool 312 kg/s of water from 43°C to 29.3°C. If the wet bulb temperature of the air is 23.4°C, estimate the size of a fully-packed tower.

An Answer:
1. Based on the liquid loading of 1.7 L/m²s, the tower cross-sectional area is about 312 kg/s/[1 kg/L × 1.7 L/m².s] = 183 m².
2. Check the gas flow; 183 m² × 1800 dm³/m².s = 330000 dm³/s. This is about 330 m³ × 1.2 kg/m³ = 396 kg/s of air. The liquid-to-gas ratio is 312/396 = 0.79. Hence, probably need less air, i.e., closer to 312

TABLE 3-20. Typical Conditions for Cooling Towers					
Location	**Wet Bulb Temperature, °C**	**Approach Temperature, °C**	**Cooling Range, °C**		
			Oil Refineries Steel Mills	**PowerPlants**	**Refrigeration Air Conditioning**
Southern Ontario	23.8	5.5	14 – 36	5.5	2.75 – 5
Gulf Coast, U.S.A.	28.2	4.2		8.3	

kg/s/1.2 kg/m^3 = 260 m^3/s. The fan requirement would be about 260 000 dm^3/s/500 dm^3/s or 520 kW.

3. Check the temperatures. The cooling range is 43°C − 29.3°C or 13.7°C. The approach temperature is 29.3 − 23.4 = 5.9°C. Hence, from Table 3–21 the fill height should be about 8 m.

Comment: Ludwig (1964) p. 207 works out this example. His configuration was a cross-sectional area of 130 m^2 and a fill height of 6.4 m.

A second approach is to use the volumetric heat transfer coefficients given in Table 3–19 to estimate the height of fill required. Table 3–14 illustrates that for cooling tower operation, the heat and mass transfer are in the same direction, the total heat load is the sensible heat to cool down the liquid which equals the sensible heat to heat up the gas and to evaporate the liquid into the gas.

1. Ensure that we know what is going on. Usually we are given the amount of water, the inlet water temperature, and the desired exit water temperature. We need to specify the wet bulb conditions pertinent to our geographical location and check that the goals and target cooling range are reasonable.

2. Since only about 4% of the liquid evaporates, we can estimate the heat duty by multiplying the liquid flowrate by the liquid heat capacity and the cooling range.

3. Estimate the temperature conditions and hence the LMTD. Evans (1974) reports that the exit air temperature is the arithmetic average of the water inlet and outlet temperatures. From this we can estimate the LMTD.

4. Estimate the percentage of the heat load that is sensible heat. The latent heat required is the gas flowrate times the uptake in humidity times the latent heat of vaporization (2256 kJ/kg). The total heat exchanged is the gas flowrate times the change in enthalpy. The percentage is the ratio and is independent of the flowrate. Thus, from the gas inlet and our estimation

based on Evans's correlation for the gas outlet conditions, we can obtain this humidity to enthalpy ratio from a humidity chart. The percentage of sensible heat is 100 less the latent heat percentage.

5. Estimate the cross-sectional area for the device based on the liquid flowrate of 1.7 L/s.m^2 for a fully packed cross-section or 2.4 L/s.m^2 for cross-flow configuration (or 1.7 and 2.4 kg/s.m^2 respectively).

6. Estimate the gas flowrate required from a variety of checks and cross-checks. These include:
   - the ratio of the liquid to gas mass flowrate should be about 0.9 to 2.7 with the usual values being about 1.
   - the gas mass flowrate is the total heat load divided by the gas enthalpy change.
   - the gas flowrate should be about 1.7 kg/s. m^2.

   From these calculations, estimate the F_G gas mass flowrate and the F_L liquid flowrate.

7. Estimate the Overall Volumetric heat transfer coefficient, the total volume required and hence the height of the fill.

8. Check that this is reasonable.

Example 3–26: Estimate the cooling tower requirement for Example 3–25 from the volumetric heat transfer coefficient.

An Answer:

1. Ensure that we know what is going on. The amount of water is 312 kg/s, the inlet water temperature is 43°C and the desired exit water temperature is 29.3°C. The wet bulb temperature is 23.4°C. The cooling range is 43 − 29.3 = 13.7°C. The approach temperature is 29.3 − 23.4 = 5.9°C. These values are OK for process operations.

2. The estimated heat duty = 312 × 4.2 kJ/kg.°C × 13.7 = 17. 95 MW.

3. The exit gas temperature is (43 + 29.3)/2 = 36.15°C.

   Hence the LMTD is
   hot water cools 43 → 29.3
   cool gas heats 36.15 ← 23.4
   $$\frac{6.85}{} \quad \frac{5.9}{}$$
   LMTD 6.38°C.

TABLE 3-21. Height of Cooling Tower Fill For the cooling range 14 to 20°C

Approach temperature, °C	Height of Fill, m
8.3 to 11	4.6 to 6
4.5 to 8.3	7.6 to 9
2.2 to 4.5	10.7 to 12

4. Estimate the percentage of the heat load that is latent heat from the gas entrance and exit humidities and enthalpies. From **Data** Part C,

	Humidity kg/kg d.a.	Enthalpy, kJ/kg d.a.
exit, 36.15 wb	0.039	138
inlet, 23.4 wb	0.0183	70
	0.021	68

Latent percentage = 0.021×2256 kJ/kg $\times 100/68 = 70\%$.

Thus, the sensible load is 30%.

5. Estimate the cross-sectional area for the device based on the liquid flowrate of 1.7 L/s.m². This was done in the previous example: 312 kg/s/1.7 = 183 m² for a fully packed cross-section.

6. Estimate the gas flowrate required.
   - the gas mass flowrate is the total heat load divided by the gas enthalpy change: 17,950 kW/68 kJ/kg d.a. = 264 kg/s.
   - check that the ratio of the liquid-to-gas mass flowrate should be about 0.7 to 2.7 with the usual values being about 1; hence the ratio is 264/312 or 0.85. This is reasonable.
   - the gas flowrate should be about 1.7 kg/s.m². Based on 264 kg/s and an area of 183 m² this is 1.44 kg/s.m².

   Thus, the F_G gas mass flowrate is about 1.44 kg/s.m²; the F_L liquid flowrate is 1.7 kg/s.m².

7. Estimate the Overall Volumetric heat transfer coefficient. From Table 3–19, the overall coefficient for packing is about:

$$h_G = 0.5 \, (1.44)^{0.7} \, (1.7)^{0.5}$$
$$= 0.84 \text{ kW/m}^3.^{\circ}\text{C}$$

Since the sensible heat load is 30%, the overall coefficient becomes 0.84/0.3 = 2.8 kW/m³.°C

Hence, the total volume required is $q_T/U \times$ LMTD

$V = 17950/2.8 \times 6.38 = 1005$ m³.

Since the cross-sectional area is 183 m², the height is 5.49 m of fill.

8. Check that this is reasonable. The results are reasonable.

Comment: This approach ties the fundamentals of heat and mass transfer in to the operations. However, because the sensible heat load is so small, the coefficient one obtains is very sensitive to that value. Furthermore, the correlations for cooling tower packing show great variability. In general, the overall volumetric coefficients for cooling towers tend to be in the range of 3 to 8 kW/m³.°C. For more Cheremisinoff (1986), Ludwig (1964), Evans (1974), and Kern (1950).

Indirect Contact Thermal Wheel. A thermal wheel, heat wheel, or Ljungstrom heater is an indirect contact device. A large hollow wheel that rotates with about a horizontal axis is filled with corrugated metal/solid strips that will serve as a heat sink. About 500 to 650 m² of heat sink area per m³ of wheel is provided. The wheel is positioned so that hot gas flows through half of the wheel in the axial direction; the corrugated strips are heated up and the gas cools down. The other half of the wheel has cold gas flowing through it; here the strips give up their heat and this gas heats up. The wheel rotates so that the net effect is that heat is transferred from the hot gas to the cold gas by means of the metal strips in the rotating wheel. Usually we try to have equal mass gas flowrates to both sides of the wheel; the face velocity is usually in the range 1 to 5 m³/s.m² or a design value of about 3.5 m/s. We obtain better heat transfer coefficiencies if the gases flow countercurrently through the wheel. Figure 3–49 shows the relationship between approximate wheel diameter and gas flowrate.

In general the heat effectiveness, ε, is defined as

$$\varepsilon = \frac{F_1(H_1 - H_2)}{F_{min}(H_1 - H_3)} \qquad (3\text{-}29)$$

where $F_1 =$ mass flowrate of supply entering as shown in Figure 3–50
$F_3 =$ mass flowrate of exhaust entering the unit
$H =$ dry bulb temperature or sensible heat, humidity ratio (or latent heat) and/or the total enthalpy, H, of the streams.

From this the exit conditions can be determined

$$H_2 = H_1 - \varepsilon \frac{F_{min}}{F_1} (H_1 - H_3) \qquad (3\text{-}30)$$

$$H_4 = H_3 - \varepsilon \frac{F_{min}}{F_3} (H_3 - H_1). \qquad (3\text{-}31)$$

In general, $\varepsilon \simeq 80\%$ on sensible heat and 60 to 80% on total heat for countercurrent operation. Table 3–22 summarizes different media that can be used in a thermal wheel.

Example 3–27: Estimate the size of a thermal wheel if the supply air flowrate is 4250 dm³/s.

An Answer: For the usual design face velocity of 3.5 m³/s·m² the required area would be:

$$4250 \frac{dm^3}{s} \times \frac{10^{-3}m^3}{dm^3} \times \frac{1 \, s \cdot m^2}{3.5 \, m^3} = 1.21 \, m^2$$

This amount of face area would be needed for the supply side of the wheel. If the exhaust side of the wheel had the same flowrate, i.e., 4250 dm³/s, then the total wheel face area needed would be 2.42 m². (This is *not* the area of the heat sink strips.) The diameter of this wheel is 1.75 m.

Comment: In practice the wheel installed was 2.1 m diameter.

Example 3–28: If the supply air from outside is 32°C dry bulb, 24°C wet bulb and the discharge air is 23.8°C and 50% relative humidity, what are the rest of the conditions if the wheel is to handle 4250 dm³/s of both supply and exhaust air? The system is shown in Figure 3–50.

An Answer: The first task is to estimate the value of the heat exchange effectiveness and to assess whether the transfer of heat involves sensible heat only or sensible and latent heat. Since the operating temperatures are low, a desiccant impregnated wheel can be used. This will have a

capital cost that is about 20% greater than an aluminum or metal wheel but one might expect that the additional heat transfer would counterbalance this cost. Hence, try a hygroscopic media.

Next, charts are available for different wheels that provide performance values for ε. Thus, normally we would select the wheel next (following the procedure used in Example 3–27) and then look up the effectiveness. As an estimate we assume ε = 80%.

$$T_{db2} = T_{db1} - \varepsilon \frac{1}{1} (T_{db1} - T_{db3})$$
$$= 32°C - 0.80 (32°C - 23.8)$$
$$= 25.4°C$$

Hence, the air entering the building will have been cooled down to 25.4°C.

Because this is a hygroscopic wheel, the humidity will also be changed. The incoming supply humidity is 16 g/kg dry air; the incoming exhaust humidity is 9.2 g/kg.

Hence $\Delta H = 16 - 9.2 = 6.8$ g/kg.
$$H_2 = 16 - 0.8(6.8)$$
$$= 10.6 \, g/kg.$$

The other equation can be used to determine the conditions in stream 4.

For details see Wing Company Bulletins WE-86, WC-86, and WCM-85, and ASHRAE handbook, Chapter 34.

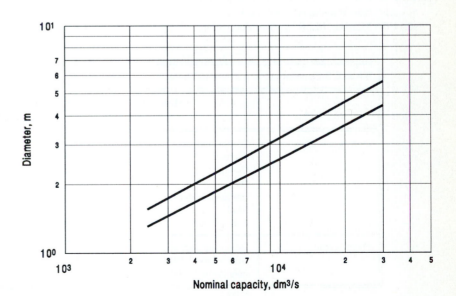

Figure 3–49 Approximate Diameter of Thermal Wheels (Typical Widths of 0.3 to 0.74 m)

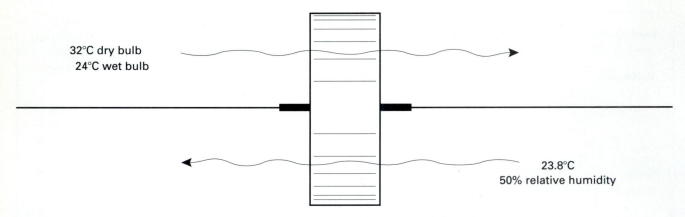

32°C dry bulb
24°C wet bulb

23.8°C
50% relative humidity

Figure 3–50   An Example of a Thermal Wheel Application

3.3 SUMMARY

Energy can be exchanged very efficiently mechanically and electrically. A general chart was given for the selection of the type of mechanical or electrical to use.

Thermal energy can be exchanged via indirect or direct contact and can be at very high, usual, or very low temperatures. For no direct contact and high temperatures, the furnace configuration is chosen, based on the amount of temperature variation that can be allowed in the stream. Then the tube area in the radiation and convection sections is estimated.

For usual temperatures, the general selection and sizing procedure is outlined. Basically, an estimate is made of the overall heat transfer coefficient based on general rules of thumb. For direct contact devices, sizing procedures are outlined for kilns, multiple hearth furnaces, liquid-liquid and gas-liquid contactors, and thermal wheels.

3.4 REFERENCES

ALBRIGHT, L. F. 1974. *Processes for Major Addition-Type Plastics and their Monomers.* New York: McGraw Hill, and 2nd ed., 1985 Malabar, FL: R. Kreiger Publishing.

ALFA-LAVAL, "Heat Exchanger Guide."

AHRAE. 1979. *ASRAE Handbook and Product Directory:* 1979 Equipment, Chapter 34, Am. Soc. Heating, Refrigerating and Air-conditioning Engineers, Inc., New York.

BERGOGNOU, M. A., C. L. BRIENS, and D. KUNII. 1986. "Design Aspects of Industrial Fluidized Bed Reactors: Research Needs and Selected Subjects" in *Chemical Reactor Design and Technology,* ed. H. I. deLasa. Dordrecht, the Netherlands: M. Nijhoff Publishers, 305–347.

BIRD, R. B., W. E. STEWART, and E. N. LIGHTFOOT. 1960. *Transport Phenomena.* New York: J. Wiley and Sons.

BOHN, M. S. 1985. "Air Molten Salt Direct Contact Heat Exchange." *J Solar Energy Engineering,* Trans ASME **107**:208–214.

BOLLES, W. L. 1968. *Case Study 6: Ethylene Plant Design and Economics,* ed. B. D. Smith, St. Louis, MO: Monsanto-Washington University.

BOTTERILL, J. S. M. 1975. *Fluid Bed Heat Transfer.* New York: Academic Press.

BROWN, T. R. 1986. "Use These Guidelines for Quick Preliminary Selection of Heat Exchanger Type." *Chem Eng* (Feb. 3): 107.

BUFFINGTON, M. 1975. "How to Select Package Boilers" *Chem Eng* **82,** 23 (Oct. 27): 98.

CHEREMISINOFF, N. P. 1986. "Cooling Tower Operations," in

TABLE 3-22. Media For a Thermal Wheel				
		Materials	**Transfer latent heat as well as sensible heat?**	**Other Comments**
Non-Hygroscopic		Aluminum foil Aluminum wire Ceramic Felt Inorganic sheet Steel Stainless steel	Probably Not	<150°C
Hygroscopic		Desiccant impregnated silicate baked asbestos impregnated with lithium chloride	Yes	

Handbook of Heat and Mass Transfer vol 1, *Heat Transfer Operations,* ed. N. Cheremisinoff. Houston, TX: Gulf Publishing Co.

DAMAN, E. L. "The Technology and Economics of Fluidized Bed Combustion" *Heat Engineering* (April-June 1979) Foster Wheeler Corporation, Livingston NJ, 27–33.

deCROOCQ, D. 1984. *Catalytic Cracking of Heavy Petroleum Fractions.* Paris: Editions Technip.

DE VRIES, R. J., et al. 1972. "Design Criteria and Performance of the Commercial Reactor for the Shell Chlorine Process" in Second International Symposium on Chemical Reaction Engineering" Amsterdam, May. Preprints published by Elsevier, Amsterdam, pp. B 9–59—9–69.

ELLWOOD, P., et al. 1966. "Process Furnaces." *Chem Eng* **73,** 8 (April 11): 151–174.

EPA. 1973. *Process Design Manual for Carbon Adsorption.* U.S. Environmental Protection Agency, EPA 625/1-71-002a.

EVANS, F. L. 1974. *Equipment Design Handbook for Refineries and Chemical Plants.* Houston, TX: Gulf Publishing Co.

FAIR, J. R. 1972. "Designing Direct-Contact Coolers/Condensers." *Chem Eng* (June 12): 91–100.

FAIR, J. R. 1972. "Process Heat Transfer by Direct Fluid-phase Contact." *Chem Eng Prog Symp Series* **68,** 118: 1–11.

FRANK, O. 1974. "Estimate Overall Heat Transfer Coefficients." *Chem Eng* **81,** 10 (May 13): 126.

FOXALL, D. H., and P. T. GILBERT. 1976. "Selecting Tubes for CPI Heat Exchangers." *Chem Eng* **83,** 6 (March 15): 99–104.

FRIED, J. R. 1973. "Heat Transfer Agents for High Temperature Systems." *Chem Eng* **80,** 12 (May 28): 89–98.

FROMENT, G. F., and K. B. BISCHOFF. 1979. *Chemical Reactor Analysis and Design.* New York: J. Wiley and Sons.

GUNDER, P. F. 1969. "How to Specify Process Heaters and Evaluate Bids." *Hydrocarbon Process* **48,** 10: 117–120.

HAUSBRAND, E. 1929. *Evaporating, Condensing and Cooling Apparatus.* (Tr. from German by A. C. Wright) 4th English ed., revised and enlarged by B. Heastie. New York: van Nostrand.

HOFFMAN, E. J. 1980. *Heat Transfer Rate Analysis.* Tulsa, OK: PenWell Books.

INST. OF CHEM. ENGINEERS. (UK) *User Guide to Process Integration for the Efficient Use of Energy*, p. 132, Rugby, UK.

JEFFREYS, G. V. 1961. *The Manufacture of Acetic Anhydride.* London: Institution of Chemical Engineers.

KABEL, R. L. 1985. "Selection of Reactor Types" in *Scaleup of Chemical Processes,"* ed. Attilio Bisio and R. L. Kabel. New York: J. Wiley and Sons.

KERN, D. Q. 1950. *Process Heat Transfer.* New York: McGraw-Hill.

KNEIL, L., et al. 1980. *Ethylene: Keystone to the Petrochemical Industry.* New York: Marcel Dekker Inc.

KUNII, D., and O. LEVENSPIEL. 1969. *Fluidization Engineering.* New York: J. Wiley and Sons.

LASKOWSKI, J. J., and W. E. RANZ. 1970. "Spray Quenching." *AIChE Journal:* 802–816.

LUDWIG, E. E. 1964. *Applied Process Design for Chemical and Petrochemical Plants: Vol 1.* Houston, TX: Gulf Publishing Co.

LUDWIG, E. E. 1965. *Applied Process Design for Chemical and Petrochemical Plants: Vol 3.* Houston, TX: Gulf Publishing Co.

MATSEN, J. M. 1985. "Fluidized Beds" in *Scaleup of Chemical Processes,* ed. Attilio Bisio and R. L. Kabel. New York: J. Wiley and Sons.

MOL, A., and J. J. WESTENBRINK. 1974. "Steam Cracker Quench Coolers." *Hydrocarbon Process* **53,** 2: 83–87.

MOSTOWY, T. 1988. Romatec, Montreal, personal communication.

MUTSAKIS, M. et al. 1986. "Advances in Static Mixing Technology." *Chem Eng Prog* **82,** 7: 42–48.

NELSON, W. L. 1949. *Petroleum Refinery Engineering,* 3rd ed. New York: McGraw-Hill.

PICCIOTTI, M. 1977. "Design Quench Water Towers." *Hydrocarbon Process* (June): 163–170.

——— 1978. "Dynamic Program Aids Quench Water Optimization" *Oil and Gas J* (Jan 16): 58–63.

——— 1977. "Optimize Quench Water Systems" *Hydrocarbon Process* (Sept.): 179–189.

——— "Cooling, Quench Water Systems" in *Encyclopedia of Chemical Engineering and Design,* ed. J. McKetta. New York: Marcel Dekker, 344–389.

POTTS, R. A. 1978. "Carboxylic Acids, (Manufacture)" in *Kirk-Othmer Encyclopedia of Chemical Technology,* 3rd ed. New York: J. Wiley and Sons, Vol 44, p. 835.

REH, L. 1979. "Fluid Bed Combustion in Processing, Environmental Protection and Energy Supply." International Fluidized Bed Combustion Symposium, Boston MA (April).

ROSE, L. M. 1981. *Chem Reactor Design in Practice.* Amsterdam: Elsevier Publishing.

ROSSI, R. A. 1984. "Indirect Heat Transfer in CPI Fluidized Beds." *Chem Eng* (Oct 15): 95–102.

RUSSO, G. and L. MASSIMILLA. 1972. "Acetylene Production by Quenching Methane Flames in a Fluidized Bed," in *Second International Symposium on Chemical Reaction Engineering.* Amsterdam (May). Preprints published by Elsevier, Amsterdam, pp. B9–47—9–57.

RYANS, J. L., and S. CROLL. 1981. "Selecting Vacuum Systems" *Chem Eng* **88,** 25: 72–90.

RYANS, J. L. and D. L. ROPER. 1986. *Process Vacuum Systems Design and Operation.* New York: McGraw-Hill.

SAUTER, J. R., and W. E. YOUNTS, III. 1986. "Tower Packings Cut Olefin-Plant Energy Needs." *Oil and Gas J* (Sept 1): 45–50.

SAWISTOWSKI, H. and W. SMITH. 1963. *Mass Transfer Process Calculations.* New York: Interscience Publishers.

SCHIFFTNER, K. C., and H. E. HESKETH. 1983. *Wet Scrubbers.* Ann Arbor, MI: AnnArbor Science Publishers.

SHAH, G. C. 1979. "Trouble Shooting Reboiler Systems." *Chem Eng Prog* **75,** 7 (July): 53–58.

SIDEMAN, S. 1966. "Direct Contact Heat Transfer between Immiscible Liquids." *Advances in Chemical Engineering,* Vol. 6. San Diego, CA: Academic Press, 207–286.

SITTIG, M. 1978. *Vinyl Chloride and PVC Manufacture: Process and Environmental Aspects*. Park Ridge, NJ: Noyes Data Corp, p. 78.

SLY MANUFACTURING CO. (undated) "Dust Collection Data Study Data No. 64–7" W. W. Sly Manufacturing Co., 4700 Train Ave, Cleveland OH 44101.

SUNDSTROM, D. W., and R. L. DEMICHIELL. 1971. "Quenching Processes for High Temperature Chemical Reactions." *Ind. Eng. Chem. Process Des. and Dev.* **10**, 1: 114–122.

SWERN, D. 1964. *Bailey's Industrial Oil and Fat Products*. New York: J. Wiley and Sons.

TREYBAL, R. E. 1968. *Mass Transfer Operations*. New York: McGraw-Hill.

UHLERR, M., and I. VOGL. 1968. "Section 6: Design and Operation of Plants producing Sulfuric Acid by Fluidized Bed Roasting of Pyrites" in *Sulfur in Australia: a Set of Design Studies*. Dept of Chemical Engineering, the University of Sydney, Sydney, Australia.

WING CO. (undated), Wing Co. Bulletins WE86, WC-86, and WMC-85, Aeroflow Dynamics Inc., Linden, NJ.

ZDONIK, S. B., et al. 1974. "How Feedstocks Affect Ethylene." *Hydrocarbon Process* **53**, 2: 73–82.

ZDONIK, S. B., and G. L. HAYWOOD. 1975. "Olefins Production by Gas Oil Cracking." *Hydrocarbon Process* **54**, 8: 95–98.

ZENZ, F. A. and D. F. OTHMER. 1960. *Fluidization and Fluid Particle Systems*. New York: Reinhold.

3.5 EXERCISES

3-1 To drive the 400 kW compressor at 1800 rpm, what alternative drives are possible?

_____

_____

3-2 If a steam turbine is used, from Table 3–1, approximately how much steam is needed? Compare this with the projected estimates from Figure 3–3. Note, in Table 3–1 a condenser is hooked up to the exit of the turbine to extract the steam at a lower value than a "back pressure" turbine.

3-3 From Figure 3–3 select a drive for 100kW blower to be run at 1800 rpm. What other possibilities are there?

3-4 If a reaction is to require 10^7 kJ/h total, approximately how much total area is needed?

Estimate the volume of the radiant section.

Select a configuration assuming tube wall temperature variation of a factor of 4 is allowed.

3-5 From Figure 3–12, if we need to find a heating medium to provide 250°C temperature, what names and pressures are required to supply this? What would you choose?

3-6 From Table 3–8, rough-size a coil in a box of cold water to condense gasoline vapor at a rate of 500 kg/h.

Assume "gasoline" is octane and use the vapor pressure data from **Data** Part Db to estimate temperatures if the gasoline condenses at atmospheric conditions.

3-7 We need to select a cracking furnace. The cracking reaction requires a very uniform heat flux on all the tubes. In other words, unwanted side reactions occur in different tubes if they receive different heat fluxes. If we require an average heat flux in the radiant section of about 10^5 kJ/m^2.h and the total load in the furnace is 10^7kJ/h,

   a. select the configuration.

   b. estimate the *total* area needed in both the radiant and the convection section.

   c. estimate the size of the radiant section (the volume).

3-8 We wish to air-condition the room so that the temperature is 20°C, when outside in the summer it is 32°C. From Figure 3–12,

   a. select the liquid to remove sensible heat in the "air-conditioning" coils; and

   b. select a fluid to evaporate to extract the heat from the fluid chosen in (a). Indicate the operating pressure.

3-9 We wish to size a kiln to destroy the equivalent of 100 Mg/d of PCB. Assume this "behaves-like" cement. What volume and diameter kiln do you need?

3-10 Steam at 1.825 MPa absolute, saturated is reduced in pressure to 0.965 MPa through a pressure control valve. The steam then goes to a heat exchanger,

   a. how many degrees of superheat has the steam entering the exchanger?

   b. how will this affect your design calculations compared to if you assumed that the steam was saturated at 0.965 MPa?

3-11 Sketch the type of condenser configuration and identify the condensing medium for the following duty. To condense the overhead vapors from a column operating under a vacuum (pressure 20 kPa). The vapors are wax-like monoglycerides. The vapors must not solidify and plug the tubes.

3-12 Steam is produced in the boiler at 235°C, 3.5 MPa. It is superheated to 290°C. If this is delivered to the bonegrease section of the plant where it is reduced in pressure across a pressure reducing valve to 0.8 MPa, what is the condition of the steam when it is used on the bonegrease plant?

3-13 Size the reboiler for the depropanizer distillation column given in PID-2A.

3-14 Size the pyrolysis furnace for the ethylene plant given in Figure 1–1. You might wish to consider this as eight separate reaction trains.

3-15 The reactor effluent from each of the six ethylene pyrolysis reactors (given in Figure 1–1 and shown in Figure 1–2) is 1.81 kg/s at 834°C. The gas density is about 0.46 kg/m^3. We want to quench this to 65°C by using a deluge of 16.5 kg/s of water entering at 43°C and leaving at 110°C. The operating pressure is 380 kPa. If we assume that the net heat exchange involves no net humidification or dehumidification, rough-size a spray tower quencher.

3-16 For a slightly different design of an ethylene pyrolysis scrubber, consider the treatment of 0.83 kg/s of pyrolysis gas at 593°C that is to be water quenched to 37.7°C. The heat load is 1.275 MW with about 1.16 MW of sensible heat load. The gas density varies between 0.43 and 1.25 kg/m^3 as it goes up the quencher. The water enters at 29.3°C and the temperature is constrained so as not to exceed 46.5°C. The overall operation is dehumidification. Rough-size a quencher.

3-17 For a solar energy operation we wish to heat up air by contacting it with molten salt that runs down over pall ring packing. The salt enters at 330°C. The molten salt flowrate is about 1 kg/m^2.s. The air flowrate is about 0.6 kg/m^2.s.

Estimate the overall volumetric heat transfer coefficient. There is no humidification or dehumidification.

3-18 As a followup to Example 3–20, we wish to cool the gas from 510°C to 150°C in a packed tower quencher. The total heat duty is 2050 kW with 86% of the load being to dehumidify the gas. The quenching liquid is 20.9 kg/s of a mixture of acetic acid and acetic anhydride. This quench liquid enters the top at 35°C and leaves the bottom at 100°C. Rough-size the quencher. The gas density is about 0.9 kg/m^3 at these conditions.

3-19 As a followup to Example 3–21, we wish to cool the cupola gas from 260°C to 72°C with scrubber water entering at 15°C and leaving at 95°C. The overall heat load is 4378 kW which is 46% sensible heat cooling and the remainder is dehumidification. Rough-size the unit if we want to use spray.

3-20 A cooling tower is to cool 1800 kg/s of water from 48°C to 31°C. The local wet bulb temperature is 26.8°C. Rough-size the cooling tower.

3-21 *Terry Sleuth and the Case of the Revised Exchangers*
"Hurrah!" reverberated off the walls. Terry Sleuth walked over to Don's desk; Terry's curiosity was aroused. "What's all the excitement about?"

"I've just found a way to save steam on the overhead condenser of the fatty acid distillation column. Let me show you. Recall that we use hot water as the coolant on that condenser to keep the fatty acids from solidifying in the tubes. The present scheme is . . . " and Don drew the sketch shown in Figure 3–51.

"But we can save all that new steam by a simple repiping job," Don continued. "I don't see why we didn't think of this before. Here's all we have to do . . . ," and he sketched as shown in Figure 3–52.

Don turned to Terry and waited for Terry's praise. But Terry pondered and then said, "I think we had better rethink that one." What did Terry see?

3-22 *Terry Sleuth and the Case of the New Piping*
"Save Energy"—the sign blurted out its message from above the drinking fountain in the engineering complex. "That's what I've done," beamed Herb. "Come on back to my desk and see my latest brainstorm." Terry Sleuth followed with anticipation. Herb usually had some real winners. "This is my scheme to eliminate the additional pump and tidy up the pipework on the vinyl acetate unit," announced Herb, and he proudly produced BEFORE and AFTER sketches in Figure 3–53 for Terry's approval.

"I've a few questions, Herb," said Terry. "What are the pressure drops across the two exchangers in lines A and B BEFORE?" "In line A it is 10 kPa; in line B it is 38 kPa," replied Herb with a questioning look. "Did you calculate the heat exchanger ratings for your AFTER conditions?" persisted Terry. "No, this is changing the pump system on the water, not the flowrates or heat exchange." Terry paused, frowned, and said . . . What did he say?

3-23 For the sulfuric acid plant given in **PID-3**, the Marketing Department has indicated a need to sell 77% acid to be shipped in tankcar lots. Modify, adjust, and size whatever has to be changed, adjusted, or added to the PID-3 so that we could meet a demand for 50 Mg/d of this concentration of acid.

3.6 PROCESS & INSTRUMENTATION DIAGRAM FOR SULFURIC ACID ILLUSTRATES APPLICATIONS

The sulfuric acid process was chosen to enrich this chapter because of the wide range of heat exchanger type used, because this process is as much a plant to produce steam as it is to produce sulfuric acid, and because a variety of turbine and electrical drives are used.

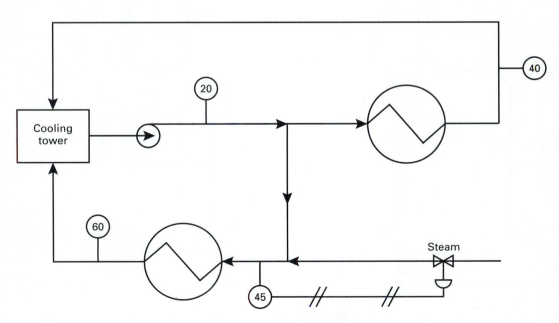

Figure 3–51 Don's Sketch of the Present System

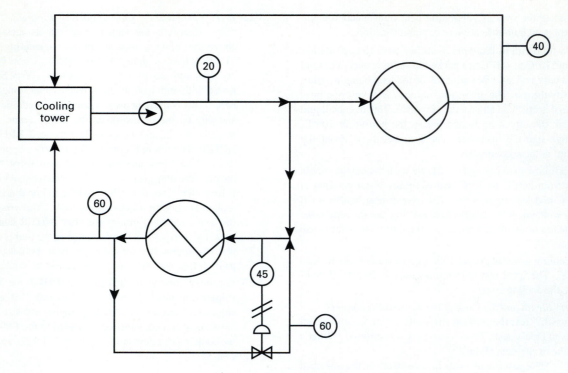

Figure 3–52 Don's Proposal

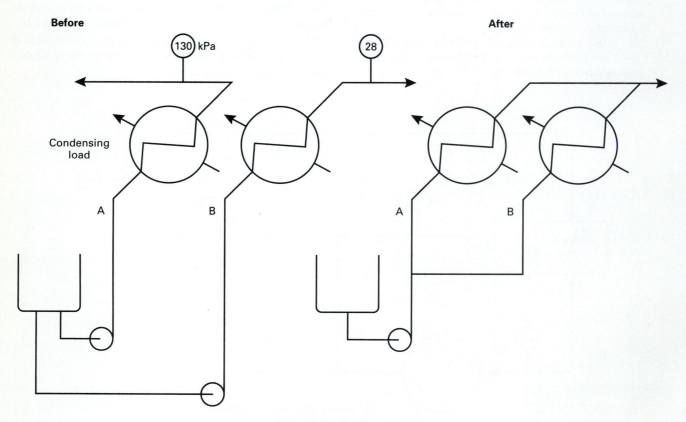

Figure 3–53 Herb's Sketches

PID-3: Sulfuric Acid

One indication of the degree of development of a country is the amount of sulfuric acid it produces. In terms of annual tonnes of production, sulfuric acid usually is #1; ammonia usually ranks third. Sulfuric acid is a critically important industrial chemical.

The production can be from sulfur-containing "off-gas" from sulfur bearing coal power plants, from sulfide ore roasting, or from elemental sulfur. This process uses solid yellow sulfur as its feed. The process has about four parts:

- feed preparation: where sulfur is burnt to produce a concentrated gas stream of SO_2;
- reaction or conversion: where the SO_2 is converted to SO_3;
- absorption: where the sulfur trioxide is absorbed in "water" (more correctly in dilute acid) to produce the products; and
- energy recovery: to produce steam.

A sketch of the flow sheet and its relationship to photographs of an actual plant are given in Figure 3A–1. The PID-3 provides some details about the equipment sizes for a 300 Mg/d plant. While this PID offers a lot of application of heat exchange principles, it has also different types of liquid pumps and some gas moving equipment. The latter complements our study of Chapter 2.

Now for the details in **PID-3**. In the feed preparation section of the process, air is blown through a drying tower where the gas contacts sulfuric acid which absorbs moisture, **T300**. The sulfur is melted in the sulfur pits, **V305**, and pumped into the sulfur burner, **T301**. For startup, this furnace is heated up via oil combustion until the sulfur burns spontaneously in the air to produce SO_2. The reaction is

$$S + O_2 \rightarrow SO_2 \qquad (3A-1)$$

The exit temperature is controlled at 971°C. Heat is extracted from the exit gas and any particulate carryover is removed in the dust filter, **V303**.

In the reaction section, sulfur dioxide is catalytically converted to sulfur trioxide:

$$SO_2 + 1/2\ O_2 \quad <=> \quad SO_3 \qquad (3A-2)$$

Again, heat is recovered in exchangers **E301** and **E302**.

The sulfur trioxide is absorbed in dilute acid to make 99% acid (Tower **T302**) or to make oleum (Tower **T303**). This is shown as an absorption in "water" in Equation 3A-3; in practice, this is absorption in dilute sulfuric acid.

$$SO_3 + \text{"}H_2O\text{"} \rightarrow H_2SO_4 \qquad (3A-3)$$

Since the reaction is exothermic, again, heat is removed, **E305, 306, 307,** and **308.**

The other section of the plant produces steam from the steam drum **V302**.

The cost contributions to the unit cost of producing sulfuric acid are given in Table 3A–1. The Battery Limits capital cost of a 300 Mg/d plant is about $13 million +/– 30% (MS = 1000). MS is the Marshall and Swift construction cost inflation index. Its value for construction in the process industry was 100 in 1926, was about 300 in 1970 and is 1000 in the early 1990s. Current values of the index can be found in *Chemical Engineering* magazine. The MS index works well for relating capital costs from one time to another. To estimate the capital cost for the time of interest, multiply the cited cost by the ratio of the MS index at the time of interest/1000. For example, in the first quarter of 1992, the MS index value was 932.9. Thus, the cost of a sulfuric acid plant at that time would have been about 932.9/1000 × $13 million.

For more details about this process, see C. M. Crowe et al. 1971. *Chemical Plant Simulation.* (Englewood Cliffs, NJ: Prentice Hall).

TABLE 3A-1. Cost contributions to the cost per tonne of sulfuric acid product (based on sulfur as the feedstock)

Contribution from	Approximate Unit Cost Breakdown, %
Raw material	53.5
Utilities	1
Labor	9
Credit for byproducts	-3.5
Maintenance	6
Supervision	1
Depreciation	12
Indirectly attributable costs	12
General expense	9
Total	100

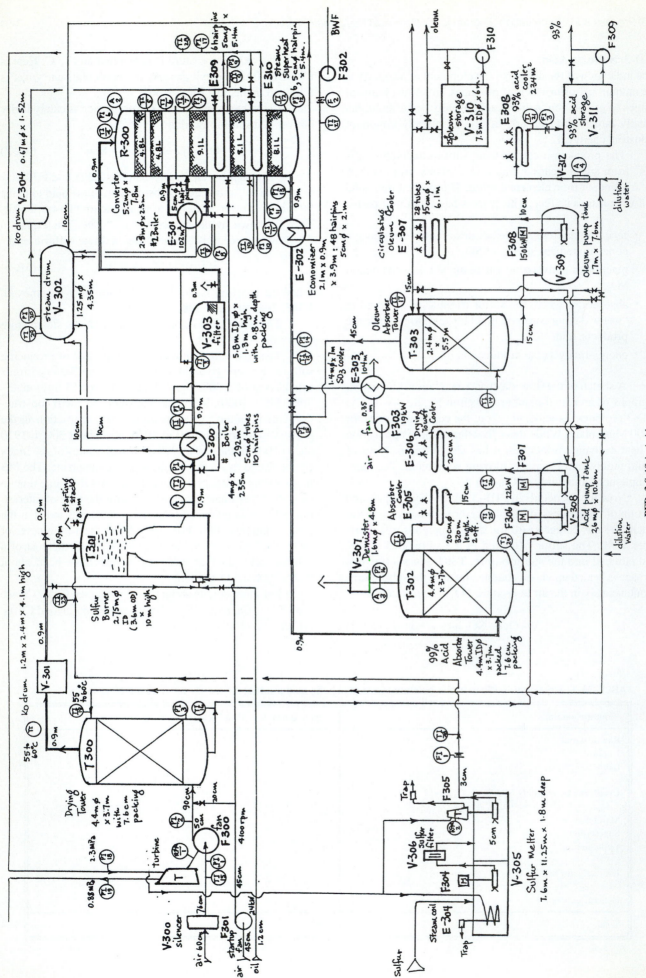

PFD-3 Sulfuric Acid

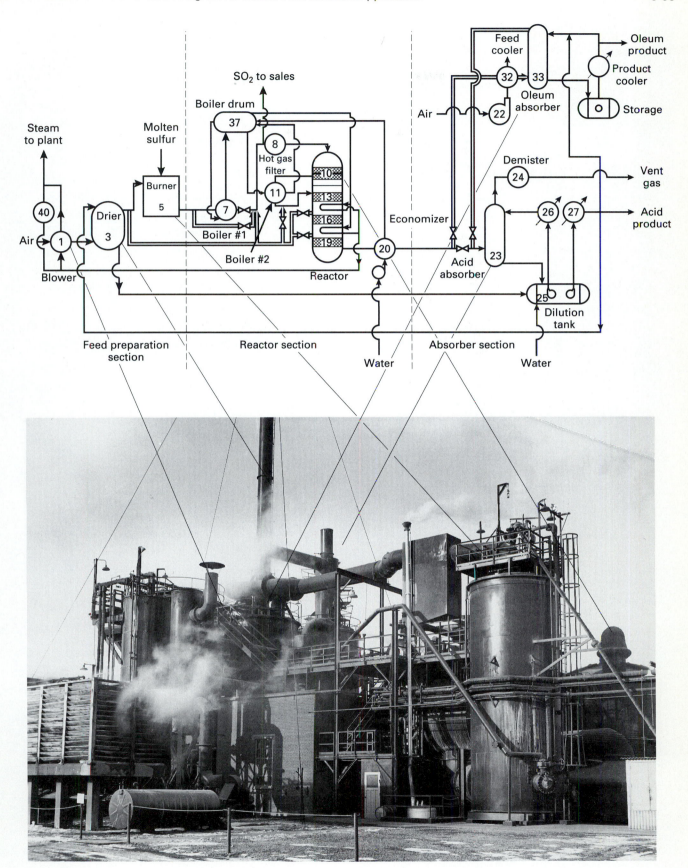

Figure 3A–1   Flow Diagram and Photograph of a Contact Sulfuric Acid Plant
(Crowe et al., CHEMICAL PLANT SIMULATION, © 1971. Reprinted
by permission of Prentice Hall, Inc., Englewood Cliffs, NJ)

Minichapter 4

Selecting Options for the Separation of Components from Homogeneous Phases

We often need to separate or concentrate species. They may occur in a single homogeneous phase (like a glass of Coke that is to be separated into components); they may occur as an intimate mix of two or more different phases (like a mixture of sand and gold or a mixture of sand and water). If the starting mixture is a homogeneous phase, we start with the ideas given in this chapter. If the initial mixture is heterogeneous, we start with the ideas in Chapter 5.

4.1 SELECTING OPTIONS FOR SEPARATING A HOMOGENEOUS PHASE

To separate a homogeneous mixture, we primarily exploit the differences in properties among the different species. In actually selecting the type of equipment, we also study the feed concentration, the product requirements, and unique characteristics of the equipment itself. In this chapter we focus on how to select feasible equipment options. How to rough-size the options is given elsewhere (Woods, 1993a).

4.1-1 Overview to Selecting Options for Separations

In selecting separation options, we should keep an open mind. Tradition, as exemplified by this text, may suggest certain options. However, new developments are occurring

rapidly. Individuals and companies have their "favorite" separation techniques and thus may prefer to work with the familiar. The purpose of this textbook is to provide short-cut methods that will allow you to rough-size many options before deciding. Hence, in selecting options, choose about three to consider in more detail for any separation.

Next think of combinations of options; perhaps ion exchange followed by solvent extraction followed by distillation is an appropriate choice.

The general criteria for selecting options include the properties of the species, the feed and product constraints, and the characteristics of the equipment.

A. Exploit the Properties of the Species. If the original mixture appears as one homogeneous phase, a guideline to the properties that can be exploited and the name of the technique or unit operation that might be used to separate the species are listed in Table 4–1. If the original mixture is heterogeneous (such as solid particles dispersed in liquid), the property's differences that might be exploited and the name of the unit operation for physical separation are listed in Table 5–1.

Thus, the first main consideration for any separation is *there must be a difference in physical or chemical properties that can be exploited.*

Usually the mixtures we are trying to separate have a

Table 4-1. General criteria for selecting options for separation

| If there is a difference in . . . | and its phase is . . . | Equilibrium — create two physical phases to give different equilibrium concentrations in two | | | Rate — Exploit differences within a single phase | |
		by ± Agent	by ± Agent — Agent	by ± Agent	by adding barrier	by using nonuniformities in concentration
Vapor pressure (G-L) — T_b	L	Evaporation Distillation Molecular distillation Cryogenic distillation	Liquid solvent or Steam	Azeotropic distillation Steam distillation Extractive distillation		
Freezing point and solubility (L-S) — T_f, δ_h	L	Melt crystallization Freeze concentration				
δ_p	S	Zone refining				
Solubility L-S — K_{sp}	L	Solution crystallization Precipitation				
G-L — δ_h, δ_p	L		Steam/gas	Desorption		
V, T_b	G		Solvent	Absorption		

Property	Symbol	State	Process	Medium / Sorbent	Membrane process	Other process
Partition Coeff. L-L	$\delta_h\ \delta_p\ V$	L	Solvent Extraction	Immiscible solvent		
Exchange Equilibrium G-S	$\delta_h\ \delta_p\ V\ T_b$	G	Adsorption	Solid absorbent		Foam Fractionation / Ion Flotation
L-S	M	L	Adsorption	Solid absorbent		
L-S	K_a, valence	L	Ion Exchange	Resin		
Surface Activity L-G	M,*	L	Ion Exchange / Adsorption			
Molecular Geometry G-S	M	G	Adsorption	Zeolites		Size Exclusion Chromatography
L	M	L			Membrane permeation	Dialysis / Ultrafiltration
	M K_a	L			Reverse Osmosis / Electrodialysis	
Electromigration	K_a	L			Electrodialysis	Electrophoresis
Molecular K.E.						Thermal Diffusion
Reactivity	BOD/COD	L	Biological reactions	Micro-organisms		

large number of components. For example, a cup of coffee contains water, dissolved sugar, cream, and the host of components that have been "perked" from the coffee bean to give it a taste. The water contains dissolved oxygen and some iron, sodium, chloride, calcium, and magnesium. Where do we stop in listing the species? In keeping with the engineer's Principle of Optimum Sloppiness, we focus first on the two major things that we are trying to separate; we call these the key components. In the above coffee example, we might want to separate the "sugar" from all of the rest. We would characterize all the rest as "water." Hence, sugar and water become the key components.

King (1971) defines the separation factor α_{ij}, which is the ratio of the concentration ratio of the two key components in one separated product stream to the concentration ratio in the other product stream. That is

$$\alpha_{ij} = (x_i/x_j)_2/(x_i/x_j)_3 \qquad (4\text{-}1)$$

where x = mole or mass composition
     i = one key component
     j = other key component

For any differences in properties, we can calculate the separation factor. This definition is illustrated in Figure 4–1. Naturally, the larger the separation factor, the more attractive the separation process and the easier the separation. Figure 4–2 relates the separation factor to the feed concentration and shows the usual regions of application for most of the equipment options given in Table 4–1. Because different equipment exploits differences in *different* properties, the separation factor has different definitions. Table 4–2 provides the appropriate definition of the separation factor for the different unit operations.

Three other important property guidelines are the actual size of the species, the temperature sensitivity of the species (will the species decompose at high temperature?),

and whether it is organic or inorganic. The size may be expressed as the ionic or molecular "size," as the molar mass, or as the molar volume (density divided by the molar mass). The approximate relationship between the size and the molar mass is given in Figure 4–3a. Thus, a triglyceride oil with a molar mass of about 900 would have a "size" of about 1.3 nm. Concerning temperature sensitivity, the bigger the molecule, the easier it is to break down at higher temperatures. In general, for organic species, as the size increases the more sensitive the species is to temperature. Most materials degrade when heated to a high temperature, and especially if they are kept at a high temperature. Hickman and Embree (1949) suggest that the rate of degradation varies linearly with time and doubles for each 10°C rise in temperature. Thus, a decomposition hazard index can be defined as the product of the absolute temperature and the number of seconds the species is held at that temperature. They found that it was more convenient to express the temperature effect in terms of the operating pressure at which the material is boiling. Thus, they define a decomposition hazard index as the product of the residence time and the absolute operating pressure when the species is boiling. Some example data are given in Figure 4–3b. Thus, the triglyceride oil (whose boiling point at 0.7 Pa is about 300°C) could not be kept boiling at 0.7 Pa pressure for more than about 10 minutes before it would start to decompose. We note that the decomposition index is not directly related to the molar mass; some small, fragile molecules decompose more readily than larger, more robust species. A guideline is that pharmaceutical and biological species are usually very fragile. Many proteins denature if the temperature is raised to 80°C; certain polymers undergo transitions if the temperature exceeds 60°C and so on. The data in Figure 4–3c can be used as a guide and a reminder of how sensitive species are to temperature. This becomes a major guideline because one of the more popular options for separation is evaporation or distillation,

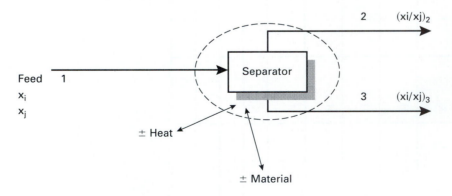

$$\alpha = \frac{(x_i/x_j)_2}{(x_i/x_j)_3}$$

Figure 4–1   Definition of the Separation Factor, α

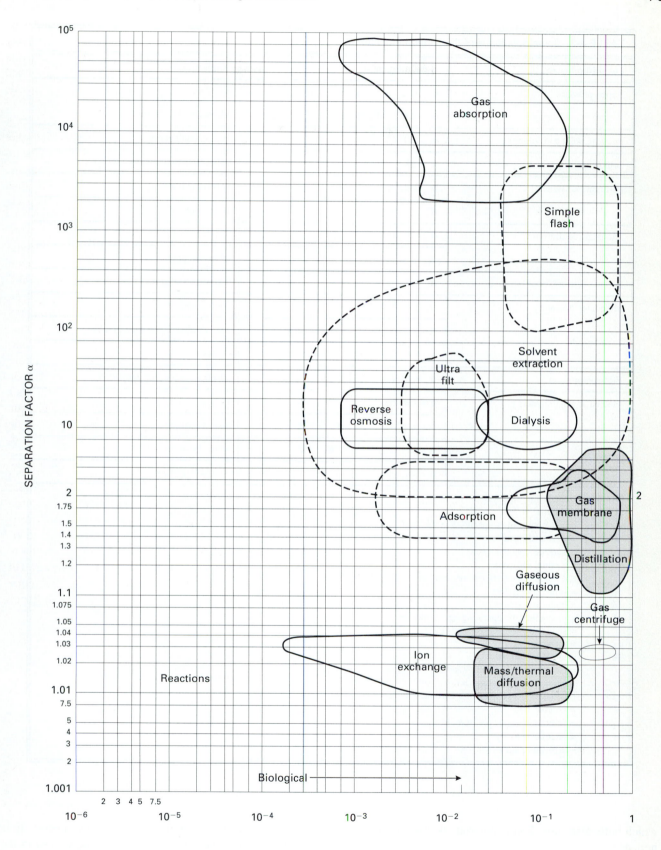

INITIAL CONCENTRATION, mass fraction

Figure 4–2   General Regions of Applicability (from Kodatsky, 1978)

TABLE 4-2. Some definitions of the separation factor α

For difference in . . .	A definition of α . . .	Application
Vapor pressure	$\alpha_{vp} = vp^o_1/vp^o_2 = p^o_1/p^o_2$	distillation; evaporation
	$\alpha_{vp} = p^o_2 \sqrt{M}_1 / p^o_1 \sqrt{M}_2$	molecular distillation
Freezing point and solubility	$\alpha_{MC} = T_2 / T_1$	melt crystallization
Solubility	$\alpha_s = pK_{sp2} / pK_{sp1}$	solution crystallization
Partition coefficient	$\alpha_{SX} = \kappa_2 / \kappa_1$	solvent extraction
	α_{Ad}	adsorption
Exchange equilibrium	$\alpha_{IX} = \Gamma^+_B(1-c^+_B)/(1- \Gamma^+_B)c^+_B$	ion exchange
	α_{Cr}	chromatography
Surface activity	$\alpha_{SA} = \Gamma A/cV$	foam fractionation
Molecular geometry	$\alpha_M = \kappa_2 \mathcal{D}_2 / \kappa_1 \mathcal{D}_1$	membrane
	$\alpha = <t_2>/ <t_1>$	size exclusion chromatography
	$\alpha_D = c_{21} \mathcal{D}_2 / c_{11} \mathcal{D}_1$	dialysis
Electromigration	$\alpha_{El} = \mu^+_2 / \mu^+_1$	electrophoresis, ion migration
Molecular kinetic energy	$\alpha = (M_2/M_1)^{1/2}$	
	$\alpha = \exp \{(M_2 - M_1) \omega^2 r_2^2/ 2 RT$	gas centrifuge
	$\alpha = \exp\{(z N_s RT/\rho)(1/\mathcal{D}_2 -1/\mathcal{D}_1)\}$	mass diffusion
	$\alpha = <k_T> \ln (T_{hot}/ T_{cold})$	thermal diffusion

Symbols:

A	= area
c_2	= concentration of species 2 in the feed
$\mathcal{D}$	= diffusivity; cm^2/s
Γ	= surface concentration
k_T	= thermal diffusivity
κ	= distribution coefficient or partition coefficient; mol/mol; or mol/L/mol/L
K_{sp}	= solubility product
M	= molar mass
N_s	= molar flux of separating agent; mol/cm^2.s
p^o	= saturation vapor pressure, kPa
r_2	= outer radius, cm
R	= ideal gas constant
μ^+	= electric mobility, cm^2/V.s
T	= absolute temperature, K
$<t>$	= peak residence time in a chromatograph
V	= volume
vp^o	= saturation vapor pressure, kPa
ω	= angular velocity
z	= diffusion path length, cm

which boils materials. Some materials degrade if they are boiled.

Thus, to obtain a separation there must be some difference in property between the species, and we ask about the size, the temperature sensitivity, and whether the species are organic or inorganic. We look to the physical properties listed in Table 4–1 (and **Data** Part D).

When considering the properties of species, it is useful to list the species *in the order* of the change in the property. Thus, if we decide to exploit vapor pressure difference to separate four species, we would list them in the order of decreasing volatility. If we elect to consider ion exchange, we would list the species in the order of valence.

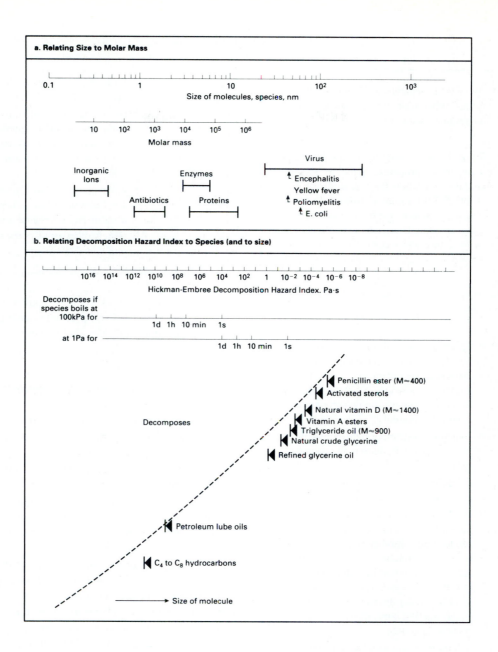

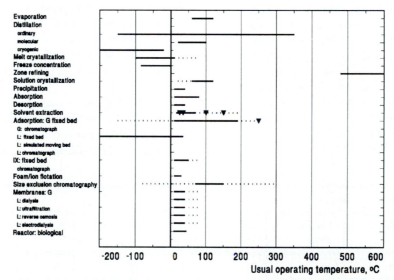

Figure 4–3   Size, Molar Mass, Decomposition Hazard, Operating Temperature
to Choice of Options

Example 4-1: We wish to separate acetic acid from water. What might the options be? The beginning state is a homogeneous liquid.

An Answer: From Table 4–1 we see that it is helpful to compare the properties of the two species and then see where they are different. Table 4–3 is a summary from **Data** Part Da. We also note that both species are relatively small; as such, it is unlikely that temperature sensitivity or decomposition would be an issue for either of these species.

Hence, from Table 4–1 some options might be:

- vapor pressure? difference in boiling point; converting this to vapor pressure difference we obtain an $\alpha = 1.4$. This is a relatively low value. Although distillation is feasible we should certainly consider other options.
- freezing point? modest difference in freezing point. Perhaps melt crystallization.
- solubility? no K_{sp} values, so solution crystallization not considered.
- solubility in absorbing solvent? reasonable differences in molar volumes and Hildebrand solubility parameters. However, this difference, combined with the relatively high boiling point do not suggest that desorption is a strong possibility. Usually for desorption, the species in the smallest concentration is a gas at room temperature.
- solubility parameters for partition coefficient? reasonable differences in molar volumes and Hildebrand solubility parameters. We should be able to exploit this difference via solvent extraction.
- exchange equilibrium? adsorption is a possibility because of the differences in molar volumes and Hildebrand solubility parameters.
- dissociation constant and valence? although differences occur, dissociation

- for acetic acid is relatively small. Perhaps ion exchange.
- surface activity? with only two carbons in the organic chain length, acetic acid is not surface active. Acetic acid dissociates to some degree. Perhaps, ion flotation.
- molecular geometry? the molar masses are different and so perhaps dialysis and ultrafiltration. Because acetic acid ionizes, reverse osmosis and electrodialysis might be possible.

The guidelines of Table 4–1 are very general. Later we will see how feed and product concentration and characteristics of the equipment help refine these initial considerations.

B. Feed and Product Conditions. A second major consideration is the *feed and product conditions*. Some separation devices work best for dilute feed concentrations less than 1%; others thrive on concentrated feeds of greater than 1%. Sometimes we need to recover both components into two separate streams; sometimes we focus on only one product stream. The amount of species recovered (or the recovery) and the purity required are other considerations. Figure 4–2 relates the feed concentration to the separation factor; Figure 4–4 shows how the purity requirements relate to equipment selection.

Another way of expressing this information is with a feed-product diagram such as illustrated in Figure 4–5. In this diagram, the feed concentration is given on the diagonal. The product concentrations in the exit streams (the overhead and bottoms) identify a point on the diagram that is joined to the feed point. Thus, in Figure 4–5a, line "A" represents a separator operating on a feed concentration of the species of 50% that produces an exit overhead stream of concentration 90% with an exit underflow concentration of 2%. The slope of the line is approximately the ratio of the mass of the total stream taken overhead to the mass in the underflow.

The diagram is slightly different when a separating agent is added, as for example in solvent extraction where an immiscible solvent is added. Now, the concentration in the exit "solvent stream" depends on the amount of the

TABLE 4-3. Exploiting property differences								
	Molar Mass	Freeze Temp.	Boil Temp.	Molar Volume	δ_p	δ_h	pK_a	K_{sp}
Acetic acid	60.06	16.6	117.9	57.2	12.2	18.9	4.76	—
Water	18.	0	100	18	22.8	40.4	14	

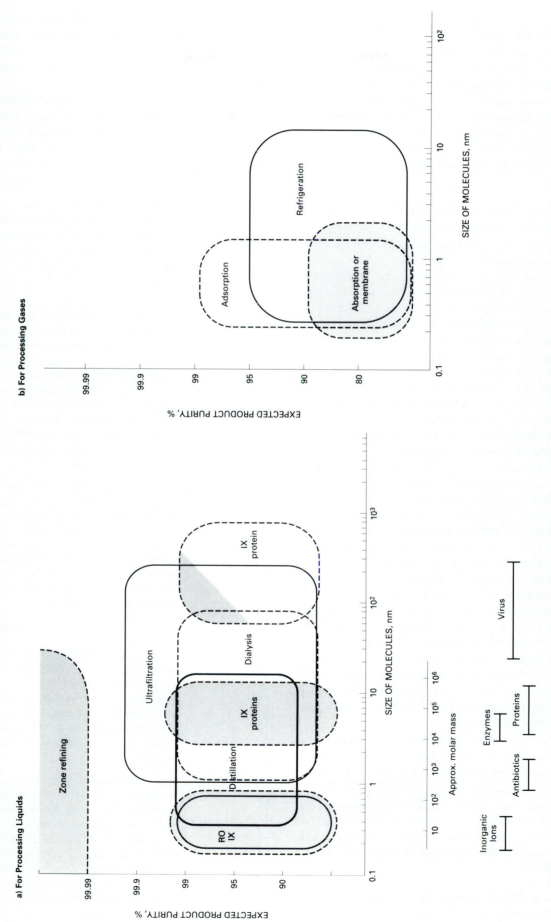

Figure 4-4 How Expected Product Purity Affects Choice

agent added. The more that is added, the smaller the concentration will be. This is shown in Figure 4–5b. As the ratio of separating agent to feed increases, the concentration decreases. This can mean that **both** the two exit streams might become more dilute than the feed. This is not unusual in many options; the diagram reminds us that streams are both dilute and asks us to question whether this is indeed what we want. Ideally, we would like the "line" to be downward to the right.

This form of presenting the information is useful because we can:

- quickly see the approximate range of feed concentrations traditionally used for a piece of equipment.
- use the two output concentrations to be the feeds to other separation devices. Thus, later we can see how to hook up separation options. This is illustrated in Figure 4–5c, where membrane separation is followed by distillation. Here, an initial feed of 20% is upgraded by the membrane to produce a membrane retentate concentration of 80%. This becomes the feed to the distillation column where the stream is concentrated further to 99%. There also are two other streams, the permeate from the membrane and the bottoms from the distillation, with solute concentrations of 1 and 5% respectively. Thus, options can be combined with each other. Figure 4–5d illustrates the combination of ion exchange with solvent extraction to provide a systems approach. In this illustration, feed of concentration 0.05% is fed to an ion exchange unit. The exit eluant, after recovery, is 5% solute. This stream then becomes the feed to a solvent extraction unit, which yields an extract of concentration 60% solute. In this illustration, we followed only the solute. The other exit streams (0.001% from the ion exchange unit and 0.01% from the solvent extraction unit) could also be considered. Indeed, in this illustration, we might wish to operate the solvent extraction unit so that the raffinate concentration is 0.05%. In this way, the raffinate could then become part of the feed to the ion exchange. Thus, this form of diagram helps to link together combinations of options and to take a systems view of all of the streams. We can use this graphical form to help us visualize the big picture or the separation system.
- relate performance qualitatively with expense. The longer the line, the more expensive the operation and the capital equipment. For example, Figure 4–5e shows a distillation column to obtain 90% purity. This might need 20 trays and a reflux ratio of 1.2. If we then went to 99% purity, we might need 30 trays and a reflux ratio of 1.4. If the target was 99.9% purity, we might need 50 trays and a reflux ratio of 2.

Care is needed in selecting the variables for the axis. In general, the axis should be chosen so that the diagram is

downward to the right. In other words, the concentration of the species that becomes more "concentrated" should be on the abscissa. With this choice, the concentration of the same species (shown on the ordinate) should become more dilute.

The diagram cannot be used when the separation option selected is reaction. This is unfortunate, because the reaction option is often forgotten in our search. To illustrate both the usefulness of reactions and the implications of specifying purity, consider the following example.

Example 4–2: We need to remove about 1 mol % H_2S and 18 mol % CO_2 from a hydrocarbon gas stream, recover the sulfur and produce a tail gas stream that we can release to the atmosphere that has less than 500 ppm sulfur. How might we do it?

An Answer: Although we should think of many options based on the property differences, this example is to illustrate the role of reactions as a separation option. Hence, we use existing technology. Often the first step is to use an absorption/scrubbing process to remove the "sour" components from the hydrocarbon stream. First, the H_2S and CO_2 might be absorbed by a liquid agent and recovered in the recovery still, as illustrated for CO_2 removal/recovery in the ammonia plant in PID-4.

The stream might now be 30% H_2S. What to do now? Part of the "acid gas" might be fed, together with molten sulfur, to a furnace where both are burned in air to produce SO_2. This in turn goes to a catalytic reactor, where elemental sulfur is produced by the reaction between H_2S and SO_2. This "process" includes reaction and separations and is called a Claus plant.

However, even a three-stage Claus plant can remove about only 95% of the sulfur in the feed. Thus, the tailgas from the Claus plant needs to be processed further, by say a Shell Claus Offgas Treater, SCOT, which will make the overall recovery of sulfur over 99%. The exit gas stream has less than 500 ppm. The SCOT process also includes a reaction step as part of the separation scheme.

Thus, three complete processes are added in sequence (and integrated) to achieve 95% and then 99% recovery. The separations illustrated in this example are "absorb, react, and react" as viable separation options. The challenging fact is that the capital cost of each of these three plants is about the same. Thus, to move from 95% to 99% we are doubling the cost.

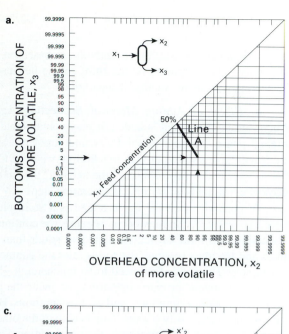

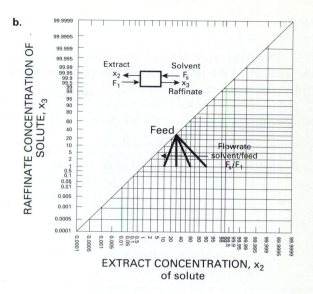

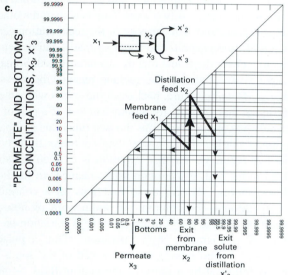

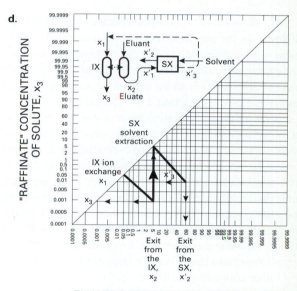

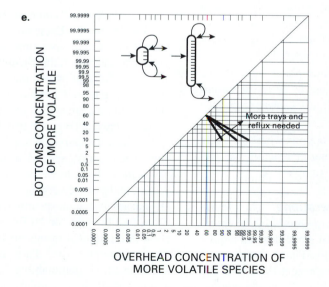

Figure 4–5 Feed-Product Diagrams

Thus, once again through this example, we see that the selection of the recovery or the purity of the product will have a major impact on the costs.

In summary, the feed concentration and the expected product specifications affect the options available.

Example 4–3: Reconsider case Example 4–1 for the separation of acetic acid from water. Assume the feed concentration is 15% and that we want 90% acetic acid as product.

An Answer: From Figure 4–4, ultrafiltration seems inappropriate because it works on larger sized molecules.

C. Beware of Trace Materials and Too Narrow a View.

In the previous sections we illustrated the usefulness of identifying a pair of key components when considering any separation. However, we must not forget the effects of other species on the separation; other species can completely rule out an option. For example, we could focus on separating two key components of ethanol and water and forget that salt is present—salt would wreck some very possible options. Or, we might be exploiting the difference between the valence in ions (like removing divalent calcium ions via ion exchange in water "softening") and neglect to realize that small quantities of trivalent would dominate the ion exchange option. This reminds us that we really need to understand the fundamental principles upon which each option is based and ensure that we consider the effect of species **other than the two key components** on the option.

Other common pitfalls are:

- to focus on finding **one** option to do the task. Sometimes we should consider a systems approach and team up sequences of different options. This has been illustrated in Figures 4–5c and d, where membrane-distillation combinations and ion exchange-solvent extraction combinations are considered.

- to consider only homogeneous phase options; we might find it more appropriate to **create a second phase** and use the techniques from Chapter 5 to give the separation.

- to consider only **separation** options when we might be better served by **destroying** or reacting the species. This was illustrated in Example 4–2.

D. Equipment Considerations.

A fourth major factor is the *special concerns imposed by the equipment*. From Table 4–1 we note that some devices require the addition or subtraction of energy. That is usually relatively easy and so these usually represent the easier and most economical devices to use. Others based on equilibrium considerations require the addition of agents. This can

cause complications. Still other methods are based on rate phenomena. These tend to be our third choice because of the economics and the quality of separation.

1. Effect of Adding Material Separating Agent.

Whenever another agent is added, it may contaminate the original phase (because of solubility and vapor pressure) and it will have to be removed. In addition, we have to either recover/regenerate or discard the agent. If the agent is recovered, this adds an additional separation step.

In general, when the agent added is a solid phase, more expensive difficulties occur because for continuous processing, it is more difficult to transport solids than it is fluids and we must separate the solids after the contacting is complete. The solids then need to be regenerated. If we use a discontinuous or batch operation, we incur the penalty of regeneration equipment and cyclic operations. Basically we need at least two units. One is active; the other is being regenerated. The purpose of the regeneration can be two-fold: to regenerate the agent so that it can be reused, and to recover the solute that the agent attracted. Some example ways of doing this include change the pressure, change the temperature, backwash the solids with an eluting agent that recovers the solute. The use of a pair of fluidized bed columns performs the two functions in separate columns but has the solids moving continuously between the two columns. Another variation is to keep the solids fixed in the bed, but through astute changes to the liquid flow patterns, perform these functions as a continuous sweep through the column. An impression of how this might be done is given in Figure 4–6. A variation on this is a chromatographic style operation where a pulse of the process fluid is sent through a bed of the solids and this is followed by a pulse of the eluting fluid. Some of these ideas are illustrated in Figure 4–6. All in all, handling solid agents means more complex and expensive operations.

2. Effect of Temperature, Pressure, and Materials of Construction.

Excursions in temperature and pressure away from "usual" operating conditions can be very expensive. Thus, although the chemistry says that a separation is possible, the equipment may cost too much to operate under the required conditions. Similarly, if solvents or other materials are added, the materials of construction required to prevent corrosion may be prohibitively expensive.

Hazard excursions require the addition of fireproofing, water spray, special instrumentation and monitors, ventilation, blowdown or removal systems, safety garments, and possibly diking or explosion barriers. To assess the hazard, two semi-quantitative measures are available in **Data** Part D. First, the NFPA ratings are on a scale from 0 to 4. A rating of 0 means negligible hazard; a rating of 4 means extreme hazard. Three different aspects are considered: health H, flammability F, and stability or likelihood to react, explode, or combust spontaneously, S. Thus, acetone has a rating of 1,3,0; acetylene, a rating of 1,4,3 and

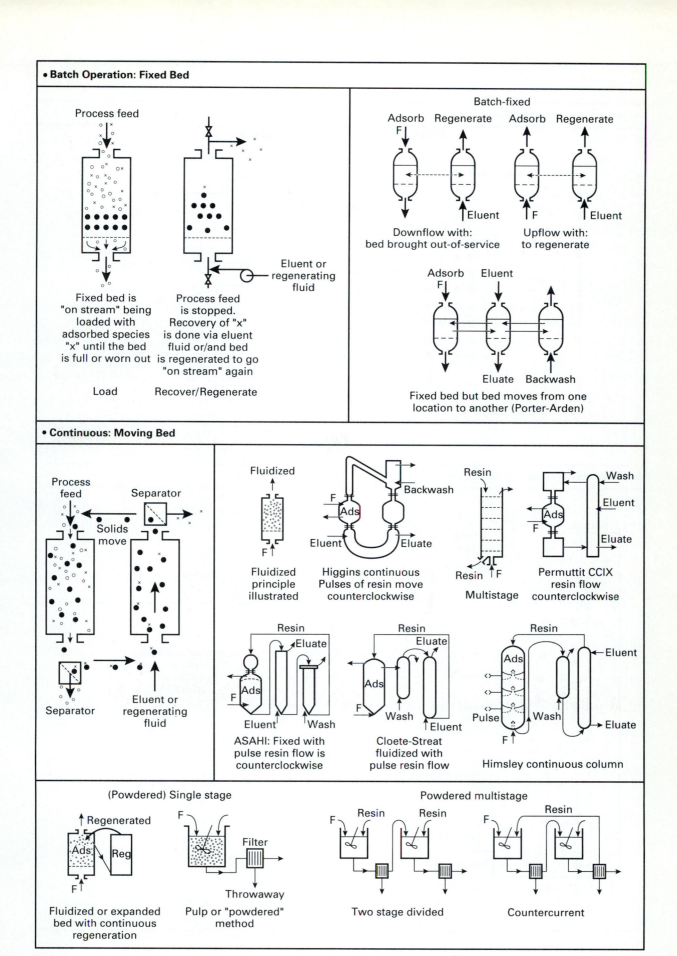

Figure 4–6 Options for Handling Solid Agents

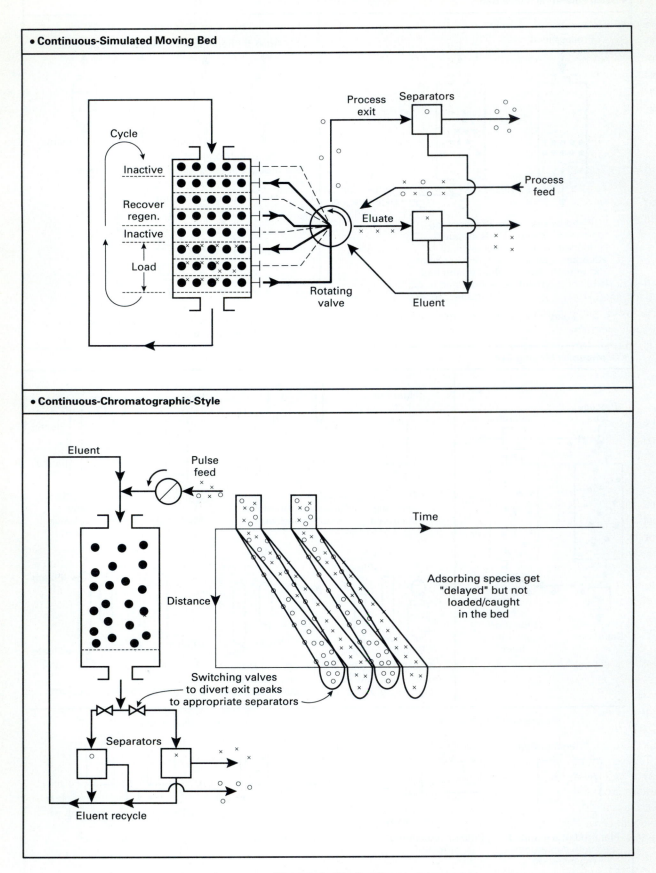

Figure 4–6   (Continued)

butyronitrile, a rating of 3,3,0. A second measure has been developed by Dow Chemical Co. (1966) and pertains only to flammability. They assign a "material hazard" rating to each species. The values for all the species processed on one section of the plant are summed and the total provides guidance as to the safety components that should be put in place.

The implications for the four excursions (temperature, pressure, materials of construction, and hazard) are illustrated for the cost of purchasing the equipment and for operating the equipment in Figures 4–7a, b, c, and d. It is not that the excursion may not be worth it; rather, we should anticipate the cost implications.

3. *Ease in Increasing the Purity by Multistaging.* Sometimes we cannot obtain the degree of separation we want by using the separator once. This is particularly true of the equilibrium type separations listed on the LHS of Table 4–1, but it is also applicable to some of the rate separations described on the RHS of this table. Often we send the process stream through a whole series of units; we call each unit a *stage*. Some methods are easy to stage; others are not. Figure 4–8 illustrates multistaging.

4. *Energy and Overall Economics.* Other rules of thumb based on the economics of processing and sequencing options affect the ultimate choice of separating device. Details will be discussed in Section 4.1–3.

5. *Economies of Scale.* Sometimes bigger is better; sometimes small is beautiful. Some of the equipment options are best suited for large-scale applications; others for small. Figure 4–9 illustrates some idea of how well different options are suited for the amount of material processed. This information pertains to the type of equipment. In general, most options have three regions: the largest or industrial scale, the intermediate or pilot plant scale, and the small or lab scale. Even as we approach the small scale, there often is a limit that we get around by operating the unit batchwise and only part of the time. For adsorption, for ion exchange, and for precipitation/crystallization the capacity range is continuous. For others, the size range occurs in bands. For example, for the large-scale operation for distillation, the size range is relatively narrow. The diameter of the device might range from 1 cm to 12 m for continuous operation (corresponding approximately to capacities of 10^{-4} to 10^2 kg/s). The "usual" continuous operation is, however, a narrower band of about 1 to 50. Similarly, membranes are built in modules that have different capacities. The "usual" are 0.1 to 1 kg/s. Some options apply primarily to small capacities. For such options, larger capacities are achieved by doubling and tripling the small-size unit. For example, zone refining processes a very small amount, about 10^{-6} kg/s. If a capacity of 1000 kg/a is required, we achieve this by installing five units. Sometimes increasing

the capacity by using multiples of a base size can be prohibitively expensive. On the other hand, we note that some options seem most applicable at larger sizes (for example, cryogenic distillation, electrodialysis, and simulated moving-bed adsorption). This is because these require a collection of specialized equipment that is extremely expensive to put together on a small scale. Some options are extensions of laboratory scale equipment and can generally be built for any capacity.

Another form of information related to this is the production scale for different types of products. Table 4–4 shows the "usual" size of a plant that is constructed.

Example 4–4: Reconsider the case of Example 4–3, where we wish to separate a 15% acetic acid, water mixture. The capacity is 10 kg/s. What options might be pertinent?

An Answer: All options still remain pertinent. At this stage we should not restrict the options. However, melt crystallization, dialysis, reverse osmosis, and electrodialysis might be less appealing because, from Figure 4–9, this capacity is greater than that for single-size process units. If we had limited time, we might consider distillation, ion exchange, and solvent extraction as the "prime" candidates.

Example 4–5: A colleague has the task of evaluating cryogenic distillation, pressure swing adsorption, and gas membranes as possible options for producing 98% purity oxygen from air. The oxygen production rate is to be 0.05 kg/s at 400 kPa. Time is limited. She asks which option might be most appropriate if only one can be considered in the time available. What do you say?

An Answer: The first caution is to repress the haste to consider only one option at this time. Several should be considered. If one *must* be considered, then peruse Figure 4–9 for the "feed" to the unit. The amount of air that must go to produce an oxygen output of 0.05 kg/s is 0.05/0.2 or 0.25 kg/s (since air is about 20% oxygen). Cryogenic distillation seems to be more applicable at higher feedrates. Either adsorption or membrane separation might be the options to consider. More details about the subtleties of the equipment options are needed before any distinction between adsorption and membrane separation can be made.

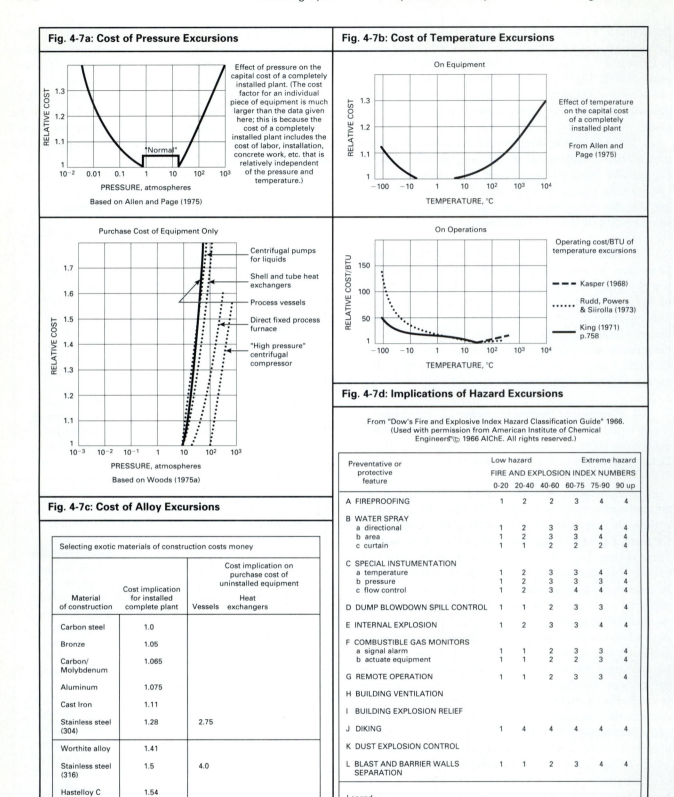

Fig. 4-7a: Cost of Pressure Excursions

Effect of pressure on the capital cost of a completely installed plant. (The cost factor for an individual piece of equipment is much larger than the data given here; this is because the cost of a completely installed plant includes the cost of labor, installation, concrete work, etc. that is relatively independent of the pressure and temperature.)

RELATIVE COST vs PRESSURE, atmospheres — "Normal"

Based on Allen and Page (1975)

Purchase Cost of Equipment Only

- Centrifugal pumps for liquids
- Shell and tube heat exchangers
- Process vessels
- Direct fixed process furnace
- "High pressure" centrifugal compressor

RELATIVE COST vs PRESSURE, atmospheres

Based on Woods (1975a)

Fig. 4-7b: Cost of Temperature Excursions

On Equipment

Effect of temperature on the capital cost of a completely installed plant

From Allen and Page (1975)

RELATIVE COST vs TEMPERATURE, °C

On Operations

Operating cost/BTU of temperature excursions

- – – Kasper (1968)
- ⋯⋯ Rudd, Powers & Siirolla (1973)
- —— King (1971) p.758

RELATIVE COST/BTU vs TEMPERATURE, °C

Fig. 4-7c: Cost of Alloy Excursions

Selecting exotic materials of construction costs money

Material of construction	Cost implication for installed complete plant	Cost implication on purchase cost of uninstalled equipment	
		Vessels	Heat exchangers
Carbon steel	1.0		
Bronze	1.05		
Carbon/Molybdenum	1.065		
Aluminum	1.075		
Cast Iron	1.11		
Stainless steel (304)	1.28	2.75	
Worthite alloy	1.41		
Stainless steel (316)	1.5	4.0	
Hastelloy C	1.54		
Monel	1.65	6.3	
Nickel/Iconal	1.71	8.2	
Titanium	2.0	8	

Fig. 4-7d: Implications of Hazard Excursions

From "Dow's Fire and Explosive Index Hazard Classification Guide" 1966. (Used with permission from American Institute of Chemical Engineers © 1966 AIChE. All rights reserved.)

Preventative or protective feature	Low hazard				Extreme hazard	
	FIRE AND EXPLOSION INDEX NUMBERS					
	0-20	20-40	40-60	60-75	75-90	90 up
A FIREPROOFING	1	2	2	3	4	4
B WATER SPRAY						
a directional	1	2	3	3	4	4
b area	1	2	3	3	4	4
c curtain	1	1	2	2	2	4
C SPECIAL INSTUMENTATION						
a temperature	1	2	3	3	4	4
b pressure	1	2	3	3	3	4
c flow control	1	2	3	4	4	4
D DUMP BLOWDOWN SPILL CONTROL	1	1	2	3	3	4
E INTERNAL EXPLOSION	1	2	3	3	4	4
F COMBUSTIBLE GAS MONITORS						
a signal alarm	1	1	2	3	3	4
b actuate equipment	1	1	2	2	3	4
G REMOTE OPERATION	1	1	2	3	3	4
H BUILDING VENTILATION						
I BUILDING EXPLOSION RELIEF						
J DIKING	1	4	4	4	4	4
K DUST EXPLOSION CONTROL						
L BLAST AND BARRIER WALLS SEPARATION	1	1	2	3	4	4

Legend

Feature optional...1
Feature suggested..2
Feature recommended..3
Feature required...4

Figure 4-7   Implications of Pressure, Temperature, Materials of Construction, and Hazards Excursions

Multistaging

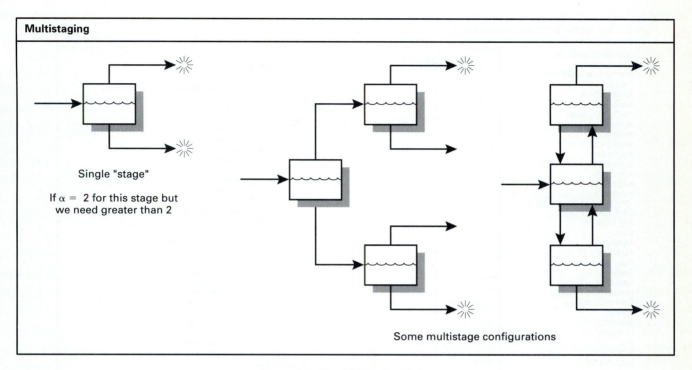

Single "stage"

If α = 2 for this stage but
we need greater than 2

Some multistage configurations

Figure 4–8 How Multistaging Works

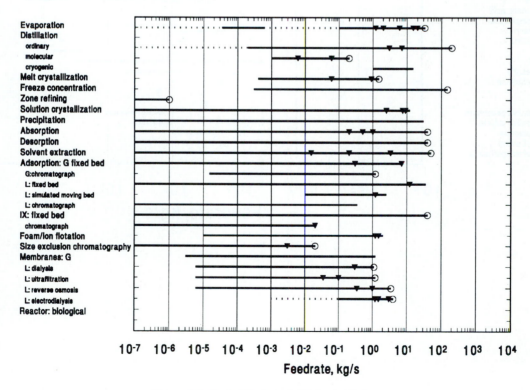

Figure 4–9 Typical Feedrates to Equipment Options

TABLE 4-4 Some production capacity data for single plants

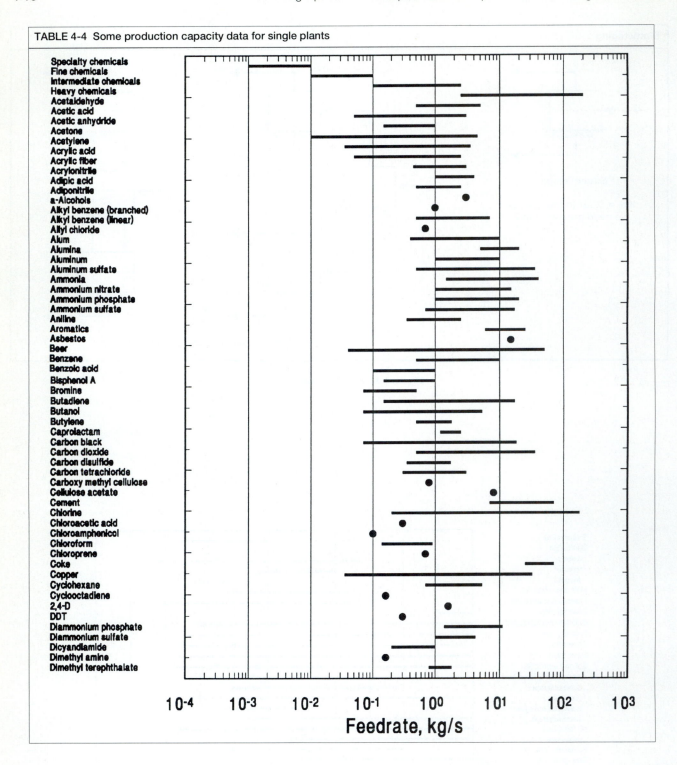

4.1-2 Equipment Options

So far we have seen that in considering equipment options, we should keep in mind the unique differences in properties between the species, typical feed/product concentrations, the effect of trace components, and the need to keep the overall system in mind. Next, we should be aware of the unique characteristics of the equipment.

In this section, each of the classes of options is described briefly. The emphasis is on the key property differences that should exist for that option to work and the principle of operation. Sketches are given. (No details are presented about selecting from among the options within a class. For example, enough detail is given so that we can decide on evaporation as a candidate. However, we will not explore which *type* of evaporator to select. Details of selecting and sizing the type are given by Woods (1993a).

The number, purity, and form of the product streams are described. For each option, we also summarize characteristics that might affect our willingness to select this option as an initial trial.

Table 4–5 is an overall summary of the characteristics of each option.

A. Exploiting Differences in Vapor Pressure.

Two options are in this category: evaporation and distillation.

1. Evaporation. The key physical property difference must be a difference in the **vapor pressure** (or more simply in the normal boiling temperatures) between the two species.

This starts with a liquid phase. The principle is that heat is used to volatize the more volatile component and thus create a vapor that has a composition rich in the more volatile component. Some sketches of equipment are given in Figure 4–10. This is basically a vessel with a boiler for the liquid in the vessel and a condenser for the vapor. (Details about how to select the best configuration of an evaporator is explored elsewhere (Woods, 1993a).)

The *product* may be either the overhead phase or the bottoms. For saline water treatment, the product is the overhead vapor; for glycerol concentration, the product is the bottoms. Figure 4–10 shows the feed-product diagram. For evaporators, we usually want to know the enrichment in the **non-volatile** species that concentrates in the underflow stream. Usually pure solvent is taken overhead. Because we do not know the purity of that stream, the lines on this figure are shown as dotted lines.

Usually about one stage is used; however, to conserve energy, sometimes up to three to five multistages are used.

The design procedures are well developed. Large-scale units can be built; the larger the amount to be processed, usually, the more attractive this option is. Large

scale usually means processing 1 to 40 kg/s of feed. Smaller capacity specialized evaporators, such as agitated falling film evaporators, have feedrates in the range 0.1 to 10 kg/s. Lower feedrates are usually handled by batch evaporation. Evaporation is an extension of a simple laboratory operation.

The main drawback to evaporation is the amount of energy required. To try to minimize the energy used, sometimes the overhead vapor is used as the heating medium for the boiler. To provide the required temperature driving force, the vapor is compressed. This approach is called vapor recompression.

Evaporation may be a questionable option if solids are present, the liquid foams readily, the species degrades at high temperatures, is thermally unstable, or if there is a low concentration of the less volatile species so that we have to boil just about all of the feed. Another concern is if the species form an azeotrope. An azeotrope is a mixture that, as far as vapor pressure is concerned, appears to be a "pure" compound. When this occurs, we cannot separate these species further. For evaporation, this is not usually a concern; for distillation it can be a major challenge.

2. Condensation. Condensation is a possible option if the feed is a gas.

3. Distillation. The key physical property difference is in the **vapor pressure** (or more simply in the normal boiling temperatures). The phase distribution occurs between the liquid and the vapor created.

The principle is that heat is used to volatize the more volatile component and thus create a vapor that has a composition rich in that more volatile component. The equipment is sketched in Figure 4–11. Each column consists of many "stages" that each represent a new liquid-gas equilibrium condition. The vapor is created at the bottom of the column in the reboiler. The overhead vapor is condensed at the top. Part is returned at the top to flow down through the stages as "reflux." The overall purity of the overhead and bottoms products can be increased by increasing the number of stages and by increasing the reflux. Usually, we express the reflux as a ratio of the reflux flowrate to the flowrate of the overhead product. Trays or packing are placed inside the column to promote the gas-liquid contacting. High performance or structured packings are often being used in place of the traditional trays.

Usually the feed enters about the middle of the column with stages above and below. The stages above the feed are called the enriching or rectification stages because their purpose is the ensure that the less volatile species do not go overhead; the stages below the feed are called the stripping stages because the more volatile species are being stripped to go overhead. For some separations, where the feed has **very** volatile species, we may need very few rectification stages and vice versa.

TABLE 4-5. Specific criteria and comments for selecting options for separations

	Usually consider when . . .	Yield Recovery	No. of Prod.	Purity	Comments	Multistaging Ease in		Scale Size	Design Know-how	Trade-off
						Ease in	No.			
Evaporation Condensation	$\alpha_{vp} > 20$; product <u>not</u> temperature sensitive; negligible solids; liquid phase; high conc. of less volatile species		1		High energy required especially when α_{vp} is small and purity needed	P	<10	L	WK	
Distillation	$\alpha_{vp} > 1.5$; product <u>not</u> temperature sensitive, negligible solids		2	99+	High energy required especially when α_{vp} is small and high purity needed; difficult if azeotropes form	E	100	B	WK	
Melt Crystallization	Heat sensitive species; lower energy; difference in freezing temperatures		1	99	Cannot obtain complete separation; watch for eutectic formation; washing to remove liquor	P		S	D	Washing to remove liquor vs. redissolve
Freeze Crystallization	Primarily to concentrate by freezing out water		1	99						
Zone Refining	High purity solid required; the contaminant to be removed lowers the freezing point	95	1	99.999						
Solution Crystallization	Heat sensitive; α_{vp} small want solid product; no need for multicomponent separations		1	99	Cannot recover all because of solubility limits; low energy required; difficult to avoid contamination	P				Washing to give purity decreases yield
Precipitation	Good, sharp first cut to remove species from liquid. Species is temperature sensitive; other options do not seem viable				Move into solid-liquid separation technology					

TABLE 4-5 (Continued)												
	Usually consider when . . .	Yield Recovery	No. of Prod.	Purity	Comments	Added Agent	Recovery regeneration	Multistaging		Scale Size	Design Know-how	Trade-off
								Ease in	Number			
Absorption Desorption	Species is difficult to condense (soluble gas); low concentration				Cross contamination of vapor from absorbing liquid must be acceptable	Add liquid Add gas or steam	Distillation or Pressure Reduction	E	<40	B		
Solvent extraction (SX)	Prime option for dilute concentration feeds; prime alternative for distillation		2	90 to 99	Solids should be removed to <100 ppm to prevent crud formation; difficult to avoid product contamination (extension: supercritical extraction)	Added solvent	Distillation or SX wash stages	E				Selectivity vs capacity
Adsorption (Gas)	Feed conc. of more volatiles <10%; αad>2. product difficult to condense; or large amount needs to be boiled		1			Add solid adsorbent	Discard or thermal regeneration or temp. swing press. swing inert purge or displacement					Selectivity vs capacity
Adsorption (Liquid)	Prime option for dilute conc. feeds especially when need more selectivity than SX		1		If particles present, use fluidized bed or pulp	Add solid adsorbent						
Ion Exchange (IX)	Feed conc. small <0.5 %w/w, high valence ions; have ions for desired solute. Relative permittivity of solution is >35. The higher the atomic number, the more economical it is to use IX				Need to control bacteria and algae growth; sensitive to chlorine; can't remove all the desired ions. IX costs increases proportional to ionic feed concentration	Add solid IX resin	Batch -elute with higher conc. of lower valence or change pH about zpc			S	D	The more selective the resin the more difficult it is to elute or regenerate
IX Chromato-graphic	Proteins with different charge depending on pH					Add solid IX resin	Change pH and elute					

TABLE 4-5 (Continued)

	Usually consider when ...	Yield Recovery	No. of Prod.	Purity	Comments	Added Agent	Recovery regeneration	Multistaging Ease in	Number	Scale Size	Design Know-how	Trade-off
Foam Fractionation	Trace quantities of organics with carbon chain length >12		1		Recovery limited by solubility and drainage of foam. Very sensitive to ionic strength	Add bubbles						
Ion Flotation	Solute is only cation or anion present		1		Add countercharged surfactant	Add surfactant and bubbles	Recover surfactant via SX plus distillation					
Chromato-graphic	Need short residence time and have temperature sensitive species; α_{vp} <1.2; α_{Cr} >2 and temp <300°C. Usually last step in purification because of small volume relative to carrier and column volume (1 to 2% column) recommended M differ by >20% for success	96 but 60 to 80% in single pass	2	99.9 9	Need to eliminate flow maldistribution and capacity-limiting thermal gradients; prevent bacterial contamination; usually use dilute buffer as carrier to minimize IX effects (0.02 mol/L). Perhaps keep concentration <10% to minimize viscosity effect	Add solid adsorbent and liquid carrier		P	500; use about 10 times the theoretical trays for distillation	S	D	
Membrane (Gas)	When product species difficult to condense; a sharp cut is not needed and moderate purity and recovery required. Gas feed conc.> 5%; gas is available at pressure; if a_M >20, then pressure ratio dominates cost	80 to 95	1	86 to 96	Need to minimize the fouling of membrane. Functions best when product is retentate. Best if (permeate/ retentate) <0.4. Economic permeate rates = 0.25 to 2.5 g/s.m²				<3	S		Selectivity vs permeability; Power use increases when improve purity via multistaging
Membrane (Liquid: Dialysis)	high concentrations of small sized species or >5% concentration of small species in a system that is fragile and damaged by shear					Add membrane plus dialysate liquid						

TABLE 4-5 (Continued)

	Usually consider when . . .	Yield Recovery	No. of Prod.	Purity	Comments	Added Agent	Recovery regeneration	Multistaging Ease in	Multistaging Number	Scale Size	Design Know-how	Trade-off
Membrane (Liquid: Ultrafiltration)	liquid with low solute concentration of species with molar mass $> 10^3$ to 10^5 and temperature 4 to 100°C	50 to 90; 90 to 95 with recycle	1	95	limited retentate concentration cannot exceed gel concentration which is usually 20 to 30%; fouling of membrane	Add membrane		P		S	D	
Membrane (Liquid: Reverse Osmosis, RO)	Feed concentration 2 ppm to 10%; If ionic solute, then reasonable selectivity or "rejection". Otherwise: fair; not clean cuts; cannot eliminate all ions. OK for organics M >300 and temperature <45°C. The higher the valence, better the rejection	50 to 90	1	99	limited pressure as the retentate concentration increases, because the pressure must be 10% greater than the osmotic pressure = 2.4 to 3.6 MPa; eliminate fouling and chlorine; costs independent of ionic feed concentration for concentration <1500 ppm., retentate concentration <10%	Membrane that has hydrophobic and hydophilic parts		P		S	D	
Membrane (Liquid: Electrodialysis)	small ionic feed concentration (ca 1000 ppm); and when need to separate ionized from unionized species		1		cannot remove organics; eliminate fouling; usually <700 ppm is economic	Add membane plus power		P		S	D	
Biological Reactions	biodegradable organics (BOD/COD >0.5) in concentration range 100 to 1000 mg/L; pH 6 to 8; temperature 35 to 65°C		1		destroys species	Add microorganisms	species not recoverable			S	D	

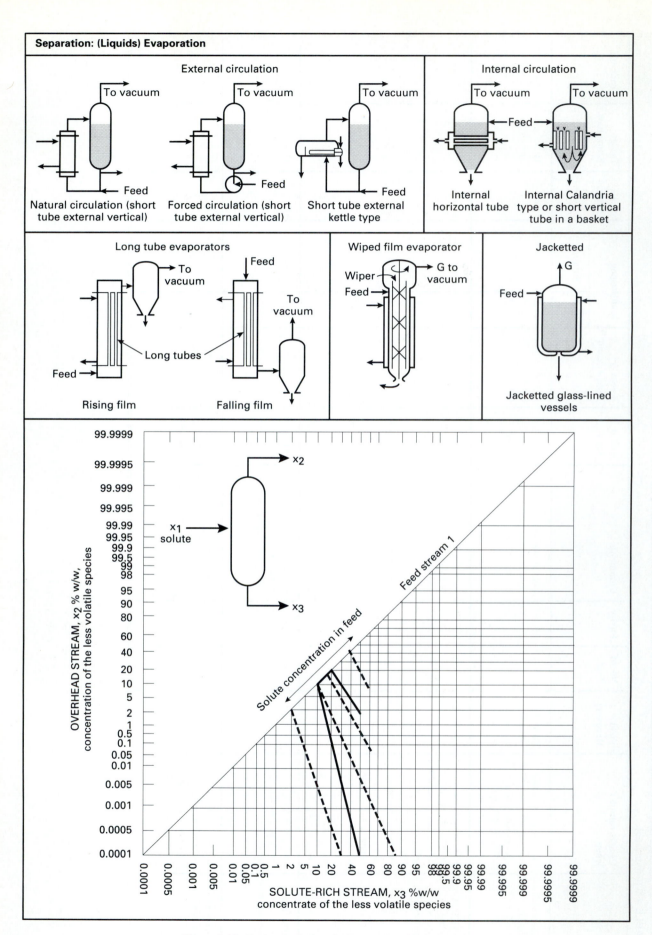

Figure 4–10 Equipment Options for Evaporation: Configurations and Feed-Product Diagram

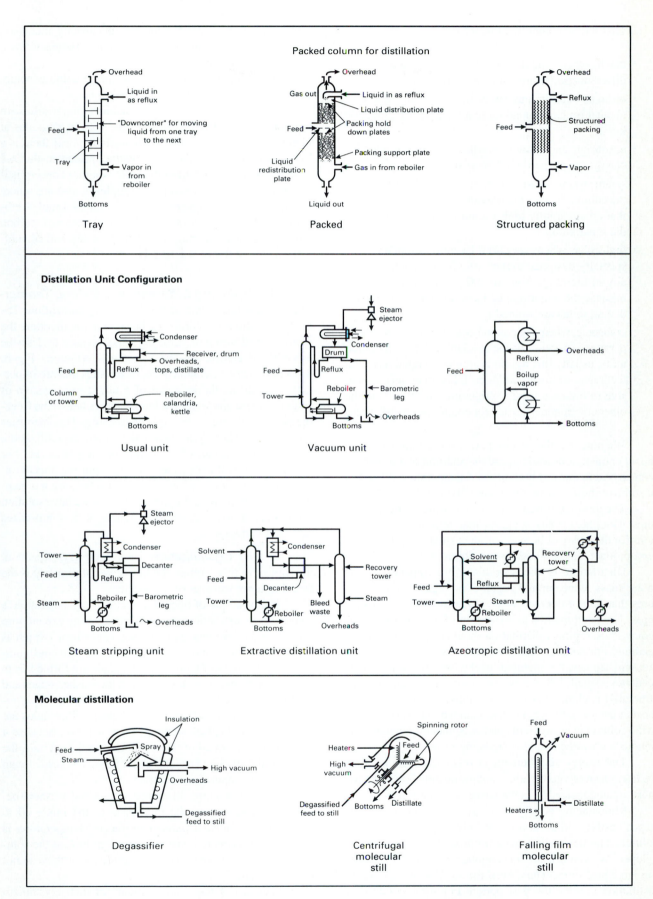

Figure 4–11   Equipment Options for Distillation: Configurations

Distillation comes in various forms:

- batch, where small amounts are processed and the different species are collected as time progresses;
- ordinary, continuous (with the maximum diameter about 12m corresponding to a feed rate of about 300 kg/s);
- azeotropic and extractive, where a solvent is added to change the properties of the system;
- steam, where steam (or an inert) is added directly to the column to alter the pressures within the system so that an apparently high vacuum can be achieved for the separation;
- molecular, where very high vacuum is applied to the specially designed system (the feed capacity is usually in the range 5 g/s to 200 g/s; the upper limit is because the column must have no leaks. This means that all joints are welded.)
- cryogenic, where the column operates at very low temperatures (because of the availability of refrigeration units, usually this option is applied to relatively high feedrates. At lower feedrates, it becomes very expensive to obtain a small refrigeration package and the other components needed for continuous distillation.)

We note that the first two operations require energy; the other operations also require the addition of a solvent or inert, or excursions in pressure and temperature. Sketches of some of these configurations are given in Figure 4–11.

The *product* is usually both the overhead and the bottoms. Douglas (1988) suggests that reasonable expectations for the purity of both the product streams be 99%. The feed-product diagram is shown in Figure 4–12.

Distillation is the workhorse for separating low molar mass, concentrated organic, or organic/water mixtures that do not degrade when exposed to high temperatures. The equipment is an extension of a laboratory technique. Thus, ordinary distillation should be possible for any feedrate. For smaller feedrates in the range 0.1 to 5 kg/s, columns are usually operated batchwise. For even smaller feedrates, small laboratory-size units are used. Before structured packings became so popular, for tray columns the diameter usually was not less than 0.6 m to allow access to the column in between the plates. Smaller diameter, continuously operated columns are being installed.

The energy requirement is relatively high because the latent heat of vaporization (approximately 500 kJ/kg for most organic materials) is the energy driving force.

Multistaging is very easy; we can obtain about 100 stages easily; a limitation is the height we can build one column. The design and scale-up procedures are well developed. We can achieve an economic advantage by building large-size units. Thus, even for small-scale operation, distillation is often the best option. For some conditions, another separation technique might be preferred at small

scale; however, as the throughput or capacity increases, distillation may become the favored option. Hence, we usually start by considering distillation.

As with evaporation, vapor recompression is sometimes used to try to minimize the energy use.

Distillation may be a questionable option if solids are present, the liquid foams readily, the species degrades at high temperatures or is thermally unstable, or if there is a low concentration of the less volatile species so that we have to boil just about all of the feed. Another concern is if the species form an azeotrope. Examples of systems where azeotropes form include ethanol-water (at 89.4 mol % ethanol), acetone-chloroform (at 34 mol % acetone), carbon dioxide-ethane (at 65 mol % carbon dioxide), and ethanol-benzene (at 44.8 mol % ethanol).

B. Exploiting Differences in Melting Temperature or Liquid-Solid Solubility.

Crystallization: The next five options are confusing because in the literature the word *crystallization* or "freezing" may be used to describe any of these. Hence, we start with a few definitions. Figure 4–13 illustrates the chemical characteristics that are important. This shows the behavior of a liquid-solid system of sodium sulfate and water. At temperatures below the freezing point, solid forms because of freezing. This identifies region "A," where processes called melt crystallization, freeze concentration, or zone refining might occur. For temperatures well above the freezing point but concentrations near the solubility limit, solid forms because it precipitates at location "B" in Figure 4–13. Here we have solution crystallization or precipitation. Table 4–6 summarizes other key differences.

1. Melt Crystallization.

For melt crystallization, the subject of this section, the key physical property is the **freezing temperatures** of the species.

The principle of operation for melt crystallization is analogous to distillation except that the energy expenditure relates to the freezing and melting latent heat (which is about 20% of the vaporization latent heat). The configuration is illustrated in Figure 4–14. Crystals are formed from the melt at the bottom, move through the washer stages and the "pure, washed" crystals are melted to create the overhead pure product and the reflux. Typical reflux ratios are about 2 to 4/1. Typical washing stages are five. Because it is difficult to wash all of the solution from the crystals, the purity of the stream is usually about 99%. The feed stream leaves as an "impure" concentrate.

The temperature difference across the washer (between the freezer and the melter) is in the range 20 to 100°C. Although the difference in freezing temperatures is the major criterion to consider, we might look at the solubility products of other species to ensure that they do not precipitate out.

Sketches of the configurations for melt crystallization, given in Figure 4–14b, illustrate the wide range of

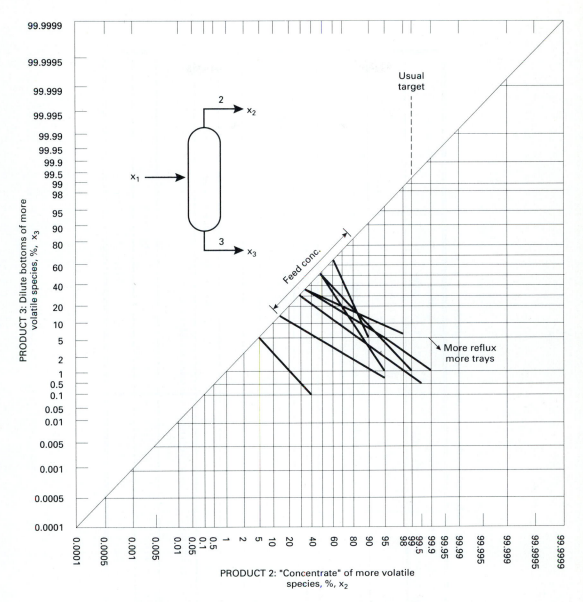

Figure 4–12 Equipment Options for Distillation: Feed-Product Diagram

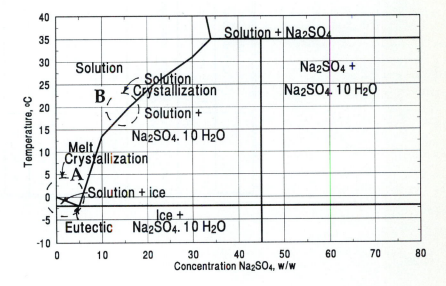

Figure 4–13 Melt and Solution Crystallization

TABLE 4-6. Options that are called "crystallization"

	Melt Crystallization	Freeze Concentration	Zone Refining	Solution Crystallization	Precipitation
Start with	liquid: melt or molten solute		solid	liquid: solute in a solvent "mother liquor"	liquid: potential solute in a solvent
Key	solidify by reducing the temp.		temp. increase/decrease to melt/solidify	cause solute to become insoluble by producing supersaturation condition	cause solute to be insoluble by chemical reaction
Main control	Heat Transfer			Mass Transfer	
Adjust	Temperature $\Delta T \cong 50$ to $100°C$ from freezing usually stop before eutectic			Temperature $\Delta T \cong 2$ to $3°C$ away from solubility line	Temp., Pressure, salt conc., or chemical composition in solvent
Key Parameter		$T_{freezing}$		K_{sp}	K_{sp}
Driving Force		ΔT		Degree of supersaturation relative to solubility	
Action	solute solidifies	solvent solidifies	solute melts then solidifies	solute solidifies as crystal	
Limited	by Eutectics			by solubility	
	location "A" in Figure 4–13			location "B" in Figure 4–13	

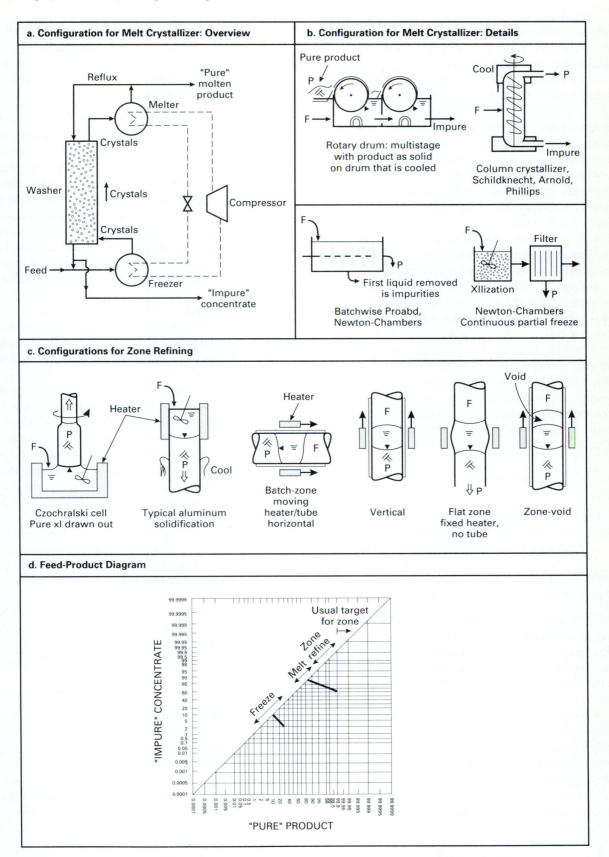

Figure 4–14 Equipment Options for Melt and Freeze Crystallization and for Zone
Refining: Configurations and Feed-Product Diagram a) Configuration for Melt
Crystallizer: Overview b) Configurations for Melt Crystallizer: Details c) Configurations
for Zone Refining d) Feed-Product Diagram

equipment configurations that have been developed. On these sketches the "pure" product is given the symbol "P" and the concentrate is called "impure."

The *product* is usually the single, "overhead" phase. The feed-product diagram is given in Figure 4–14d. Usually the feed concentration is relatively high.

Multistaging is possible although usually fewer than 10 stages are installed. The design and scale-up procedures are developing. This used to be a small-feed capacity option. With the development of direct contact refrigeration for the freeze cycles, feed capacities as large as 3 kg/s can be handled. Using indirect refrigeration, the feed capacity is about one-fifth of this or 0.6 kg/s.

Melt crystallization is an attractive option when the species are temperature sensitive and/or when the boiling points of the species to be separated are similar.

For melt crystallization, one of the main limitations is that many mixtures form eutectics, as illustrated in Figure 4–13. Once the melt has been cooled to the eutectic temperature (−1.1°C on Figure 4–13), all of the species begin to solidify together. For the system in Figure 4–13, this means that the solid is a mixture of ice and sodium decahydrate. Although for some systems the individual species will retain their identity and can be physically separated, for most systems the eutectic condition is to be avoided. The eutectic condition can be estimated by the use of van't Hoff's law:

$$\ln x_2 = \frac{\Delta H_f}{RT}\left(\frac{T}{T_f} - 1\right) \qquad (4\text{-}2)$$

For a binary, this law is applied for both of the species and will result in the freezing temperature-concentration curves that are projected until they intersect at the eutectic.

Because of the eutectics, separation of the feed into two "pure" phases is usually impossible.

2. Freeze Concentration.   Freeze concentration is closely related to melt crystallization in that the principles and the equipment are similar. The key physical properties differences for separation are that there must be a difference (for the species to be separated) in the **freezing** temperatures.

We start with a liquid phase; the principle is that heat is removed to freeze or solidify one of the components (usually the solvent) and leave behind a liquid phase rich in the other components. Thus, in melt crystallization we usually freeze the *solute;* in freeze concentration, the *solvent.* In melt crystallization the product is usually the "crystals"; in freeze concentration, the product is usually the unsolidified liquid (or what was called the "impure" concentrate in Figure 4–14).

Figures 4–14a and b show the equipment used. Here the overhead melt is usually water. The underflow is usually the desired product.

The *product* can be either of the phases; fruit juices are concentrated by removing the frozen "water." Potable water is created by selecting the frozen phase.

The feed-product diagram is given in Figure 4–14d. The equipment configurations and the size applications are similar to those for melt crystallization.

3. Zone Refining.   In zone refining, as with the previous two options, the key physical properties differences for separation are that there must be a difference (for the species to be separated) in the **freezing** temperatures.

The fundamental principle used in zone refining is that any contaminant will concentrate in the liquid phase if its addition to the solid lowers the freezing temperature of the solid. Zone refining is very similar to melt crystallization except that we start with a contaminated *solid* (rather than a liquid melt). The approach is to add or subtract heat to the solid. How this is done depends on the configuration.

Two basically different devices are used, as illustrated in Figure 4–14c. In the Czochralski cell, or progressive crystallization or "crystal-pulling" approach, a starting crystal of the pure species is attached to a vertical rotating shaft. The crystal touches the feed and the "pure" species crystallizes out on the end of the shaft. The shaft gradually is drawn upwards, drawing the pure crystal with it. In the second option—the "progressive freezing" or "batch zone melting" approach—a heater moves slowly over the cylindrical solid. Within the band of the heater the solid melts. As the heater moves along, it leaves behind the freezing or solidification front. The impurities are swept out ahead of the front and concentrate in the molten material in the heater band.

The *product* is the solid phase. The feed-product diagram is given in Figure 4–14d. The purity and the recovery are very high. The purity is 4N or 4 nines or 99.99%. The purity can be improved by repeating the process 20 to 70 times so as to obtain 6N to 8N purity.

Zone refining is the top choice when very high purity solid product is required. However, as can be seen from Figure 4–9, the capacity is about 200 kg/a because the speed of crystal pulling is in the order of 40 μ/s. It may take 30 to 70 days to complete one pass.

The major limitation is the formation of eutectics.

4. Solution Crystallization.   For solution crystallization, the key physical properties differences for separation are that there must be a difference (for the species to be separated) in the **solubility product.**

The feed is a liquid. The principle is that conditions are altered (usually by using heat) so that solid will come out of a supersaturated solution to solidify but solidify under controlled conditions so that crystals grow. This is illustrated in Figure 4–13.

For this to work, we add or subtract heat, or we use energy to pull a vacuum. Figure 4–15 shows some sketches of equipment.

The *product* is usually the solid phase; in the removal

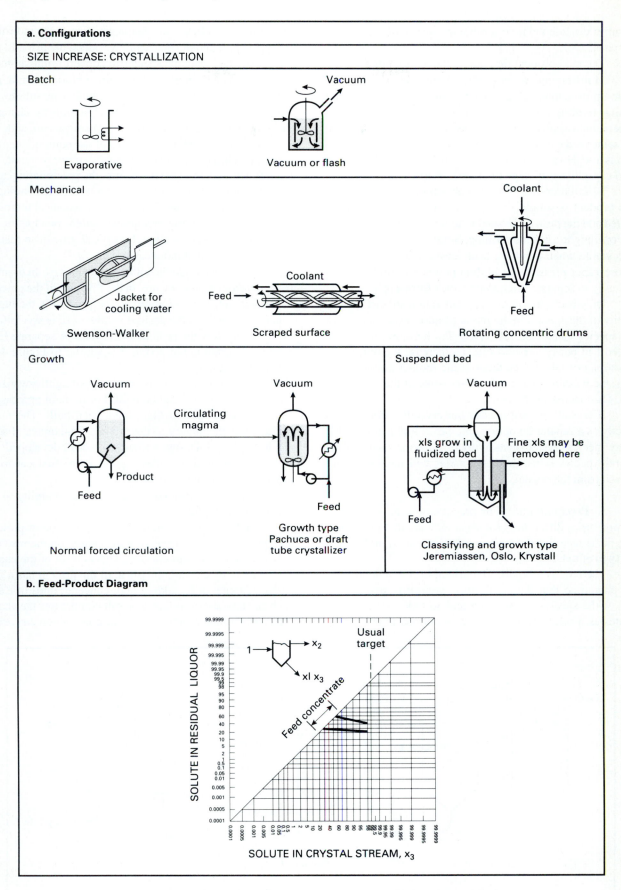

Figure 4–15 Equipment Options for Solution Crystallization: Configurations and Feed-
Product Diagram a) Configurations b) Feed-Product Diagram

of sulfur dioxide via lime scrubbing, the product is the liquid phase.

The feed-product diagram is shown in Figure 4–15.

Multistaging is possible; design know-how and scale-up principles are evolving. Usually crystallizers are similar in design to evaporators. Thus, in Figure 4–9, the general industrial, continuous operation devices have about the same feed capacity as evaporators. Because crystallization is a scale-up of laboratory techniques, the full range of feed capacities can, in general, be constructed.

Solution crystallization is an attractive option when one product is sought that is solid; when crystals are required and the purity expected is about 99%. For higher purity one might use zone refining. Solution crystallization is also viable when handling heat sensitive species, or ones whose vapor pressures or boiling points are similar. Often it is used to separate inorganic species from water.

Drawbacks to solution crystallization are that the solubility of the species in the mother liquor means that all of the species cannot be recovered. Furthermore, there is a tradeoff in purity versus yield. To increase the purity, we wash the crystals to free them of the mother liquor. However, the washing process dissolves some of the solid crystals. This decreases the yield.

If crystals are not needed but crystallization is used strictly as a separation option, another complication is that many species crystallize as hydrates. This means that if the "pure" species is required, the water will have to be removed from the crystal.

5. *Precipitation*.   For precipitation the key physical properties differences for separation are that there must be a difference (for the species to be separated) in the **solubility products.**

The feed is a liquid. The principle is that conditions are altered so that the equilibrium **solubility product** of one of the species is exceeded and so that species precipitates as a solid. To change the conditions to cause pre-

cipitation, we often add chemicals. For most industrial, inorganic precipitations, usually we adjust the pH. Alternatively, we may react the species to convert it into an insoluble form. For example, soluble zinc chloride can be precipitated by converting it into an insoluble zinc sulfide. For some species, we might cause precipitation by changing the temperature. Usually, however, chemicals are added.

Figure 4–16 shows sketches of the equipment; since most "precipitation" reactions are extremely rapid, the main function of the vessel is to mix the precipitating agent with the fluid. A variety of mixer can be used depending on the volume of the vessel. The vessel has a conical bottom to collect and remove the precipitate. Often, precipitation is followed by vessels to allow for crystal growth so that the solid can be handled more easily.

The *product* may be either of the phases. In uranium recovery from brannerite ore, in the first part of the process, unwanted gangue precipitates so the product is the liquid. Later in the process the desired yellow-cake is precipitated as the solid. Now the product is the solid. A separate feed-product diagram is not given; it is similar to Figure 4–15 for crystallization.

Because the design is relatively straightforward, almost any size device can be built. For mineral processing, capacities of 2 million Mg/a have been built. The major limitation is the quality of the mixing. Economies of scale can be exploited. Design know-how is well developed.

Precipitation provides a good sharp first cut to remove species from liquid.

The yield is limited because of the solubility of the species in the liquid.

(In the context of biotechnology, sometimes proteins are "precipitated" by removing the charge on them so that they coagulate to form clusters. This control of the charge can be done by altering the pH until the protein has a zero surface charge, or by adding high concentrations of a salt, such as ammonium sulfate. The salt compresses the charge effect around the protein so that the coagulation can occur.

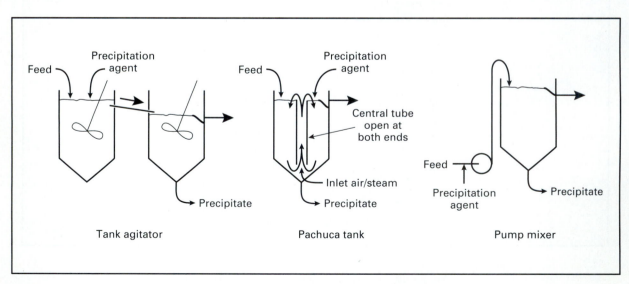

Figure 4–16   Equipment Options for Precipitation: Configurations

Regardless of which approach is taken, this "precipitation" is not instantaneous. Rather, it follows the characteristics of coagulation, a unit operation discussed in detail elsewhere (Woods, 1993b). Hence, in this text, protein precipitation is considered as a coagulation process.

C. Adding Separating Agents to Exploit Differences in Solubility, Partition Coefficient, Exchange Equilibrium, or Surface Activity.

The next collection of options are based on the addition of an agent to achieve the separation. Such addition usually costs more and offers the following additional complications.

1. The agent is effective because of its interaction with the species to be separated. Thus, if we add a scrubbing liquid, we have to select a liquid and compare the interaction between the species with that liquid. For example, so far, if we wanted to separate ethanol from water, we compared the boiling temperatures and the freezing temperatures of ethanol and water. Now, we need to:
   - select a solvent and explore the interaction between ethanol and the solvent with the interaction between water and the solvent;
   - select a scrubbing liquid and do the same;
   - select a membrane and do the same.

   Thus, we have to have some data on how the species interact with the probable agents we might add. In each of the sections, some data will be given. More details will be given later when we spend more time evaluating the option.

2. Usually, we want to recover, regenerate, and reuse the agent. This means that all of these agent-promoted options are really **two separations** in one: species-agent and agent regeneration. Another way of looking at it is that the agent is an intermediary with it achieving a first-step species separation and then a species recover. For example, to separate ammonia from air, we might absorb the ammonia in water. Thus, the air leaving the water absorber/scrubber will have negligible ammonia in it. The water has removed the ammonia. But what do we do with the water? We might take the water into a separate vessel, lower the pressure, and desorb the ammonia into a concentrated ammonia stream. This removes the ammonia from the water. Now the water can be reused in the absorber. This two-step separation is illustrated in Figure 4–17 on a feed-product diagram.

3. As seen on the feed-product diagram, the mass ratio of the agent-feed streams has a dramatic effect on the concentrations in the product streams. We often encounter a tradeoff because for a phase contacting device to work well, we may need a given operating ratio (as illustrated in Figure 3–40). On the other hand, the species-agent interaction dictates how much of the species is transferred. Hence, we could encounter a situation where the agent stream is much more dilute than the original stream. We usually can handle this. It just is a realization that adding too much of an agent is not necessarily useful.

Thus, three issues that will be explored for each of the next six options are the species-agent interaction, the agent recovery/regeneration options, and the effect of the agent-feed ratio on the feed-product diagram.

1. Absorption/Desorption to Exploit Differences in Gas-Liquid Solubility.

The key physical property difference for separation is the **Henry's law constant** for the species to be transferred. This, in turn, depends on the gas solubility or on the molar volumes and the polar and hydrogen bond solubility parameters.

In absorption, we "dissolve" a soluble gas species in a liquid. In desorption, we "strip" a volatile component from a liquid by blowing gas past the liquid or by suddenly reducing the pressure. The principle is that by adding the other phase, the volatile or soluble species will go to the other phase.

The species we commonly consider are ammonia, carbon dioxide, sulfur dioxide, sulfur trioxide, chlorine, hydrogen sulfide, methane, and oxygen. All of these are gases at room temperature. Example "species-agent" data for some commonly used absorbing agents are given in Table 4–7. Often, "package absorption systems" are available for such commonly required tasks as removal of CO_2 and H_2S from gas streams and degassifying water for steam production.

For this process, the medium added is either a liquid or a gas phase depending on the feed stream; for a gas feed stream a liquid is added and vice versa.

Sketches of the equipment are given in Figure 4–18. Usually the absorbing liquid or solvent is put in at the top of the absorption tower. Occasionally, when the preferred solvent is itself rather volatile, the absorber has stages above the feed as illustrated in Figure 4–18b. Thus, we have a two-solvent system. At the top of the column is an absorber to trap the volatile components of the "solvent"; in the lower part of the column is the workhorse solvent that removes the key species from the gas. Another variation is to add a stripping section below the gas feed to strip any unwanted components from the absorbing liquid. In both absorption and stripping at least one species is transferred from one phase to another.

The *product* can be either the exit gas phase that has had the solute removed or that has acquired the species, or the solute that leaves the system in the liquid or both (as with sulfur recovery from sour gas).

In the feed-product diagram for absorption, in Figure 4–18, the usual feed concentration is in the range 0.1 to 20% solute in the gas. Also shown are two practical cases (ammonia into water, acetone into water) where the concentration in the scrubbing liquid is very small; indeed, the concentration in the scrubbing liquid is about the same as

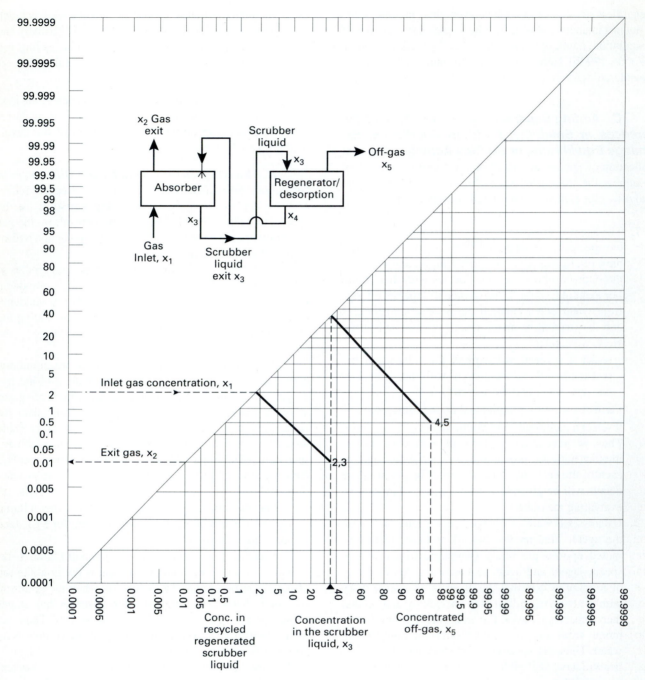

Figure 4–17   Equipment Options for Precipitation: Feed-Product Diagram

the concentration in the scrubbed gas. The example with a feed concentration of 80% is like a quench in that methyl ethyl ketone is being scrubbed from hydrogen with water.

The feed-product diagram for desorption is shown in Figure 4–19. The recovery by pressure reduction alone yields high gas concentrations.

The "recovery/regeneration" of the species and/or regenerate of the scrubbing liquid is often done by distillation or by pressure reduction desorption. This is illustrated in Figure 4–17. **PID-1B** illustrates this absorber-recovery system. Figure 4–38 provides other examples. The number of transfer units is usually about five (unless a tray or packed

tower configuration is used that would supply any number of transfer units).

Absorption columns are commonly built up to 12 m in diameter. Scale-up is usually not an issue although the liquid to gas flowrate should be such that the appropriate gas-liquid contact is achieved. Figure 3–40 gives typical ratios.

One complication is that the more selective the absorbing liquid is, the more difficult it is to regenerate it.

The major limitations of absorption are to be able to locate an acceptable solvent/stripper and for the cross-contamination between the liquid and the gas to be acceptable.

TABLE 4-7. Absorption/desorption: example species-agent interaction

Solute/. . . . . ."Species"	Agent
Absorption: O Gas System H_2S/Hydrocarbons H_2S/air	Triethanolamine "Stretford process"(sodium vanadate)
CO_2; H_2S	Monoethanolamine; hot potassium carbonate; dieth diglycolamine, sulfinol, selexol, catacarb, tributylphosate, methanol
NH_3/air HCl NO_2 SO_3 H_2O/air Ethylene/CH_4, H_2	water water water dilute sulfuric acid sulfuric acid; diethylene glyol, Lithium chloride in water lean oil C_4^+
CO_2	monoethanolamine (MEA), hot potassium carbonate, diglycolamine DGA, "Fluor solvent" sodium hydroxide; propylene carbonate; glycerol triacetate
SO_2	xylidine-water; dimethylaniline; calcium hydroxide; ammonia-water; sodium sulfite, sodium carbonate solution.
Desorption: O Liquid H_2S; NH_3/water	Gas Steam

2. Solvent Extraction to Exploit Differences in Liquid-Liquid Solubility.

For solvent extraction, the key physical properties differences for separation are that there must be a difference (for the species to be separated) in the **molar volumes and the polar and hydrogen bond solubility parameters.**

Solvent extraction is based on the principle that an immiscible liquid solvent is added to the system. The solvent is one to which the solute goes preferentially. Some example "species-agent" interactions are given in Table 4–8. The choice of solvent is a tradeoff because usually the more selective the solvent is for a species, the smaller the capacity for the species in the solvent (or the smaller the concentration of species in the solvent).

The starting feed is a liquid; the species of interest may be, but are not necessarily, ions.

Sketches of the equipment are shown in Figure 4–20.

The *product* may be either the liquid phase left behind (as the raffinate) or the solvent or "extract" phase. The feed-product diagram is shown in Figure 4–20. Solvent extraction is a major workhorse. It is especially appealing for the separation of dilute species and is a major alternative to distillation for more concentrated feeds. The temperature of operation is often room temperature. The feed-product diagram shows that it finds application over most of the range of feed concentrations. Again, the solvent to feed ratio affects the concentrations in the exit (or extract) phase, but usually the concentration is reasonably high.

The "recovery-regeneration" of the species and the regeneration of the solvent is often done by distillation or by "scrubbing" solvent extraction where the extract is con-tacted with another solvent, usually water, to produce the desired effect. The former approach tends to have been used in chemical engineering whereas the latter is common in minerals engineering.

On the one hand, this approach has great flexibility because of the wide variety of very selective solvents that have been or can be developed. On the other hand, the major limitations are the mutual solubilities of the other species, the mutual cross-contamination of both feed and solvent, solvent loss (through vaporization or solubility), and the safety hazards the solvent might introduce. Any solids that are present in the feed should be removed because they tend to collect at the solvent-feed interface and cause the buildup of "crud" that interferes with the successful operation.

The wide variety of contacting devices usually offers some option for the system under consideration. The diameter of the device can be very large. Scale-up procedures are developing.

The *type* of extraction device selected depends on the number of stages and the physical property product of the density difference and the interfacial tension. Thus, if the $\gamma\Delta\rho < 1$ Mg/m^3.mN/m, then centrifugal devices are used; mixer settlers, if > 4. This is illustrated in Figure 4–20c.

The *type* of solvent chosen is related to the properties of the "solute." Table 4–8b relates the solvent type to the solute; + means "has good potential" as an effective solvent. Figure 4–20d relates the Hildebrand solubility parameters δ_h and δ_p and the molar volumes $\tilde{V}$ to the type of solvent.

Data for these solubility parameters and molar volumes are given in **Data** Part D.

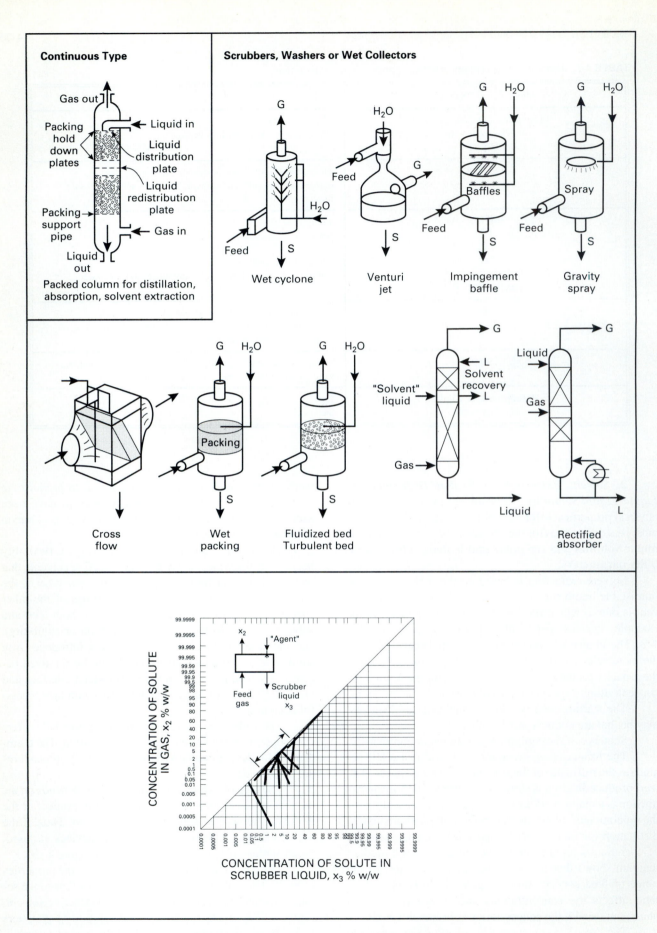

Figure 4–18 Equipment Options for Absorption: Configurations and Feed-Product Diagram

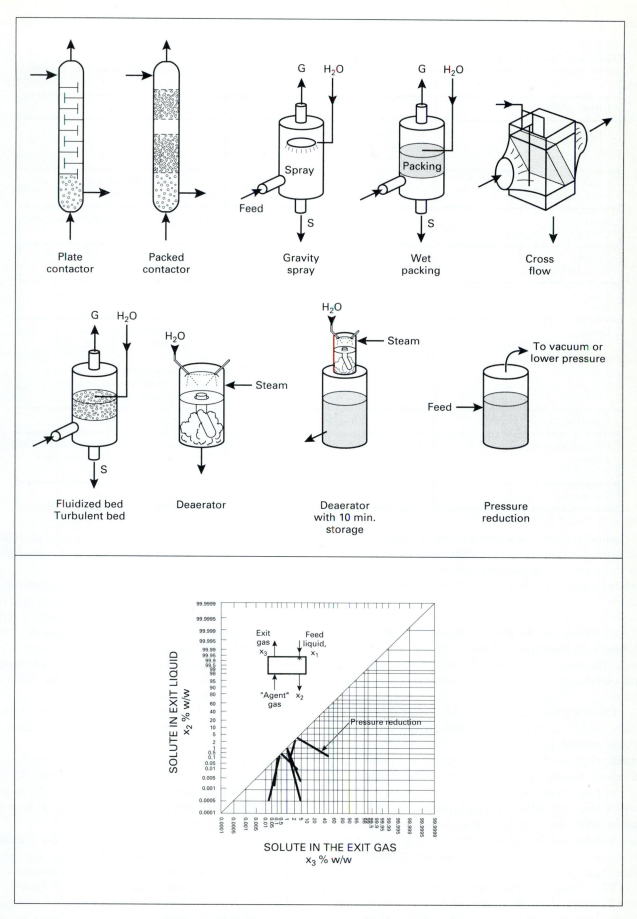

Figure 4–19 Equipment Options for Desorption: Configurations and Feed-Product Diagram

TABLE 4-8. Solvent extraction: example species-agent interaction

Solute Species	Agent	Regeneration of Agent
INORGANIC		
Co/Ni/	(tertiary amine, di-2 ethyl hexyl phosphoric acid (D2EHPA) or tributyl phosphate) in kerosene or organic carrier	
Cobalt- Nickel Ni/Co/ sulfate leach liquor	20% D2EHPA in kerosene	strip via SX with Co-rich aqueous phase to leave Co in organic and extract Ni
Copper Cu/H$_2$SO$_4$ leach liquor pH 2 to 4	10% 2-hydroxyl 5-nonylbenzophenone oxime (L1X 64N) in kerosene	strip via SX with concentrated sulfuric acid
Niobium Nb/Ta/HF leach liquor	methyl isobutyl ketone	strip via SX with HCl 12N HCl strips Ta; 6N HCl strip Nb
Tungsten W/	tricaprylylamine 12% trin-butyl phosphate (TBP) 12% plus 76% kerosene	strip via SX with ammonium tungstate solution + NH$_3$ at pH = 11
Uranium: U/sulfuric acid leach liquor	5% tertiary amine in kerosene or 3 to 10% alkyl phosphate in kerosene	strip via SX with various options: -HCl: acid chloride (1 to 1.5M, pH2) -acid nitrate solutions -neutral ammonium sulfate solutions -sodium carbonate solutions
U/nitric acid leach liquor	tributyl phosphate (TBP) in kerosene	sodium carbonate solutions
U, Plutonium/nitric acid + fission products	30% v/v TBP in kerosene or in dodecane	strip Pu via SX with aqueous solution of ferrous sulfate; strip U via SX with dilute nitric acid
Zirconium/H$_f$ nitric acid	50% TBP in kerosene	5M HNO$_3$ + strip via water

Supercritical extraction uses the principles of solvent extraction in the context of temperatures and pressures that exceed the vapor-liquid critical point for the solvent. Carbon dioxide is a popular solvent for supercritical extraction.

3. Adsorption to Exploit Differences in Surface Exchange Equilibrium. Whereas in the previous two options the species moved into the bulk phase, in the next three options, the species interacts with the surface of the media.

Like the term *crystallization, adsorption* represents a variety of systems (operating on either gas or liquid feeds) and, because the agent is usually a solid, a variety of configuration options as illustrated in Figure 4–6. Consider the options for treating a gas phase and then a liquid phase.

a) Gas phase: For gas phase adsorption, the key physical properties differences for separation are that there must be a difference (for the species to be separated) in the molar volume and the vapor pressure (or boiling temperature) although for molecular sieves the main criterion is a difference in the diameter of the species.

The principle is that solid adsorbent particles are added to the gas. The gas solute species are preferentially adsorbed onto the solid surface. Once all the adsorption "sites" are saturated with the solute, the feed to the bed is stopped. A regeneration process occurs whereby the solute is removed from the solid. Thus, usually, the configuration used is batchwise, fixed bed operation.

Sketches of the usual configurations are given in Figure 4–21. (The moving bed configuration has been used for the purification of vent gases; the chromatographic-style configuration has been used for the bulk separation of the constituents in perfumes and to separate C$_4$-C$_{10}$ normal hydrocarbons from isoparaffins. The larger scale application of chromatographic style operation is evolving. Units as large as 4.2 m diameter have been constructed. Usually in the initial selection activities, we concentrate on the probable feasibility of the option and do not consider the detail of the configuration.)

The *product* is usually the gas phase leaving the adsorber (as with air drying where moisture is removed).

The feed-product diagram, in Figure 4–21, shows that this option has been applied to a relatively wide range of feed concentrations.

Usually adsorption can be more selective than absorption and has the advantage that, because the agent added is a solid, there is negligible cross contamination. Some species-agent combinations are given in Table 4–9. Different adsorbents have different temperature ranges over which they are applicable. The advent of molecular sieves has greatly enhanced the applicability of adsorption.

TABLE 4-8a. Solvent extraction: example species-agent interaction

Soluble Species	Agent	Regeneration of Agent
ORGANIC		
acetic acid/water, formic acid, propionic acid	isoamyl acetate	azeotropic distillation
aromatics/aliphatics	liquid SO_2	distillation and desorb/strip (3 columns)
<u>aromatics/lube oil</u>	phenol + water furfural	distillation and desorb/strip (3 columns)
		distillation and desorb/strip (5 columns)
BTX/aliphatics	diethylene glycol-water	2 step distillation
	tetraethylene glycol-water	strip via SX with dodecane regenerate dodecane via distillation
	dimethyl sulfoxide (DMSO)	strip via SX with paraffinic solvent with subsequent distillation [with water wash and vacuum distillation of raffinate and extract]
	N-methyl pyrrolidone plus 12 to 20% water	distillation
	N-formyl morpholine plus water	distillation
	sulfolane (tetra hydro-thiophene-1, 1-dioxide)	extractive distillation
caprolactam/aqueous impurities	toluene or benzene or methylene chloride	strip via SX with water, purge portion of extract to distillation. Steam strip toluene from water stream
penicillina/broth, sulfuric acid	amyl acetate or n-butyl acetate	aqueous buffer solution pH=6
vitamin A and D/fish liver oil	liquid propane	
p xylene/o- and m-xylene	$HF_{(l)}$ - $BF_3(g)$	distillation

It is useful to distinguish between "bulk" application, where the adsorbed species is greater than 10% of the feed, and "purification" application, where the species has smaller concentrations. In the table, the preferred absorbed species is at the start of the list. Thus, silica gel can be used to adsorb propylene from ethylene.

Many options are available for the "regeneration/recovery." These include thermal recovery (or burning off the species), temperature swing, pressure swing, inert purge, and displacement purge. Which one is used depends on the feed concentration and the desired fate of the adsorbed species.

Large-diameter beds can be installed; the technology is reasonably well developed.

The major limitation is to locate the appropriate adsorbent that can be regenerated and to manage strongly adsorbed impurities that would interfere with the key separation.

b) Liquid phase: The key physical properties differences for separation are that there must be a difference (for the species to be separated) in the molar volume of the species and the solute solubility in the liquid phase.

The principle is that solid adsorbent particles contact the liquid; the target species is preferentially adsorbed onto the solid.

Some representative species-agent interactions are given in Table 4–10, with some of these being similar to those given for gas adsorption applications. Four completely different types of applications occur: high feed concentrations with recovery; low feed concentrations or purification with recovery; purification and throwaway the adsorbent; and purification with high temperature adsorbent regeneration. In general, carbon is difficult to chemically regenerate; usually we regenerate carbon thermally. This means that recovery of the adsorbed species is difficult. On the other hand, resins (such as styrene-

TABLE 4-8b. Solute-solvent interaction

Solvent Interaction with the Solute		Type of Solvent								
Solute		1	2	3	4	5	6	7	8	9
1. Strong potential to form H bonds	paraffin OH, alcohol, water, imide or amide with active H	–	0	+	+	+	+	+	+	+
2. some active H atoms and donor atoms	phenol, aromatic OH acid	0	–	–	–	–	0	+	+	+
3. No active H but donor atoms	ketone, aromatic nitrate, tertiary amine, pyridine, sulfone, trialkyl, phosphate, phosphine oxide or methyl isobutyl ketone	–	+	0	+	+	–	0	+	+
4.	ester, aldehyde, carbonate, phosphate, nitrite or nitrate, amide without active H, intra-molecular bonding, e.g. o-nitrophenol, ethyl acetate	–	+	+	0	+	–	+	+	+
5. water soluble	ether, oxide, sulfide, sulfoxide, primary or secondary amine or imine ethyl ether	–	+	+	+	0	–	0	+	+
6. active H atoms but no donor atoms	multihalo paraffin with active H: methylene choloride	0	+	–	–	–	0	0	+	+
7.	aromatic, halogen aromatic olefin (toluene)	+	+	0	+	0	0	0	0	0
8. no H bonding	paraffin, CS_2 sulfides, mercaptans, hexane	+	+	+	+	+	+	0	0	0
9.	monohalo paraffin or olefin, butyl chloride	+	+	+	+	+	0	0	+	0

divinylbenzene or phenyl formaldehyde resins) are more readily regenerated chemically. However, such resins usually have a lower capacity for the solute.

Some sketches of the usual equipment configurations are shown in Figure 4–22. Many options are used: moving bed, fluidized bed, simulated moving bed, "pulp" or tank, and batch, fixed bed. Chromatographic-style is possible.

The *product* is usually the liquid phase leaving the adsorber (as with waste water treatment where phenol is removed).

The feed-product diagram, Figure 4–22, shows two different regions of application: the low concentrations for purification and the high feed concentration for bulk separation. The dotted lines on the diagram are used when the solute is not recovered and the only data are for the concentrations of the feed and the purified liquid.

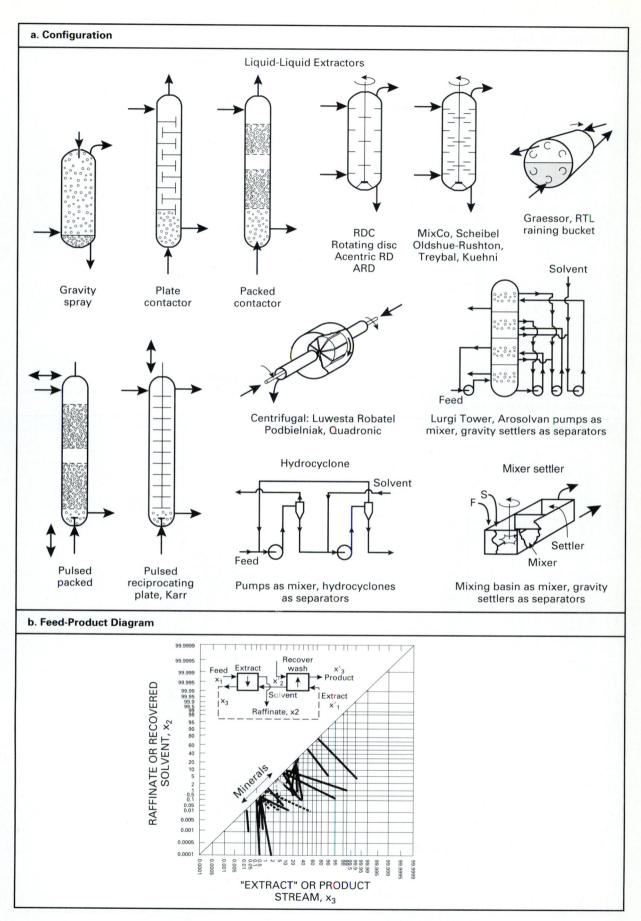

a. Configuration

Liquid-Liquid Extractors

Gravity spray

Plate contactor

Packed contactor

RDC Rotating disc Acentric RD ARD

MixCo, Scheibel Oldshue-Rushton, Treybal, Kuehni

Graessor, RTL raining bucket

Pulsed packed

Pulsed reciprocating plate, Karr

Centrifugal: Luwesta Robatel Podbielniak, Quadronic

Lurgi Tower, Arosolvan pumps as mixer, gravity settlers as separators

Hydrocyclone

Solvent

Feed

Pumps as mixer, hydrocyclones as separators

Mixer settler

F S

Settler

Mixer

Mixing basin as mixer, gravity settlers as separators

b. Feed-Product Diagram

Feed x_1 Extract x_2 Recover wash x'_3 Product

x_3 Solvent Extract x'_1

Raffinate, x2

Minerals

RAFFINATE OR RECOVERED SOLVENT, x_2

"EXTRACT" OR PRODUCT STREAM, x_3

Figure 4–20 Equipment Options for Solvent Extraction: Configurations and Feed-Product Diagram a) Configuration b) Feed-Product Diagram c) Selecting the Configuration d) Classification of Solutes and Solvent

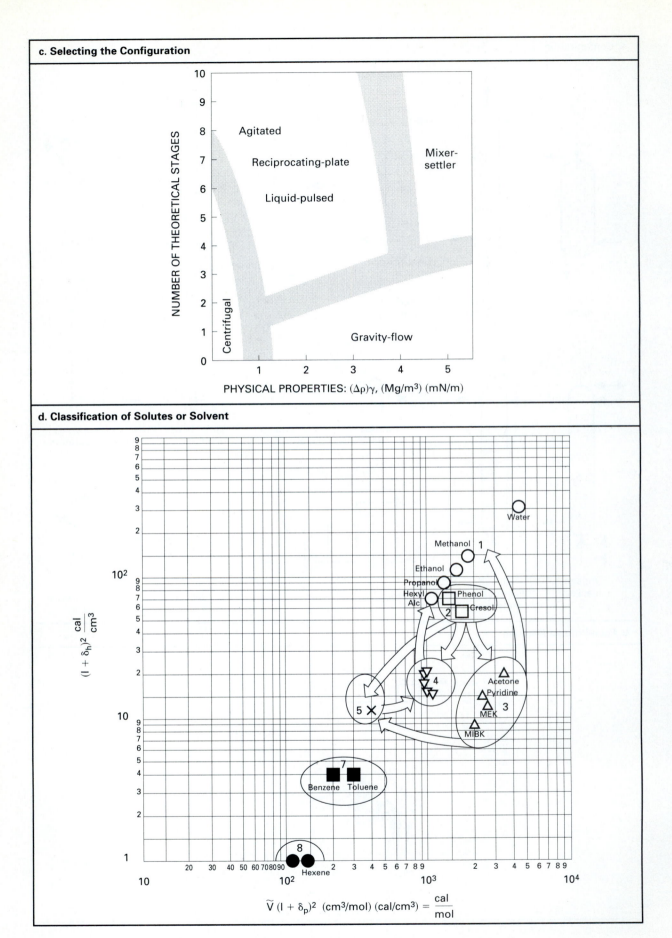

c. Selecting the Configuration

d. Classification of Solutes or Solvent

Figure 4–20 (Continued)

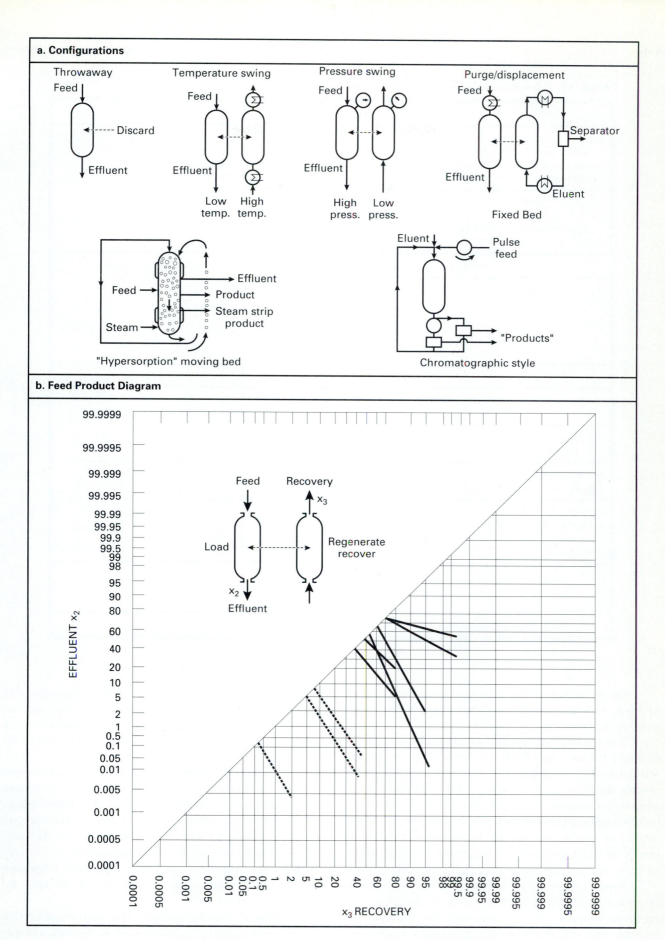

Figure 4–21 Equipment Options for Gas Adsorption: Configurations and Feed-Product Diagram a) Configurations b) Feed-Product Diagram

TABLE 4-9. Adsorption in gaseous systems: example species-agent interaction

Species		Agent
general water>propylene>n butane>isobutane>acetylene>propane> ethylene>CO_2>CS_2>NO_x>ethane>xenon>krypton>methane >CO>oxygen>Ar>N_2>Ne>H_2>He		Silica gel (-240 to 40°C)
purification H_2O> olefins, CH_4, air, synthesis gas	bulk separations	
CO>CH_4>CO_2>N_2>Ar>NH_3>H_2		Activated carbon
purification solvents > air odors > air	bulk separations CO, CH_4, CO_2, N_2, A, NH_3, >H_2 acetone	
		Activated alumina
He, Ne, Ar, CO, H_2, O_2, N_2, NH_3, H_2O/ species > 0.3 nm		Zeolites 3A. (type 5)
————————	———————— O_2 > N_2	4A. (type 4)
Kr, Xe,CH_4, C_2H_6, CH_3OH, CH_3CN, CH_3NH_2, CH_3Cl, CH_3Br, CO_2, C_2H_2, CS_2, H_2S, C_2H_5OH, C_3H_6/ species > 0.4 nm		4A. (type 4)
N_2 > O_2 n-Paraffins, C_3H_8, nC_4H_{10}, n C_7H_{16} . . . n-olefins / species > 0.5nm n-C_4H_9 OH		5A. (type 3)
————————	bulk separations ———————— IsoSiv. ; N_2/O_2	
iso-Paraffins, iso-olefins	/species > 0.8 nm	10X (type 2)

Often, we standardize on a maximum bed diameter of 3.6 m. Simulated moving beds have been built as large as 6.7 m diameter.

The major use of liquid phase adsorption is in purification or the removal of trace amounts of species. Often this means that rather than trying to recover the adsorbed species, the focus is on throwing away or regenerating the adsorbent. Common methods for regenerating carbon adsorbents are temperature swing (with temperatures of about 300°C) or combustion/multiple-hearth furnace regeneration or by steam stripping. For species recovery, resin adsorbents are used that can be eluted with solvents.

Some examples of bulk separations are the recovery of n-paraffins from hydrocarbon streams by the adsorption on zeolites with eluent recovery and the separation of p-xylene from mixed xylenes on zeolites with paradiethylbenzene as the eluent, Bravo et al. (1986).

Liquid adsorption is the prime option for dilute con-

centrations of the species in the feed. Adsorption provides more selectivity than does solvent extraction and has the advantage of minimizing the cross-contamination. The flexible configurations for the contacting also add to the appeal.

The limitations are in being able to recover the species if it is adsorbed (and, in turn, to recover the solute from the eluate). For "bulk recoveries" via chemical displacement, Keller et al. (1987) suggest that rarely will adsorption be able to compete with distillation because of the complexity, capital investment, and energy used in the recovery of the solute from the eluate. On the other hand, adsorption is attractive to purify a stream where we can "discard" the adsorbed species.

Affinity chromatography is a form of adsorption being developed in biotechnology for the separation of species from liquids. The key property difference is the ability of the target species to form a complex with a ligand that is

TABLE 4-10. Adsorption in liquid systems: example species-agent interaction

Species		Regeneration Recovery	Agent
water/hydrocarbon chlorophyll/soybean oil decolorizing oils			Silica gel Fullers' earth
purification:	recovery:		Activated carbon
phenol/water			
	p-cresol/brine	4 to 10% NaOH	
acetic acid/water		acid	
	ethylene diamine/water	NaOH	
	p-nitrophenol/water	NaOH	
	p-chlorobenzene/water	NaOH	
mineral oil/vegetable oil			
water/hydrocarbon			Activated alumina
p-xylene/C_8-hydrocarbons		toluene	Zeolites "K-Ba Y" or NaY
olefins/paraffins			CaX or SrX
n-paraffins/branched and cyclic hydrocarbons		light paraffin	5A
fructose/dextrose and polysaccharides			CaY
p or m cresol/cresol isomers p-cymene/cymene isomers			
TNT/RDX//water		acetone	Resins
phenol/water		acetone formaldehyde	

immobilized on the solid packing. Here, the solid packing should be inert. Thus, the target species becomes trapped in the bed until elution uncouples the species-ligand bond. This method has very high selectivity. The situation is illustrated in Figure 4–23.

Some example species-ligand interactions are given in Table 4–11. Also listed is the eluent which is batchwise sent through the loaded bed to break the complex and thus release the target species. The selection and care of the ligand, sometimes called the gel, offers a challenge. Only a limited amount of ligand can be immobilized; only about 1% of the immobilized ligand becomes "saturated with the target species," the ligand can be washed out of the column if care is not taken, and a high concentration of the eluent is required to free the species.

Because this is a specialized separation option, further details are not given here. However, if the task is a high purity, selective separation of biochemical systems, affinity adsorption should be kept as a possible option. One might consider it as an extended "pilot scale" operation.

The usual diameter is about 0.1 m with the maximum reported as being about 0.3 m.

4. Ion Exchange to Exploit Differences in Surface Exchange Equilibrium.
The key physical properties differences for separation are that there must be a difference (for the species to be separated) in the **valence of the ions** with the ions to be exchanged being the high valence ions. The species must be ions in the liquid phase.

The principle applied is that solid resin particles are added to the liquid. The resin has ionizable groups embedded in the surface that ionize to form a local charge when the resin is placed in the liquid. The ions in solution form an equal and opposite countercharge region close to the resin surface. It is with the "monovalent" ions in the countercharge region that higher valence ions from solution can exchange. Thus, the higher valence ions become concentrated in the liquid within the resin bed. Once the bed becomes "saturated" with the higher valence ions, the feed to the bed is stopped. A regeneration stream containing a high

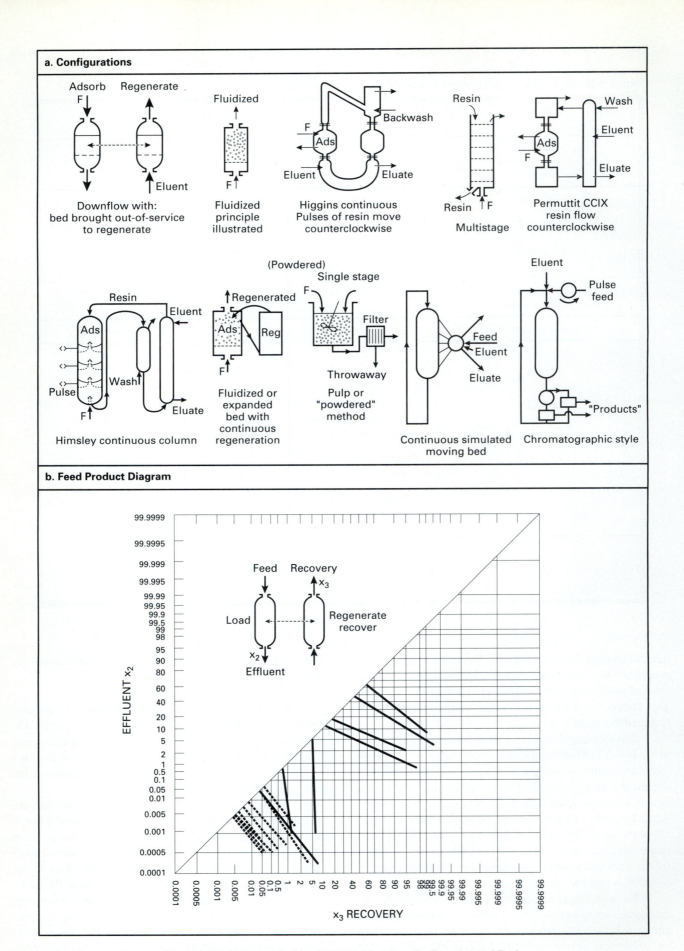

Figure 4–22 Equipment Options for Liquid Adsorption: Configurations and Feed-Product Diagram a) Configurations b) Feed-Product Diagram

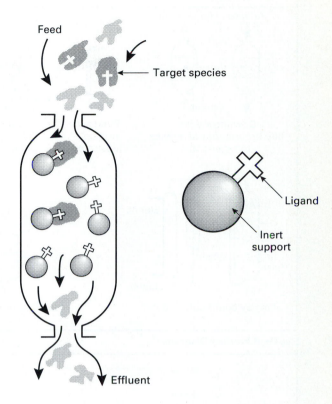

Figure 4–23 Equipment Options for Affinity Chromatography: Configuration

concentration of monovalent ions flows through the bed and now regenerates the bed and kicks off the high valence ions into the regeneration eluent.

For this to work, we add a solid resin that has the correct surface charge in the liquid stream. Sketches of the typical equipment configurations are given in Figure 4–24. The more popular options tend to be batch, fixed-bed, fluidized, moving bed, and chromatographic style. Usually a fixed-bed operation is used. As with adsorption, usually the maximum diameter of bed is about 3.6 m.

The *product* may be either the feed stream leaving the bed (in the case of water softening, where divalent calcium and magnesium ions are exchanged with monovalent sodium in the bed) or it could be the ions trapped in the bed

that are recovered in the eluent stream (in the case of recovering uranium from solution).

The feed-product diagram, Figure 4–24, suggests that usually the feed concentration is small. Higher feed concentrations mean that the bed becomes loaded or saturated so quickly that other options become feasible. Furthermore, the higher concentrations usually mean more contaminant ions are present to compete for the existing exchange sites.

The agent-species interaction is illustrated in Table 4–12. The guidelines in the table are not always straightforward. If the pH < pK_a then the species will be unionized; many species, especially metals, form complexes that can be either cationic or anionic. For more details about metals

TABLE 4-11. Affinity adsorption: examples species-agent interaction with recovery

Species	Ligand	Wash	Elute
• **specific** albumin	cibacron blue held on trisacryl		
plasminogen	lysine held on Sepharose 4B	1M NaCl	0.1M ε-aminocaproic acid
antithrombin III	heparin held on Sepharose CL-6B	0.15 M NaCl	2M NaCl
• **general** enzyme antibody lectin	inhibitor antigen cell wall		

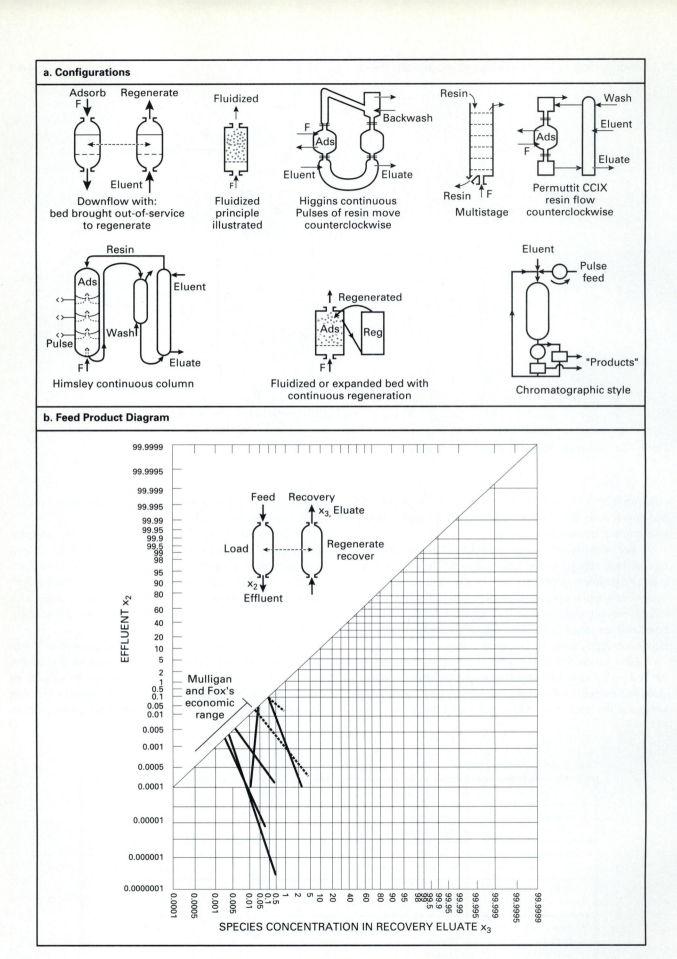

Figure 4–24 Equipment Options for Ion Exchange: Configurations and Feed-Product Diagram a) Configurations b) Feed-Product Diagram

TABLE 4–12. Ion exchange: example species–agent interaction

Solute Species	Agent
$Ba^{++} > Pb^{++} > Sr^{++} > Ca^{++} > Ni^{++} > Mn^{++} > Be^{++} >$ $Cd^{++} > Cu^{++} > Co^{++} > Zn^{++} > Mg^{++} > UO_2^{H} > Ag^+ >$ $Cs^+ > Rb^+ > K^+ > NH_4^+ > Na^+ > H^+ > Li^+$ (For sorption of metals pH>2)	Strong acid (cation exchange) resin Resin- $SO_3^- H^+$; polystyrene sulfonic acid 8% cross linked with divinylbenzene
cations in general	weak acid (cation exchange) resin Resin $NR_3^+OH^-$ polyacrylic acid crosslinked with divinylbenzene
$I^- > NO_3^- > Br^- > HSO_4^- > NO_2^- > CN^- > Cl^- >$ $BrO_3^- > OH^-(II) > HCO_3^- > CH_3COO^- > F^- > OH^-(I) >$ $SO_4^= > CO_3^= > HPO_4^=$ especially anions of weak acids (cyanides, carbonates, silicates)	strong base (anion exchange) resin Resin- $NR_3^+OH^-$ polyvinylbenzyl trimethyl ammonium hydroxide crosslinked with divinyl benzene
anions of strong acids (sulfates, chlorides)	weak base (anion exchange) resin

see Kennedy (1980). Kennedy offers the following guidelines for metal ion exchange: strong adsorption of monovalent metal ions occurs mainly below 0.5 to 1 M hydrogen ion acidity; for divalent, below 1 to 2 M; for trivalent below 2 to 3 M; and for quatravalent, at all pH. To regenerate the exchange resin, a high concentration of a low valence species is usually used. Occasionally, the resin is thrown away rather than regenerated.

Naturally, the species must be present as ions. The pH and the dissociation constant tell us if ions are present. In general, if the species is composed of ions from the LHS of the periodic table combined with ones from the RHS, then they ionize readily. For example, HCl and NaCl. However, the most interesting species are nearer the center of the periodic table where higher valence ions occur. For organic species, the dissociation constants should be considered, as given in **Data** Part D. Another guideline is that the relative permittivity of the ions in solution should be greater than 35. If it is less, than insufficient ionization has occurred.

The higher the atomic number of the elements making up the target species, the more economical ion exchange becomes. As the concentration of ions increases to about 20,000 mg/L, ion exchange may lose its appeal because there are so many competing ions. The cost of ion exchange increases proportionally to the feed concentration. Inorganic ions seem to work well because they exchange without much chemical reaction/interaction with the resin. Organic ions, such as phenol, can exchange but they may interact with the resin so that higher concentrations of eluent are required to regenerate the bed.

A challenge is in selecting the agent or exchange resin. The more selective the resin is, the more difficult it is to recover the species or to regenerate the bed. Bacteria and algae growth must be prevented; operating temperatures are usually less than 45°C. Chlorine can be very detrimental to the operation of ion exchange. Ion exchange cannot remove all of the ions.

Ion exchange chromatography is mentioned as a popular option in biochemical separations; especially for proteins. Actually, because the exchange is so rapid, "chromatography" is a misnomer. The ion exchange is done in the batch, fixed-bed mode and not in the "chromatography-style." The principles are as follows.

Proteins usually bear a charge in liquid solution. The charge is dependent on the pH and indeed each protein has a unique charge-pH diagram such as illustrated in Figure 4–25. Here, the ordinate is the "mobility" or movement of the charged species in a DC field. At pH = 4, β-lactogloblin has a positive charge; at pH = 7, it has a negative charge. At pH = 5.1, the species has zero charge. We call this condition the isoelectric point (or zero point of charge). The iep or zpc condition is very useful in selecting conditions. Some examples are given in Table 4–13; more tabulations are given by Righetti et al. (1976) and Malamud et al. (1978).

This characteristic of proteins can be exploited (just as the valence difference of ions can be exploited) through ion exchange. The agent is usually a cellulose-based resin which, at the feed pH conditions used for loading the bed, has a charge opposite to the charge on the protein. When the bed is loaded, the feed is stopped. The pH is changed to the isoelectric point of the protein and the protein is eluted out of the bed. Thus, the eluant is fluid with a pH of the isoelectric point. The maximum diameter chromatographic column is about 0.4 m.

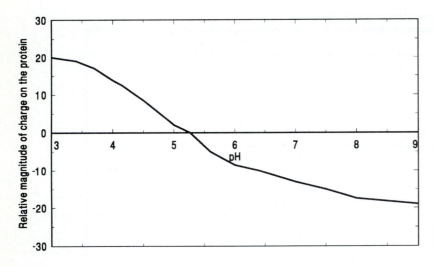

Figure 4–25 How Surface Charge on a Molecule Varies with pH

Ion exchange membranes are used in Electrodialysis, described in Section 4.1-2: D7.

5. *Foam/Ion Fractionation to Exploit Differences in Surface Activity.*   The key physical property difference is that the target species is surface active. This means that the species preferentially resides in the liquid-air surface. The feed must be a liquid. A surface active species is usually made up of two parts: one is hydrophobic, such as a hydrocarbon of chain length 12 to 18, and one is hydrophilic, such as sulfate, ammonium. Thus, sodium docecyl sulfate and cetyl ammonium bromide are surface active in water.

The principle of operation is that a large amount of gas-liquid surface is created in the liquid by blowing in bubbles. The surface active materials move to the surface and are carried out of the bulk liquid phase into a foam that builds up on top of the liquid. The foam is allowed to drain—but not break! Then the foam is swept off the top of the liquid and the two "phases" (the foam and the bulk liquid) are separated.

Sketches of the equipment are given in Figure 4–26. The product may be either the recovered surface active material in the foamate or the "purified" liquid. Usually it is the liquid except in biochemical separations where the foamate contains the target species.

The feed-product diagram, in Figure 4–26, emphasizes that this is used primarily for very dilute feed concentrations. Often the amount of liquid taken overhead in the foamate is relatively small. Thus, the line is almost vertical on the feed-product diagram.

The agent-species for foam fractionation is relatively straightforward. The agent is usually air (although where oxygen might degrade the species, nitrogen could be used). Similarly, little needs to be said about regeneration since once the foam has been collapsed, the gas is released to the atmosphere or to the vent gas system. Some examples are given in Table 4–14 (including some data for ion flotation). The technology for foam fractionation is developing.

Multistaging is possible and moderately large-size devices can be built. Vessels about 1.4 m diameter have been built.

Foam fractionation finds use for dilute concentrations of surfactant, as in the case of waste water treatment.

Ion flotation is a variation on foam fractionation. In foam fractionation, the target species is present in the feed as a surface active agent. In ion flotation, the target species is present in the feed as an ion or as a charged species (such as a protein). We make the ions surface active by adding a surface active agent of the opposite sign. Thus, the ion-surfactant are held together by coulombic forces; the ion goes to the liquid-gas surface because of the surface activity of the surfactant.

The agent-species interaction now includes the surfactant as well as the air (as illustrated in Table 4–14). The recovery-regeneration step requires that we cope with the presence of and the possible recovery of the surfactant. Solvent extraction or vacuum distillation might be options used for the recovery.

The effectiveness of ion flotation is very dependent upon the ionic strength of and the presence of other competing ions in the liquid.

D. *Exploiting Differences in Molecular Geometry.*   **Molecular geometry** is the basis for separation for the next set of options. Size-exclusion chromatography (SEC) involves the flow of a carrier fluid that carries the species through a packed bed. The length of time the species spend in the bed depends on the size of the species. **Membranes** offer another approach. Here, a barrier is placed in the system; some species pass through the barrier; some do not.

1. *Size-Exclusion Chromatography.*   The term *chromatography* has been applied to a wide range of separation processes that differ greatly in the fundamental principles of how the species are separated. On the other hand, the term is used for a particular way of operating a packed bed of agent (what is referred in this text and in Figure 4–6

TABLE 4-13. Example isoelectric point conditions for proteins

Species	Isoelectric condition pH =	Comment
Albumin, serum	5.85	
Albumin, egg	4.6	
α crystallin	4.85	
Cytochrome c	10.80	
γ Globulin	6.6	
ß Glucosidase	4.9	
Haemoglobin A	6.8	
Insulin (beef)	5.72	beef
Insulin (pig)	6.	pig
Invertase	4.24	
ß Lactoglobulin	5.26	
Lipase	3.8	
Lysozyme	10.0	
Malic enzyme pig c.	5.1	pig cytoplasm
Myoglobin	7.1	horse muscle
Pepsin	2.86	
Ribonuclease	8.7	guinea pig pancreas
Thyroglobulin	4.5	
Transferrin	6.0	
Urease	4.88	

as the "chromatographic-style" of operation). Table 4–15 compares and contrasts these. Thus, in this section we consider size-exclusion chromatography. (The use of chromatographic-style of operating other options is discussed in those sections.)

The key physical property difference is one in **molar mass** or size. Two different types of industrial applications are of interest. In "group" or bulk separations, small inorganic ions or organic species (like salt and ethanol) are separated from large macromolecules such as polymers and proteins; that is, there are large differences in molar mass. Differentiating based on small differences in molar mass (like between molar masses of 3000 and 5000) is possible as long as the difference is greater than about 20%. Thus, we can "fractionate" proteins to have a lipoprotein separated from albumin.

The principle of operation is that a carrier fluid receives a pulse of the feed, which moves the mixture through a packed bed of inert spheres. Smaller species diffuse into the interstices or "pores" between the spheres;

turn around and diffuse back out of the pores and eventually leave the bed. Thus, the passage of small species through the bed is "delayed" because they interacted with the bed. Larger species, on the other hand, are too big to diffuse into the pores and are swept straight through the bed by the carrier fluid. Thus, big molecules exit the bed first; the smallest species leave the bed last. If the size of the feed pulse, the size of the pores, the speed of the carrier fluid, and the length of the bed are adjusted appropriately, then all of the different species leave the bed as separate pulses depending on the size of the species. It remains then for us to use the "chromatographic-style" of operation, illustrated in Figure 4–6, to convert this idea into a practical option for separation. Columns up to 0.4 m diameter have been used.

Sketches of the configurations are given in Figure 4–27. The products are different mixtures of the carrier species. The feed-product diagram is shown in Figure 4–27. Usually, 100% recovery of the species is possible. Unfortunately, little information is reported about the actual concentrations of the species being separated. In gen-

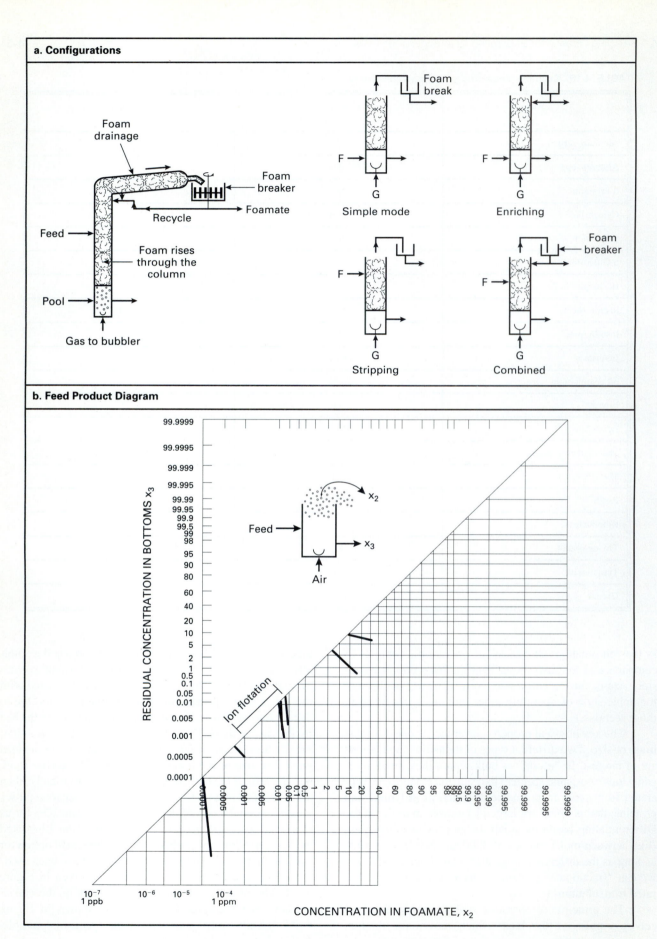

Figure 4–26 Equipment Options for Foam Fractionation: Configurations and Feed-Product Diagram a) Configurations b) Feed-Product Diagram

TABLE 4-14. Foam fractionation: example species-agent interaction

Species	Agent	Recovery
surface active dye/water	air	NA
vanadium/crude oil (containing ''natural''surfactants)	carbon dioxide	
acid oils and phenols/''spent'' caustic	0.04% surfactant added to liquid + air	
chromium ion/water	surfactant + air	surfactant recovered by solvent extraction plus vacuum distillation

eral, the target high molar mass species is diluted by the carrier fluid by a factor of about 1.5 to 10. The more band broadening occurs, the more dilution will occur. Thus, the concentration of the separated species in the product, x_3, will be the feed concentration divided by dilution. The concentration of the high molar mass species in the later pulse should be negligible. Hence, x_2 should be zero. This is shown as dotted lines on Figure 4–27.

The agent-species interaction involves the carrier and the bed of inert, solid particles. Some examples are given in Table 4–16. Because both the species and the bed may have slight charges, the carrier is often a dilute buffer (0.02 M) so that the ionic strength of the medium swamps out the ion exchange effects. The solid phase, beads or stationary phase is selected based on the range of molar masses being considered and the compressive strength needed. Sephacryl is better to resist compression but because of the cross-linking, the pore size is relatively small and hence is more applicable to "group separation" than to fractionation. On the other hand, sephadex and biogel are softer, can resist less compressive strength, but have larger pore sizes and can be used for fractionation of large molecules. The regen-

TABLE 4-15. Comparison of "chromatographic-style" options

Name		Basis of Separation	Comment	In this Text
Size-exclusion, gel permeation, gel filtration, molecular sieve, steric, restricted-diffusional;	G L	Size of the species; small size species diffuse into the pores between inert solids	As the term is used in this book	Size-Exclusion Chromatography
Adsorption, fixed-bed, "chromatography"	G L	"Adsorption" of species onto the solid surface; exchange equilibrium	The "chromatographic-style" of operation is an option that has been developed on both the large and small scale for both gases and liquids	Adsorption
Ion Exchange	L	Exchange equilibrium; valence and charge interaction between species and the solid, charged resin	In the biotech area of protein separations, because the species exchanges so rapidly, the "chromatographic-style" is not used. Rather, a batchwise, fixed bed operation is preferred. Thus, chromatography is a misnomer when used in the context of ion exchange	Ion Exchange
Affinity, immunosorbent, dye-ligand	L	Complexation between species and a ligand that is immobilized on the solid surface	In the biotech area, because the species complexes so rapidly, the "chromatographic-style" is not used. Rather, a batchwise, fixed bed operation is preferred. Thus, chromatography is a misnomer when used in the context of affinity adsorption	Adsorption: affinity

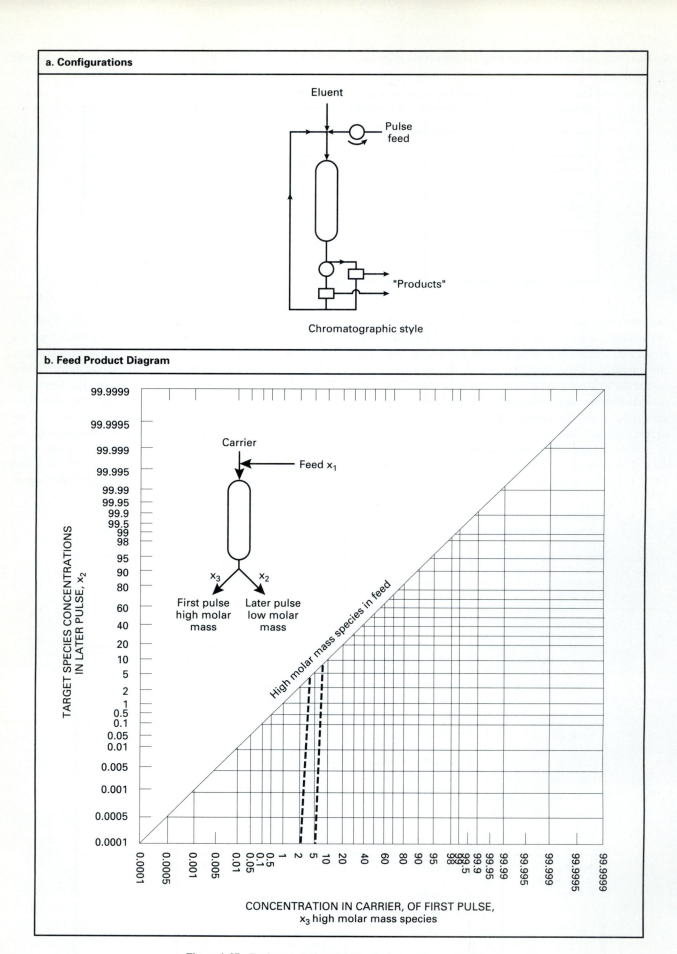

Figure 4–27 Equipment Options for Size-Exclusion Chromatography: Configurations and Feed-Product Diagram a) Configurations b) Feed-Product Diagram

TABLE 4-16. Size-exclusion chromatography: example species-agent interaction

Species	Agent (stationary phase)	Carrier	Recovery Regeneration
Tetanus vaccine/broth		sodium hypophosphate buffer solution	dialysis
Diphtheria vaccine/broth			
Whey or skim milk/ sugars and salts	Sephadex		ultrafiltration
Human serum albumin/ ethanol, water	Sephadex Sephacryl	0.025M sodium acetate water carrier/feed = 5/1 0.05M NaCl carrier/feed = 9/1	ultrafiltration
Insulin/broth		1M acetic acid carrier/feed = 50/1	ultrafiltration
Albumin/glucose	Sephadex	0.005M tris-HCl	
Lipase/microbial rennet	Sephacryl		

Sephadex: crosslinked dextran gel M < 250,000
Biogel P: crosslinked polyacrylamide gel M < 400,000
Sepharose: agarose-based gels 50,000 < M < 40 × 10^6
Biogel A:
Sephacryl: copolymer of allyldextran and bisacrylamide

eration/recovery is dictated by the carrier-species interaction.

Size-exclusion chromatography is used when we need to separate species whose molar masses differ by at least 20%. Thus, it is very useful for separating different fractions of proteins, polymers or enzymes. It also is useful for separating high molar mass species from very low molar mass: polymers from salts. Usually the feed concentration is less than 8 to 10%; larger concentrations of protein increase the viscosity of the fluids so that excessive pressure drop occurs in the column. Usually, SEC is used as one of the last purification steps because of the large volume of carrier and column needed for a relatively small volume of feed. For example, a 100 L column is used to process about 1 to 2 L of feed if used for fractionation. The space velocity relative to the feed only is about 0.003 Bed Volumes/h. The space velocity of the carrier-feed mixture is about 0.1 to 0.2 BV/h. The flowrate is about 50 mL/s.m². For "group" separation, the feed volume can be up to 40% of the bed volume with a flowrate of about 500 mL/s.m².

Size-exclusion chromatography is scaled up on the basis of the total liquid flowrate. Multistaging is relatively easy with about 500 theoretical stages being relatively easy to institute. The challenge is to eliminate flow maldistribution, to minimize thermal gradients, and to prevent bacterial contamination.

2. Membranes: Overview. **Membranes** can be applied to gaseous and liquid systems. A range of different configurations are used. These are illustrated in Figure 4–28. Thus, for dialysis and prevaporation, there are two input streams; for all of the rest, there is one input stream. Some are based on a concentration driving force, others on pressure, and another on electrical field.

Figure 4–29 illustrates how the liquid flux through the membrane relates to the size of the species being processed. In other words, reverse osmosis and dialysis apply to small molecules with a small flux of liquid into the permeate. Ultrafiltration processes larger polymers and has a correspondingly higher flux. (For comparison, also shown on this figure are the pertinent values for "filtration" operations that are considered in Chapter 5. On this figure, the ordinate represents the flux of "water" through the membrane per unit *pressure gradient,* **A.** This flux depends on the physical makeup of the membrane and such parts as the "pore" size.)

In considering membranes, the "solubility" of the species in the membrane material is a factor. Thus, the Hildebrand solubility parameters of the polymers used for the membranes are important starting data. These parameters can be expressed as a "total," or as individual components for the dispersion, polar, and hydrogen bond contribution. Since the dispersion contribution is relatively constant around 16 to 17 MPa$^{0.5}$, the values of the polar and hydrogen bond contributions are the important items. Some values for the polar, hydrogen bond, and *total* solubility parameters are given in Table 4–17 (whereas in **Data** Part D only the polar and hydrogen bond contributions are cited).

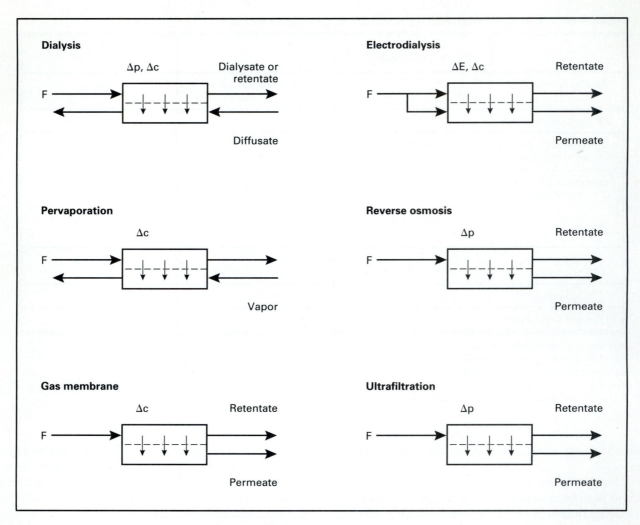

Figure 4–28   Various Membrane Applications

The square of the *total* is the sum of the squares of the dispersion, the polar, and the hydrogen bond contributions. Thus, any increase in the total value usually reflects an increase in either or both the polar or hydrogen bond contributions. In other words, the higher the total value (as given in Table 4–17) the more polar the membrane and the more likely it is to do well with polar species. For the processing of gaseous species, Figure 4–30 illustrates how the total Hildebrand solubility parameter for the polymer membrane relates to the selectivity for CO_2/CH_4 separation as applied to gas membrane permeation. Here, the species traveling through the membrane is the carbon dioxide. This plot assumes that the *physical* characteristics of the membranes are similar and that the chemical characteristics alone are the controlling factors. It also assumes that the concentration gradients are comparable for the situations given in this figure.

In general, membranes do not give a sharp separation between species and produce moderate purity and recovery.

3. Membranes: Gas or Gas Permeation.   The key physical property differences for separation are that there must be a difference (for the species to be separated) in the **diffusivity** which, in turn, depends on the molar volume of the species and the **solute solubility** in the "membrane" phase. We also need to consider how the membrane itself interacts with the species. Thus, the α_m is the ratio of the product of the diffusivity and the solubility of the species in the membrane, as given in Table 4–2.

The principle is that the feed contacts a membrane that interacts with the species and allows certain species to pass through the membrane quickly. Other species might pass through the membrane, but slowly. Thus, there is direct interaction between the membrane and the species. We obtain the retentate on the feed side of the membrane and the permeate from the other side of the membrane.

A sketch of the configuration is given in Figure 4–31.

The *product* is either the concentrated retentate or the permeate. The strength is in producing *one* product. The feed-product diagram shows that permeation is used

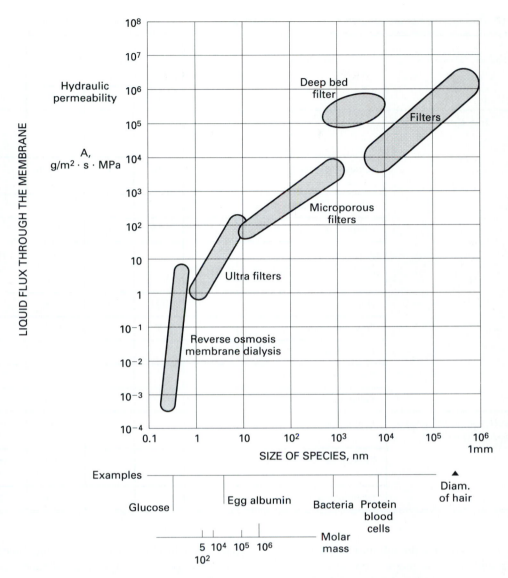

Figure 4–29   General Applicability of Separations Based on Size (based on de Filippi (1977) and used with permission from Marcel Dekker, Inc. copyright 1977)

mainly when the feed concentration of the target species is relatively high.

The agent-species or membrane-species interaction is summarized in Table 4–17 where the ">" represents the relative permeability ratios. Thus, for polysulfone, helium permeates through the membrane faster than hydrogen. The recovery/regeneration of the membrane is not usually a concern although the membrane is very sensitive to temperature and to fouling.

Membranes are used when the target species is difficult to condense and when moderate purity and recovery are required. A membrane does not give a sharp separation between species. For gas permeation, the species concentration should be above 5%. Gas membranes are especially attractive when the gas initially is under pressure and/or when the retentate product is to be at a pressure. Usually

good performance occurs when the permeate/retentate is <0.4.

Selecting a membrane is a tradeoff between the membrane selectivity and the permeation rate. Usually the greater the selectivity the less the permeation rate or the overall capacity.

Although multistaging is possible by hooking up a series of membrane units, rarely are more than 3 stages feasible because the cost of the interstage compression exceeds the increased yield of product.

The limitations include having a membrane that provides sufficient selectivity and that is not contaminated or degraded by any of the species in the feed. More specifically, both contaminant particles must be removed from the feed and conditions must be such that no free liquid enters or forms inside the unit. As a guideline, if water moisture is

TABLE 4-17. Gaseous membrane separation: example species-agent interaction and membrane solubility

Species: Gaseous	Membrane	Solubility Characteristics, MPa$^{0.5}$			T_{glass} or T_{melt} °C
		δ_p	δ_h	δ	
$H_2 > C_1^+$	cellulose	40	25	49	
	cellulose acetate	16	13	21 to 26	230 to 300
	cellulose diacetate			22	
	cellulose triacetate	15	12	25	
	Kapton (Dupont) polyinide			24.6	400 to 500
	Nylon	12.6	13.6	27.8	
	Polyacrylonitrile	22.5	7.5	31.5	
	Polycarbonate			20.6	
	Poly(phenylene oxide)			19.6	
	Polystyrene	7.6	0	18.6	
$H_2O>He>H_2>>H_2S>CO_2>NO_x> O_2>>Ar>CO>N_2>CH_4$	Polysulfone			21	190 to 230
	PES(Union Carbide)			21.3	180 to 185
	Ultem(polyetherinide)			21.4	215

present, the gas inlet humidity should not exceed 85% of the saturation humidity at the temperature and pressure of the membrane operation. If a species that might dissolve in the membrane is present, the inlet feed partial pressure should be less than 10% of the vapor pressure of the pure component (at the operating conditions). For more see Russell (1983).

4. Membranes: Liquid Dialysis. The key physical properties differences for separation are that there must be a difference (for the species to be separated) in the **diffusivity** which, in turn, depends on the molar volume of the species and the solute solubility in the liquid phase. The species are relatively small with molar masses less than 20,000. Usually for us to have sufficient difference in diffusivity we require fairly large differences in size between the species to be separated.

Dialysis applies to liquid systems. The principle is to use a membrane that emphasizes the difference in diffusivity of the species in the feed. The feed flows on one side of the membrane; a dialysate liquid flows countercurrently on the other side of the membrane. The *concentration differ-*

ence across the membrane is the driving force for the diffusion of the species through the membrane.

Sketches of equipment configurations are given in Figure 4–32.

The *product* is usually the purified feed stream. The feed-product diagram is given in Figure 4–32.

The membrane-species interaction is illustrated in Table 4–18. In this table, the species to the left of the solidus (/) is the species that diffuses through the membrane; the species to the right is the slower moving species and, in the time available, does not pass through the membrane. The membranes are generally hydrophobic.

The feed concentration of the target species should be high (usually greater than 5%) because the concentration difference across the membrane is the driving force.

Thus, dialysis is attractive when the feed concentration of the species is high, when the species are temperature sensitive or fragile so that they degrade under high pressures and high shear rates, and when the species is quite different in size from the other species.

One complication that limits the applicability in-

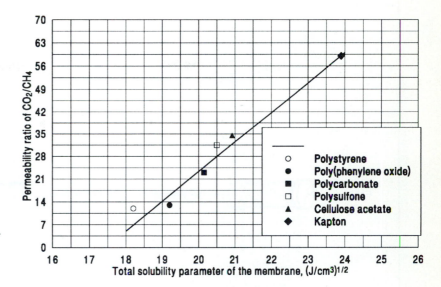

Figure 4–30 *Performance of Gas Permeability Membranes at 35°C and 2 MPa Pressure Difference (from Bravo et al., 1986, and reprinted with permission from Noyes Data Publishing Co., copyright 1986)*

cludes the additional treatment needed if the dialysate is to be regenerated and reused. A sharp cut is not possible.

5. Membranes: Ultrafiltration. The key physical properties differences for separation are that there must be a difference (for the species to be separated) in the **molar mass.** The target species should be at least two orders of magnitude larger than the solvent (in the range 1 to 10nm or with molar masses of 1000 to 100,000).

The principle of operation is that the feed, under moderate pressure (about 700 kPa), contacts the membrane. The relatively large molecules cannot pass through the membrane and so exit as the retentate. The smaller molecules pass through the membrane to form the species in the permeate.

Sketches of the configurations are given in Figure 4–33.

The *product* is usually the concentrated retentate although in rare instances it is the permeate.

The feed-product diagram, given in Figure 4–33, illustrates that usually the feed is relatively dilute. Indeed, the concentration of the target species in the retentate cannot exceed the so-called gel concentration. The latter is the concentration when the flow through the membrane is zero; usually this is about 20 to 30%, depending on the species.

The agent-species interaction, illustrated in Table 4–19, is between the molecules and the membrane. However, for ultrafiltration the big molecules are retained; the small molecules pass through. Thus, in Table 4–19, the species to the left of / passes through the membrane; the species to the right does not. Normally we do not have to worry about regeneration of the membrane although the periodic replacement of the membrane is a significant expense and concern. Other drawbacks are the need to minimize membrane degradation (by chlorine) and fouling.

Naturally, we cannot use membranes when the species interact/dissolve or chemically react with the membrane. This can place a limitation on the types of organic species where ultrafiltration can be a viable option. Temperatures normally should not exceed 100°C.

Ultrafiltration is a viable option for separating low concentrations of species with high molar mass, greater than 10^3. Units are built up to about 60 m^2; the usual size is about 1 m^2 per device.

The maximum continuous operation size corresponds to a feedrate to the device of about 2 L/s; the smaller sizes in the range 0.1 L/s. This rate is not the feedrate to the system because often we recycle retentate to the feed. This recycle can be as high as 20:1 and is included in the flow to the device given above. For laboratory-scale operation, smaller-sized, stirred-pot configurations are available.

The limitations are that the retentate exit concentration cannot exceed the gel concentration. If a sharp cut is required, do not use ultrafiltration.

The design and development procedures are evolving. Multistaging is possible but rarely does it seem economically attractive to have more than 5 stages.

6. Membranes: Reverse Osmosis. The key physical properties differences for separation are that there must be a difference (for the species to be separated) in the **osmotic pressure.** This usually occurs because of ions and can be significant in monovalent ionic solutions.

The principle is as follows. When a semipermeable membrane separates water from a concentrated solution of ions, the water moves through the membrane (by osmosis) into the concentrated solution to "try to even out the water concentration." Such movement occurs until sufficient "osmotic pressure" builds up on the concentrated solution side of the membrane to prevent further flow of the water "up the concentration gradient." We can turn this situation around by *applying* a pressure on the concentrated side of the membrane. The pressure is so large that it *exceeds* the osmotic pressure. Under the pressure gradient the species

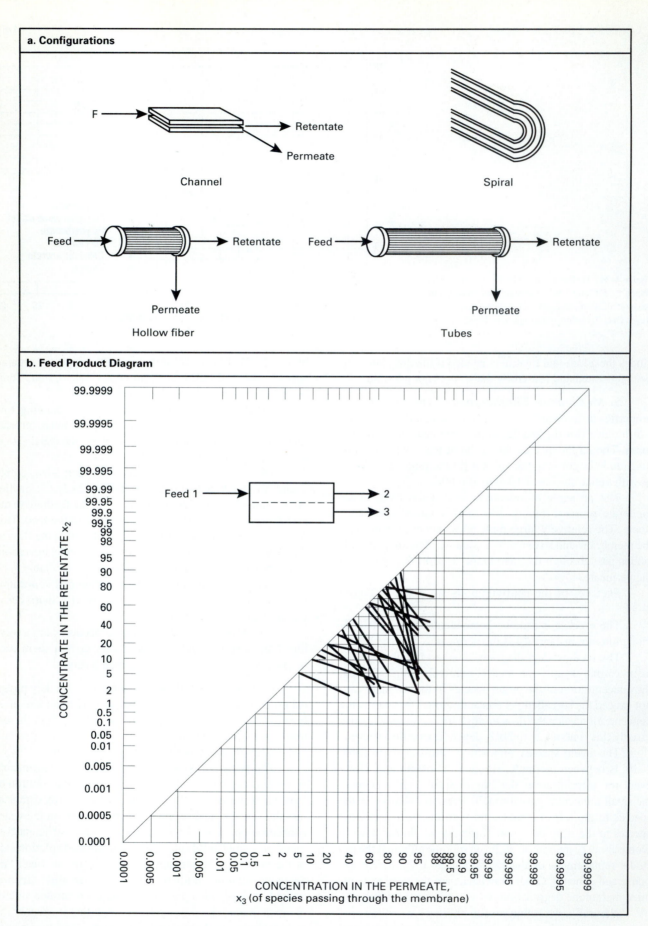

Figure 4–31 Equipment Options for Gas Permeation: Configurations and Feed-Product Diagram a) Configurations b) Feed-Product Diagram

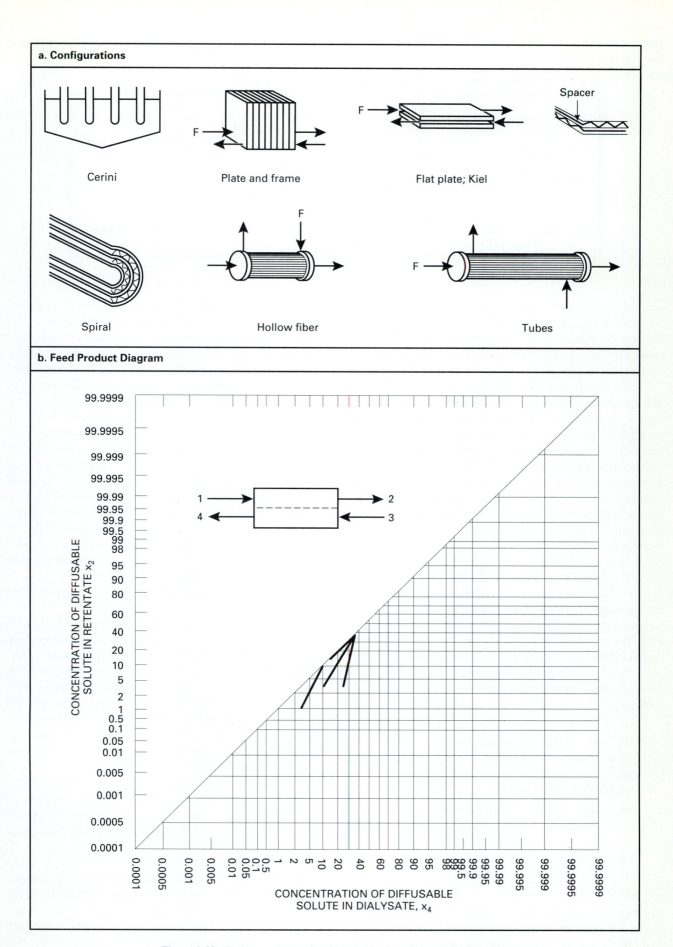

Figure 4–32 Equipment Options for Dialysis: Configurations and Feed-Product Diagram a) Configurations b) Feed-Product Diagram

TABLE 4-18. Dialysis: example agent-species interaction

Species	Agent-Membrane	Dialysate
NaOH/hemicellulose	Parchmentized cotton cloth	water
H^+/Ni^{++} in electrolytic refining fluids	Parchment	water
urea/blood proteins	cellulose polycarbonate polyacrylonitrile	saline solution
salt/raffinose		water

are pushed in the direction reverse to the flow of the solvent by osmosis. Thus, the term "reverse osmosis." Figure 4–34 illustrates the movement of the species, the resulting osmotic pressure, and the application of pressure greater than the osmotic pressure.

This discussion so far has focused on terminology. What happens in the separation? The principle is to use a membrane that allows the passage of water but rejects and prevents the passage of certain ions or species through the membrane. Usually the species are smaller in diameter than the pore size in the membrane; thus, the species-membrane interaction is the key to the separation. We obtain a concentrated retentate on the feed side of the membrane and the permeate from the other side of the membrane. The main ideas are that the operating pressure must exceed the osmotic pressure and the selectivity of the separation depends on the species-membrane interaction.

If target species are ions (the species is dissociated and the pK_a is large) then most membranes *reject* the ions; few ions appear in the permeate and reasonably good separation occurs (90% rejection). The higher the valence, the better the rejection. Thus, reverse osmosis, RO, is a popular option for creating potable water permeate from salt water.

If the species are not dissociated (pK_a is small), reverse osmosis can still be a viable option. The features to look for are:

- the "size" of the rejected species is larger than water (usually the Stoke's radius is used as the criterion); organic molecules of molar mass greater than 300 often satisfy this criterion;
- the rejected species is less polar than water (check the polar Hildebrand solubility parameter in **Data** Part D) but not so non-polar that the molecules stick to the membrane.

This leads to the importance of both the chemical makeup and the fabrication technique used for the membrane. Usually the membrane has both hydrophobic and hydrophilic characteristics. How the membrane is fabricated is also important. Figure 4–35a shows the perform-

ance of membranes that illustrates the variability in performance and also illustrates the characteristic tradeoff between selectivity and capacity. The greater the selectivity (or rejection), the smaller the permeate flux or the capacity. Here, the different shaded regions refer to different feed concentrations and to different types of membranes. (The performance is relatively independent of the feed concentration so the main features illustrated in Figure 4–35a are the effect of the physical and chemical properties of the membrane.) Figure 4–35b shows how the "movement" of sodium chloride by membranes depends on the membrane's Hildebrand solubility parameter. In this situation, the desired condition is for the electrolyte to have the *smallest* mobility. That is, the membrane would "reject" the electrolyte. These data are given at a constant flux of water through the membrane as a permeate (of about 1 $g/m^2.s.MPa$). Thus, this tries to have comparable physical characteristics for the membranes.

Sketches of reverse osmosis configurations are given in Figure 4–36.

The *product* is either the concentrated retentate (in the case of concentrated tomato juice) or the permeate (in the case of saline water feed with potable water as the permeate). The feed-product diagram is shown in Figure 4–36. Also shown on this diagram is Mulligan and Fox's (1976) range of economic feed conditions for reverse osmosis. We note that although RO is used primarily for dilute concentrations (because of the pressure limitations) for some applications, the feed concentration can be very high for organics (such as ethanol/water separations).

The species-membrane interaction is illustrated in Table 4–20. Some of the species are not ions; yet, these do yield an osmotic pressure and thus this approach is applicable. The species to the left, in this table, passes through the membrane; that to the right, is "rejected."

Reverse osmosis function best with ions. It works well with monovalent ions (whereas ion exchange works for ions with higher valence). RO works well for feed concentrations between 2 ppm and 10%.

The limitations of RO are that it does not yield high selectivity, even for ions. A sharp cut is not possible. Nor can it recover or exclude all of the target species.

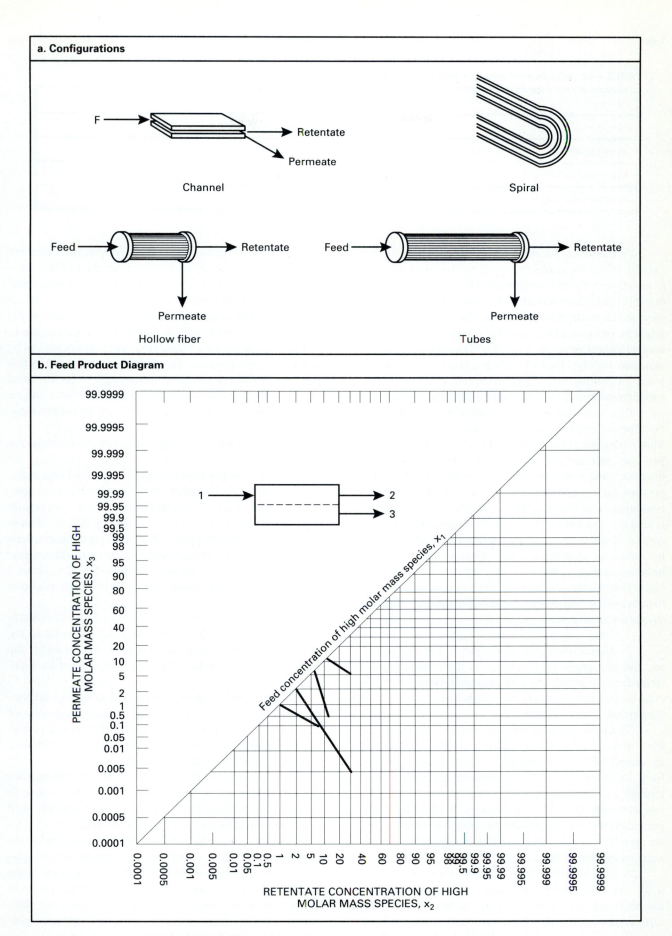

Figure 4–33 Equipment Options for Ultrafiltration: Configurations and Feed-Product Diagram a) Configurations b) Feed-Product Diagram

TABLE 4-19. Ultrafiltration example species—agent interaction

Species	Agent-Membrane
whey/protein	Anisotropic cellulose acetate
water/egg albumin	Polyvinylidene fluoride
water/feta cheese	
water/polyvinyl alcohol	

Furthermore, the exit concentration in the retentate is limited. When the concentration of the retentate increases, to say above 15%, the osmotic pressure can become so high that it becomes impractical for the operating pressure to exceed the osmotic pressure. Thus, the operating pressure limits the exit concentration of the retentate. Practical operating pressures are about 2.4 to 3.6 MPa. Operating temperatures are kept less than 45°C. Major challenges are to minimize the fouling of the membrane and to develop robust membranes capable of withstanding chlorine environments.

Pervaporation is a variation in which a purge gas flows on the permeate side of a reverse osmosis membrane. This is illustrated in Figure 4–28. The conditions on the permeate side are such that all species that pass through the membrane evaporate. The driving force here is **not** the pressure (as in traditional RO); rather it is the species concentration (as in dialysis). In general, the permeate pressure is kept low and indeed is an important operating variable to control. Since the permeating species evaporates, the latent heat must be supplied from the preheated feed or a preheated inlet purge gas.

The challenges include the complication that the membrane may swell at the higher temperatures required to evaporate the species.

7. Membranes: Electrodialysis.

For electrodialysis, the key property difference is the **ionic mobilities (transference numbers) and concentrations** of the species. Electrodialysis works only if the feed is a liquid and the desired species are ions.

The principle is to use a pair of membranes that allows passage of ions but not the passage of water. The movement of the ions through the membrane is facilitated by a DC field across the membrane: the cations move through a cationic permeable membrane toward the negative electrode (cathode); the anions, through an anionic permeable membrane toward the positive (anode). Thus, the membrane stack consists of alternating anionic and cationic membranes placed in a DC field.

A sketch of the configuration is shown in Figure 4–37. The ion-exchange membranes alternate between cation-permeable and anion-permeable such that a "cell" has a different membrane on either side. Feed flows to all cells. The effluent streams from neighboring cells are different. One is rich in electrolyte; the other, depleted. The DC field is applied across the stack of cells. Industrial size operation uses about 70 m² of membrane area per stack. This corresponds to about 1 to 2 L/s feedrate. The feed-product diagram is given in Figure 4–37.

The species-agent interaction is given in Table 4–21. The usual application is to pass sodium and chloride ions to desalinate water or other feeds. However, other ions can be separated from solution. All ions are not removed in this process. Which ones migrate fast enough to be separated from the stream depends on their ionic mobilities and on the concentrations. Electrodialysis cannot be used to separate organics.

The product is usually the retentate (namely the water).

The power required increases as the feed concentration of electrolyte increases, as the concentration of electrolyte in the product decreases, and as the temperature decreases. Thus, in general, we operate with feed concentrations less than 1000 ppm, with electrolyte concentrations in the product water greater than 250 ppm and with temperatures above 10°C. In general, the power load is about 5 MJ/1000 L of product water for every 1000 ppm reduction in electrolyte.

The usual current density is about 10 to 20 mA/cm² of membrane area. Higher current densities facilitate the movement of hydrogen and hydroxyl ions through the membrane (thus defeating the whole objective of the separation).

The general challenges are to find an appropriate ion-exchange membrane (that has an exchange capacity of about 1 to 3 milliequiv per g of dry membrane and a transference number of about 0.9 or above), to prevent fouling, and to cope with the electrode reactions.

To prevent fouling and membrane degradation we need to have low concentrations of less than 0.1 ppm of such contaminants as ferric ion, manganese ion. The feed concentration of organics should also be kept low.

For the electrode reactions, at the cathode, reduction occurs. For example, water may be reduced to form gaseous hydrogen. Thus, the pH will increase because of the buildup of hydroxyl ions. This can cause other species to precipitate. At the anode, oxidation occurs and often chlorine will evolve. Thus, consideration must be given to minimize the hazard and interference that occurs because of this.

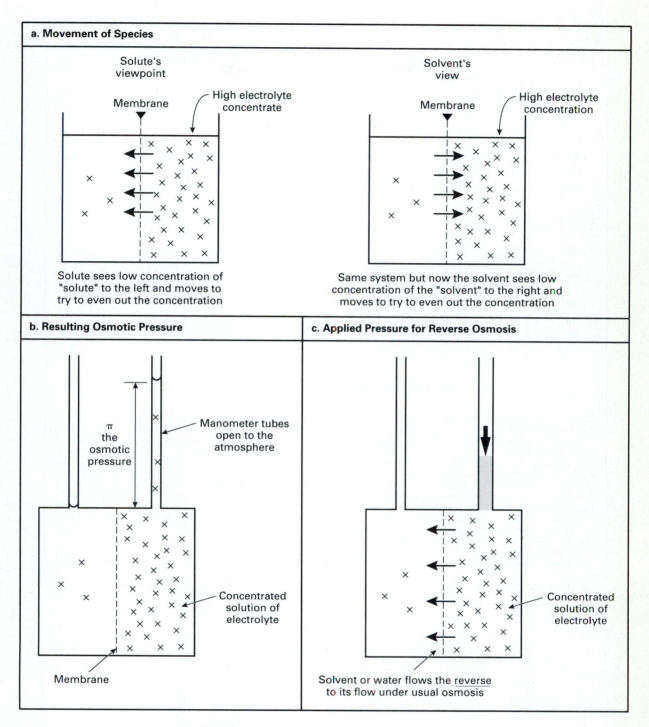

Figure 4–34 Illustrating Osmotic Pressure

E. Exploiting Reactivity.
Sometimes, especially for small concentrations, it is economically attractive to convert the target species to another "preferred" species by reaction or to biologically convert the target species.

1. General Reactions.
We have an extensive arsenal of conversion-removal techniques. For example, many gas absorption separations have been tailor-made to remove certain "preferred" species: carbon dioxide, sulfur di-

oxide, hydrogen sulfide, and water. Thus, if we can convert a target species into these preferred species, we can use off-the-shelf technology to complete the removal. Table 4–22 summarizes some example reactions. Figure 4–38 and PID-1B illustrate some of these "packaged processes."

The results of the reactions can be related to the feed-product diagram by first converting the diagram to mol fractions and then working with "the reaction coordinate" introduced by Smith and van Ness (1975). The reaction co-

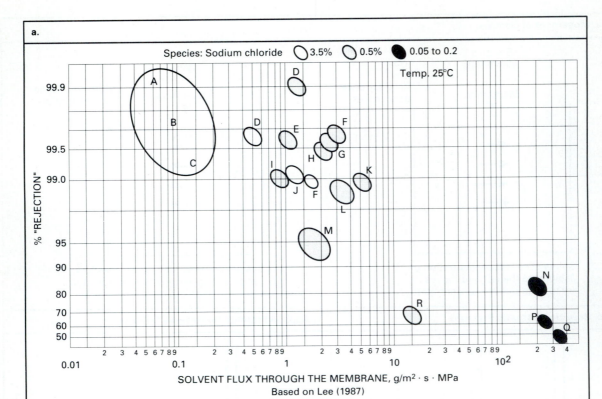

a.

Membranes from various manufacturers: A, Hollo-sep—cellulose triacetate hollow fiber membrane (Toyobo); B, sulfonated polysulfone composite hollow fiber mem-brane (Albany International); C, B-10—aromatic polyamide hollow fiber membrane (Du Pont); D, PEC-1000—composite flat-sheet membrane (Toray); E, NS-200—composite polyfurfuryl alcohol membrane; F, FT-30—composite polyamide flat-sheet membrane (FilmTec/Dow); G, NTR-7199—composite polyamide/polyurea flat-sheet membrane (Nitto Denko); H, TFC-803 (PA-300)—composite polyeth-eramide flat-sheet membrane (Fluid Systems/Signal; I, NS-100—composite polyurea flat-sheet membrane; J, TFC-801 (RC-100)—composite poluetherurea flat-sheet membrane. (Fluid Systems/Signal); K, NTR-7197—composite poly-amide/polyurea flat-sheet membrane (Nitti Denko); L, B-15—asymmetric polyamide flat-sheet membrane (Du Pont); M, asymmetric cellulose acetate flat-sheet mem-branes (generic); N, NF-70—composite flat-sheet mem-brane (FilmTec/Dow); P, Romembra SU-composite polyamide flat-sheet membrane (Toray); Q, NF-50—composite flat-sheet membrane (FilmTec/Dow); and R, NTR-7250—composite polyvinyl alcohol flat-sheet membrane (Nitto Denko).

b.

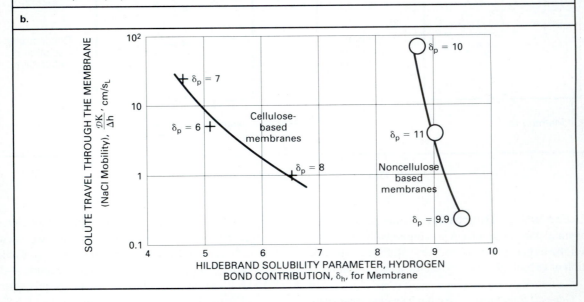

Figure 4–35 Reverse Osmosis: Illustrating Species-Agent Interaction:
a) Effect of Membrane Characteristics on Solute "Rejection" by the Membrane (adapted
from Lee, 1987, and published with permission from Academic Press and from E. K. Lee)
b) Effect of the Membrane Solubility Parameters on Solute Travel

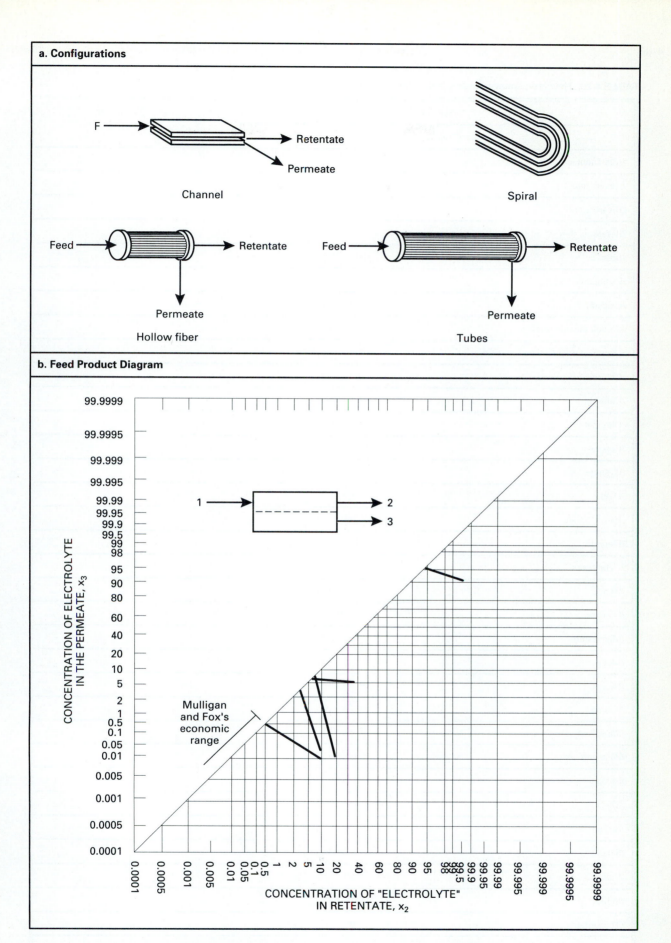

Figure 4–36 Equipment Options for Reverse Osmosis: Configurations and Feed-Product Diagram a) Configurations b) Feed-Product Diagram

TABLE 4-20. Reverse osmosis: example species-agent interaction

Species	Agent-Membrane
water/"ion reject" %	
/Aluminum ion (99%)	
/Ammonium (88 to 95%)	
/Bicarbonate (95 to 96%)	
/Borate (35 to 70%)	
/Bromide (94 to 96%)	
/Cadmium (95 to 98%)	
/Calcium (96 to 98%)	
/Carbon dioxide (diss) (30 to 50%)	
/Chemical oxygen demand (COD) (80 to 95%)	
/Chlorine (dissolved) (30 to 70%)	
/Chloride (94 to 95%)	
/Chromate (90 to 98%)	
/Copper (96 to 99%)	
/Cyanide (90 to 95%)	
/Dyes (100%)	
/Ferrocyanide (99+%)	
/Fluoride (94 to 96%)	
/"Hardness" (96 to 98%)	
/Iron (98 to 99%)	
/Glucose (99.9%)	
/Magnesium (96 to 98%)	
/Manganese (98 to 99%)	
/Nickel (97 to 98%)	
/Nitrate (93 to 96%)	
/Phosphate (99+%)	
/Potassium (94 to 96%)	
/Silicate (95 to 97%)	
/Silver (94 to 97%)	
/Sodium (94 to 96%)	
/Stontium (96 to 98%)	
/Sulfate (99+%)	
/Sulfide (98 to 99%)	
/Thiosulfate (99+%)	
Phenol/water	

TABLE 4-20. (Continued)

Species		Agent-Membrane
water/"ion reject" %		
Acetic acid/water		
water/ethanol	(60%)	ethanol rejecting membrane
ethanol/water		water rejecting membrane
water/tomato juice		
/whey		
acetone/methylacetate		cellophane
acetone/chloroform		saran

ordinate, ε, is the mol of reaction or amount of reaction that occurs. Thus, if the feed concentration of the limiting reaction is 10% and all of it reacts, the reaction coordinate is 10. If the reaction is methanation to convert a target species of CO to the preferred species methane, then the reaction is:

$$CO + 3H_2 \rightarrow CH_4 + H_2O$$

with the corresponding stoichiometric coefficients of

−1   −3   1   1   net reaction stoichiometric coefficient = sum of products less reactants or −2.

Thus, if we want the concentration of methane after the reaction, the composition is:

$$\frac{\text{Initial concentration} + \text{stoichiometric coefficient} \times \varepsilon}{\text{of methane} \qquad \text{of methane}}$$

$$100 + \text{net reaction stoichiometric coefficient} \times \varepsilon$$

Thus, the mol % methane would be $0 + 1 \times 10/(100 + (−2 \times 10) = 13\%$.

2. Biological Reactions. For biological reactions to treat dilute waste waters the general criteria are that:

- the BOD/COD ratio should be greater than 0.5 or that the rate of COD biodegradation should be relatively high; (these data are given in **Data** Part D)
- the pH should be between 6 to 8;
- the temperature should be between
  8 to 25°C for aerobic, low mesophilic range,
  30 to 40°C for the anaerobic mesophilic range,
  45°C for the anaerobic thermophilic range of operations.

The feed concentration of BOD is usually between 100 and 1000 mg/L.

F. Summary. The general applicability of each equipment option is summarized in Table 4–5 and on the feed-product diagrams given for each. The emphasis was on identifying the key property difference that is exploited, and the unique characteristics.

4.1-3 An Overall Selection Procedure

The overall selection procedure considers first the separation of two species. Often one option will dominate. Then multicomponent systems are considered where more than one species have to be separated. Now, we need to take a systems approach and try to evolve a sequence of separation options.

A. Initial Selection for Separating Binaries. We follow the principle of successive approximation and consider how to select options when the task is to separate a binary. The general procedure is:

1. List the species and their properties; compare and identify the characteristics that can be exploited. Use Table 4–1 to suggest options. Consider Figure 4–2 to identify usual differences in properties required for the separation factor for different options. For dilute concentrations, identify which species is the "solute" species. Use the standard tabulation format, **ST,** given on page 4–94.
2. Note the temperature sensitivity of the species. In general, inorganics are robust; foods, pharmaceutical and biological species tend to be very temperature sensitive. Figures 4–3b and c may help to quantify both the effect on the species being processed and the conditions under which the option can operate.
3. Identify the feed conditions and the product requirements. Consider the feed-product diagrams for the different options, and Figure 4–4 for the effect of purity on the options.
4. Note the size of the plant or operation. Use Figure 4–9 as a guideline.
5. Use the data of Table 4–5 to provide guidance; enrich this with the detailed look given to equipment characteristics in Section 4.1-2.

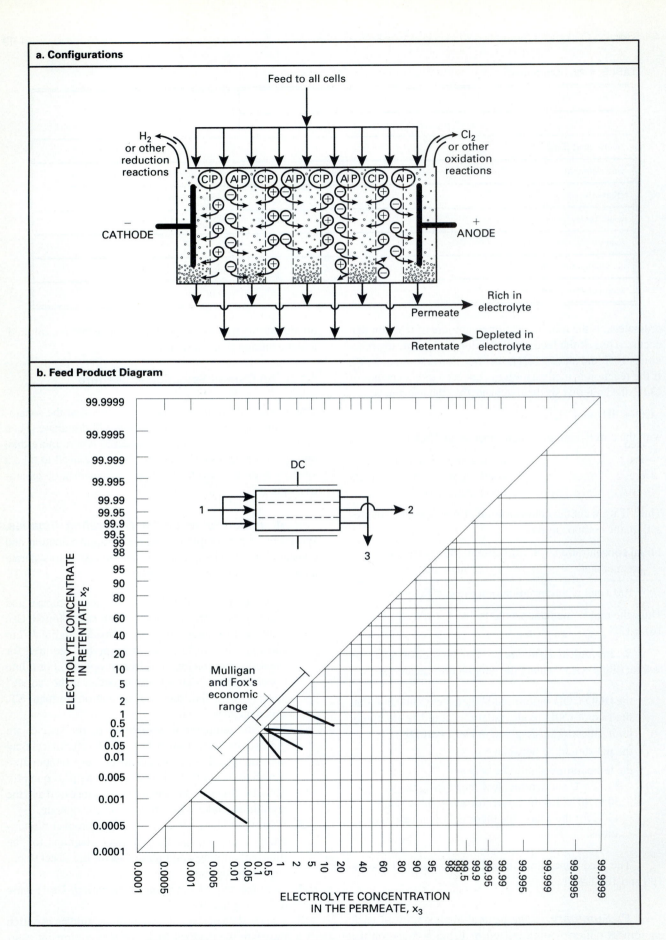

Figure 4–37 Equipment Options for Electrodialysis: Configurations and Feed-Product Diagram a) Configurations b) Feed-Product Diagram

TABLE 4-21. Electrodialysis: example species-agent interaction

Species	Agent-Membrane
Na⁺ Cl⁻/water	cationic and anionic
Na⁺ Mg⁺⁺/pulp waste	exchange membranes or
NaCl/whey	cationic exchange plus neutral membranes
NaCl/milk	
NaCl/dextrin	
Strontium/milk	
Citric acid/fruit juices	
"desalting" in general	

6. If a separating agent is added, consider whether a second separation system is needed to regenerate the agent or to recover the target species.

Based on this approach we should have about half a dozen options. These might include a variety of combinations. Some of the options will apply to special regions and there may be only one possible. Sometimes there will be overlap with many technically acceptable options; here economics will be used as the basis of selection.

How do we decide from among the different options? The answer is on the basis of economics, safety, environment, and risk in developing the technology. For example, whereas distillation technology is well developed and relatively risk-free, the design of membranes is evolving.

B. More Detailed Selection of Binaries.
In this section, we compare based primarily on economics. Which option would cost the least to operate?

1. For Dilute Gases <1%. The prime options include adsorption, absorption, reaction/combustion, and perhaps condensation or cryogenic distillation. We note that both adsorption and absorption require the addition of a separating agent. Hence, we may need two separation sequences.

The choice from among these depends strongly on the species involved, the capacity and the purity-recovery requirements. It is important to identify which species is the "solute"; ideally, the solute species will be the dilute species. Figure 4-39 illustrates the general applicability, based on economics, with absorption extending into higher concentrations as the capacity increases. Similarly, as the capacity increases, some of the options shift over to cryogenic distillation. This is very generally summarized for some example applications (for both small and larger concentrations) in Table 4-23.

Example 4-6: We need to dry air so that the dew point is −20°C. What might we do? The capacity is 1000 dm³/s.

An Answer:
1. List the species and their properties; compare and identify the characteristics that can be exploited. Consider Figure 4-2 to identify usual values required for the separation factor for different options. The species are water and air.

The properties are: **ST**
water 18   0°C   100°C   18.1   22.8   40.4   15.74
air 29   —   −183

Thus, there is modest difference in molar mass, a large difference in boiling and freezing points. Assume the air is 50% relative humidity at room temperature. Thus, from **Data** Part D and the humidity chart, the feed concentration is about 9 g/kg dry air; the product is less than 1 g/kg dry air. The "ideal solute species" is the water. The feed concentration is very dilute. Figure 4-2 suggests that an alpha of 2 would be needed for adsorption; an alpha of about 10^4 would be expected for absorption. From Table 4-22, reaction/combustion is not a probable option.
2. Note the temperature sensitivity of the species. The air is robust; however, for temperatures below 0°C, water would freeze if the pressure is atmospheric.
3. Identify the feed conditions and the product requirements. From the feed-product diagrams for the general options, all seem to be applicable.
4. Note the size of the plant or operation. This is 1000 dm³/s or about 1.2 kg/s.

TABLE 4-22. Example reactions (and packaged processes) to separate species

Target Species	Phase	Convert to . . . by	Preferred Species	Removal of Preferred Species by
CO, CO_2 (trace)	Gas	"Methanation" with H_2 $CO + 3H_2 \rightarrow CH_4 + H_2O$ $CO_2 + 4H_2 \rightarrow CH_4 + 2H_2O$	CH_4	use as fuel
Acetylene (trace)	Gas	hydrogenation $C_2H_2 + H_2 \rightarrow C_2H_4$ $C_2 + H_2 \rightarrow C_2H_6$ $C_2H_6 + H_2 \rightarrow$ "green oil" presence of CO promotes reaction 1	C_2H_4	as feedstock
S compounds (trace)	Gas	hydrogenation (Beavon)	H_2S	absorption in liquid
CO (trace)	Gas	water gas shift reaction $CO + H_2O \rightarrow CO_2 + H_2$ $C + H_2O \rightarrow CO + H_2$	CO_2	absorption in liquid
Organics (trace)	Gas	catalytic oxidation	H_2O CO_2	absorption in liquid if necessary
H_2S <1%	Gas	oxidation liquid phase (Stretford, Cataban, Takahax processes)	S	filtration of precipitate
H_2S >50%	Gas	Claus redox gas phase $3/2\ O_2 + H_2S \rightarrow SO_2 + H_2O$ $2\ H_2S + SO_2 \rightarrow 2\ H_2O + 3S$	S	product
SO_2/ H_2S	Gas	Claus redox gas phase $2\ H_2S + SO_2 \rightarrow 2\ H_2O + 3S$	S	product
SO_2 <1%	Gas	usual absorb or absorb-react in liquid phase. wide variety of processes available		
SO_2 >9%	Gas	catalytic oxidation $SO_2 + 1/2\ O_2 \rightarrow SO_3$	SO_3	absorb to make H_2SO_4
organics	Gas/ Air	oxidize (burn) ensure that the hydrocarbon content is at least <25% of the explosive limit	CO_2, H_2O	
sulfides	Liquid	oxidation	sulfates	
ammonia	Liquid	chlorination/oxidation $2\ NH_3 + 3\ Cl_2 \rightarrow N_2 + 6\ HCl$	N_2	strip gas neutralize acid
H_2S	Liquid	$H_2S + 3\ Cl_2 + 2\ H_2O \rightarrow SO_2 + 6\ HCl$	SO_2	strip gas neutralize acid
CN^-	Liquid	$2\ CN^- + 5\ Cl_2 + 4\ H_2O \rightarrow 2\ CO_2 + N_2 + 8\ HCl + 2\ Cl^-$	CO_2, N_2	strip gas neutralize acid
ferrous sulfate	Liquid		ferric compounds	precipitate
acids base	Liquid	neutralization	H_2O + salt	
ammonia	Liquid	biological nitrification/denitrification		
nitrates	Liquid	biological denitrification		
proteins	Liquid	denature via temperature		

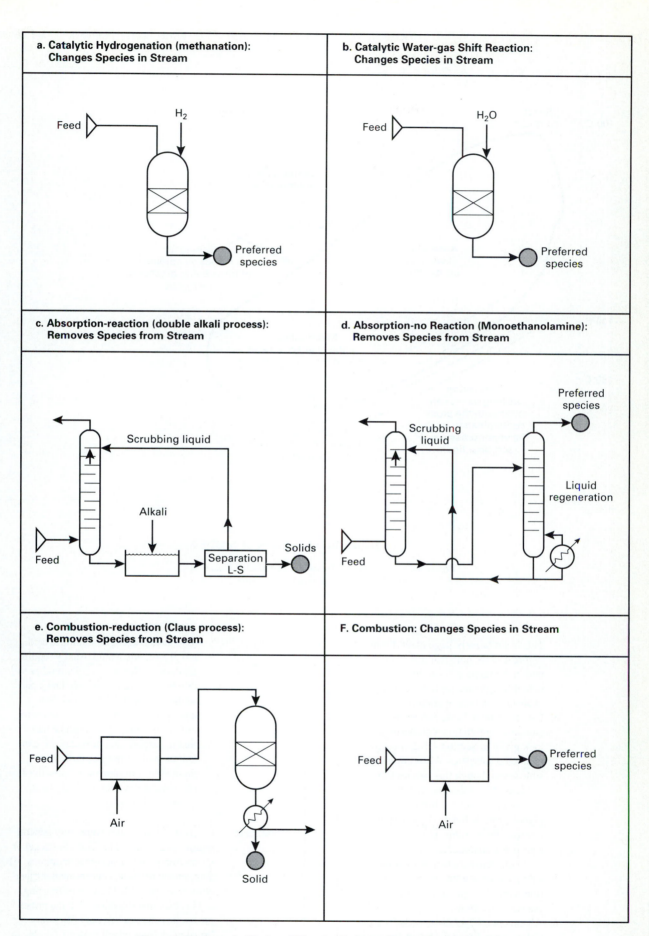

Figure 4–38 Example, "Packaged" Reaction Options to Separate Species: Gas Phase

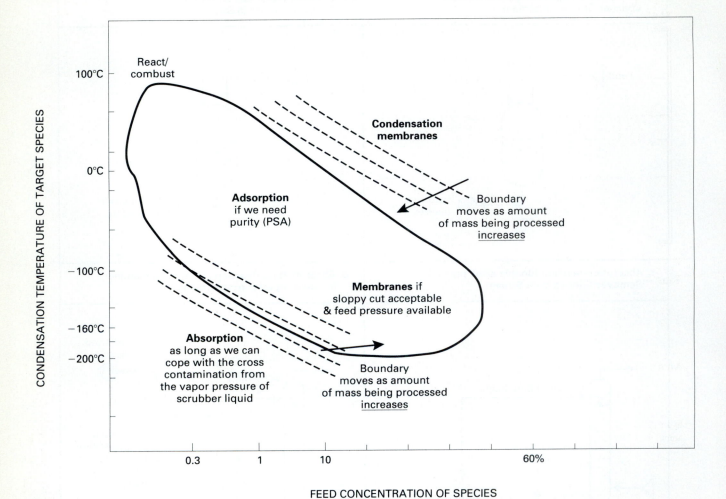

Figure 4–39   Treatment of Gases by Condensation, Reaction,
Membranes, Adsorption and Absorption

This is a relatively large plant. From Figure 4–9 we note that this is still within the range of a single adsorption unit although one might expect a shift to absorption type operation.

5. Use the data of Table 4–5 to provide guidance; enrich this with the detailed look given in Section 4.1-2 to equipment characteristics. Absorption is possible (even though this species can be condensed); we assume that the absorbing liquid might be sulfuric acid or a glycol (from Table 4–7). Adsorption is feasible. From Table 4–9, silica gel is a probable adsorbent.

   Hence, refrigeration, absorption in sulfuric acid or with glycol, or adsorption with silica gel or molecular sieves seem to be options.

6. All except refrigeration require a separating agent. Since the target species is water, we would not try to recover it. The main consideration would be to re-

generate the agent. For sulfuric acid and glycol, this would usually require distillation. Hence, glycol would probably be preferred unless we had a particular use for the dilute acid. We should also be concerned about corrosion. Concentrated acid can be handled in carbon steel provided the concentration does not decrease below about 90%. For more see Appendix I. The silica gel would probably be regenerated thermally.

Comment: Figure 4–40 illustrates some very general regions of applicability. Thus, silica gel or absorption might be acceptable options. At atmospheric pressure, condensation/refrigeration is not possible because of freezing.

   This example illustrates that the principles in this chapter can only offer very broad guidelines which may not be able to discriminate between options.

TABLE 4-23. Effect of throughput capacity on options for separating gases

Options	Species		
	H_2/purge gas	CO_2/natural gas	air O_2/N_2
Cryogenic distillation	> 650 Mg/d of purge gas	probably combine with membranes	> 25 Mg O_2/d
PSA Adsorption	small scale		small capacity
Absorption	> 650 Mg/d of purge gas	cheapest at > 1000 Mg/d	
Membranes	small scale < 650 Mg/d and pressure > 3 MPa	can concentrate a stream if < 200 Mg/d	

Example 4–7: We wish to remove hydrogen from a purge gas. What might we do? The hydrogen concentration is 1%; assume the purge gas is ammonia and the flowrate is 700 Mg/d. The pressure is 6 MPa.

An Answer:

1. List the species and their properties; the species are hydrogen and ammonia, **ST**:

hydrogen   2 0 4 0 21 0 3 0 -259 -252 28.4 - - -
ammonia 17.03 3 1 0   0 0 0 -77 -33.4 20.8 13 11 4.75

There is a high difference in molar mass, there is hazard with both species, and there is a large difference in freezing and boiling temperatures. The options might be, from Table 4–1, membranes (to exploit difference in molar mass), adsorption (because of the differences in boiling temperature T_b and the molar mass), absorption (to exploit the differences in boiling temperature, although the molar volumes are very similar), condensation/refrigeration (to exploit the differences in boiling temperatures). One might consider reaction/combustion but we would have to find a reaction for hydrogen that does not contaminate the ammonia and that ammonia does not participate in.

From Figure 4–2 the generally expected alpha values are as listed in the previous example. The "ideal solute species" is the hydrogen. Unfortunately, most of the property differences point to the ammonia as being the solute species.

2. Note the temperature sensitivity of the species. Since both have small molar masses, from Figure 4–3, they are both relatively robust. Hence, temperature need not be a concern.

3. Identify the feed conditions and the product requirements. For condensation, we would condense the ammonia to leave the hydrogen. Since we would have to condense just about all of the stream, this is not a probable option for this inlet concentration. For absorption, from Figure 4–18 this looks like it handles the small concentrations; however, the ammonia is what would be absorbed. Ammonia is the solute species. The reverse is what would be desirable, namely for the smaller concentration species to be absorbed. Hence, absorption is probably not economically feasible because the hydrogen is so dilute. For adsorption, Table 4–9 suggests that the ammonia would be adsorbed preferentially to the hydrogen (for bulk separations using activated carbon). Again, the wrong species is the solute species. With zeolite 3A, it is unclear whether hydrogen would be preferentially adsorbed. It would be worth exploring with the supplier. For membranes, from Table 4–17 the hydrogen might permeate and leave the main flow of ammonia as retentate. This is the desired situation and therefore probably has the greatest potential. However, nothing is immediately apparent. The more promising options seem to be adsorption (if we can locate the correct adsorbent) and membranes.

4. Note the size of the plant or operation. The size of the plant operation is 700 Mg/d × 300 d/a or 210 000 Mg/a. From Figure 4–9, this is large. This is about the limit of a single adsorption unit and larger than usual single membranes. This can be overcome by splitting the feed and having several feed trains.

5. Use the data of Table 4–5 to provide guidance; enrich this with the detailed look given in Section 4.1-2 to equipment characteristics. Based on these data, the membrane system might be more appealing because of the inlet pressure. Economically, this, from Table 4–23, is larger than usual and the purity of the separation will not be high. Nevertheless, because the concentration of hydrogen is so small, membranes and adsorbers are probable ones to consider. Perhaps a combination of membranes and distillation might be useful at the large scale.

6. The membrane, although it is a separating agent, may degrade. Nevertheless, with a membrane we would obtain two streams: one rich in hydrogen and one depleted. This is advantageous. However, for an adsorber, we would have to regenerate the adsorbent and recovery of the hydrogen might not be easy. We might try Pressure Swing Adsorption.

Comment: Hydrogen might have a dramatic effect on the choice of materials of construction.

2. For Concentrated Gases >1%.   The prime options include adsorption, absorption, gas membranes, and perhaps condensation or cryogenic distillation. One might condense the whole thing and consider liquid phase separation options. Thus, for the larger concentrations, the same general possibilities—except react-combust—are available and, as illustrated in Figure 4–39 and Table 4–23, the same general vagueness applies.

Example 4–8:   The effluent from the reactor to produce MEK from acetone is condensed. The resulting uncondensed gas is

Hydrogen	13.9% w/w
2-Butanol	5.38
Methyl ethyl ketone	80.72 (saturating the non-condensable hydrogen)

The total flowrate is 269.7 kg/h; the temperature is 62.7°C. What might we do to "separate" the MEK from the hydrogen? Assume that the 2-Butanol can be considered to be MEK.

An Answer: 1. List the species and their properties; compare and identify the characteristics that can be exploited.

Here the species are, **ST**:

hydrogen	2	0	4	0	21	0	3	0	-259	-252	28.4	-	-
MEK	72	1	3	0	16	1	1	1	-86	79.6	98.5	9.3	8.5

There are large differences in molar mass, boiling and freezing temperatures, and solubilities (as seen by T_b and molar volumes). From Table 4–1, absorption, adsorption, and membranes are the main options. Combustion/reaction is not feasible because the concentration is relatively high and all of the gas would burn. Figure 4–2 shows that a large difference in solubility is needed for absorption to apply. Such a large difference does exist because hydrogen is relatively insoluble.

2. The molar masses of both species are so small that, from Figure 4–3, this should not be a concern. From Figure 4–3a, the temperature of the stream is within the usual range of temperatures when absorption is used. The temperature is a little too high for most gas membranes; hence, we would have to cool the gas down first. Temperature is not an issue for adsorption.

3. Identify the feed conditions and the product requirements. Here the target species for absorption would be the methanol. The feed concentration is about 86%. From Figure 4–18, this is above the "usual" range but data are shown for a scrubber operating about this feed concentration. The product concentration in the scrubber liquid depends on the solubility and the mass flowrate of scrubber liquid.
For adsorption, the target species would probably be the species in the smallest concentration, namely the hydrogen. From Table 4–9, a 3A zeolite might be appropriate. From Figure 4–21, a feed concentration of 15% is reasonable.

For membranes, Table 4–17 and Figure 4–31 suggest that this is feasible provided the feed temperature was reduced.

4. For a flowrate of 270 kg/h or 0.075 kg/s, all three options seem feasible for a single unit.

5. From Table 4–5, for absorption, we have already condensed as much MEK as we can without altering the conditions. Absorption is very feasible except that we note that we are absorbing most of the gas stream. Also, some of the water evaporates into the hydrogen gas stream. This probably would have to be removed before the hydrogen would be processed further.

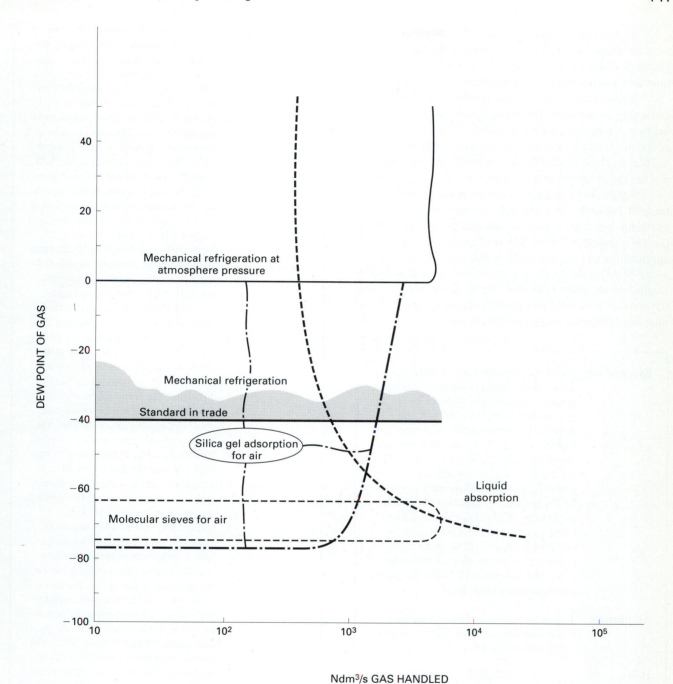

Figure 4–40   Options for Drying or Dehydrating Gases: Removal of Water

For adsorption, the hydrogen gas concentration is higher than one usually tries to adsorb (15% instead of less than 10%). This would mean very short loading times.

For membranes, the gas pressure is not high. Hence, we would have to compress the gas and cool it before this might be attractive.

Hence, adsorption is attractive. Because of the large polar and hydrogen

solubility parameters for MEK, water is probably a good solvent.
6. Since we have added a separating agent we now have to recover the MEK from the solvent. Options to do this are considered in the next sections.

Comment: Austin and Jeffreys (1979) describe this case problem. They selected absorption into water. The water flowrate was 1935.05 kg/h. The exit concentration in the absorber liquid was 10% w/w MEK.

3. For Dilute Liquids <1%.

For dilute liquid systems the prime options are ion exchange, reverse osmosis, electrodialysis, solvent extraction, adsorption, foam fractionation, ion fractionation, ultrafiltration, and reaction.

Some of these options may be eliminated simply because they do not work. However, if there is a lot of overlap between the options, then Figure 4–41 illustrates a general economic tradeoff between ion exchange and solvent extraction and how this is affected by the solids concentration and the target species concentration.

Figure 4–42 illustrates for an ionic separation the tradeoff between electrodialysis, reverse osmosis and ion exchange based on expected feed and product concentrations (Eckenfelder, 1976). Electrodialysis is for large scale applications and not for small scale. Here the general regions of applicability (based on economics and technical feasibility) are given in terms of the feed concentration and the desired "concentrated" product concentration. This applies only when the target species are ions.

Example 4–9: The stream from the CCD (counter current decantation system) in a uranium recovery process contains 1 g/L uranium ions in the quatravalent form, U_3O_8. What might we do to recover the uranium (or concentrate it to about 20 g/L)? Assume that the other species is sulfuric acid of pH about 2. This is the exit from a sulfuric acid leach circuit. However, this has been partially neutralized to precipitate insoluble inert ore that might be represented as calcium sulfate. Ferric sulfate is also present in solution; in this example we will be aware of its presence, but assume this is a binary separation. The flowrate is 11.3 kg/s. The solids concentration is about 26 g/L.

An Answer:
1. List the species and their properties; here we have uranium "ion," and sulfuric acid-water, **ST**:

H_2SO_4	98	- - - -	3	3	3	10.5	340d		−3
U_3O_8	842					high	high		ions

We could exploit the difference in boiling temperatures, precipitate, extract, and exploit the ionic form, especially because it is such a high valence. We might consider reverse osmosis, ion exchange, and electrodialysis. We might consider reaction.

Consider Figure 4–2 to identify usual values required for the separation factor for different options. Values are given for the boiling (evaporation) and for solvent extraction. The latter is about 2 to 500.

2. Note the temperature sensitivity of the

---

species. For this separation, the species are insensitive to temperature. Although the size of the molecule is relatively large, and hence from Figure 4–3b we might expect temperature sensitivity, inorganics are robust.

3. Identify the feed conditions and the product requirements. The feed concentration is relatively dilute at 1 g/L or about 0.1%. The desired product is 20 g/L or about 2%. For evaporation, from Figure 4–12, the feed concentration is very dilute. We would have to evaporate almost all of the feed. Hence, evaporation might not be attractive. Figure 4–20 for solvent extraction suggests that this option is viable; similarly, Figure 4–24 suggests that ion exchange is feasible. Reverse osmosis and electrodialysis, Figures 4–36 and 4–37, suggest this is in the correct range of application.

4. Note the size of the plant or operation. For the flow of 11.3 kg/s shows that this is a relatively large-scale operation. However, single units to handle this size of operation are feasible for solvent extraction, precipitation, and ion exchange. For electrodialysis and reverse osmosis, the capacity would be handled by using multiple units in parallel. (This means that we might be at a slight cost disadvantage with these options because we cannot take advantage of the economy of scale.)

5. Use the data of Table 4–5 to provide guidance; for evaporation, we do *not* have a high concentration of the less volatile species. This confirms our earlier rejection of this option. For precipitation, we would have to look carefully at the other species present. In particular, we would have to explore the possible coprecipitation of ferric sulfate. For solvent extraction we would have to be concerned about the solids concentration since it is larger than 100 ppm and we might have a lot of crud formation.

Ion exchange seems to be ideally suited based on the criteria. The species has a high atomic number and a high valence.

From Figure 4–41, the solids concentration seems to be a factor in choosing between ion exchange and solvent extraction. However, we could run the ion exchange as a pulp slurry or as a fluidized bed.

Hence, based on this analysis, ion exchange and solvent extraction appear to be very attractive options.

4. For Concentrated Liquids >1%. For concentrated liquid systems the prime options are distillation, evaporation, melt crystallization, solvent extraction, adsorption, desorption, dialysis, solution crystallization, and precipitation. Reaction is usually not a viable option for concentrated target species.

Desorption, dialysis, and solution crystallization each are unique. Desorption—the species is very volatile. Dialysis—the species is fragile. Solution crystallization and precipitation—the target species is a solid. With precipitation—we might select it because all of the other properties seem to be too similar.

For many separations, the usual sequence of options to be considered is distillation, solvent extraction, crystallization, and then adsorption (to distinguish between isomers). Figure 4–43 shows one criteria for distinguishing between solvent extraction and distillation. If the relative volatility of the two species to be separated is about 1.2, then if we can find a solvent to give a distribution ratio of about 2.6 or better, solvent extraction might be economically more attractive than distillation. Alternatively, we might search for a solvent for extractive distillation. For extractive distillation to be attractive, the relative volatility would have to be about 1.5 or better.

An approximation to Figure 4–43 is that the alpha value for solvent extraction should be about double that for distillation if solvent extraction is to be appealing; extractive or azeotropic distillation is a contender if the alpha value is 40% higher than that for ordinary distillation (Bojnowski and Hanks, 1979).

Example 4–10: We wish to separate a 50–50 mixture of styrene, ethylbenzene to produce 99.5%w products. What might we do?

An Answer: The relative volatility of styrene/ethylbenzene, from Figure C-9 in **Data Part C**, is about 1.3. For diethylene glycol as a solvent, the alpha value is about 2.7. From Figure 4–43, this suggests that probably distillation would still be more attractive because the distribution ratio, although high, is not above the 3.5 suggested by Figure 4–43.

Another way of exploring the relative attractiveness of options is to consider the amount of energy costs needed for each. Assuming that the cost of energy is the sole distinguishing feature, then Null (1980) and Rush (1980) have compared adsorption, solvent extraction, melt crystallization and reverse osmosis. Some data are given in Figure 4–44.

Figure 4–44a compares the distillation reflux ratio, R; the fraction of the feed that is taken overhead and the temperature of the steam available to identify conditions

where solvent extraction has less energy cost than distillation. (In this analysis, the assumption is that in the solvent extraction process the solvent is recovered by distillation. For many processes the solvent is recovered by scrubbing. Hence, this analysis would not apply for such a scrubbing recovery system.) Thus, if the required reflux ratio is high and the amount of feed recovered as an overhead product is high, solvent extraction becomes viable. This figure assumes that the solvent is nonvolatile and that the required heat for the solvent stripper is at 320°C. If the solvent is volatile, the solvent recovery still needs reflux. As the amount of reflux required in the solvent recovery distillation increases, the line shifts upward to the right as illustrated in Figure 4–44b. Thus, if the steam temperature for both stills is 320°C and the distillate is 40% of the feed, and if the required reflux ratio in the solvent recovery column is 1:1, solvent extraction is favored from an energy requirement viewpoint if the distillation separation requires a distillation reflux ratio of about 5:1.

For melt crystallization, the tradeoff depends on the distillation reflux ratio, the crystallization reflux ratio, and the temperatures of the melt crystallization, the steam temperature to the melter. If we can generalize that the heat of vaporization is about 5 times the heat of fusion and that the temperature of steam to the melt and to the bottoms of the distillation column are the same, then Figure 4–44c provides some guidance. Thus, if the distillation requires a reflux ratio of 2:1 and the steam temperature to both the distillation column and the melter is 320°C, then melt crystallization would be preferred from an energy viewpoint if the melt crystallization reflux ratio is less than about 3.5:1 and the freeze refrigerant temperature is about −50°C or hotter.

For gas adsorption that is regenerated by temperature swing, Figure 4–44d shows the energy comparison. If the least volatile species is more readily adsorbed, then the average rate of adsorption is about the ratio of the bottoms to the overhead split in the distillation separation. If we assume that the heat to regenerate the bed is about 5 times the latent heat of vaporization, then the tradeoff as a function of the steam temperature to the reboiler of the distillation column is as shown in the figure. This assumes that the regeneration of the adsorber occurs at 320°C.

The data comparing reverse osmosis to evaporation only consider the energy component, namely the pressure required for reverse osmosis vs the thermal energy needed for the evaporation. As with all of the data shown in Figure 4–44, the energy cost is *one* cost component; others to be considered include maintenance, solvent/membrane replacement costs, depreciation, operating labor, and so on. Nevertheless, these give an idea of how the energy component can be a guideline. Figure 4–44e, for reverse osmosis, suggests that (for the low concentrations where the osmotic pressure is still reasonably low and where membranes can be fabricated) reverse osmosis is very attractive relative to evaporation.

"Usual" reflux ratios are in the order of 1.2 to 2.0;

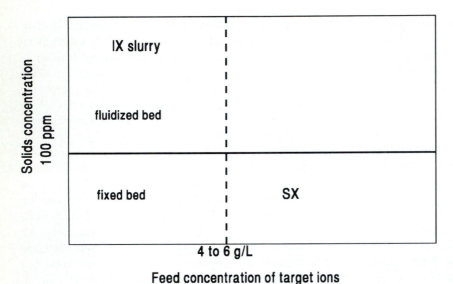

Figure 4–41 General Regions of Applicability for Ion Exchange and Solvent Extraction

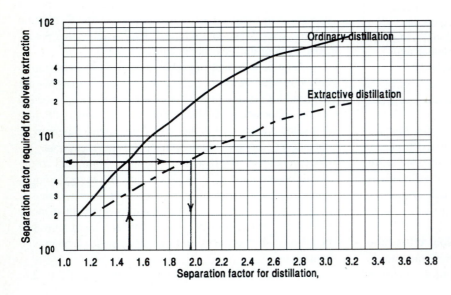

Figure 4–42 General Regions of Applicability for Dilute Liquid Systems (based on Eckenfelder, 1976, and reprinted courtesy of Chemical Engineering, McGraw Hill publication, copyright 1976)

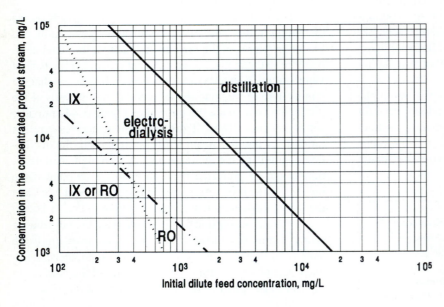

Figure 4–43 When To Use Distillation, Solvent Extraction, or Extractive Distillation (from Souders, 1964 and reprinted courtesy of the American Institute of Chemical Engineers, copyright 1964 AIChE)

usual splits delta are about 0.5. Thus, we can see that usually distillation is more energy efficient than these options. Where distillation becomes questionable is when the required reflux ratio is high, the amount taken overhead is large. Other conditions that are not shown on these figures are when refrigeration cooling is needed (instead of cooling water).

Figure 4–44, in general, is more illustrative than conclusive criteria because only energy costs are considered. Nevertheless, this provides a starting basis for exploring the economic implications of some of the options.

Example 4–11: Consider the concentration of acetic acid-water mixtures. This is a common consideration in the pharmaceutical industry, for example. If the concentration is 3 wt% acetic acid, what options are most applicable? Assume we have 15 kg/s of feed.

An Answer: 1. List the species and their properties; compare and identify the characteristics that can be exploited, **ST**:

Molar

Water	18	0	0	0		0	0	0	100	18	23	40	15.7	
Acetic Acid	60	3	2	1	10	4	3	2	16.6	118	57	12.2	18.9	4.7

Thus there is a reasonable difference in boiling temperatures and in freezing temperatures, the solubility parameters are quite different, and we might be able to exploit the difference in molar mass. Hence, we have quite a range of possibilities. From Table 4–1, distillation, solvent extraction, melt crystallization, and membrane reverse osmosis are possible candidates. Figure 4–2 illustrates the type of alpha values that would be needed.

2. From Figure 4–3, these species should be relatively robust. Temperature should not be a limitation. This is interesting because in most pharmaceutical operations, the species are very temperature sensitive. These two species, although found within such a process, are not.

3. Identify the feed conditions and the product requirements. For distillation, the feed concentration is very low (Figure 4–12). This combined with the alpha value of about 1.4 suggests that distillation might be very expensive. For solvent extraction, the feed concentration is in the usual range. For melt crystallization freeze concentration might be reasonable;

for melt crystallization we have to look carefully at which species freezes first. Here acetic acid freezes first and hence would appear as the pure molten product. The feed concentration, however, does not seem to be high enough. Since the species ionizes, membrane separation with reverse osmosis seems to be in the appropriate feed concentration (Figure 4–36). Ion exchange might be feasible from the feed concentration diagram and because the species ionizes.

4. Note the size of the plant or operation. Single units for distillation, solvent extraction, and ion exchange could handle this capacity. Multiple units of melt crystallizers and membrane units would be needed.

5. Use the data of Table 4–5 to provide guidance; enrich this with the detailed look given in Section 4.1-2 about equipment characteristics. Based on this, solvent extraction and perhaps reverse osmosis might be possible.

Comment: Busche (1985) analyzes this problem and considers simple distillation, melt crystallization, solvent extraction, supercritical extraction, and electrodialysis (when acetic acid has been converted to a salt form).

C. Sequencing Heuristics for Separating Multicomponents.
When multicomponent mixtures are to be separated, we should take a systems view. Now, not only are we concerned about the option; we must consider the sequencing. Do we use three distillation columns in series? What is the function of each? Many have explored what general guidelines or heuristics can be applied.

1. Remove the most hazardous, unstable and corrosive species early. (Use the data in **Data** Part D to get an idea of the hazard rating and the corrosive rating of the species.)
2. Consider if mixing/blending of streams can accomplish the goal (rather than separation).
3. Save the most difficult separations until last (or do the easy ones first).
4. Separate the most plentiful components early.
5. Leave high specific recoveries until last.
6. Try to avoid having to add an agent to achieve the separation. If an agent must be added, recover immediately the agent. Try not to have to use *another* agent to recover the first agent. In other words, try to

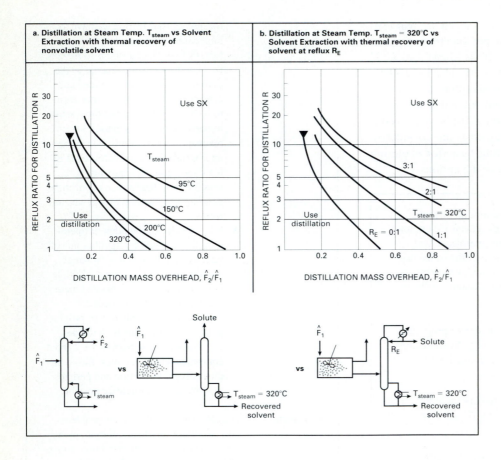

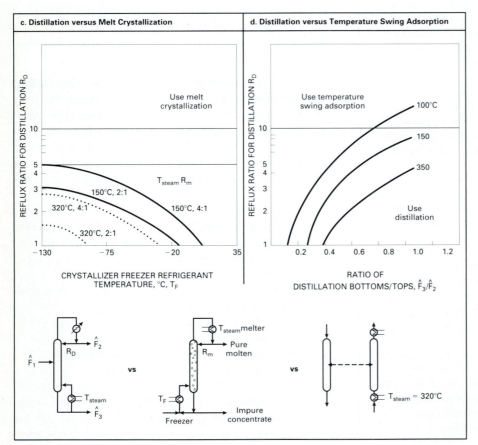

Figure 4-44 Energy Criteria for Selecting Options (adapted from Null, 1980, and used with permission from the American Institute of Chemical Engineers, copyright 1980 American Institute of Chemical Engineers)

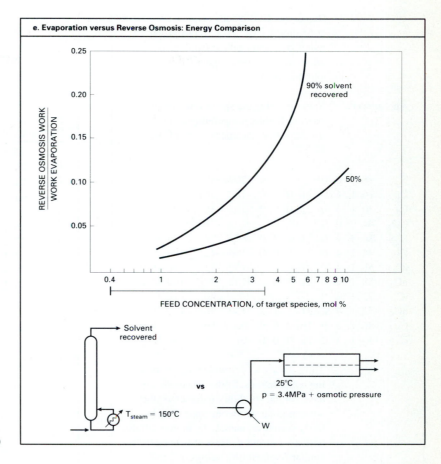

e. Evaporation versus Reverse Osmosis: Energy Comparison

Figure 4–44   (Continued)

solve the separation by adding or subtracting heat, as illustrated in Table 4–1.

7. Avoid extremes in operating conditions (as illustrated in Figure 4–7).

8. When mapping out some options, consider safety, controllability, materials of construction implications, and energy integration. (Appendices I and II address some of these considerations.)

Since distillation is often a first consideration, it is useful to list the species in order of increasing boiling temperature.

If a series of *distillation* columns are chosen, then:

1. Prefer to take the species overhead one by one.
2. Prefer separations that give 50–50 splits.
3. Prefer to take the most valuable product as an *overhead* product (because it is easier to control).

These are not always mutually exclusive. Indeed, some of these may be contradictory for the separation under consideration.

More details of these are given by King (1971), Rudd Powers and Sirrola (1973).

For energy integration considerations see Moore (1987) and Douglas (1988).

The general procedure is to apply the heuristics to block out the probable separation sequencing. Then apply the "binary" approach listed above. Check out operability and process control, energy integration, and safety.

Example 4–12: Carbon dioxide is injected into an oil well to enhance the recovery of oil from the reservoir. The composition of the gas coming out of the well is:

H_2S	0.05
CO_2	90.25
N_2	0.16
C_1	4.76
C_2	1.75
C_3	1.52
iC_4	0.22
nC_4	0.58
iC_5	0.21
nC_5	0.22
C_6	0.16
C_7+	0.12

The flowrate is 45 Nm^3/s at 40°C and a pressure of 2 MPa. Devise alternative schemes to create a methane rich fuel, recycle CO_2, plus a liquefied natural gas, which is C_2^+ hydrocarbon stream that is free of H_2S and CO_2.

Recall that CO_2-ethane form an azeotrope at 65 mol % CO_2. This azeotrope can be handled in azeotropic distillation by the addition of a solvent-agent C_4^+ hydrocarbon stream.

An Answer: Start by listing the species in increasing order of boiling temperature and list the data about the species from **Data** Part D, **ST**.

```
N₂    28  0 0           0 -209 -196
C₁    16  1 4 0 21 0 0 0  -182 -161
C₂    30  1 4 0 21 0 0 0  -183 -88   93
CO₂   44  0 0              -78(s)
H₂S   34  3 4 0 21 3 3 0  -82  -59   28 - - 7
C₃    44  1 4 0 21 0 0 0  -189 -42   90 0 0
iC₄   58  1 4 0 21         -159 -11  106 0 0
nC₄   58  1 4 0 21 0 0 0  -138 -0.5  100 0 0
iC₅   72  1 4 0                  30  116 0 0
nC₅   72  1 4 0 21 1 1 2  -130 36   115 0 0
C₆    86  1 3 0 16 0 0 0  -95  69   131 0 0
C₇+  146  1 3 0 16 0 0 0  -91  98   147 0 0
```

Thus, we see a very wide range in boiling temperatures, from −196 to 98; all have similar solubility parameters; they are non-polar. Only one appears to be fairly corrosive, namely hydrogen sulfide with a 3 3 0 rating. Most of the rest have similar flammability ratings of 3 to 4 (except CO_2 and N_2, which we suspect would have a rating of 0).

Next we recall the specific requirements; separate and recycle the CO_2; a fuel basically of methane and a C_2^+ stream. Both of these must be free of CO_2 and H_2S.

Apply the heuristics:

1. Remove the most hazardous, unstable and corrosive species early. Here, since most have similar hazard ratings, but hydrogen sulfide has a high corrosion rating, we might remove hydrogen sulfide first.
2. Consider if mixing/blending of streams can accomplish the goal (rather than separation). Not appropriate here.
3. Save the most difficult separations until last (or do the easy ones first). The easiest ones would be where there is the greatest difference in molar mass or between boiling temperatures or between freezing temperatures (since most of the other dimensions are similar). Molar masses:

similar with a cluster around the 30 to 45 range. For boiling temperatures, the line in the table represents where there is a large difference. All the others have a difference of about 30°C:

```
N₂    28  0 0            0 -209 -196
C₁    16  1 4 0 21 0 0 0  -182 -161
─────────────────────────────────────
C₂    30  1 4 0 21 0 0 0  -183 -88   93
CO₂   44  0 0              -78(s)
H₂S   34  3 4 0 21 3 3 0  -82  -59   28 - - 7
C₃    44  1 4 0 21 0 0 0  -189 -42   90 0 0
iC₄   58  1 4 0 21         -159 -11  106 0 0
nC₄   58  1 4 0 21 0 0 0  -138 -0.5  100 0 0
iC₅   72  1 4 0                  30  116 0 0
nC₅   72  1 4 0 21 1 1 2  -130 36   115 0 0
C₆    86  1 3 0 16 0 0 0  -95  69   131 0 0
C₇+  146  1 3 0 16 0 0 0  -91  98   147 0 0
```

The most difficult ones will be between the "normal" and the "iso" species, where the temperature difference is about 6 to 10°C. Thus, the nitrogen and the methane can be separated early; the C1-C2 split should be relatively easy. CO_2 from C_2 looks difficult; save this one until later. It also is difficult because the two form an azeotrope. Hence, separate the C_1 out early. If we want to isolate a C_4^+ stream to be used to break the azeotrope, then C_3/C_4 cut is wanted. This has a 30°C difference and so should be not too difficult.

4. Separate the most plentiful components early. Here, the most plentiful species is carbon dioxide, which should be removed early. It should also be separated early because it sublimes (and could cause problems by having solid formation that blocks lines, and because it forms an azeotrope that would affect subsequent distillation).
5. Leave high specific recoveries until last.
6. Try to avoid having to add an agent to achieve the separation. There seems to be a reasonable difference in boiling temperatures except for the most plentiful species, CO_2, which sublimes. According to hint 4, we want to separate this early. Thus, distillation might be a viable option for all separations except the separation of CO_2.
7. Avoid extremes in operating conditions. Atmospheric pressure operation is OK for iC₅ and higher. How-

ever, for the other species we would have to operate under pressure to bring their boiling temperatures into the usual range. We note, however, that the inlet pressure is 2 MPa, so it already is at a high pressure.

8. When mapping out some options, consider safety, controllability, materials of construction implications, and energy integration.

We note that some contradict: #1 says remove H_2S first; #4 says separate CO_2 early; #3 says separate the C_1 vs C_2 early, or separate CO_2 vs C_2 last.

A dominant theme is to separate the corrosive one first if we can. H_2S can be removed by absorbing H_2S preferentially in an absorbing liquid. This violates hint #6 because we add an agent first thing. On the other hand, such a packaged scrubber-regeneration plant (similar to that in PID-1A) works well. We do have to remove any moisture that has been introduced because of the absorbing scrubber liquid. Thus, as a basic start we will remove the H_2S (and the subsequent moisture).

Now come some of the options:

Idea 1: separate the methane cut first: use simple distillation with some reflux with C_4 "solvent" stream added near the top of the tower to prevent the CO_2 from freezing. The top then is the methane; the bottoms is the rest. Use extractive distillation on the bottoms with the C_4 cut as the solvent to separate CO_2/C_2. This is illustrated in Figure 4–45a.

Idea 2: separate the methane cut first: use distillation without reflux and without the temperatures that might cause CO_2 freezing. Do not use a solvent on the first column. This allows more CO_2 to go overhead and is lost by scrubbing the methane to free it from CO_2. The bottoms goes to extractive distillation. Figure 4–45b shows this variation.

Idea 3: separate the most plentiful one first: concentrate the CO_2 via a membrane. Treat the retentate with distillation to drive off the methane. Take the bottoms and do a CO_2-C_2 azeotrope/C_3^+ cut in a second distilla-

tion column. Treat the azeotrope overhead with a membrane/distillation to remove the CO_2. Now combine the various streams. Figure 4–45c illustrates how this occurs.

Idea 4: separate the most plentiful one first by doing a C_2/CO_2 split. Such a separation is not based on distillation; rather, it depends on the use of a C_4^+ stream as an initial absorption separation. Then on the overhead gas do the CO_2/C_1 separation; on the bottoms we isolate a C_4^+ stream to serve as the absorbent. This is given in Figure 4–45d.

Comment: For more details see Schendel et al. (1984), (1983).

4.1-4 Summary of Overall Selection Procedures

To select a separation option for a binary, a systematic 6-step procedure was outlined and illustrated for dilute and concentrated gas systems and dilute and concentrated liquid systems.

For a multicomponent mixture of species, general heuristics are presented about the sequencing of the separation; specific suggestions are given for a sequence of distillations. The focus has been on **selection**; details about **sizing** are given in Woods (1993a).

4.2 REFERENCES

ALLEN, D. H., and R. C. PAGE. 1975. "Revised Technique for Predesign Cost Estimating." *Chem Eng* (Mar 3): 142–150.

AUSTIN, D. G., and G. V. JEFFRIES. 1979. *The Manufacture of Methyl Ethyl Ketone from 2-Butanol.* London: George Goodwin Ltd.

BOJNOWSKI, J. H., and D. L. HANKS. 1979. "Low-energy Separation Processes." *Chem Eng* (May 7): 67–71.

BRAVO, J. L., J. R. FAIR, J. J. HUMPHREY, et al. 1986. *Fluid Mixture Separation Technologies for Cost Reduction and Process Improvement.* Park Ridge, NJ: Noyes.

BUSCHE, R. M. 1985. "Acetic Acid Manufacture-Fermentation Alternatives," Chapter 9 in *Biotechnology: Applications and Research*, ed. P. N. CHEREMISINOFF and R. P. OUELLETTE. Lancaster: Technomic Publishing Co., pp. 88–102.

CARTER, A. L., and R. RE. KRAYBILL. 1966. *Chem Eng Prog* **62**, 2: 99.

DE FILLIPI, R. P. 1977. "Ultrafiltration," Chapter 6 in *Filtration, Principles and Practice*, ed. C. ORR. New York: Marcel Dekker Inc.

DOUGLAS, J. M. 1988. *Conceptual Design of Chemical Processes.* New York: McGraw-Hill.

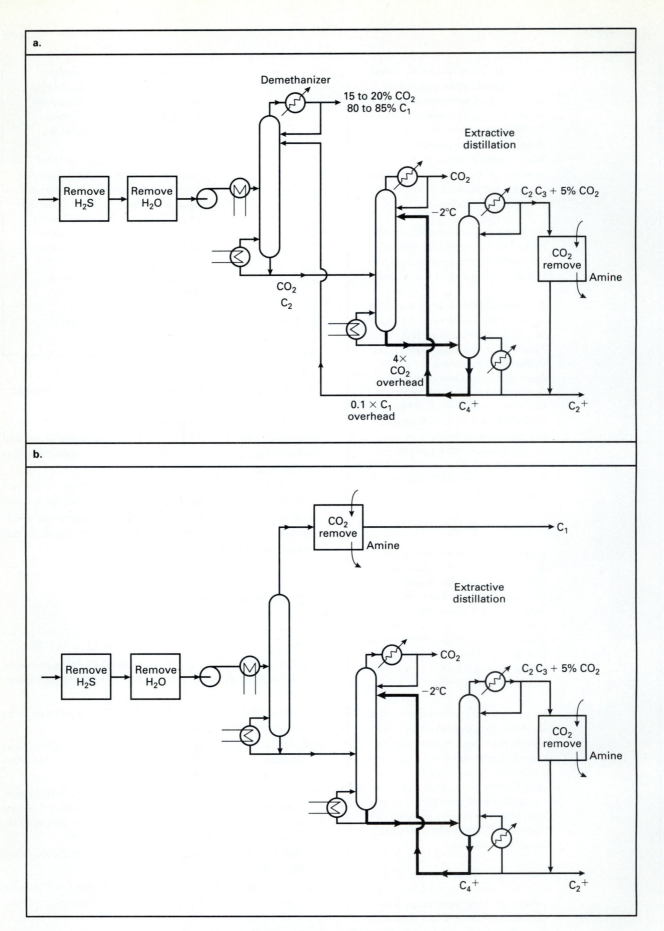

Figure 4-45 Options for Separation

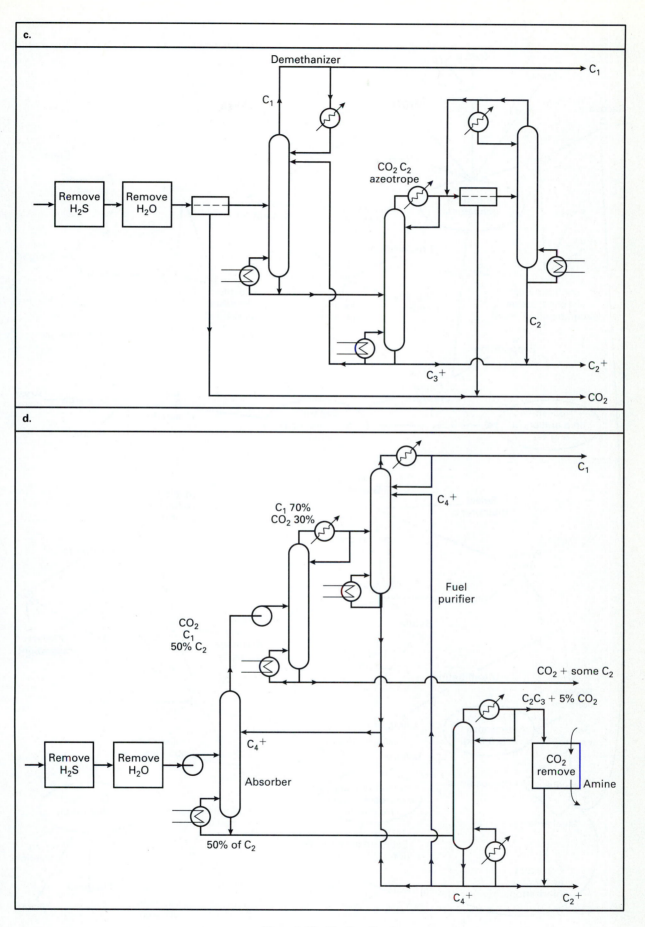

Figure 4–45 (Continued)

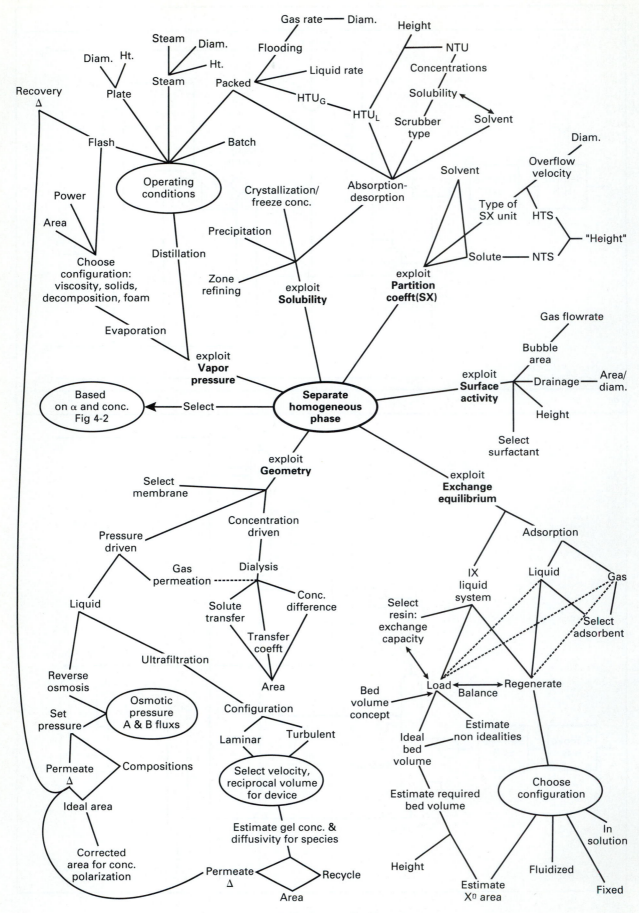

Figure 4–46 An Overview to Homogeneous Separations

"Dow's Fire and Explosive Index Hazard Classification Guide." *Chem Eng Prog* Technical Manual. 1966. New York: American Institute of Chemical Engineers.

ECKENFELDER. 1976. "Wastewater Treatment." *Chem Eng* (Oct 18): 60.

HEAVEN, D. L. 1969. M. S. Thesis, Dept of Chemical Engineering, University of California, Berkeley.

HEIST, J. A. 1979. "Freeze Crystallization." *Chem Eng* (May 7): 72–82.

HICKMAN, K. C. D. 1948. "High Vacuum Distillation." *Ind Eng Chem* **40,** 1: 16.

HICKMAN, K. C. D., and N. D. EMBREE. 1949. "Decomposition Hazard of Vacuum Stills." *Ind Eng Chem* **40,** 1: 135.

HOLIDAY, A. D. 1982. "Conserving and Reusing Water." *Chem Eng* (April 19): 117–137.

HUMPHREY, J. L., et al. 1984. "The Essentials of Extraction." *Chem Eng* (Sept 17): 94.

KASPER, S. 1968. "Selecting Heat-Transfer Media by Cost Comparison." *Chem Eng* **75,** 25 (Dec 2): 117–120.

KELLER, G. E., III, et al. 1987. "Adsorption," Chapter 12 in *Handbook of Separation Process Technology,* ed. R. W. ROUSSEAU. New York: J. Wiley Interscience, pp. 644–696.

KENNEDY, D. C. 1980. "Predict Sorption of Metals on Ion-exchange Resins." *Chem Eng* **87,** 12 (June 16): 106–118.

KING, C. J. 1971. *Separation Processes.* New York: McGraw-Hill.

KODATSKY, W. K. 1978. *Selecting Separation Options."* Hamilton, ON: McMaster University; see also WOODS, D. R., and W. K. KODATSKY. 1985. "Discovering Short Cut Methods for Equipment Sizing and Selection." ASEE Annual Conference Proceedings, Atlanta, GA paper 1616.

LEE, E. K. 1987. "Membranes, Synthetic Applications" in *Encyclopedia of Physical Science and Technology,* Vol. 8. San Diego, CA: Academic Press.

MALAMUD, D., et al. 1978. "Isoelectric Points of Proteins: a Table." *Analytical Biochemistry* **86:** 620–647.

MOORE, I. J. 1987. M.Eng Thesis. McMaster University, Hamilton, Ont, Canada.

MULLIGAN, T. J., and R. D. FOX. 1976. "Treatment of Industrial Wastewaters." *Chem Eng* **83,** 22 (Oct 18): 49–66.

NULL, H. R. 1980. "Energy Economy in Separation Processes." *Chem Eng Prog* (Aug): 42–49.

RIGHETTI, P. G., et al. 1976. "Isoelectric Points and Molecular Weights of Proteins." *J of Chromatography* **127:** 7–28, Chromatographic Reviews.

RUDD, D. F., G. F. POWER, and J. J. SIIROLLA. 1973. *Process Synthesis.* New York: J. Wiley and Sons.

RUSH, F. E., JR. 1980. "Energy Saving Alternatives to Distillation." *Chem Eng Prog* (July): 44–49.

RUSSELL, F. G. 1983. "Field Tests of DELSEP Permeators." *Hydrocarbon Process* (Aug): 55.

SCHENDEL, R. L., C. L. MARIZ, and J. Y. MAK. 1983. "Is Permeation Competitive?" *Hydrocarbon Process* **62,** 8: 58–62.

SCHENDEL, R. L. 1984. "Using Membranes for the Separation of Acid Gases and Hydrocarbons." *Chem Eng Prog* **80,** 5: 39–43.

SMITH, J. M. and H. C. VAN NESS. 1975. *Introduction to Chemical Engineering Thermodynamics.* Third ed., New York: McGraw-Hill.

SOUDERS, M. 1964. *Chem Eng Prog* **60,** 2: 75.

WOODS, D. R. 1993a. *Process Design and Engineering Practice: Selecting and Sizing Homogeneous Separations.* Hamilton, ON: McMaster University.

WOODS, D. R. 1993b. *Process Design and Engineering Practice: Reactions, Mixing and Size Change.* Hamilton, ON: McMaster University.

4.3 EXERCISES

4-1 For the following binary separations, what are some possible options:
1. 500 ppm of TNT from waste water.
2. concentrate 30% monoglycerides to 90%. (The monoglycerides are produced by the reaction vegetable fatty acids plus glycerine plus sodium hydroxide yields monoglycerides plus water.)
3. separate 5% acetophenone from 95% phenol.
4. separate benzene from cyclohexane.
5. aromatic hydrocarbons from kerosene.
6. separate 60% toluene from methylcyclohexane to yield a toluene stream that is 90% pure with a 90% recovery.
7. separate 40% styrene ethylbenzene.
8. recover ethylene from a gas stream that is 25% ethylene and 75% propane.
9. purify water by removing 50,000 ppm of sulfides present in the water from sulfidic caustic wastes.
10. 50% m-cresol and 50% p-cresol into 95% pure concentrates.

4-2 Propose a separation system to handle 5.7 L/s of aqueous waste that contains:

1 500 ppb	benzene
60 000 ppb	tetrahydrothiophene
900 ppb	vinyl chloride
20 000 ppb	chloroform
20 000 ppb	trichloroethylene
6 000 ppb	methyl chloride
27 mg/L	cyanide
430 ppb	phenol

4-3 A homogeneous liquid contains 0.1% w/w of a target component that has an exchange equilibrium given by $\alpha = 1.02$, that differs in vapor pressure by an $\alpha = 1.75$ and by solubility by an $\alpha = 1.2$. It is not surface active. What separation options might be considered?

4-4 For the ethylene plant given in Figure 1–5 and **PID-1A;** the concentration of the gas leaving the reactor is as follows:

	mol %	wt%
Hydrogen	32.7	3.52
methane	6.3	5.42
acetylene	0.2	
ethylene	33.8	50.88
ethane	24.9	40.18
propylene	1.0	
propane	0.2	
C_4^+	0.9	
	100	

Propose various separation sequences. Compare and contrast these with the one given in **PID-1.**

4.4 PROCESS AND INSTRUMENTATION DIAGRAM FOR AMMONIA ILLUSTRATES APPLICATIONS

Separations—the production of ammonia offers an intriguing example of how to separate a variety of species. Some are trace contaminants like carbon, CO, and CO_2. Some is to separate hydrogen from and some is to separate the product ammonia from unreacted nitrogen and hydrogen. Interesting.

PID-4: Ammonia

Ammonia is #3 in annual tonnage production in the world. Although one of its prime uses is in fertilizer, it is needed in the production of explosives, textiles (such as nylon), urea and other amines and amides, and as a refrigerant.

The general flow diagram and a sketch of a plant is given in Figure 4A–1. Although many different feedstocks can be used, this illustrative process uses methane. The process, shown in PID-4, consists of seven processes or miniplants to produce "synthesis feed gas" at 32 MPa for the ammonia reactor, the ammonia reactor loop including the separation and removal of the product ammonia, and the "purge" gas treatment to remove contaminants from the recycle "synthesis feed gas." Thus an ammonia plant is an integrated system of nine miniplants.

The seven miniplants to prepare the synthesis feed gas are:

- feed preparation: methane desulfurization and conditioning to create a water/stream to methane ratio of 3:1. Desulfurization may be done, for example, by hydrogenation over a Co-Mo catalyst and desulfurization over zinc oxide, **R406.** Alternatively, the desulfurization can be done by adsorption on activated carbon. Water is added to methane through the "saturator" **T400** and by direct steam addition.

- creation of a source of hydrogen to mix with nitrogen to create a 3:1 ratio in the synthesis gas: methane "reforming" in the "primary" (**R400**) and "secondary" (**R401**) reformers. Here the target reforming reaction is:

$$CH_4 + H_2O \rightarrow CO + 3H_2 \qquad (4A\text{-}1)$$
$$CH_4 + 2H_2O \rightarrow CO_2 + 4H_2 \qquad (4A\text{-}2)$$

In the "primary," the steam-methane mixture passes over nickel catalyst at about 760 to 830°C, 2.8 to 4.5 MPa. Hot air is added and the mixture fed to the secondary reformer. Here the oxygen reacts exothermically with methane to help drive the reforming reaction:

$$CH_4 + 2O_2 \rightarrow CO_2 + 2H_2O \qquad (4A\text{-}3)$$

The temperature in the secondary reformer is about 980°C.

Side reactions occur. At the end of this step we basically have the synthesis feed gas. However, it is contaminated with CO and CO_2 (and some unreacted CH_4). The next four miniplants clean up the synthesis feed gas. The "separations" use reactions and packaged plants described in Table 4–22 and illustrated in Figure 4–38; methanation, water gas shift, and absorption (option d in Figure 4–38).

- the recovery of heat energy by steam raising, **V401,** to produce steam at 10 MPa and 500°C for sale. The steam is superheated in the "primary" reformer furnace **E402** and used to drive the air compressor by the turbine drive, **D400.** Some of the turbine exhaust steam is required as feed to the reformer; some is sold.

- our strategy is to convert all of the CO into CO_2. This is done by the water gas "shift" reaction at high and low temperatures (in reactors **R402** and **R403**). The high-temperature shift reactor produces a residual CO concentration of about 2%. Because the equilibrium in this reaction is favored at low temperature, the gas is cooled and passed over Cu zinc oxide catalyst at low temperature. The output from this process is cooled further and condensed liquid is removed in knock out pot **V402.**

- next the CO_2 is removed in packaged plant **P400** by absorbing it in an appropriate solvent. An example packaged process is given in PID-1B.

- any remaining CO_2 and CO are converted to methane in the methanation reactor, **R404.** In this particular process configuration, the raw gas is compressed, via **F402,** and preheated before the methanator. (In some processes, the compressor is after methanation and drying.) The target methanation reaction is:

$$CO + 3H_2 \rightarrow CH_4 + H_2O \qquad (4A\text{-}4)$$

- the last step is to cool and "dry" (remove water from) the synthesis feed gas, Exchangers **E413** and **415,** knock out pot **V403,** and molecular sieve adsorption, **V404.** The synthesis gas feed contains the correct ratio of hydrogen to nitrogen and is relatively free from other contaminants, other than argon and methane. Argon is present in the air fed initially to the secondary reformer; methane is unreacted methane from the reformers and that generated from the CO via the methanator.

In the ammonia synthesis loop, synthesis gas is pumped around through the synthesis reactor **R405,** a series of gas cooler/condensers, **E417 to 21** (including an ammonia refrigerant condenser **E421**), and a high-pressure knock out pot that removes the condensed ammonia, **V405.** With about 25% conversion per pass, the gas is pumped round and round by the "circulating" compressor **F403.** The temperature of the synthesis gas entering the synthesis

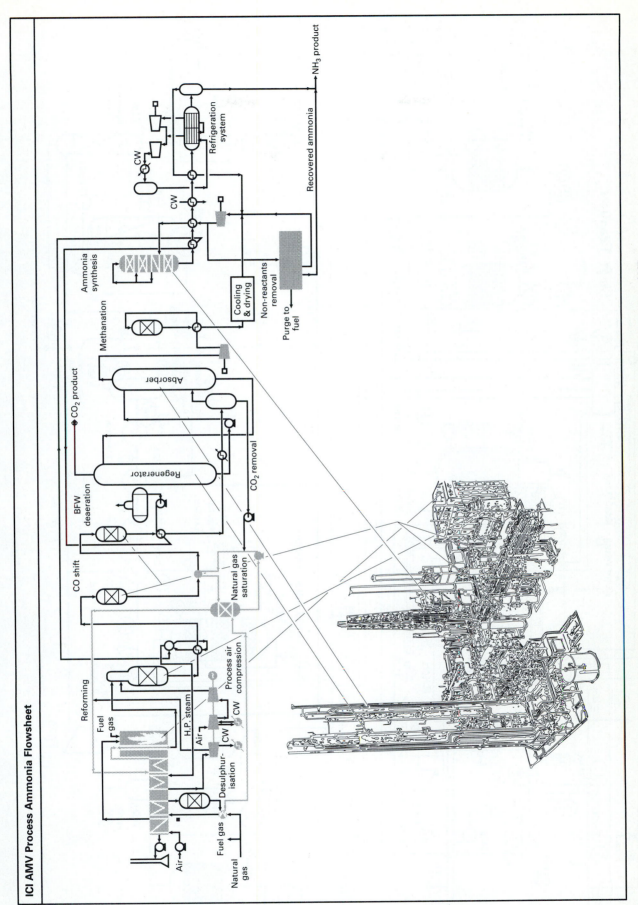

Figure 4A–1 Flow Diagram and Pictorial of an Ammonia Plant (reproduced courtesy of ICI)

4-91

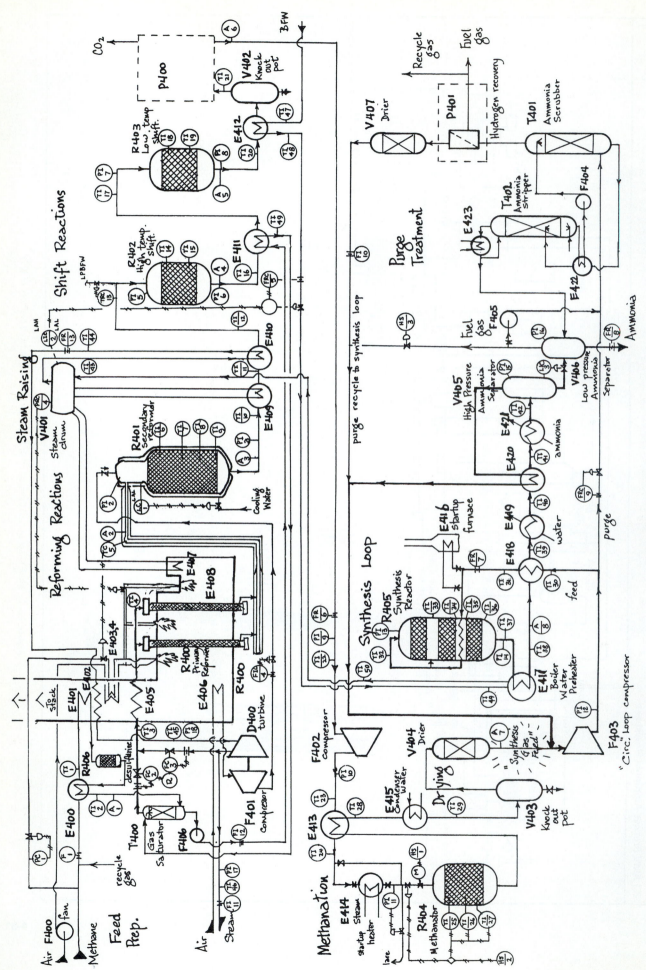

Figure PID-4 An Ammonia Process

reactor is about 35°C; the gas leaves the reactor at about 460°C. The synthesis reaction occurs at 400 to 500°C. The produce, liquid ammonia, is removed from the system to a low-pressure ammonia separator **V406.**

Contaminants argon and methane have no way to escape the synthesis loop unless some is bled off as a purge. The purge is treated to recover and recycle any hydrogen. The commonly used methods of separation, for the packaged plant **P401,** include cryogenic liquefaction of the methane, membrane separation of the hydrogen as a permeate, or pressure swing adsorption. For these options to be effective, ammonia (present in the purge steam) must be removed first. Absorption of the ammonia with water, with subsequent ammonia recovery by steam stripping, is commonly used. In this process the purge treatment is done at loop pressure. The vapor from the low-pressure ammonia separator can be processed with the purge gas by compressing it up to loop pressure by **F405.**

The general unit cost distribution and capital cost for this process are given in Table 4A-1 and Table 4A-2.

4.5 REFERENCES

CALABRESE, A. M., and L. D. KREJCI. 1974. "Safety Instrumentation for Ammonia Plants." *Chem Eng Prog* **70,** 2: 54.

ICI. "The Unique Energy Saving Ammonia Process from ICI: AMV." Billigham UK, undated brochure.

FORSTER, F. 1980. "Improved Reactor Design for Ammonia Synthesis." *Chem Eng* **87,** 18 (Sept 8): 62–63.

PINTO, A., and J. G. LIVINGSTONE. "The ICI AMV Process Reduces Cost." Paper presented at the AIChE Los Angeles Meeting, 1982.

STRELZOFF, S. 1981. *Technology and Manufacture of Ammonia.* New York: J. Wiley and Sons.

TABLE 4A-1. Cost contributions to the cost per tonne of ammonia product (based on ammonia from methane/natural gas as the feedstock)

Contribution from	Approximate unit cost breakdown, %	
Raw material	37	to 31
Utilities	31	to 9.7
Catalyst	2	to 4
Labor	1.5	to 4.3
Credit for byproducts	negligible	
Maintenance	4.5	
Supervision		
Depreciation	14.5	
Indirectly attributable costs	9.5	
General expense		
Total	100	

TABLE 4A-2. Capital cost contribution for the cost to construct a grass roots facility to produce ammonia from natural gas (A 1000 Mg/d plant costs an estimated $300 million ± 30%; MS = 1000)

Contribution from	Battery limits; approximate capital cost breakdown, %
Primary and secondary reforming	21.5
Compression	21.5
Steam generation	12
Shift reaction and methanation	3.5
CO_2 removal	16.5
Ammonia synthesis and ammonia recovery	20
Purge treatment	5
Total	100

ST, Standard tabulation of properties, to exploit differences for homogeneous phase separation. To make it easy to tabulate data, in this book (and the companion book **Data for PDEP**) we list the properties in the following standard sequence: molar mass; three items, health, flammability and stability, in the NFPA ratings; the Dow material rating; the three items in the Material of construction hazard/corrosivity rating for copper, carbon steel and 316 stainless steel, respectively; the freezing temperature ($^{\circ}$C); the boiling temperature ($^{\circ}$C); the molar volume (cm^3/mol); the Hildebrand solubility parameters for the polar and hydrogen bond contributions, respectively, (J/cm^3)$^{1/2}$; the dissociation constant, pK_a; the solubility product, pK_{sp}; the biodegradability, expressed as mg COD/h per g of target species; and the BOD$_5$/COD ratio. In the book, the code **ST**, meaning standard tabulation, is used to identify the meaning and units for the sequence of numbers presented.

Minichapter 5

Selecting Options for the Separation of Components from Heterogeneous

Often we need to separate or concentrate species. They may occur in a single homogeneous phase (like a glass of Coke that is to be separated into components); they may occur as an intimate mix of two or more different phases (like a mixture of sand and gold or a mixture of sand and water). If the starting mixture is a heterogeneous phase, then we start with the ideas given in this chapter. If the initial mixture is homogeneous, then we start with the ideas in Chapter 4.

As with homogeneous phase separation, to separate a heterogeneous mixture, we primarily exploit the differences in properties among the different species. Having said that, we have some very sloppy guidelines for initially selecting separation equipment that:

- consider the types of phases present: gas-liquid; liquid-liquid, etc.
- make an initial selection based on feed concentration and particle size.

Later, as we consider the options in more depth, we consider the types of property differences that can be exploited and the types of product required. In this chapter we focus on how to select feasible equipment.

5.1 OVERVIEW TO SELECTING OPTIONS FOR SEPARATIONS

In selecting separation options, we should keep an open mind. Tradition, as exemplified by this text, may suggest certain options. However, new developments are occurring rapidly. Individuals and companies have their "favorite" separation technique and thus may prefer to work with the familiar. The purpose of this textbook is to provide shortcut methods that will allow you to rough-size many options before deciding. Hence, in selecting options, choose about three to consider in more detail for any separation.

Next, think of combinations of options. Perhaps a screen followed by sedimentation is an appropriate choice for a liquid-solid system.

The general criteria for selecting options include the properties of the species, the feed and product constraints, and the characteristics of the equipment.

5.1-1 A First Selection: Exploit the Size and Concentration for the System

For heterogeneous systems, the type of system has a major impact on what options to consider. Hence, it is appropriate first to identify whether we are separating:

- a gas-liquid system,
- a gas-solid system,
- a liquid-liquid system,
- a liquid-solid system, or
- a solid-solid system.

Once this is established, an initial guess may be made by considering the size of the "particulate" phase and the feed concentration. To some extent this focus on the particle size *early in the screening process* is misleading because we may need to exploit differences in properties other than size in order to make an option work. (For example, a big and small particle of different materials might have the same settling velocities; hence, sedimentation could not be used.) Nevertheless, particle size and concentration give us an initial idea about the probable options for different systems.

The particles in any system can be drops, bubbles, mist, dust, gas, liquid, or solid; they can be big—like a bottle—or small—like suspended clay particles that make drinking water look murky. In the context of separation we usually think of particles as having a size greater than about 0.1 μm.

A. Separating Gas-Liquid Systems.

Figure 5–1 shows sketches of the equipment options usually available to separate gas-liquid mixtures: cyclones, knock out pots, scrubbers, and burners. Figure 5–2 illustrates the ballpark region when the different types might be applicable. Actually, since gas-liquid separations are surprisingly difficult, most of our effort is spent trying to prevent the unwanted formation of mists or bubble dispersions. For example, when liquid is drained from a tank, vortex breakers are put in the exit pipe to prevent the overhead gas vapor from being caught in the vortex and entrained. Demisters mainly function to increase the size of drops or bubbles so that the separation can occur rapidly by the action of gravity on the drops.

When a separation is required, cyclones can be used but they have not always worked satisfactorily. The most common device is a knock out pot, which provides sufficiently slow gas velocity for the drops to settle out by gravity. Thus, this exploits difference in settling velocity. For low concentrations, this separation can be achieved by baffles. The terms in brackets in this figure refer to the removal efficiency.

B. Separating Gas-Solid Systems.

Figure 5–3 lists options for this type of separation. These include cyclones, electrostatic precipitators, scrubbers, and bag filters. Figure 5–4 illustrates regions of applicability for removing a dust from a gas. This suggests that the particle size seems to be the major variable, since for both wet scrubbers and electric precipitators almost any inlet loading

can be handled. The smaller the particle size to be removed, the greater the energy requirement.

Example 5–1: A stream contains 10 g/m^3 of a 20 μm dust. What options might we consider?

An Answer: A relatively high-energy cyclone (since the point is in the lower portion of the application region), a relatively low-energy bag filter (since the point is near the top of the application region), and a low-voltage electrostatic precipitator.

Comment: The lower one is in any region, the more energy is required.

Another feature to note is that most devices are about 90 to 99% efficient in *mass* removal. The part that is not removed will be the smaller-sized particles. Thus, one can use Figure 5–4 by moving to the left *one* cycle on the abscissa to obtain the concentration and go to the limit of the device application region. Read the particle size from the ordinate.

Example 5–2: An inlet gas stream has 1 g/m^3 of dust of about 1 μm diameter. What device is appropriate and what are the characteristics of the exit gas leaving the separator?

An Answer: From Figure 5–4, we could use a high voltage electrostatic precipitator, a moderate-energy venturi scrubber or a high-energy bag filter. If we used the electrostatic precipitator, the exit stream might be about 10^{-1} g/m^3 of about 0.1 μm particles.

C. Separating Liquid-Liquid Systems.

Liquid-liquid systems can be separated by decanters, hydrocyclones, flotation, deep bed filtration. Figure 5–5 illustrates the equipment in general. Figure 5–6 shows the options. The usual size of drops in an oily emulsion or dispersion is about 200 μm. The basic device is gravity settling or decantation. This can be augmented by adding coalescence promoters, electrodes, and/or plates. A hydrocyclone or centrifugal device exploits the same general principle. As the concentrations and size become smaller, the options shift to coagulation (increasing the size of the particles by clustering the particles together to form a larger one), entrapment (such as filters, solvent extraction, and flotation) combinations. Combustion and high gradient magnetic separation (HGMS) are options for very dilute and very small systems.

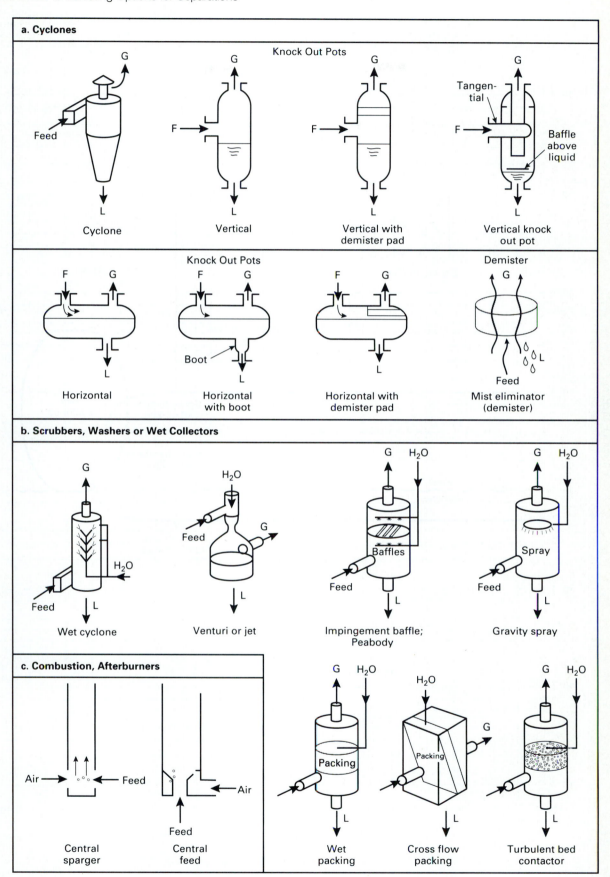

Figure 5–1 Equipment Options for Separating Gas-Liquid Systems

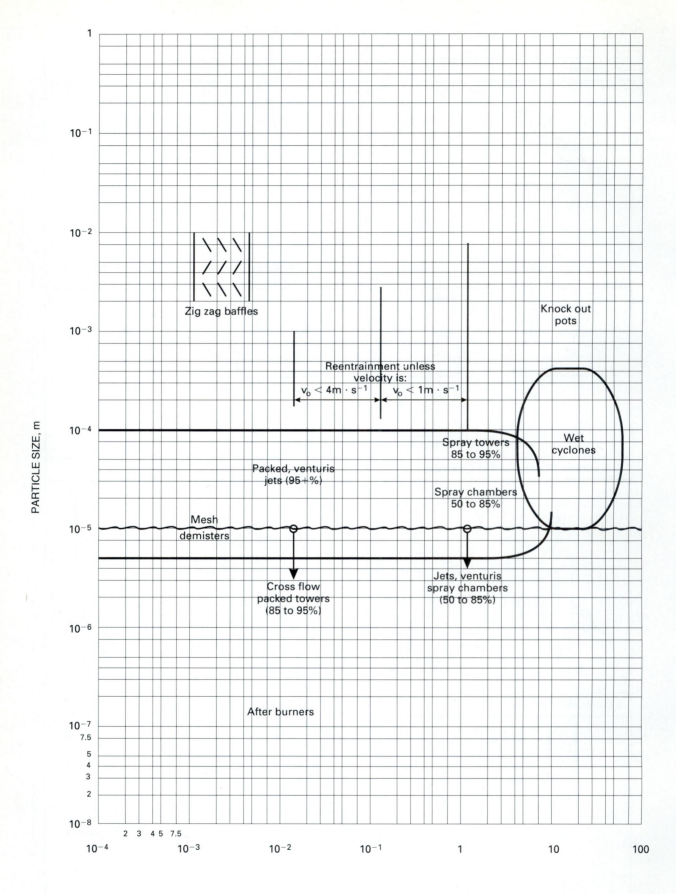

Figure 5–2 General Regions of Applicability for Gas-Liquid Separations: Based on Particle Size and Concentration (Bracketed Terms = % Removal Efficiency)

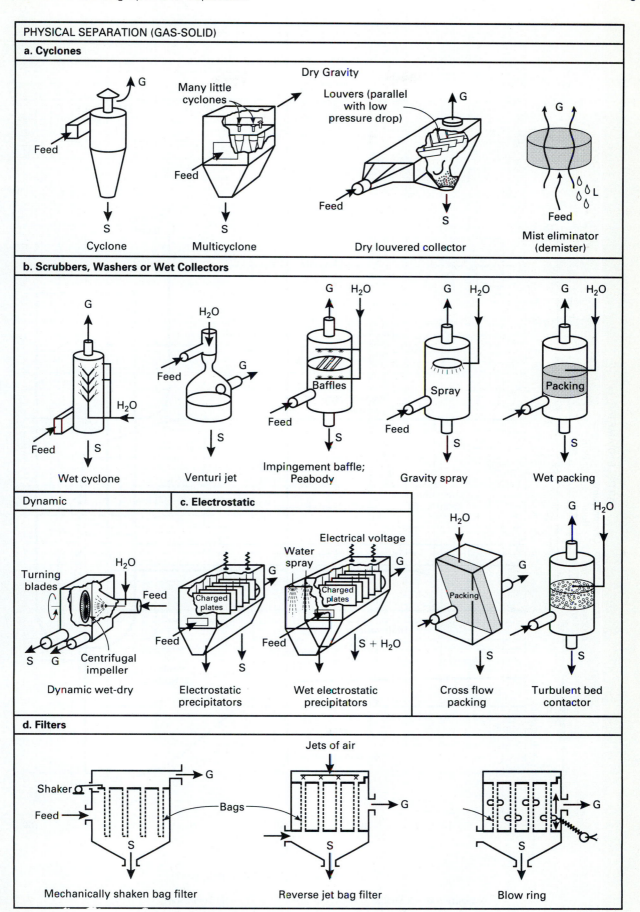

Figure 5–3 Equipment Options for Separating Gas-Solid Systems

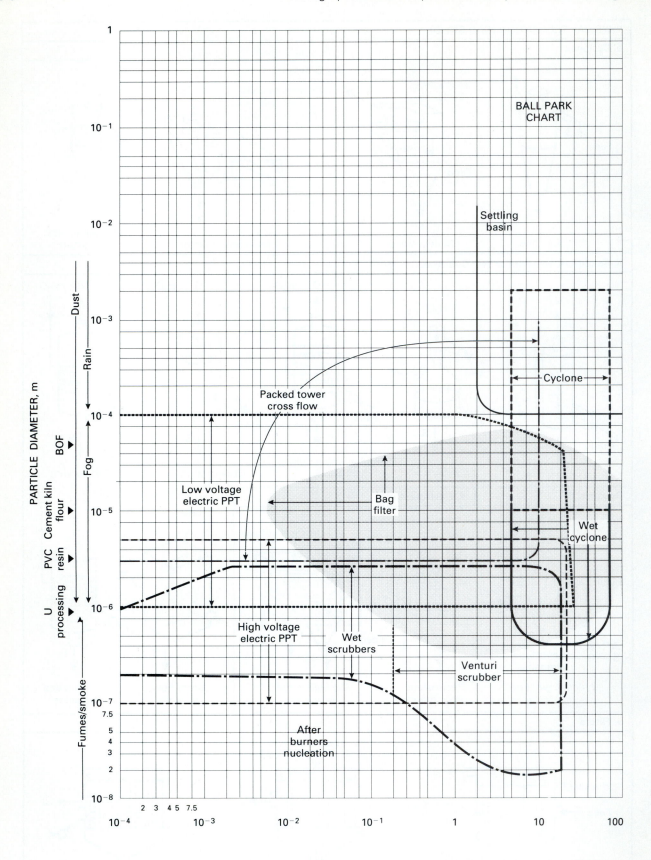

Figure 5–4　General Regions of Applicability for Gas-Solid Separations:
Based on Particle Size and Concentration

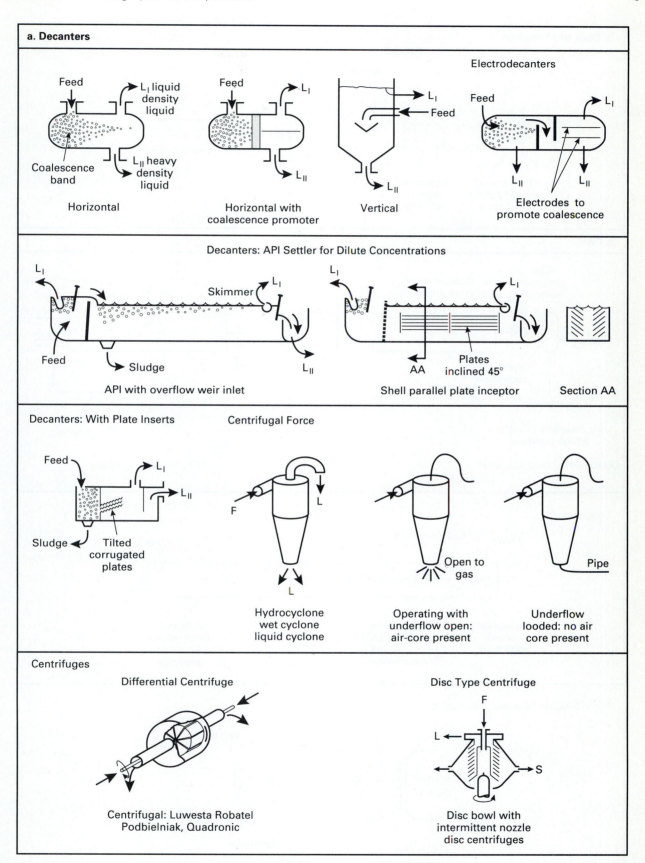

Figure 5–5   Equipment Options for Separating Liquid-Liquid Systems

a) Decanters

b) Deep Bed Filtration; Membrane

c) DAF; Coagulation; High Gradient Magnetic Separation

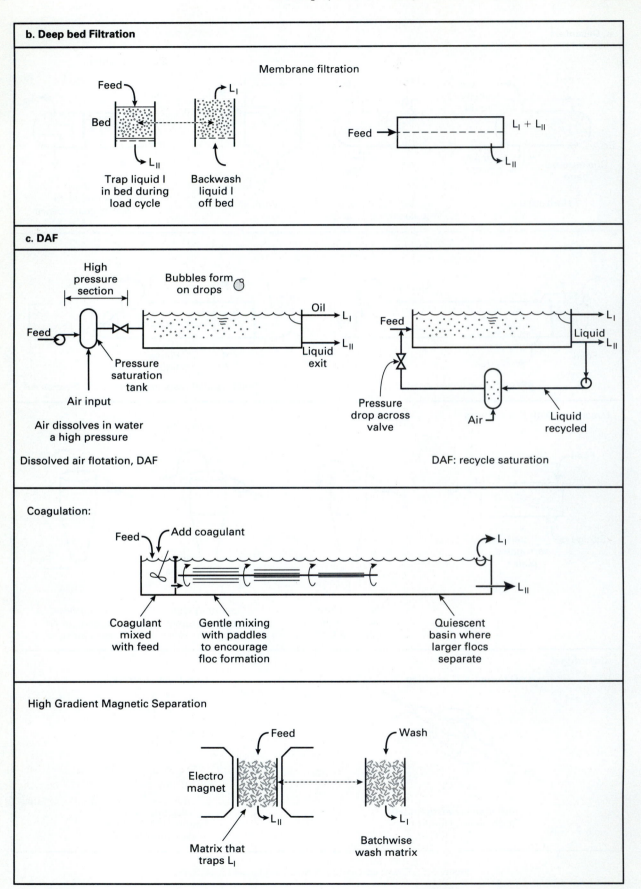

b. Deep bed Filtration

Membrane filtration

Feed

Bed

L$_I$

L$_{II}$

Trap liquid I
in bed during
load cycle

Backwash
liquid I
off bed

Feed → | L$_I$ + L$_{II}$

L$_{II}$

c. DAF

High
pressure
section

Bubbles form
on drops

Oil

Feed

L$_I$

L$_{II}$

Pressure
saturation
tank

Liquid
exit

Air input

Air dissolves in water
a high pressure

Dissolved air flotation, DAF

Feed

L$_I$

Liquid L$_{II}$

Pressure
drop across
valve

Air

Liquid
recycled

DAF: recycle saturation

Coagulation:

Feed   Add coagulant

L$_I$

L$_{II}$

Coagulant
mixed
with feed

Gentle mixing
with paddles
to encourage
floc formation

Quiescent
basin where
larger flocs
separate

High Gradient Magnetic Separation

Feed   Wash

Electro
magnet

L$_{II}$

L$_I$

Matrix that
traps L$_I$

Batchwise
wash matrix

Figure 5–5 (Continued)

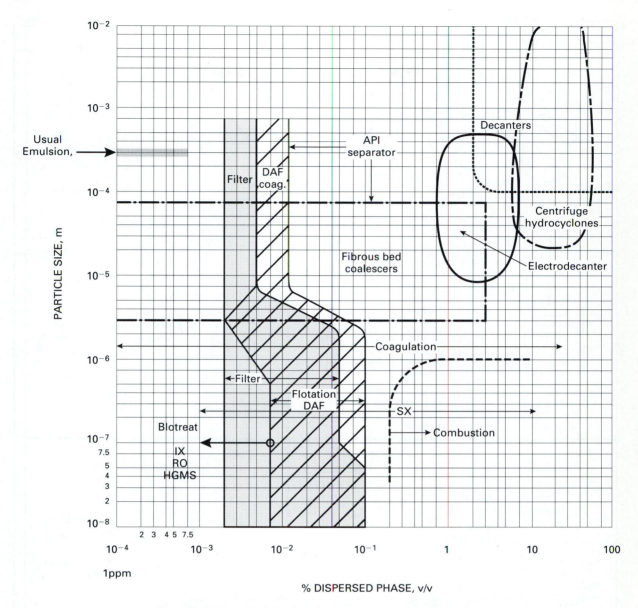

Figure 5–6 General Regions of Applicability for Liquid-Liquid Separations: Based on Particle Size and Concentration (reprinted from Woods and Diamadopoulos, 1988, p. 464, by courtesy of Marcel Dekker, Inc.)

D. Separating Liquid-Solid Systems. For liquid-solid separations, a very wide range of options exist: screens, dryers, leachers, expellers, filters, and settlers. These are illustrated in Figures 5–7, 5–8, and 5–9. Some options exploit differences in vapor pressure; Figure 5–7 shows dryers, evaporators, or devolatizers. Some options are based on differences in settling characteristics; Figure 5–8 illustrates these. Some options exploit the "size" of the particles relative to the size in a mesh. Figure 5–9 shows trommels screens and filters. If the size is very small, we often increase the size via flocculation or coagulation. Sometimes we add a bed of coarse particles or bubbles to try to capture the small particles. These latter options have been sketched in Figure 5–5. Now, however, the phase L_I should be replaced by the "Solid" phase.

Figure 5–10 shows the general regions of application for these options.

For the settling devices illustrated in Figure 5–8, Figures 5–11 and 5–12 provide more detail about where slightly different types of settlers are best suited.

For screens and filters, the overall *class* of filter is selected based on the size of particle (screens versus filters versus filtering centrifuges); within a class, the concentration helps suggest the *type*. For example, Figure 5–12 shows the type of centrifugal filter to use. Figure 5–13 gives the type of filter or screen to use.

Larger-sized particles in liquid-solid systems may also be processed by "dewatering presses or expellers" or by leaching. This type of equipment is illustrated in Figures 5–14 and 5–15. Presses and expellers are like filters in the

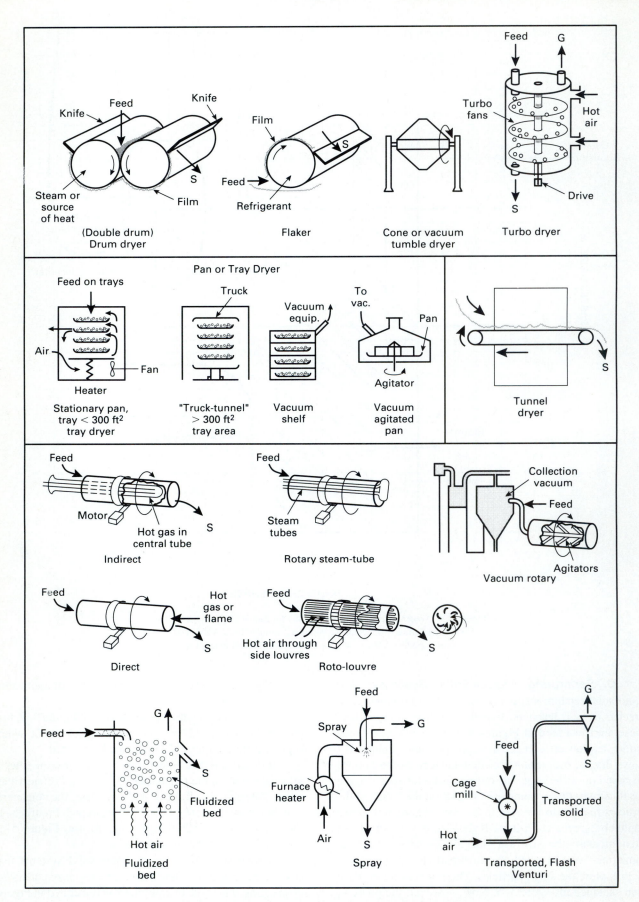

Figure 5–7 Equipment Options for Separating Liquid-Solid Systems: Dryers

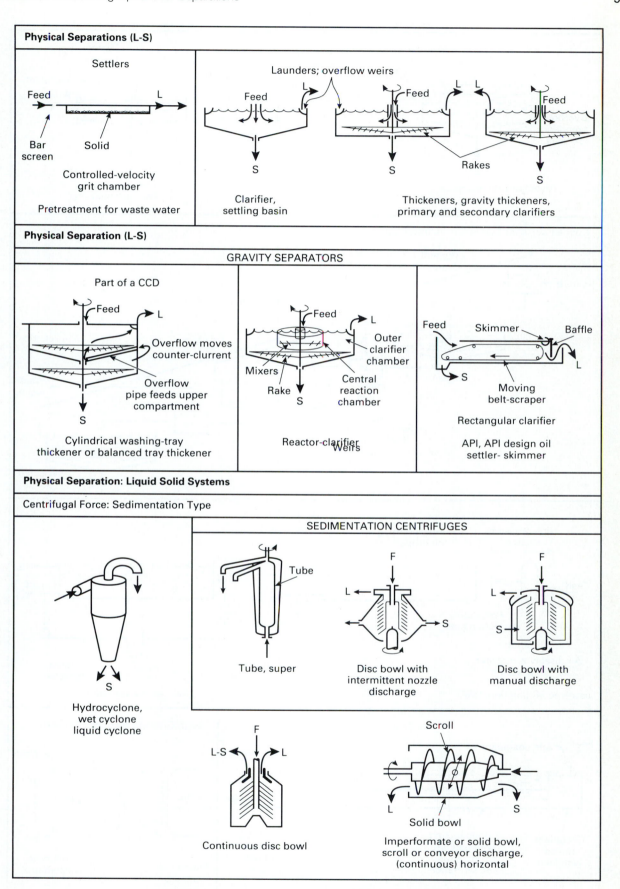

Figure 5–8 Equipment Options for Separating Liquid-Solid Systems:
Settler, Thickeners, Hydrocyclones and Centrifuges

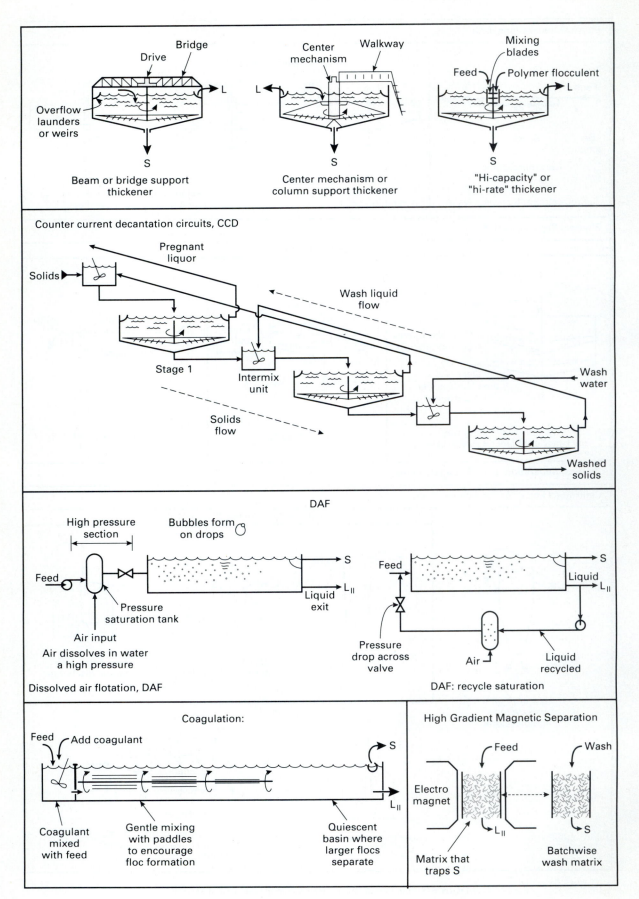

Figure 5–8 (Continued)

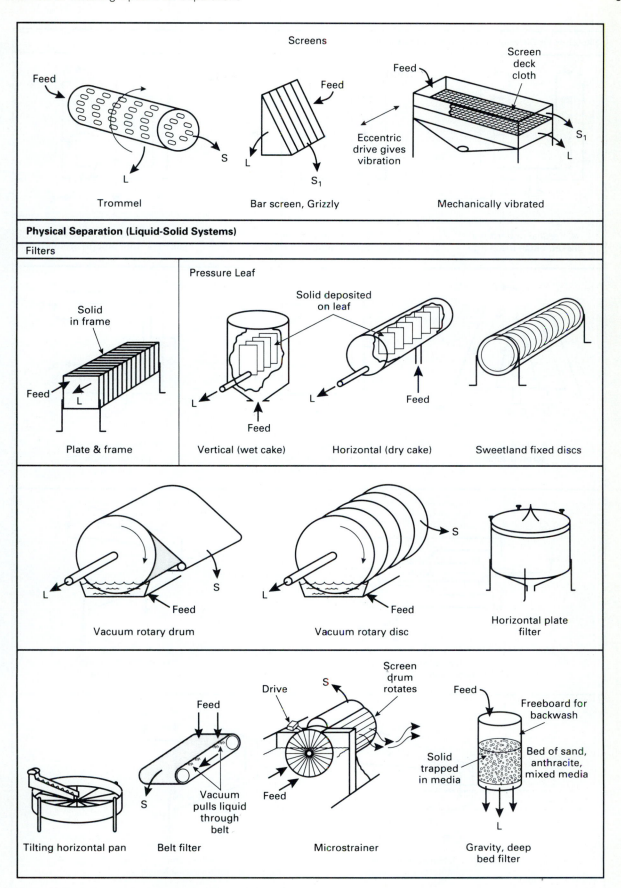

Figure 5–9   Equipment Options for Separating Liquid-Solid Systems:
Screens and Trommels, Filters and Filtering Centrifuges

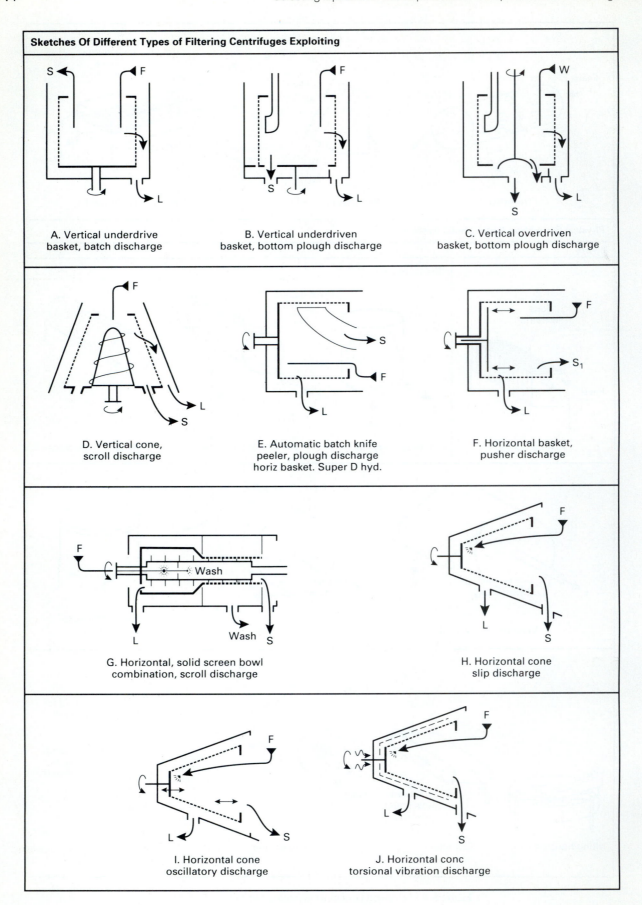

Figure 5–9 (Continued)

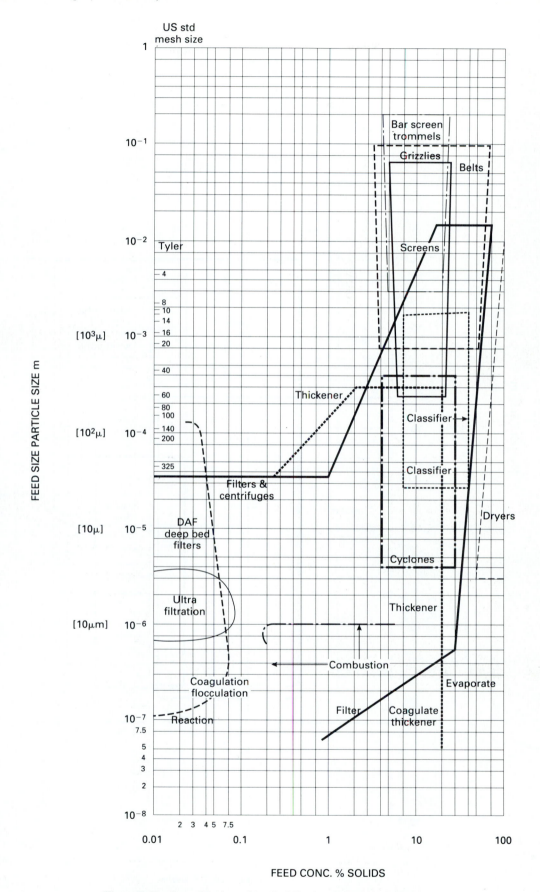

Figure 5–10 General Regions of Applicability for Liquid-Solid Separations: Based on Particle Size and Concentration (adapted from Dahlstrom and Cornell, 1971)

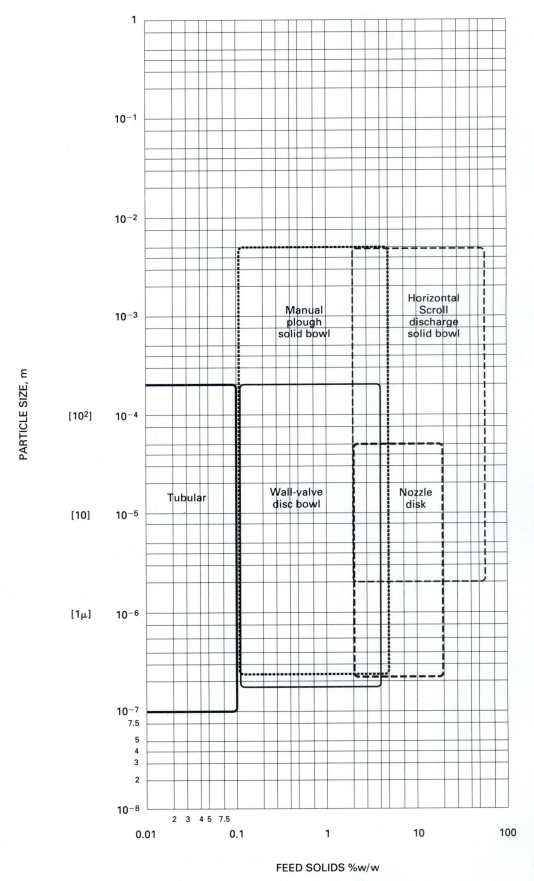

Figure 5–11 General Regions of Applicability for Liquid-Solid Separations:
Based on Particle Size and Concentration for Sedimentation Centrifuges

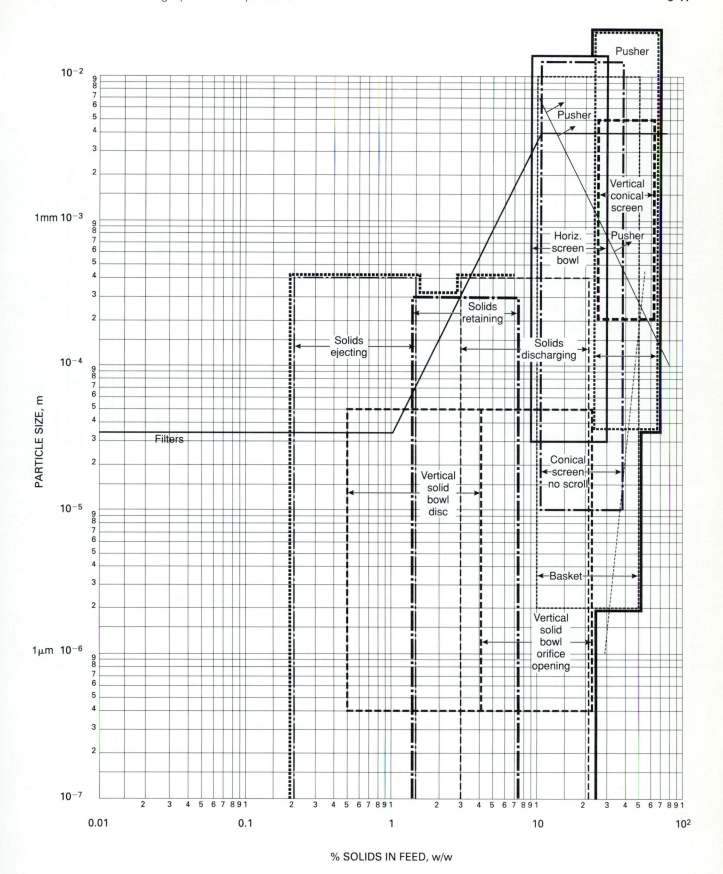

Figure 5–12 General Regions of Applicability for Liquid-Solid Separations:
Based on Particle Size and Concentration for Filtering (and Sedimentation) Centrifuges

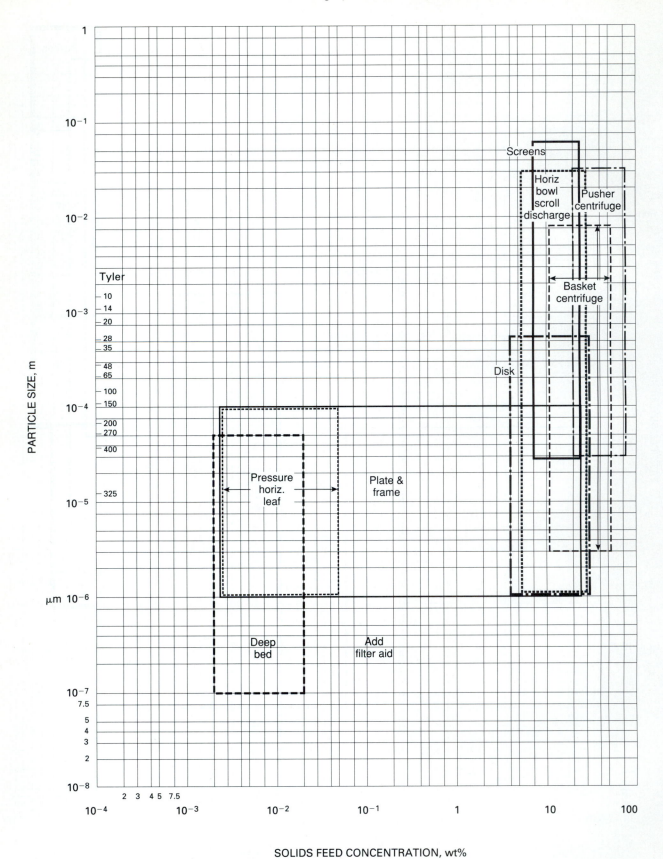

Figure 5–13   General Regions of Applicability for Liquid-Solid Separations:
Based on Particle Size and Concentration for Filters and Centrifuges

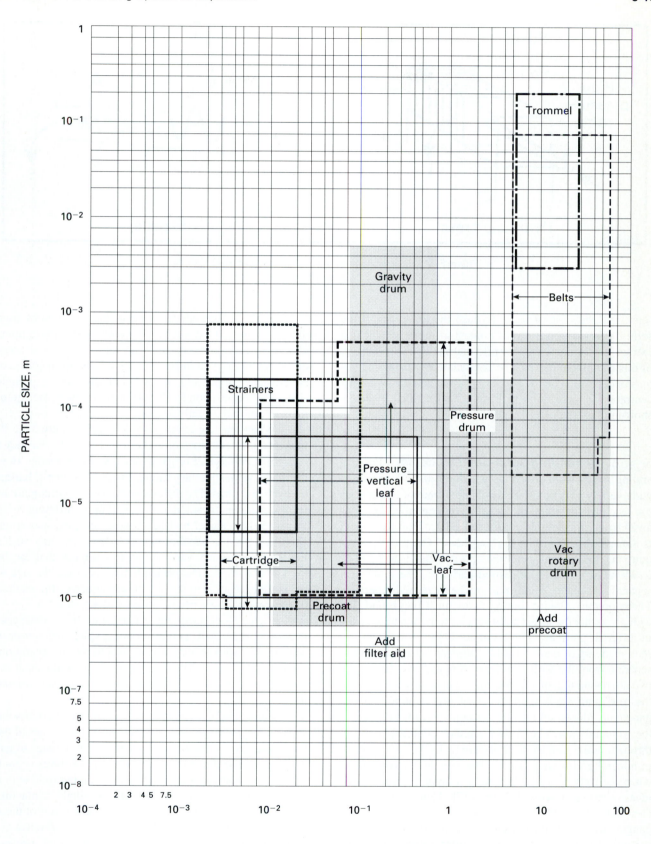

Figure 5–13 (Continued)

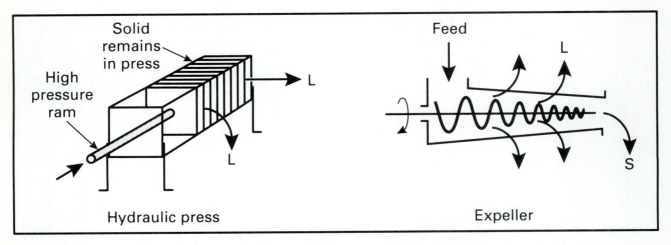

Figure 5–14 Equipment Options for Separating Liquid-Solid Systems:
Presses and Expellers

sense that they use pressure to push the liquid out of the solid. In leaching, the solids contact a liquid solvent that extracts the liquid from the solid matrix. The applicability of this equipment can best be appreciated by using the *liquid* concentration on the abscissa (rather than the *solid* concentration as was done in Figure 5–10). Such a plot is shown in Figure 5–16.

E. Separating Solid-Solid Systems.

For separating solids from solids, as with liquid-solid systems a wide variety of options are available. Some of the options use leaching (in which the desired component is *dissolved* away from an insoluble, undesired solid). This equipment is illustrated in Figure 5–15, "leachers." Other options require that the desired species is a particle that is to be *physically separated* from undesired particles. These physical separations are the options that are the focus of this section. These options require that particles must be ground fine enough so that the target solid species to be separated exist in different particles. For example, Figure 5–17 shows a rock made up of dark, valuable "mineral" that we want to separate from the white, useless "gangue." As shown, the two are bound together as one rock. Thus, we need to crush the rock until we have particles of just mineral and just gangue as illustrated ideally in Figure 5–17b. In practice, the mineral may be finely spread inhomogeneously throughout the gangue, or, if we are lucky, be concentrated in mineral grains embedded in the gangue. Unfortunately, when the rock is crushed, it doesn't break to produce the ideal situation shown in Figure 5–17b. Hence, to get particles that are predominantly mineral or predominantly gangue, we usually have to crush to about $1/100$th the size of the mineral grain. Such a size is often referred to as the "liberation" size because the two species are now in separate particles. Even with this fine a particle, usually only about 60 to 70% of the mineral would be as a unique, uncontaminated particle.

Table 5–1 introduces some of the terminology used to help distinguish between the physical separation options: classifiers, concentrators, and "separators." Although some of the options exploit differences in density (or magnetic or electrostatic properties), we will initially use the particle size and concentration to identify regions where the equipment options are usually applied.

What does the equipment look like? Figure 5–18 illustrates different types of classifiers. Figure 5–18a shows classifiers that use air or gas as the conveying media. Here the media whooshes the particles through a series of baffles or twists and turns. The different density-size combinations for the particles means that they end up at different exits. Figure 5–18b shows hydrocyclones, and rake and spiral classifiers that use water or liquids as the conveying media. Figure 5–18c shows various types of screens that can be used as classifiers. The principle here is to use the size of the particle relative to the size in the screen as the method of separation. These can be operated dry or wet.

Figure 5–19 shows jigs, tables, and sluices that serve as concentrators. These concentrators exploit differences in settling velocities of the particles to achieve the separation. Figure 5–20 shows DMS or dense media separators that exploit differences in density between the target species and the rest.

Figure 5–21 illustrates the workhorse of solid-solid separations: the froth flotation unit. Here, the surface of the target species is conditioned so that the solid clings to gas bubbles that rise through the liquid. The gas bubbles or froth can then be skimmed off, bringing the clinging solid with it into the "float." Figure 5–21a shows single contacting stages or cells. Figure 5–21b illustrates how series of these cells are hooked together. The fundamental difference exploited here is the surface wettability of the target species.

Figure 5–22 illustrates electrostatic separators, which are effective because some particles will acquire an electrostatic charge; others will not. This difference can be exploited, as illustrated in Figure 5–22, to achieve separation.

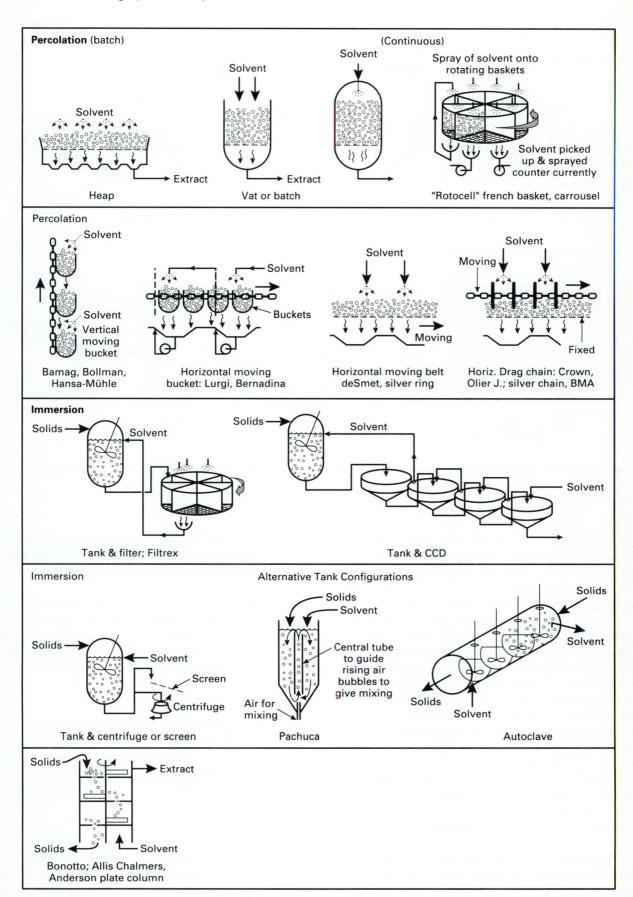

Figure 5–15 Equipment Options for Separating Liquid-Solid Systems: Leaching

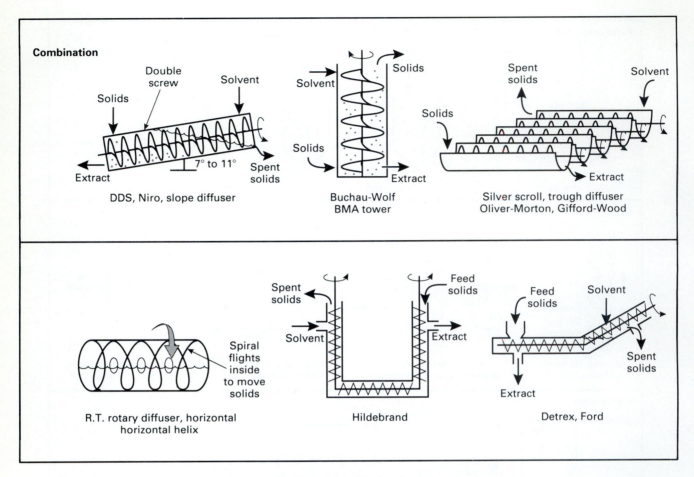

Figure 5–15   (Continued)

Figure 5–23 shows how materials with different conductivity can be separated via "eddy-current." Similarly, some target materials are sufficiently magnetic that a magnetic field can be used to obtain a separation. Equipment based on magnetic separation is illustrated in Figure 5–24.

Figure 5–25 shows the usual regions of application for options to physically separate solids from solids. Figure 5–16 shows, for percolation and immersion leaching, the usual regions of application for separating solids from solids by chemical leaching. Figure 5–25 is a very busy diagram. Many options are available, and many overlap. Nevertheless, this is useful for a "first approximation" for reasonable options. On the left-hand ordinate is summarized the size of the particles expressed in terms of *mesh* size.

Another way to consider solid-solid separations is analogous to that given for homogeneous phase separation in Figure 4–1. This is shown in Figure 5–26, where now the separation factor is defined for solid-solid options in Table 5–2.

In summary, for each combination of phases, a convenient, initial choice of option can be made based on the size of the particles, drops, or bubbles and the concentration.

5.1-2 A More Detailed Consideration of Options: Exploit the Properties of the Phases and Consider the Product Requirements

So far we have considered a very sloppy method of selecting possible separation options based on only the feed concentration and the particle size. If more details are needed, then by successive approximation, we need to consider the differences in properties that can be exploited and the product requirements.

A. Exploiting the Property Differences.   Despite what the screening charts given in the previous section suggest about the importance of particle size, the key fact is that for any separation *there must be a difference in physical or chemical properties that can be exploited.*

The options include:

1. a difference in **vapor pressure.** This is especially useful for a liquid-solid mixture. By adding heat or by pulling a high vacuum we can convert the liquid trapped in the pores between the solid particles into a gas that can be separated easily from the solid. This

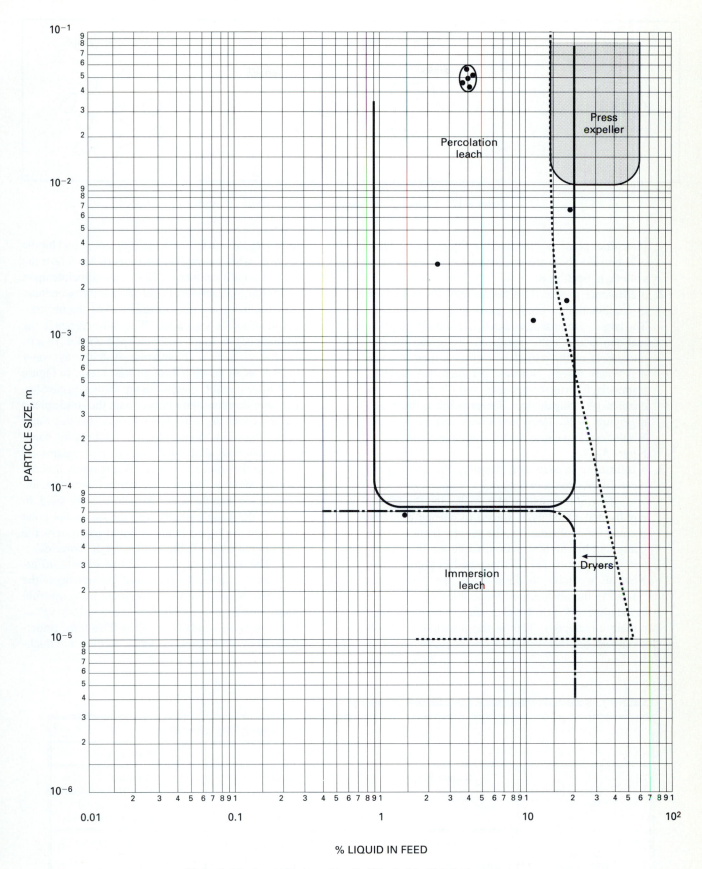

Figure 5–16 General Regions of Applicability for Liquid-Solid Separations: Based on Particle Size and Concentration for Presses, Expellers, and Leachers

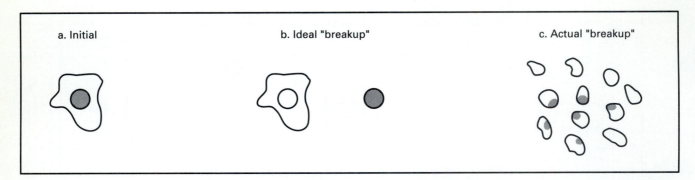

Figure 5–17 "Liberating" the Target Species for Solid-Solid Separation

interstitial phase is illustrated in Figure 5–27. More will be said about the importance of the interstitial volume in Section B.

2. a difference in **size.** This can be used for every type of combination of phases: gas-liquid, gas-solid, etc. The larger-sized particles can be separated by a physical barrier that either mechanically blocks the free movement of the particles or attracts them via van der Waals forces. Screens, grizzlies, bar screens, trommels, filtering centrifuges, and filters all are mechanical blocking devices. The principle is illustrated in Figure 5–28. Deep bed filters rely on van der Waals forces to cause the separation as illustrated in Figure 5–29.

3. a difference in **size and density.** We can exploit these differences in two ways.

First, if the mixture of particles and fluid flows rapidly and suddenly is forced to change direction because of an obstruction, the heavier particles keep going and hit the obstruction, while the fluid bends and flows past it. This is illustrated in Figure 5–30. Now all we have to do is make the particulates "stick" to the obstruction. Sometimes we use water as the obstruction (to which particles will impact and remain enveloped). This is the principle of a wet scrubber. Another variation on the same theme is when large-diameter particles keep going and hit the obstacle, whereas small diameter particles follow the gas and miss the obstacle. This is the principle upon which some classifiers of sold-solid systems operate. Sometimes we use the "wettability" of the obstruction to effect the separation. Wettability refers to the behavior of two fluids in contact with a solid. That is, we always have a three-phase system: say water droplets in air flowing through wire mesh. In Figure 5–31 three possible conditions are shown. In case (a), the water droplets upon impact with the solid spread out and cover the wire. Here we say that the water "wets" the solid. In case (b), the water retains much of its spherical shape and, although it remains attached to the solid via van der Waals force, it is not strongly attached because the contact area is small. Here we say that the water does *not* wet the solid. In case (c), the water partially wets the solids. For good separation of a liquid-liquid or liquid-gas system, the target liquid must wet the solid mesh or obstacles.

The second way that size and density difference can be exploited is in the settling velocity of the particles. We can use this to separate particles from liquids or we can use the *difference* in settling velocities to separate solids from solids. This is the principle used in centrifuges, cyclones, settling and thick-

TABLE 5-1. Solid-solid separations: classification of options

		Exploit Particle Size Difference	
		Narrow range	Wide range
Exploit Particle Density Difference	Narrow Range	"Separators" Froth flotation, Electrostatic Magnetic	"Classifiers" Air classifiers, Hydrocyclones, rake classifiers, spiral classifiers, screens, trommels
	Wide Range	"Concentrators" Jigs Tables Sluices Dense Media	Use concentrator then a classifier or vice versa

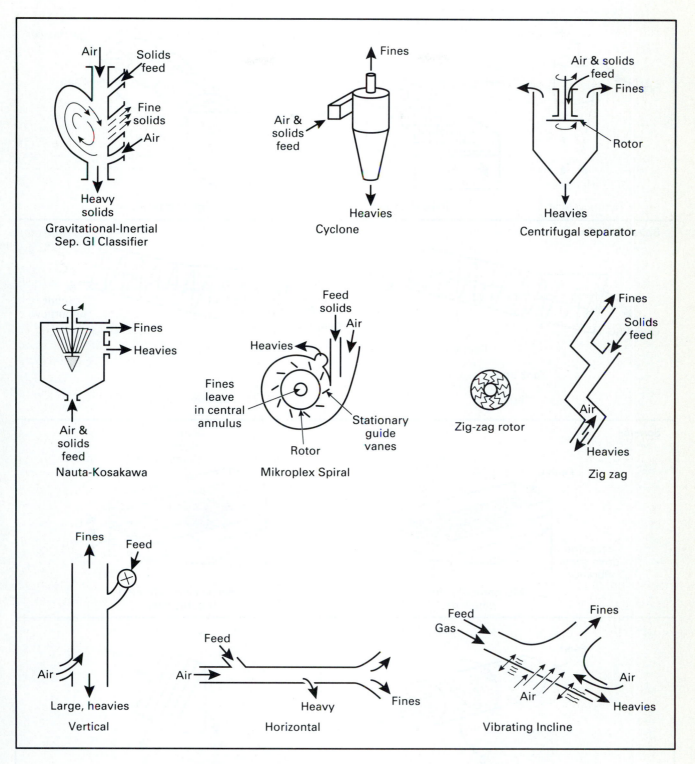

Figure 5–18 Equipment Options for Separating Solid-Solid Systems: Air and Liquid Classifiers, Screens and Trommels

a) Air or Gas Classifiers

b) Hydroclones and Rake and Spiral Liquid Classifiers

c) Screen Classifiers

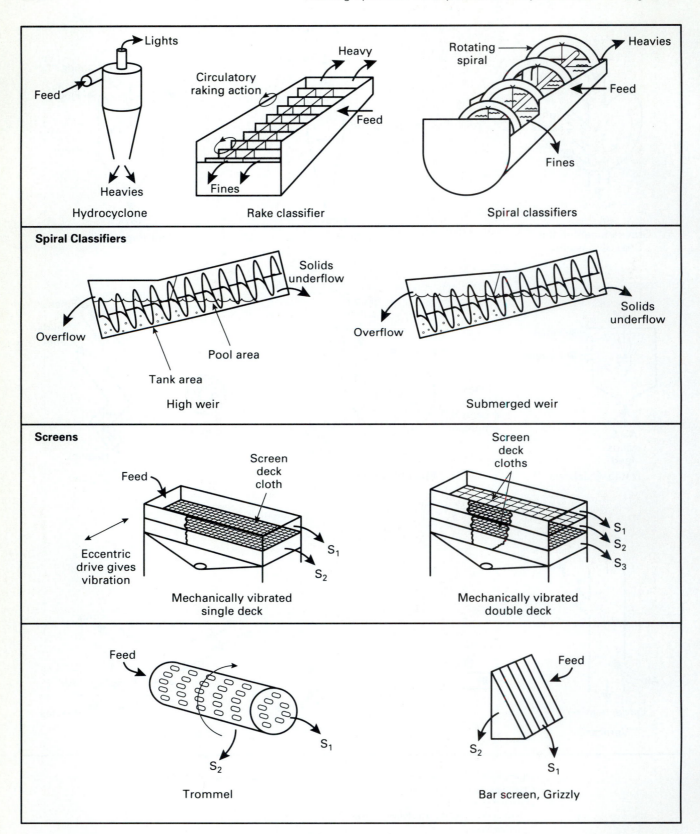

Figure 5–18 (Continued)

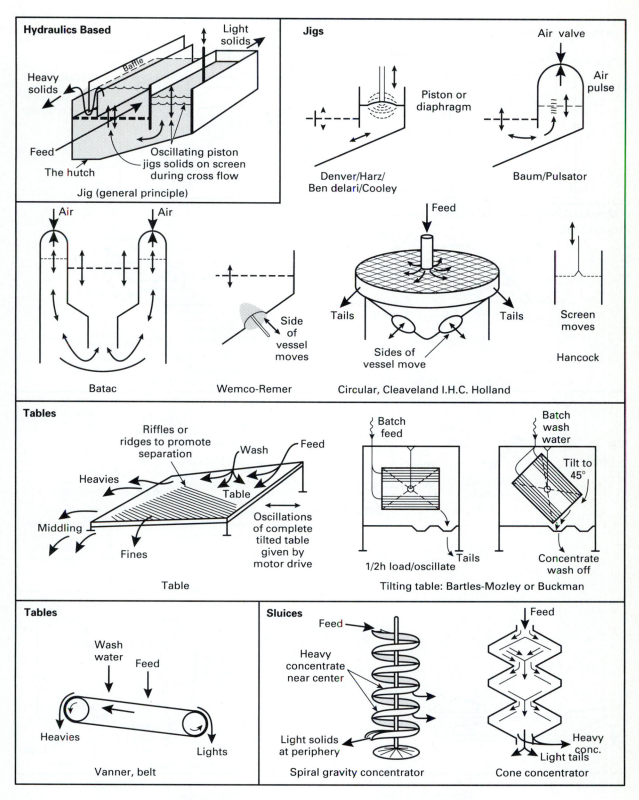

Figure 5–19   Equipment Options for Separating Solid-Solid Systems: Concentrators

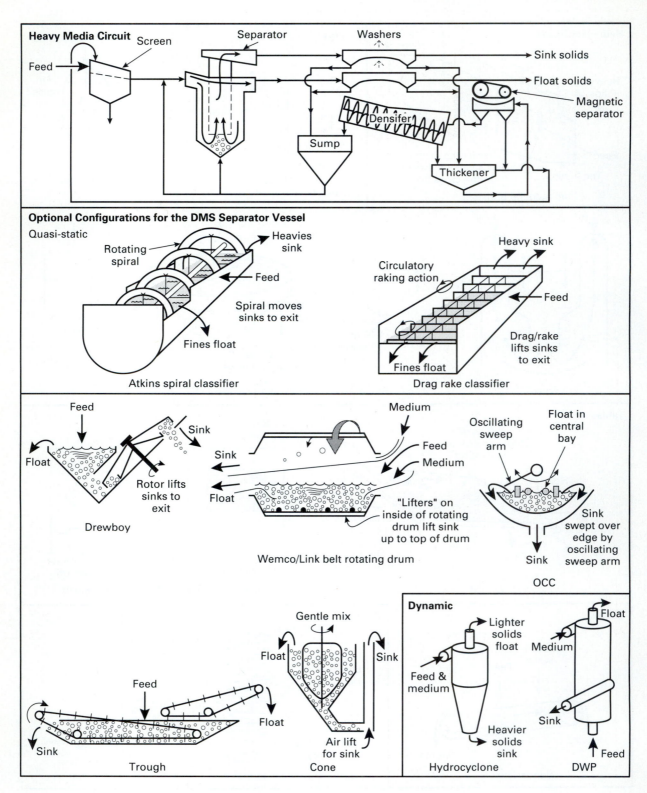

Figure 5–20 Equipment Options for Separating Solid-Solid Systems:
Dense Media Separators

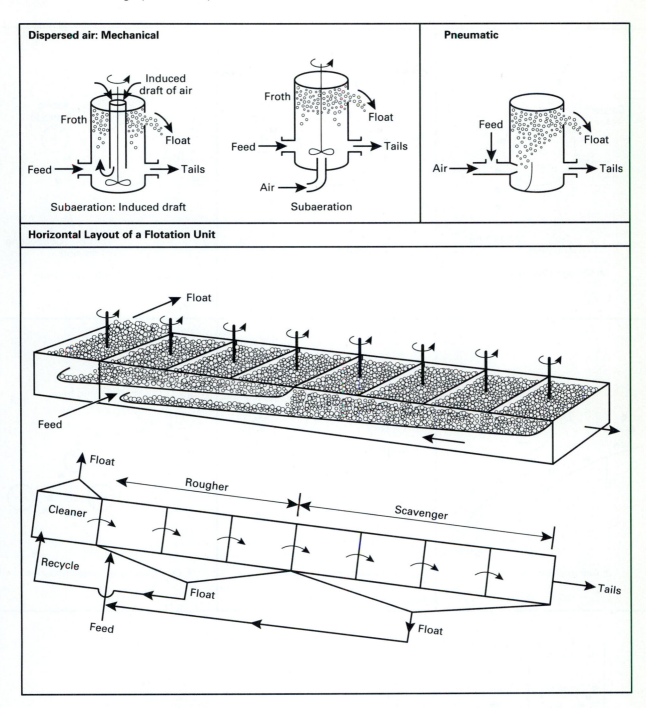

Figure 5–21 Equipment Options for Separating Solid-Solid Systems: Froth Flotation
a) Dispersed Air: Mechanical and Pneumatic
b) Horizontal Layout of a Flotation Unit

ening basins, hydrocyclones, decanters, jigs, rakes, tables, and spirals.

4. a difference in **density.** We may be able to separate solid-solid systems by carefully selecting the density of the continuous phase, or the surrounding media. This is called "dense media separation."

5. a difference in **wettability.** The concepts of wettability are illustrated in Figure 5–31. Here a liquid spreads out over or wets a solid, or the liquid may retain a spherical shape and *not* wet the solid, or it may partially wet it. We can use this to "capture" liquids on solid wires or surfaces.

   Displacement or "solvent drying" is based on wettability. For a solid bed of particles or for capillaries or cracks in rocks that are filled with liquid, we can use wettability to displace the liquid. A displacing solvent is pushed into the system. The solvent preferentially wets the solid and displaces the previ-

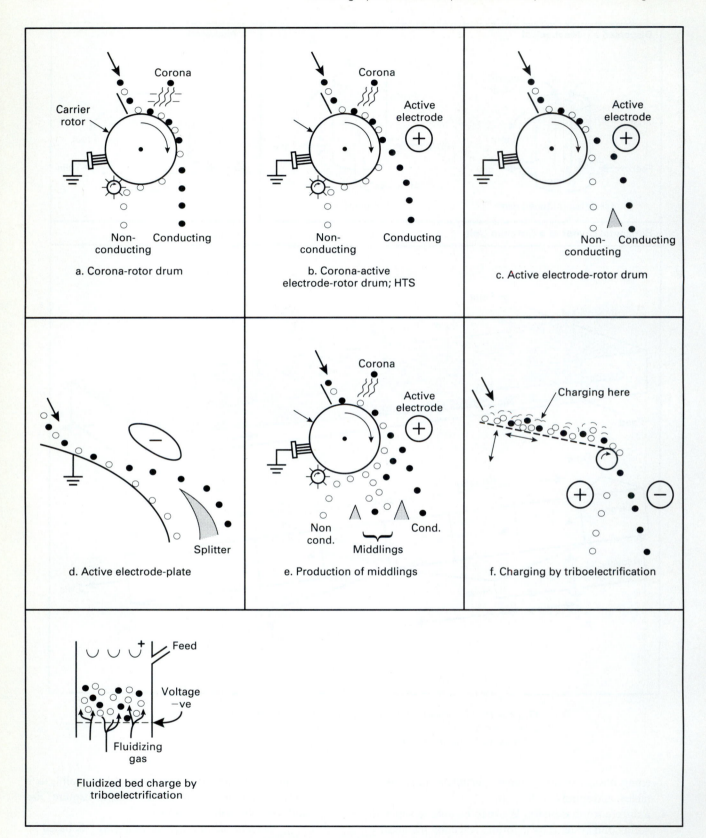

Figure 5–22 Equipment Options for Separating
Solid-Solid Systems: Electrostatic Separators

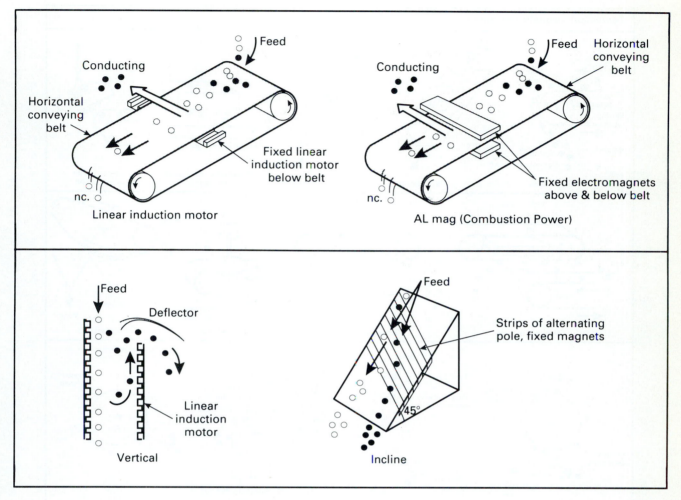

Figure 5–23 Equipment Options for Separating
Solid-Solid Systems: Eddy-Current Separation

ous liquid that "peels away from the solid" as drops. Solvent "drying" is an analogous process in which water is *displaced* (not dissolved) by a solvent.

In *froth flotation,* air bubbles are introduced into a liquid pool that contains two different types of solids. One of the solids is *not* wet by the liquid; it prefers to attach to the bubble and leave the liquid with the bubble. Thus, one type of solid is separated from another.

In *dissolved air flotation,* liquid supersaturated with gas is added to the pool of liquid containing the suspended solids. The gas nucleates to form bubbles on the solid particles present in a pool of liquid. The bubbles rise and lift the solids free from the liquid. Thus all of the solid particles are separated from the liquid.

6. a difference in **solubility.** In leaching, a solvent is added to which the liquid component (of a liquid-solid mixture) is preferentially absorbed. For example, hexane is added to ground-up peanuts to extract the peanut oil from the fibrous protein peanut cake.

In *dissolution,* a solvent or heat is added to convert one solid in a solid-solid mixture into a liquid. For example, sodium hydroxide is used to dissolve solid alumina from solid (but insoluble) silicates in bauxite ore.

Alternately, we can dissolve one of the phases in a two-phase system and thus create a homogeneous phase. Techniques introduced in Chapter 4 could then be used for the separation.

7. a difference in **electrical conductivity.** Particles with electrical conductivities different from those of the surrounding media can acquire a charge that can, in turn, be used to cause the particles to migrate in a DC field. This is the principle of electrodecantation, precipitators, and electrostatic separators.

8. a difference in **magnetic permeability.** This can be used to separate two solids by placing them in a magnetic field. For example, magnets can be used to keep metals from getting into boxes of breakfast cereal or to prevent metals from getting into crushing machines in cattle-food processing.

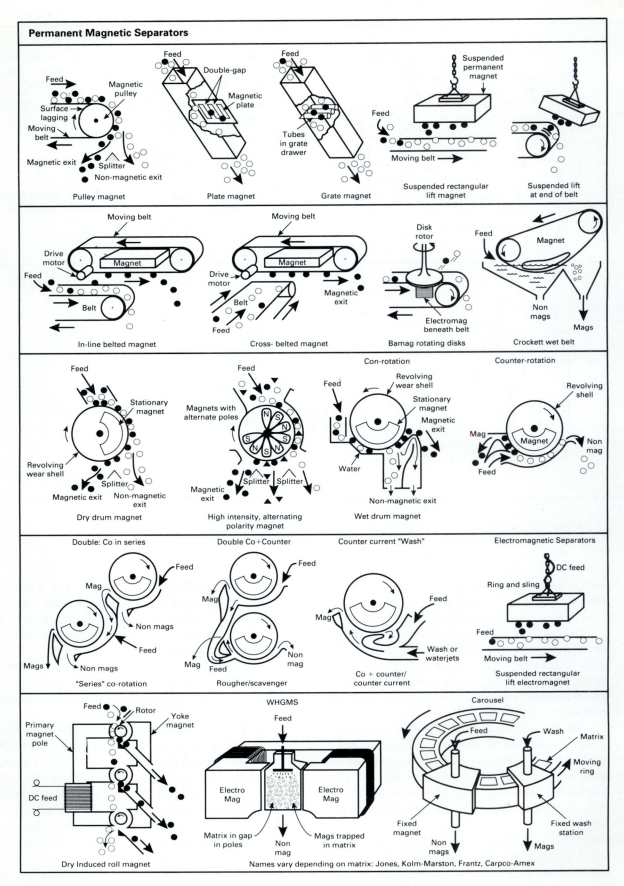

Figure 5–24 Equipment Options for Separating
Solid-Solid Systems: Magnetic Separators

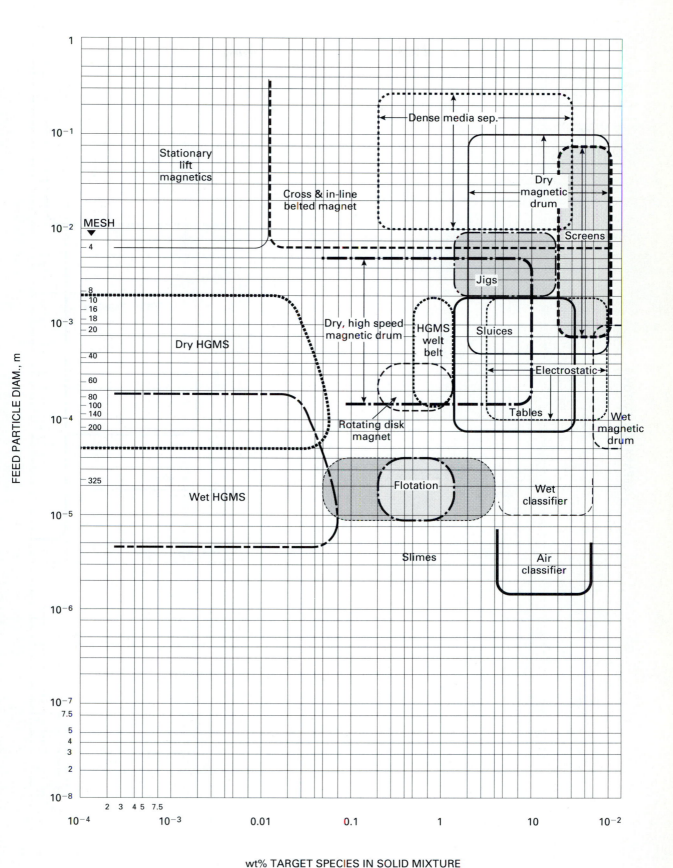

Figure 5–25 General Regions of Applicability for Solid-Solid Separations:
Based on Particle Size and Concentration

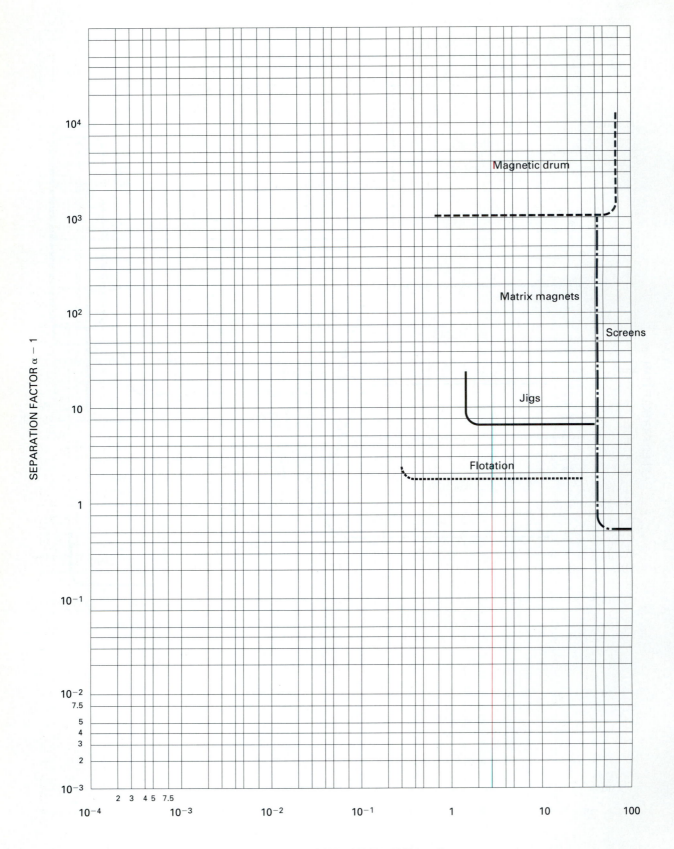

Figure 5–26 General Regions of Applicability for Solid-Solid
Separations: Based on Separation Factor

TABLE 5-2. Definitions of separation factor a for solid-solid separations

Exploiting Difference in . . .	α	Definition and Example
Size	$\alpha = \dfrac{D_{p1}}{D_{p2}}$	D_p = diameter Screens, trommels, grizzlies
Size and density	$\alpha = \dfrac{v_1}{v_2}$	v = settling velocity rakes, tables, jigs, spirals
Electrical conductivity	$\alpha = \dfrac{\varepsilon_{r1}}{\varepsilon_{r2}}$	ε_r relative permittivity electrostatic separators
Wettability	$\alpha = \dfrac{S_{c1} - S_{c2}}{-\gamma}$	θ = contact angle induced draft flotation S_c = spreading coefft. γ = surface tension liquid
Solubility	$\alpha = \dfrac{K_1}{K_2}$	K = solubility in phase
Density	$\alpha = \dfrac{(\rho_1 - 1)}{(\rho_2 - 1)}$ or Bird no.	ρ = density
Magnetic permeability	$\alpha = \dfrac{\kappa_1}{\kappa_2}$	κ = magnetic susceptibility

9. a difference in **electrical charge at the surface of the particle.** Whenever two *neutral* phases are mixed, a small charge usually develops at the surface separating the phases. This can be because ions preferentially adsorb at the surface or are preferentially soluble in one of the phases. The overall system is neutrally charged; it must be! Therefore to provide this neutrality, locally near the surface will be an equal and opposite countercharge. Usually these charges go unnoticed because they are so small. However, if the size of the particles is below 1000 μm, then the charge can indeed dominate. An example of this occurs when we mix clay with water. The fine clay particles usually acquire negative charges because of the adsorption of OH$^-$ ions. Because all the particles have similar charges, the clay particles remain suspended and do not clump together and settle out.

The presence of the charges can be exploited by placing the charged particles in a DC field. Electroosmosis, electrophoresis, and electrodialysis are centered around this principle.

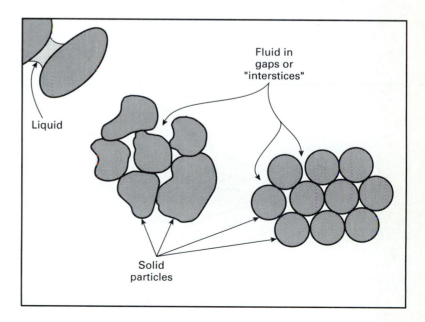

Figure 5–27 Fluids Fill Gaps Between Solid Particles

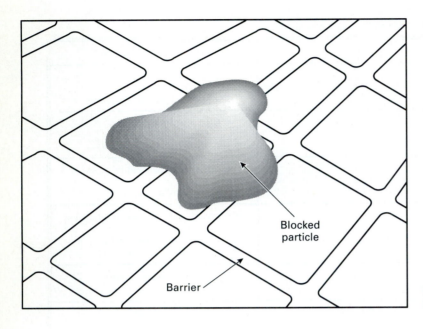

Figure 5–28 Physical Barrier Prevents Large Particles from Moving Through

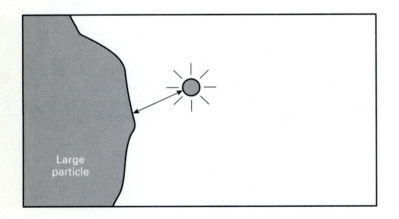

Figure 5–29 Large "Filter" Particle Attracts (Via van der Waals Force) Small Particle

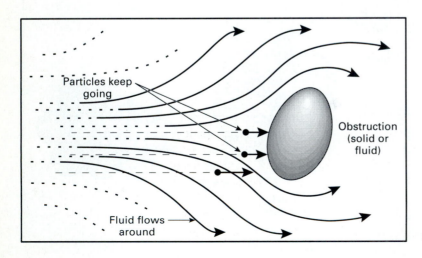

Figure 5–30 Particles Hit Obstacle

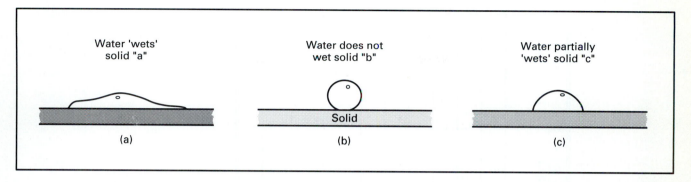

Figure 5–31 Water "Wets" Different Solids Differently

We can also recognize that this surface charge is the reason that particles might not separate out on a filter or by gravity. Hence, some of the other differences may not be able to be exploited until the surface charge has been controlled. This is the basis for coagulation and flocculation as a precursor to sedimentation or filtration.

10. a difference in **reactivity.** We can react one component to create new properties to be exploited; so we can react a two-phase system to create a homogeneous phase system. Then the principles of Chapter 4 can be exploited.

These then are the property differences that can be exploited. Tables 5–3 through 5–7 relate the names of the options to the property to be exploited for each of the different phase combinations. Actually, for many of the phase systems only a few options are available, and they may all exploit the same property difference.

B. Consider the Desired Characteristics of the Product.
Another important distinguishing feature among options is based on the desired characteristics of the product. Three different classes of product specifications are common. For example, if we have a liquid-solid system we might have a goal:

1. to separate the solids from the liquid to create a "solid-free" liquid; this is illustrated in Figure 5–32a.
2. to separate the liquid from the solid to create a "liquid-free" solid; this is shown in Figure 5–32b.
3. to do **both** of the above. Figure 5–32c illustrates this.

The difficulty in easily getting both phases separated sharply occurs because the spaces or interstices between a bed of particles is filled with the other phase, as was illustrated in Figure 5–27. Because we are dealing with particles often we must appreciate three complications:

1. Not all of the particles may be the same size or shape. We account for this by identifying an average size, or a critical size pertinent for the separation. A "distribution" of particles is usually represented graphically as the number (or mass) of particles of a size less than a given size. This is called a cumulative plot. We can characterize the distribution by stating the mean or average and the standard deviation. The standard deviation represents the spread of the particles' sizes about the mean and is related to the "slope" of the cumulative plot. This is illustrated in Figure 5–33, where at the top is shown a *frequency* plot showing the "number" of particles of each size, where all the particles are the same size and where they are different. Below is shown the corresponding cumulative plot. The key point is that the larger the differences in particle size, the larger will be the standard deviation.

2. All beds of particles contain void volume, ε, between the particles. Even in the most tightly packed "bed" of particles, about 30 to 40% of the total bed volume consists of the interstitial volume, $\varepsilon = 0.3$ to 0.4. Figure 5–34 illustrates pictorially the void volume of liquid in a gas-liquid-solid system. Also shown are the typical product conditions for liquid-solid separations performed by different types of separators. Thus, gravity thickening will give a liquid porosity of between 0.8 to 0.9.

The interstitial or void volume for a bed of particles depends on the degree of packing, the shape of the particles, and the size distribution, as represented by the standard deviation. Figure 5–35 illustrates this variation. Thus, a bed of tightly packed spheres with a standard deviation of 2 would have a porosity or interstitial volume of about 0.19.

The key point is that with particulates, we do not easily get complete separation of the two phases. The "continuous" phase occupies the interstices. Some separation options can expel this fluid relatively easily; others, with difficulty.

3. Because of the porosity of beds of particles, the "density" of particles may be ambiguous. We can report the "solid" density of the pure solid (coal is 1.4 Mg/m^3); we can report the "bulk density" in air of the bed of particles (coal is 0.6 to 0.8 Mg/m^3).

TABLE 5-3a. Gas-liquid separations: Exploiting properties

Properties	Gravity	Centrifugal Field		Filter Media +	Agent +	Heat ±	Power or Mech. Agitation ±
		via Mechanical	via Pressure				
Vapor Pressure							
Size							
Size and density	knock out pots		cyclones		wet scrubbers		
Density							
Wettability					demisters		
Solubility							
Electrical conductivity							
Magnetic permeability							
Electric charge							
Reactivity					Combustion		

TABLE 5-3b. Gas-liquid separations: Effect of product recovery requirements

Recover Gas	Combustion, wet scrubbers
Recover both	Knock out pots, cyclones, demisters
Recover Liquid	

TABLE 5-4a. Gas-solid separations: Exploiting properties

Properties	By Gravity	Centrifugal Field		Filter Media +	Agent +	Heat ±	Power or Mech. Agitation ±
		via Mechanical	via Pressure				
Vapor Pressure							
Size	knock out pots			filters			
Size and density			cyclones		wet scrubbers		
Density							
Wettability							
Solubility							
Electrical conductivity							Electrostatic precipitator
Magnetic permeability							
Electric charge							
Reactivity					Combustion		

TABLE 5-4b. Gas-solid separations: Effect of Product recovery requirements

Recover Gas	Combustion, wet scrubbers
Recover both	Cyclones, precipitators, filters
Recover Solids	Wet scrubbers with subsequent L-S separation

TABLE 5-5a. Liquid-liquid separations: Exploiting properties

Properties	By Gravity	Centrifugal Field		+ Filter Media	+ Agent	± Heat	± Power or Mech. Agitation
		via Mechanical	via Pressure				
Vapor Pressure							
Size				Deep bed filters			
Size and density	decanters	centrifuges sedimentation type	hydrocyclones				
Density							
Wettability	dissolved air flotation				Fibrous bed coalescers Dissolved air flot.		
Solubility							Solvent extraction
Electrical conductivity							
Magnetic permeability							
Electric charge							
Reactivity							

TABLE 5-5b. Liquid-liquid separations: Effect of product recovery requirements

	No coalescence occurs	Complete coalescence occurs
Recover the dispersed phase	Solvent extraction	
Recover both phases	decanters, hydrocyclones, centrifuges (but there is interstitial cross contamination)	decanters, solvent extraction, with no interstitial cross contamination
Recover continuous phase	Dissolved air flotation, deep bed filters, solvent extraction, fibrous bed coalescers	dissolved air flotation, deep bed filters, hydrocyclones, centrifuges and fibrous beds

TABLE 5-6. Liquid-solid separations: Exploiting properties

Properties	By Gravity	Centrifugal Field		+ Filter Media	+ Agent	± Heat	± Power or Mech. Agitation
		via Mechanical	via Pressure				
Vapor Pressure						drying freeze drying	
Size	deep bed filter	filtering centrifuge		expeller, hydraulic press filter			screens, grizzlies, bar screens trommels
Size and density	settling basin thickener elec.	sedimentation centrifuge	hydrocyclones				
Density							
Wettability	dissolved air flotation				dissolved air flotation		
Solubility					leaching		
Electrical conductivity							
Magnetic permeability							
Electric charge							electroosmosis
Reactivity							

TABLE 5-7. Solid-solid separations

Properties	By Gravity	Centrifugal Field		Filter Media +	Agent +	Heat ±	Power or Mech. Agitation ±
		via Mechanical	via Pressure				
Vapor Pressure							
Size				screens, trommels, grizzlies			screens, trommels, grizzlies classifiers
Size and density							rakes, jigs, tables, sprials
Density					dense media separation		
Wettability					induced draft flotation		
Solubility					leaching	dissolution	leaching
Electrical conductivity							electrostatic eddy current
Magnetic permeability							magnetic
Electric charge							
Reactivity							

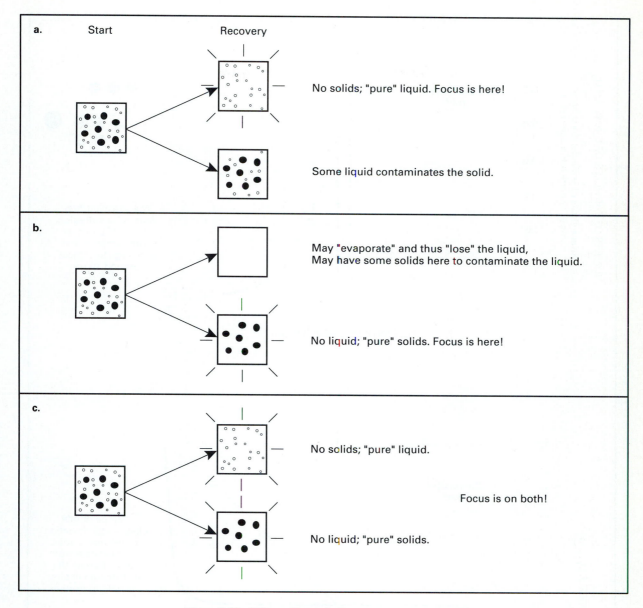

Figure 5–32 Different Targets for Recovery
a) Creating a "Solid-Free" Liquid
b) Creating a "Liquid-Free" Solid
c) Creating Both a "Solid-Free" Liquid and a "Liquid-Free" Solid

Thus, for particulate systems we must clearly identify what products we are trying to obtain: recover "pure" solids, recover "pure" liquids, or recover both. Different equipment options are suited for different specifications. This is qualitatively illustrated for liquid-solid systems in Figure 5–34, where the target solid concentration possible from different options is presented. In the next section, the effect of product specifications is outlined for each type of phase combination, where appropriate.

C. Putting It All Together: Exploiting the Property Difference and the Product Specifications. Now that the principles have been introduced in

sections *A* and *B,* we consider the application of these principles to the selection of separation options. We consider each combination of phases in turn.

1. Gas-Liquid. For **gas-liquid** systems, the options are predominantly based on size-density differences as illustrated in Table 5–3a. Table 5–3b illustrates that many of the options recover both the liquid and the gas. Combustion and wet scrubbers would recover the gas but would destroy the liquid in the process. Hence if we wanted to recover the liquid, neither of these two options would be attractive. No method is designed to recover only the liquid. Essentially Figure 5–2 is an acceptable basis for selection.

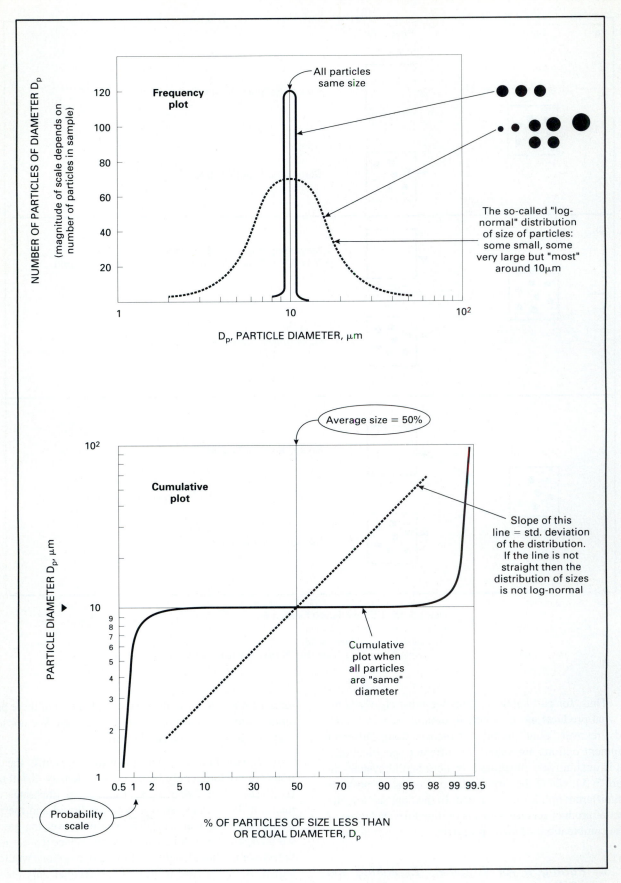

Figure 5–33 Reporting Particle Size Characteristics When
There Is More Than One "Size"

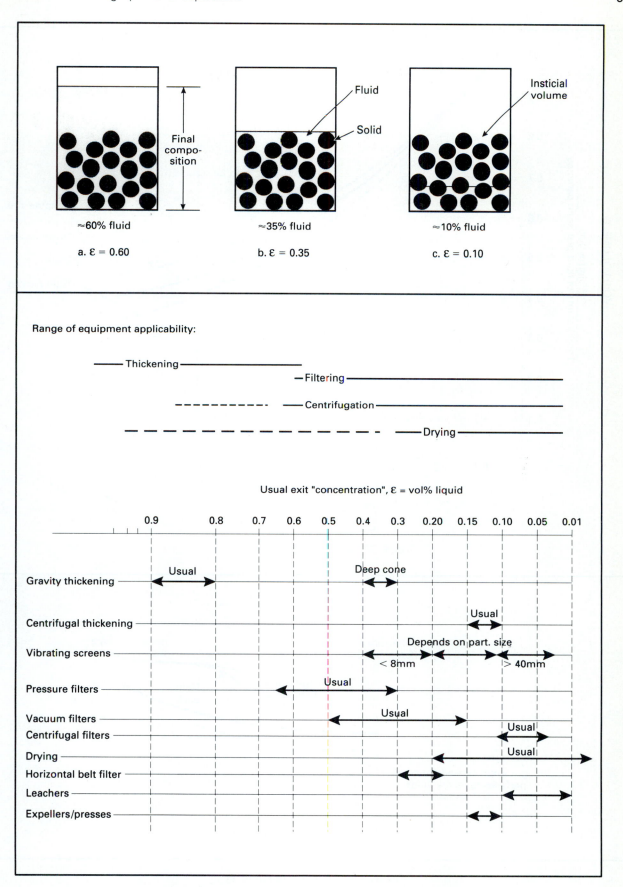

Figure 5–34 Relating Solid, Fluid, and Interstitial Volumes to the "Usual" Exit Concentrations for Different Equipment Alternatives

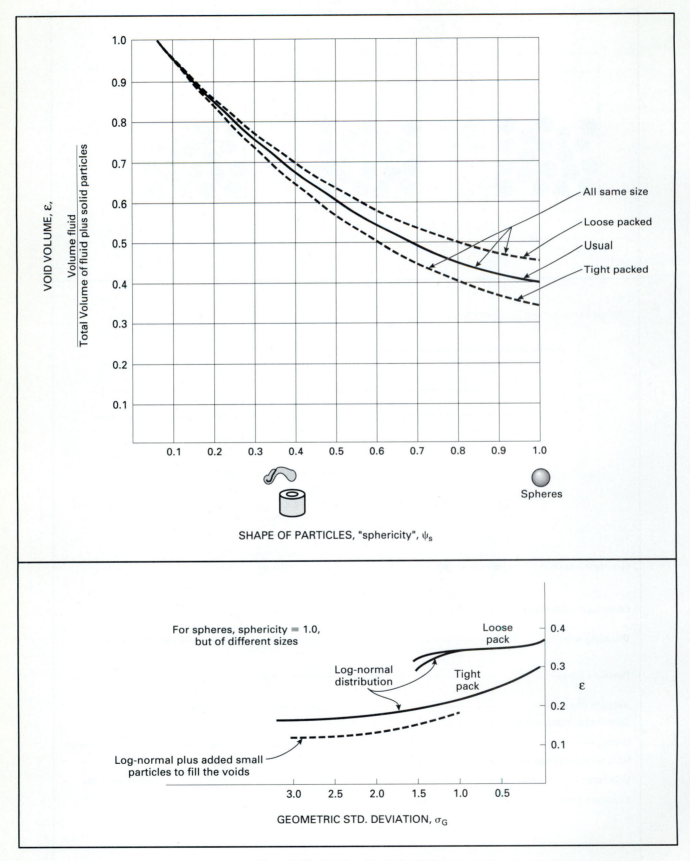

Figure 5–35   Porosity of Particulate Beds

2. Gas-Solid. For **gas-solid** systems, size-density differences offer the major basis as shown in Table 5–4a. However, size or electrical conductivity offer other choices. For electrical conductivity to be attractive, the conductivity of the solid particulates should be greater than 10^{-10}/ohm-cm. Thus, damp ferrous oxide or damp aluminum oxide meet this criterion, but these same particulates *dry* do not.

Table 5–4b shows the effect of recovery expectations; wet scrubbers and combustion make it difficult, if not impossible, to recover the solids (although they can remove the solids from the gas). Some options provide a separate gas and a recovered solid stream. (For the solids, the interstices would likely be contaminated by the gas). This information, together with Figure 5–4, provides an acceptable basis for initial selection.

3. Liquid-Liquid. **Liquid-liquid** systems are challenging. A variety of options are available as illustrated in Table 5–5. What can have a great impact on the recovery of the two separate phases are:

- the mutual solubility of the liquids for each other. This can place constraints on the purity of the products before we even consider separation options;
- whether the dispersed phase "coalesces" into one large layer or whether the drops remain as drops and thus we are limited by the concentrations illustrated in Figure 5–34 and by interstitial cross-contamination.

The property being exploited in Table 5–5 assumes that *coalescence* is not important. Table 5–5b shows that if coalescence is made to occur, a different set of options is available. Consider this table in more detail. When negligible coalescence occurs, only solvent extraction can yield a relatively pure recovered phase from the dispersed liquid. Decanters, hydrocyclones, and centrifuges can give a relatively pure continuous phase and can separate the uncoalesced "band of drops." However, the band of drops will be cross-contaminated with the continuous phase in the interstices. Hence, we cannot get a recovered "dispersed phase" that is very pure. On the other hand, dissolved air flotation, deep bed filtration, and coalescers will produce a "clean" recovered continuous phase; however, it is extremely difficult to recover the dispersed phase.

When complete coalescence does occur or is required, both decantation and solvent extraction will yield relatively pure streams of both recovered fluids. However, hydrocyclones and centrifuges usually have such short residence times that coalescence is unlikely to occur even in rapidly coalescing systems. Furthermore, for liquid-liquid systems, care is needed in selecting centrifugal devices because the forces acting on the dispersed drops might *break up* the drops and more intimately disperse the phases rather than achieve the goal of separating the phases. (A brief word about coalescence. Although the coalescence between isolated drops may take only milliseconds, the coalescence of groups and clusters of drops usually takes on the order of 2 to 5 min. Furthermore, coalescence is extremely sensitive to the presence of trace quantities of contamination and to surface charge. Thus, for screening studies it is more conservative to assume the coalescence *does not* occur. On the other hand, when the target recoveries require that coalescence occur, procedures are available to size options based on coalescence as the controlling mechanism. [For details see Woods, 1993] The approach at the "selection" stage is to select options where complete coalescence is possible. This is more likely to be achieved in gravity devices than in centrifugal devices.) Electrodecantation facilitates drop migration and perhaps coalescence.

The dominant characteristic to be exploited for liquid-liquid separation is the *size-density* that affects the velocity of the droplets. Figure 5–36 shows the settling or rising velocity of drops, bubbles, or particulates as a function of size and density. This is based on Stokes's law for single droplets rising through water at 20°C. (This figure is analogous to Figure 3–26 which showed Stokes's law for particles in air.) The "usual" drop size for emulsions is for liquid-liquid surface tensions of about 30 mN/m and where the liquid mixture has not been subjected to pumping or excessive shear. The drop size decreases with a decrease in surface tension and increase in shear. Figure 5–37 shows the regions where different types of devices are used depending on the settling velocity of the dispersed phase and the flowrate of the continuous phase. Thus, if the droplets are very small we might add coalescence promoters; if the density difference is very small, then centrifugal devices should be considered. The overall result of this analysis yields information that is similar to that given on the size-concentration chart in Figure 5–6. Thus, for screening, Figure 5–6, combined with the product recovery expectations in Table 5–5b, is the prime basis for selection. The choices are sensitive to the need for coalescence to occur.

Example 5–3: We have a hexane-water mixture that is 43% v/v water. The temperature is 60°C and the density difference is 0.33 Mg/m³. The feed flowrate is 0.3 L/s. Our goal is to recover the "pure" hexane so that it can be reused in the process. The recycled hexane should not be cross-contaminated with anything other than the mutually soluble water because this is a food-processing operation. We also want "pure" water so that we can dispose of it through our waste treatment facilities. Assume that hexane is the dispersed phase. What options might we select? The feed flows from a condenser.

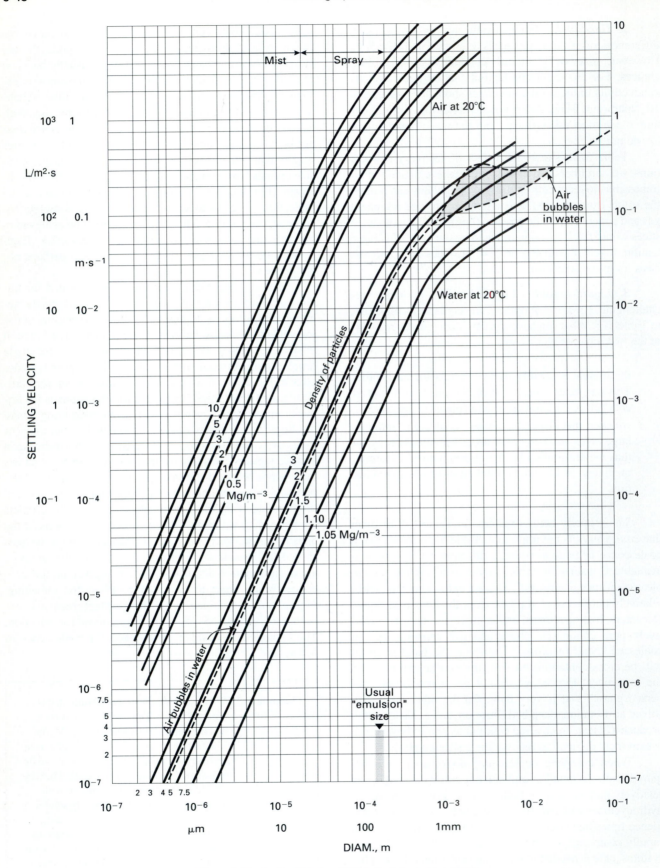

Figure 5–36 Settling Velocity of Single Particles in Air or Water

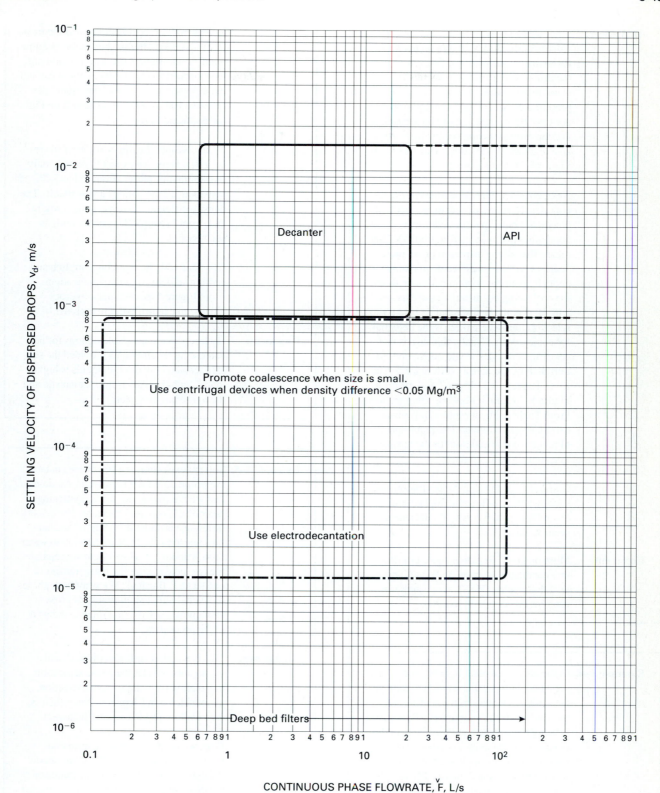

Figure 5–37 General Regions of Applicability for Liquid-Liquid Separations: Based on Settling Velocities and Flowrates

An Answer: First, we need to estimate reasonable values for the drop size. We do not know the diameter of the drops. The source of the drops is condensation and the hexane-water mixture flows by gravity into the proposed device. Hence, the liquid should not have been subjected to excessive shear and the drop size is probably "the usual" size of about 100 to 200 μm, as shown on Figure 5–36 and Figure 5–6.

Next, consider the implications of the recoveries. Here, we expect to recover both phases as "pure" phases (apart from cross-contamination by mutual solubility). This means that all the dispersed phase must coalesce. At this stage we do not know enough about the contamination or about the coalescing characteristics of this system. Hence, the best we can do is to try to select options where there is the potential for coalescence. Usually this means that centrifugal devices would not be options. Fortunately, the density difference is relatively large, at 0.33 Mg/m^3, and, from Figure 5–37, gravity options seem viable. From Table 5–5b, decanters and solvent extraction are possible. The requirement of no new contamination for the hexane rules out solvent extraction.

Consider now the size-concentration and settling velocity-flowrate implications. From Figures 5–6 and 5–37, decanters seem applicable.

Comment: More on predicting drop size from flowrate conditions is given by Woods and Diamadopoulos (1988). A decanter was used for this option in industry.

Example 5–4: In the alkylation plant light hydrocarbon is scrubbed with caustic solution to remove the acidity. The resultant liquid-liquid mixture needs to be separated. The caustic is being recirculated and may be a mixture of caustic drops in hydrocarbon. The continuous phase, the hydrocarbon, must be recovered as a "pure" phase. The total feed flowrate is 7.6 L/s and is 22% caustic. Assume caustic is the dispersed phase. What options might we consider? The density difference is about 0.58 Mg/m^3. The viscosity of the continuous phase is 0.1 mPa.s.

An Answer: First, we need to estimate reasonable values for the drop size. The source is from a contactor but we do not know if the mixture is pumped after the contactor. Assume that it is not. Hence, the liquid should not have been subjected to excessive shear and the drop size is probably "the usual" size of about 100 to 200 μm, as shown on Figure 5–36 and Figure 5–6.

Next, consider the implications of the recoveries. Here, we expect to recover only the continuous phase. From Table 5–5b, just about all the options are available. The density difference is large, at 0.58 Mg/m^3, and, from Figure 5–37, gravity options seem viable.

Consider now the size-concentration and settling velocity-flowrate implications. From Figure 5–6, decanters, hydrocyclones, and centrifuges are options.

Comment: Hydrocyclones are viable options for non-coalescing requirements provided the surface tension between the liquids is higher than 10 mN/m. This is true for caustic and the hydrocarbon system.

Example 5–5: Solvent extraction is a stagewise contacting of the liquid with a solvent. Assume that we are contacting water containing penicillin F with amyl acetate. The density difference is about 0.11 Mg/m^3; the surface tension is 9.2 mN/m. Solvent to water ratio is about 1:1. How might we separate the two phases after the mass transfer in one stage is finished? We want both phases recovered with the minimum of cross-contamination from one stage to the next in the processing.

An Answer: This is an interesting situation. Usually we do not consider the phase separation in isolation from the solvent extraction, as considered in Chapter 4. Nevertheless, let's explore just the separation aspect.

First, we need to estimate reasonable values for the drop size. We do not know the diameter of the drops. The source of the drops depends on the type of contactor used. For the usual mixer settler or for pulsed systems the drop size is about 1 to 2 mm. In other words, these are usually about 10 times larger than shown in Figure 5–6. However, for this system the surface tension is 9.2 mN/m, whereas for most systems it is between 20 and 30 mN/m. Thus, the drop size will be much smaller than 1 to 2 mm. Let's assume that they are about 100 μm.

Next consider the implications of the recoveries. Here, we expect to recover both phases as "pure" phases (apart from cross-contamination by mutual solubility). This means that all the dispersed phase must coalesce. At this stage we do not know enough about the contamination or about the coalescing characteristics of this system. Hence, the best we can do is try to select options where there is the potential for coalescence. Usually this means that centrifugal devices would not be options. Unfortunately, the density difference is relatively small, at 0.11 Mg/m^3, and, from Figure 5–37, perhaps centrifugal devices may be required.

Consider now the size-concentration and settling velocity-flowrate implications. From Figures 5–6 and 5–37, centrifuges and decanters seem applicable. Perhaps because of the low density differences we might have to go to a centrifugal device despite the challenges of getting complete coalescence.

Comment: The product of the surface tension and the density difference is about 1 Mg/m^3. mN/m. From Section 4.2-9, this suggests that a centrifugal device is appropriate for solvent extraction.

Example 5–6: A waste water stream from an API separator contains 130 ppm of immiscible oil. We wish to recover the water only. What might we do? The flowrate is 4 L/s. Assume the density difference is 0.20 Mg/m^3 and that the surface tension is about 30 mN/m.

An Answer: First, we need to estimate reasonable values for the drop size. We do not know the diameter of the drops. The source and pumping history of the stream are not given. Assume that the drop size is probably "the usual" size of about 100 to 200 μm, as shown on Figures 5–36 and 5–6.

Next consider the implications of the recoveries. Here, we expect to recover only the water of continuous phase. Hence, from Table 5–5b most of the options are available. Consider now the size-concentration and settling velocity-flowrate implications. From Figures 5–6 and 5–37, deep bed filters or dissolved air flotation may be feasible options.

Comment: For more on the separation of dilute oil-water mixtures and the surface phenomena that affect these separations see Woods and Diamadopoulos (1988). From Figure 5–41, given later in this section under liquid-solid separations, we might expect about 80% removal for dissolved air flotation. The data for deep bed filtration suggest that this concentration, at 100 ppm or mg/L, is higher than is often used in deep bed filters. However, data summarized by Woods and Diamadopoulos (1988; p. 512) suggest that for this concentration about 60 to 70% oil is removed by deep bed filtration.

---

4. Liquid-Solid. For **liquid-solid** systems we have a wide range of choice. Table 5–6 illustrates the type of property difference that can be exploited. From Figures 5–7 through 5–16 we note that there is a wide range of equipment—even within the same class—and a lot of apparent overlap on the size-concentration diagrams of Figures 5–10 and 5–16. Consider the properties that we wish to exploit in the same sequence as given in Table 5–6.

• For *vapor pressure* differences, drying is the option. This is appealing when we wish to recover the *solid.* Usually the liquid, which has been vaporized, is not recovered.

The specific type of dryer is chosen based on the condition of the feed, the expected form of the final product, and the temperature sensitivity of the solid. Table 5–8 illustrates how the type of feed material affects the choice. (The descriptions may be a little misleading because, for example, "thin liquids" sounds as though there are no solids present when indeed solids are or will be present.) Table 5–9 shows how the form of the product affects the choice. Table 5–10 illustrates qualitatively that the temperature sensitivity of the solid also affects the choice of configuration. Especially for temperature sensitive solids, leaching and expellers may be appropriate, as given by the general regions shown in Figure 5–16. Thus, Figures 5–10 and 5–16 can be used to identify the general region when drying might be appropriate; Tables 5–8 through 5–10 guide us in selecting the special type of dryer. Dryers recover the solids.

• For *size* differences, basically a screen or filter is the choice. However, the particles also might have a density different from the liquid and so will have a difference in settling velocity. Let's focus first on exploiting size difference. Options are basically "filters." However, the terminology can be confusing. For large-size particles, the term *filter* is replaced with such terms as *trommel* (a rotating filter), *screen* (a filter cloth on a frame), and *grizzly* (a filter cloth made up of rods and bars). We can also place a screen in a centrifugal field to obtain a "filtering" centrifuge. These are distinct from "sedimentation" centrifuges

TABLE 5-8. How feed condition affects choice of dryer

Type of Feed Material	Indirect Contact		Direct Contact	Vacuum	
	Conduction	Convection	Convection	Conduction	Convection
Thin liquids	single drum, pan, dielectric	spray	spray	freeze, pan	
Thick liquids (slurries)	double drum, pan, dielectric	spray	spray	freeze, pan	
Soft, paste-like	double drum, pan, dielectric			freeze, pan	
Stiff, paste-like	double drum, pan, dielectric		tunnel, disk, screen, belt, trough	shelf, belt, conveyor, freeze, pan	belt, conveyor, shelf
Moist, crumb-like	pan, screw, vibrating tray, steam rotary, dielectric	steam rotary, vibrating tray	agitated, paddle, rotating drum, tunnel, disk, fluid bed, pneumatic	kneader, shelf, freeze, pan	
Powdery, grained, grit-like <150 μm	pan, screw, vibrating tray, steam rotary, dielectric	steam rotary, vibrating tray	agitated, paddle, rotary, tunnel, disk, fluid bed, pneumatic	shelf, freeze, pan	
Granular lumps >150 μm	pan, screw, vibrating tray, steam rotary, dielectric	steam rotary, vibrating tray	shelf, tunnel, belt	shelf, freeze, pan, steam rotary	steam rotary
Coherent sheets	cylinder			freeze	
Discontinuous sheets		shelf, belt, tunnel	shelf, belt, tunnel	shelf, freeze	shelf
Shaped pieces lumber, pottery	dielectric	shelf, belt, tunnel	shelf, belt, tunnel		

TABLE 5-9. Effect of product temperature sensitivity on options

Temperature Sensitivity	Temperature Difference °C	Some Examples	Options	
			Dryers:[*] Classes	Other Options
Extremely sensitive	<5	Pharmaceuticals, antibiotics, penicillin, some foods	Indirect vacuum conduction, vacuum conduction, freeze dry	Expellers, leachers, "solvent extraction"
Very sensitive	5 to 10		Most drying options possible except direct convection	Expellers, leachers, "solvent extraction"
Sensitive	10 to 40		Direct or indirect convection plus indirect conduction	
Moderately insensitive	40 to 100	Polymers, dyes, organic pigments, general organics	Direct and/or indirect convection	
Insensitive	100 to 150	General inorganics	Direct convection	

[*]For examples of devices in each class, see Table 5-8

that do not have a filter cloth or screen inside. Often the term *imperforate* or *solid* bowl will be used to describe sedimentation centrifuges. The pictures in Figure 5–8 for sedimentation centrifuges should be contrasted with those in Figure 5–9 for "filtering" centrifuges. In this section, we focus on filtering centrifuges.

For screens and filters, select the class of screening device based on the size of particles; then within a class, select the type based on the concentration. For example, Figure 5–13 shows the type of filter to use; Figure 5–12 shows the type of "filtering" centrifugal filter to use. (Precaution: only the centrifuges in Figure 5–12 operating at concentrations greater than about 10% are filtering centrifuges. They are described by the terms *basket, conical screen, pusher, horizontal screen bowl,* and *vertical conical screen.* Figure 5–11 is for sedimentation centrifuges.)

If the size of the particle is too small for any of the options, we can add a filtering aid or use a precoat. One approach is to add a filtering aid to the feed. A filtering aid is a coagulant or flocculant (such as starch, added to in-

crease the size of the species in the feed) or diatomous earth or perlite particles (which are bigger particles that collect on the filter cloth together with the target species). Filter aids adjust the "permeability" of the filter media to be equivalent to filtering particles of about 1 to 8 μm. Figure 5–38 illustrates how the particle size is related to the "permeability" of a cake of such particles and to different filtering conditions and cake conditions. Filter aids can be used with almost any type of filter.

Another approach is to leave the feed alone and to alter the filter cloth with a precoat. Such a precoat deposits, on the filter cloth, a "mini-deep bed" of larger-sized particles, say about 1 to 10 μm. Usually a precoat is added to a drum or rotary filter so that as filtration proceeds, a thin layer of the product plus the precoat can be shaved off each revolution with a "doctor knife." We refer to this as precoat filtration.

Recovery considerations for "filters" (excluding screens, trommels, and bar screens) are summarized in Table 5–11. Depending on how the filter is backwashed and run, the solids may be reasonably dry. Figure 5–34

TABLE 5-10. How product specifications affect dryer choice

Product Expectations	Type of Dryer
Powder	Spray
Flakes	Drum
Solid cake	Pan

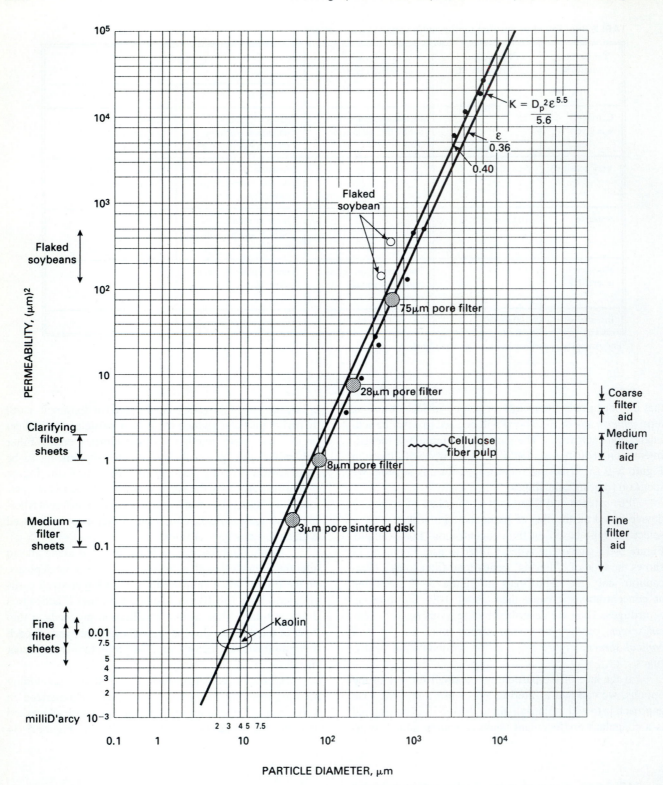

Figure 5–38 Permeability of Porous Media

TABLE 5-11. Effect of recovery on choice of filters

	Gravity or Pressure	Centrifugal
Recover Liquid	1. Deep bed, horizontal vacuum, pressure leaf, gravity flat table. 2. Gravity flat table 3. Cartidge 4. Precoat drum 5. Plate and Frame	Vertical basket (A,B)*
And high liquid viscosity	1. Plate and frame	Horizontal (H)

Recover Solid	For large sized particles:
	Expellers and presses
	
	(drying?)
		Usually the solids are "dry."
		Basket centrifuge
		Batch automatic, auto
		    vertical basket (const. speed)
		    horizontal basket (var. speed)
		
		Continuous:
		    conical screen, scroll conveyor; conical screen,
		    oscillating cylindrical screen with
		    pusher conveyor
		
		Horizontal solid screen scroll conveyor

With good washing	1. Pressure, vacuum, gravity table/pan 2. Horizontal pressure or vacuum 3. Horizontal belt 4. Vacuum drum 5. – – – – – – – – – – – – – –> 6. Plate and frame	Vertical basket overdriven (C)* Vertical basket underdriven (A,B) Horizontal batch (E) Cylindrical screen scroll discharge (F)
And crystals break easily	1. Gravity, vacuum table/pan 2. Vacuum, pressure, gravity drum 3. Plate and frame	
And cake compressible	1. Low pressure rotary vacuum drum	

*(Letters refer to the filtering centrifuge type shown in Figure 5-9)

provides guidelines. In Table 5–11, priority is shown, where appropriate, by the numerical sequence listed with the options. Thus, to recover solids and provide good washing, the first choice is usually a pressure, vacuum, gravity table, pan filter, or an overdriven vertical basket centrifuge. Beside some of the descriptions of filtering centrifuges is a code letter that refers to the sketches shown in Figure 5–9. Expellers and presses apply to large-size particles where the main function is to recover either solid or liquid. The range of application is given in Figure 5–16.

To sum up, the initial selection of filters depends on the size/concentration (as given in Figures 5–12 and 5–13), the expected recovery and complicating conditions (as given in Table 5–11), and the target concentration or "dryness" of the solids (as given in Figure 5–34).

• For *size and density* differences (or settling velocities), Figures 5–11 and 5–12 show the general regions of

applicability for centrifugal devices, based again on particle size and concentration. Figure 5–36, analogous to Figure 3–26, shows the settling velocity of different size and density of particles in water. This is based on Stokes's law, which considers only single particles. Figure 5–39 is more specific in exploiting the settling velocity in a simple gravitational field versus the flowrate of recovered, clarified liquid. This helps us to distinguish between different types of gravity and centrifugal devices. Table 5–12 summarizes how the different expected recoveries affect the choice. Figure 5–40 is a rework of Figure 5–39 in terms of solids throughput for those situations where the emphasis is on recovering the solid.

When the size of the particles is too small, coagulation and flocculation can often be used to increase the size. Dissolved air flotation is an option to remove the solids from the liquid; usually little concern is for the recovery of the solids.

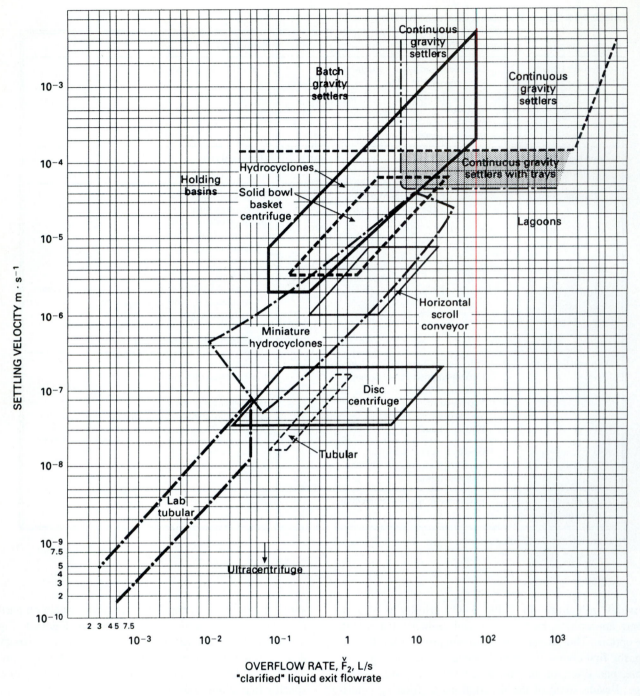

Figure 5–39 General Regions of Applicability for Liquid-Solid Separations:
Based on Clarified Liquid Recovery

• For *solubility*, the "liquid" phase should preferentially dissolve in a solvent or be "washed out" by the solvent. Which mechanism applies depends on how the liquid is "bound" to the solid. If the liquid is intimately mixed within the solid (like soya oil inside the soya bean) then the oil needs to be dissolved out. On the other hand, the liquid might just occupy the interstitial spaces between the particles and need to be washed out. Table 5–13 illustrates how the size and condition of the solid and the condition of the

liquid affect the configuration of the leacher. Figure 5–15 illustrates the configurations.

In summary, the initial screening charts of Figures 5–10 and 5–16 are but starting guidelines. Usually many overlapping options are possible. To resolve these, we can systematically work through the tables and charts described here to narrow the selection. Figure 5–41 gives some guidance as to the effectiveness in removing the solids from the liquid.

TABLE 5-12. Effect of recovery on choice of "settlers"

Purpose	Gravity Force	Centrifugal Force
To recover liquid	Clarifier, settler, washing tray thickener reactor-clarifier	Hydrocyclone, Batch, tubular bowl centrifuge Batch automatic: horizontal or vertical bowl, disc with intermittent nozzle discharge Continuous disc bowl centrifuge with nozzle discharge with/without recycle
To recover both	Continuous countercurrent decanter circuit, CCD	Horizontal, solid bowl centrifuge with scroll discharge
To recover solid	Thickener, Deep thickener Rake thickener Tray thickener	Hydrocyclone Batch automatic imperforate bowl centrifuge Continuous conical bowl centrifuge Continuous contour bowl vertical or horizontal centrifuge

Example 5–7: We want to dewater a coal slurry that has a particle size "B." This slurry is being conveyed in a pipeline. The density of the particles is 1.4 Mg/m^3 and the concentration is 50% wt. We are only interested in recovering the coal. We do not want water left in the interstices if it can be helped.

An Answer: First, the size of the particles is, from Figure 2–40, between about 200 μm and 4 mm. Thus, from Figure 5–36 the settling velocity is in the range of 0.01 to 0.3 m/s.

Consider next the size-concentration diagrams, Figures 5–10 and 5–16. This suggests that belt filters are appropriate. The size is a bit too large for hydrocyclones; the concentration of solids a little too large for a stationary screen.

The goal of having little water left in the interstices may be achieved if vibration or vacuum can be applied to the last portion of the belt filter.

Example 5–8: We want to separate the antibiotic erythromycin from the broth used to make this product. The feed concentration is 15 g dry product per L. The target is to recover the solid. What might we do?

An Answer: First, the size of the particles is, from Figure 4–4, about 1 nm. This is extremely small. We could add a filtering aid or use a precoat filter. These might alter the "diameter" of the solids to be about 1 to 5 μm.

Consider next the size-concentration diagrams, Figures 5–10 and 5–16. The feed concentration is about 1.5%. This suggests that a plate and frame or pressure leaf filter might be effective if a filter aid is used.

Table 5–10 adds little new information to aid in the selection.

Comment: Belter et al. (1989) describe the filtration of this system in their example 2.3-1, p. 25.

Example 5–9: We wish to create powdered milk. We wish to remove the water from milk. The feed concentration is about 87% water, 3.5% protein, 3.7% fat, and the rest milk sugar. The fat globules are about 0.1 to 20 μm. What options do we have?

An Answer: First, the size of the particles: proteins are, from Figure 4–4, about 10 nm; the sugars are also small. We are given the size of the fat globules. This is a wide range. This also is interesting because with this small size of particle we could treat this as a solid-liquid separation or a homogeneous phase separation. It really comes down to how small something must be before we cease to call it a particle. For continuity, let's consider these as particles. For powdered milk we would want to have everything except the water.

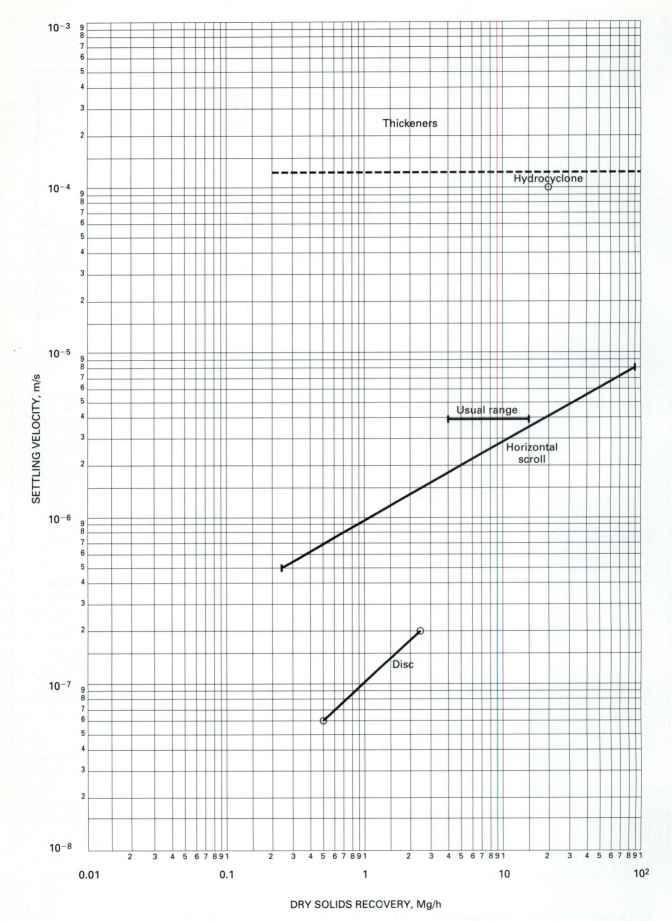

Figure 5–40 General Regions of Applicability for Liquid-Solid Separations: Based on Dry Solids Recovery

TABLE 5-13. Three types of contact for leachers

	Percolation or vat leach or heap leach or insitu leach	Immersion or stirred tank	Combination
			Liquid
Size of inert solid	large, coarse: try for uniform size (so that fines don't plug the interstices)	fine, <0.07 mm easy to suspend; or have broad size range that yield low bed porosity	
Strength of solid	relatively fragile	relatively strong; does not break up into fines because of mixing	should be strong so that it doesn't break up into fines and plug the leacher
Initial solute concentration	low and very low concentrations can be processed		high concentrations needed
For "bound" or soluble species	high rates of diffusion are needed if contact times are to be minimized	OK for low rates of diffusion because long contact times are easy to achieve	
For "unbound," washing conditions e.g. inorganics	often used for "washing" or pull a vacuum to draw liquid free of bed	usually not used	usually not used

Consider next the size-concentration diagrams, Figures 5–10 and 5–16. The feed concentration of solids is about 13%. This corresponds with filter-aid filtration and coagulation-thickening.

However, filtration would only capture the larger-sized molecules unless a precoat were used. Here we want to recover the solids and the precoat would contaminate the solids. Similarly, the particle size is so small that coagulation would have to be used to increase the size of even the fat globules. This would contaminate the product. Thus, the route would be to evaporate some of the liquid to preconcentrate the milk until an appropriate drying option can be used.

Consider Tables 5–8, 5–9, and 5–10.

5. Solid-Solid. For **solid-solid** systems, the general classes of separating devices depend on the particle size and density and how much each vary. This is illustrated in Table 5–1. The terms *classifiers, concentrators,* and *separators* help us to focus on the fundamental principle being exploited in the separation.

Classifiers exploit differences in size. Thus, an important consideration is the "cut" size of the particles that represents the boundary between the two different prod-ucts. Screens, trommels, and grizzlies are related to the solid size range and to the concentration in Figures 5–10, 5–25, and 5–26. The "cut" size of interest must be in the size range. Usually these operate dry. Other classifiers that operate dry (or in a gas or air) are shown in Figure 5–18a. The general region of application for the different types depends on the "cut" size and on the mass of solid mixture to be processed, as illustrated in Figure 5–42.

Classifiers can also be operated "wet" or using a liquid to convey the solid mixture. These "liquid" classifiers are given in Figure 5–18b and the general region of applications of these are given in Figure 5–43.

Concentrators exploit the differences in the densities of the particles about a given "cut" density that separates the lower-density particles from the higher-density particles. We can express this two ways: the Bird Number (that is, the mass of solids that is within ± 0.1 Mg/m^3 density variation from the cut density) or the density ratio between one phase and the other. Table 5–14 uses the Bird number as a criterion; Figure 5–44 uses the density ratio. Sketches of the options of rakes, jigs, tables, spirals, and dense media separators are given in Figures 5–19 and 5–20. Some general regions of applicability are also given in Figure 5–25.

Solid-solid *separators* exploit features other than the size or the density. The options include flotation, dissolution/leaching, electrostatic or magnetic separators. Sketches of the options are given in Figure 5–15 (for leachers/dissolution), Figure 5–21 for flotation, Figure 5–22 for electrostatic separations, Figure 5–23 for eddy-current separators to remove aluminum, and Figure 5–24 for magnetic

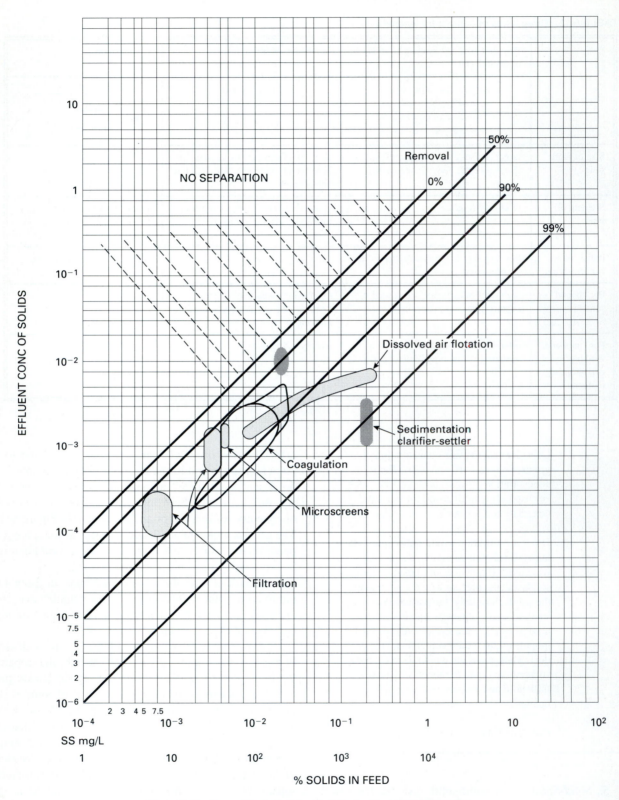

Figure 5–41   General Regions of Applicability for Liquid-Solid Separations:
Based on Effluent Concentration (with Removal Efficiencies) Based on EPA 1975

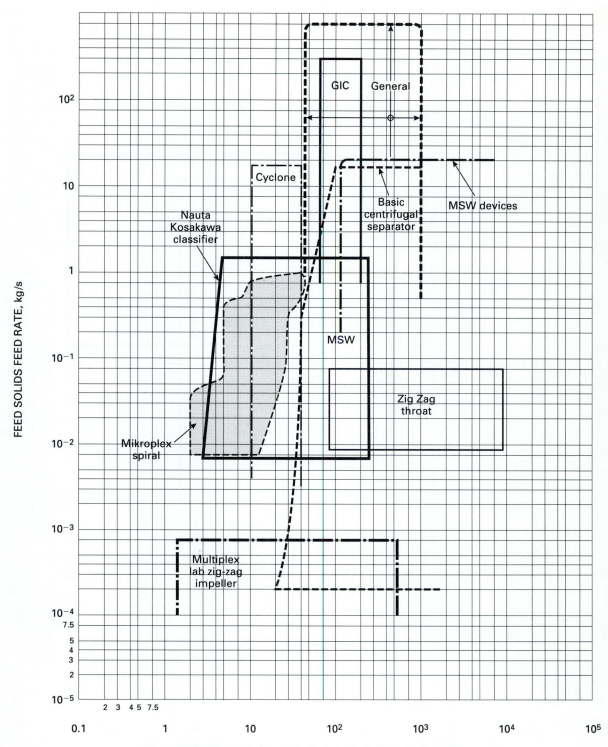

Figure 5–42 General Regions of Applicability for Solid-Solid Separations:
Based on Feedrate and Cut Diameter for Air Classifiers (from Maier, 1982)

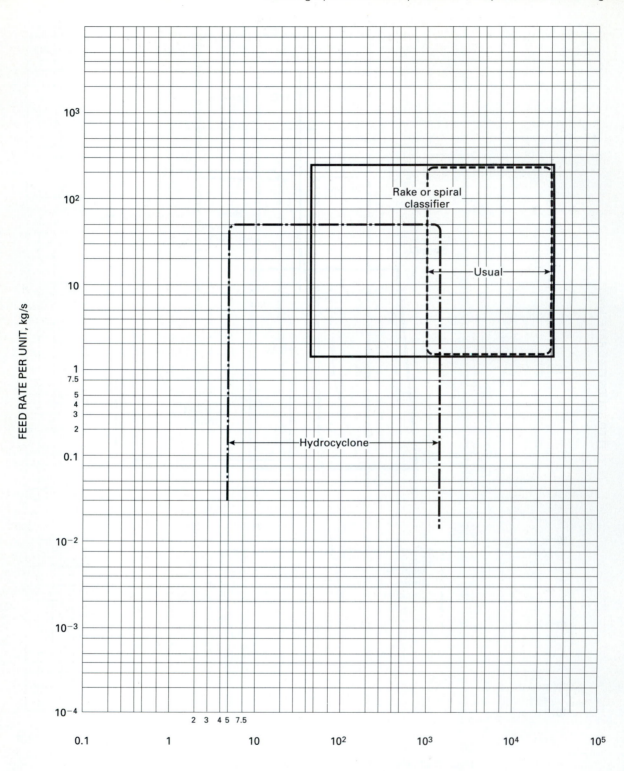

Figure 5–43   General Regions of Applicability for Solid-Solid Separations:
Based on Feedrate and Cut Diameter for Liquid Classifiers

TABLE 5-14. Criteria for selecting concentrators

Bird Number	Ease of Concentration	Type of Device
0 to 7	easy	jigs, tables,
7 to 10	moderate	sluices
10 to 15	difficult	
15 to 20	very difficult	
20 to 25	exceedingly difficult	DMS
>25		

separators. Figure 5–25 (and to some extent Figure 5–26) illustrates the general regions of applicability.

Exploiting the solubility of the target species is done via leaching/dissolution. Table 5–13 is applicable except that "unbound" washing operations are not pertinent. Here for a solid-solid separation, one of the species is soluble. Usually the solids are ground very fine to liberate the desired species, to provide surface area for the mass transfer, and to minimize the distance through which the species must diffuse. Figure 5–45 illustrates that the size of the feed varies greatly and depends on the liberation size of the target species.

The workhorse for solid-solid separation using a wet medium is flotation, which exploits the difference in wettability of the species. Flotation has a relatively narrow range of particle size over which it is applicable. Basically, to exploit flotation, the solids need to be ground to this size range. The general regions are shown in Figure 5–25 and reproduced in Figure 5–45. We note from Figure 5–40 that only a small difference in properties can produce an effective separation. Sketches of a flotation unit and individual "cells" are given in Figure 5–21. In this section, the details of which method of introducing the air are not distinguished. Such details are given by Woods (1993).

Differences in particle electrical conductivity can be exploited by electrostatic "concentrators," which operate in a non-liquid or gaseous medium. Particles with relative permittivities greater than 10 are usually classed as conducting; those with less than 10, as nonconducting. Some example values for different solids are illustrated in Figure 5–46. The general region where electrostatics are applied is given in Figure 5–25, which shows that, in general, the species concentration and the particle size are usually larger than those expected for flotation. The size, however, should be relatively uniform and less than 1 mm. Although Figure 5–46 emphasizes a difference in electrical permittivity, electrostatic separation can be effective if two species have similar conductivities but have different threshold voltages at which they become charged or if they respond with different signs when they encounter an active electrode. Details about these properties are given in **Data** Part C.

The three basic charging options are by corona, by active electrode, or by bouncing and rubbing together or "triboelectrification." These are illustrated in Figure 5–22. An illustrative guide for the applicability of different types of electrostatic separators is given in Table 5–15.

To separate aluminum from other solids, eddy current separators, illustrated in Figure 5–23, can be used.

Differences in magnetic properties can be exploited using some of the configurations illustrated in Figure 5–24. The general regions of applicability are given in Figure 5–25 (and to some extent in Figure 5–26). In general, solids can be ferromagnetic, ferrimagnetic, paramagnetic, nonmagnetic, and diamagnetic to represent very strong attraction, strong attraction, weak attraction, no attraction, and weak repulsion when the solid encounters a magnetic field. These are measured as rationalized magnetic susceptibilities or by comparing their behavior relative to iron that has been assigned a value of 100. The data on Figure 5–47 are expressed on the relative basis; the data in Figure 5–48 and in **Data** Part C are expressed in terms of the rationalized mass susceptibilities. The periodic table, in **Data** Part C, gives magnetic characteristics of elements. Figure 5–49 illustrates the regions of applicability of different options based on the magnetic characteristics.

5.1-3 Sequencing Heuristics for Separating Multiphase, Multicomponent Systems

The focus so far has been on the separation of one phase from another. Sometimes, even for that situation, a sequence of options might be used simply because no one single option is possible. For example, we might separate a liquid from a solid by preconcentrating via sedimentation and then finishing with filtration. Sometimes, size reduction is included between some of the options. At other times, more than two target species are the expected products.

Few heuristics have been developed specifically for heterogeneous separations. We might start by using the same general heuristics about the sequence of options that were developed in Chapter 4 for homogeneous separations. These are:

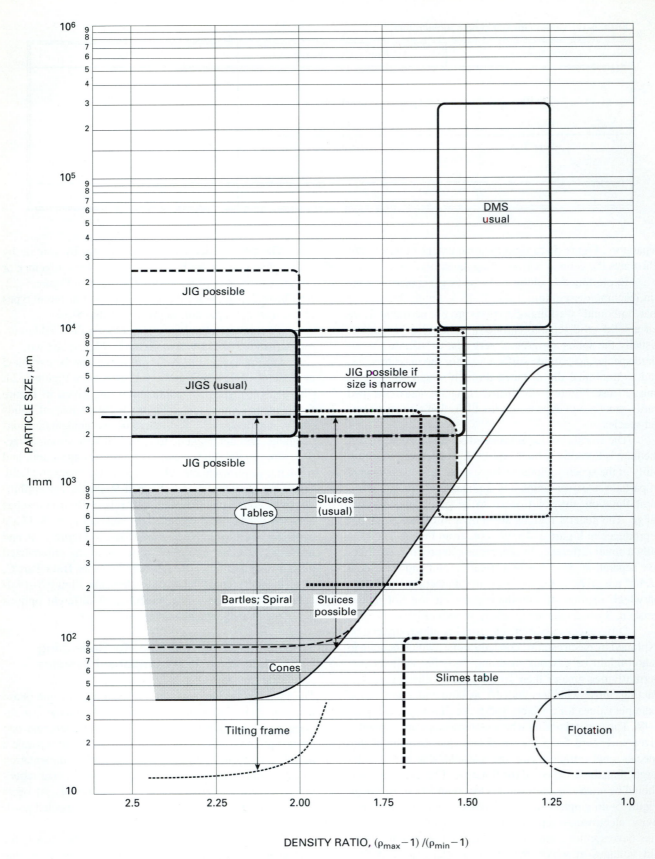

Figure 5–44 General Regions of Applicability for Solid-Solid Separations:
Based on Particle Size and Density Ratio for Concentrators

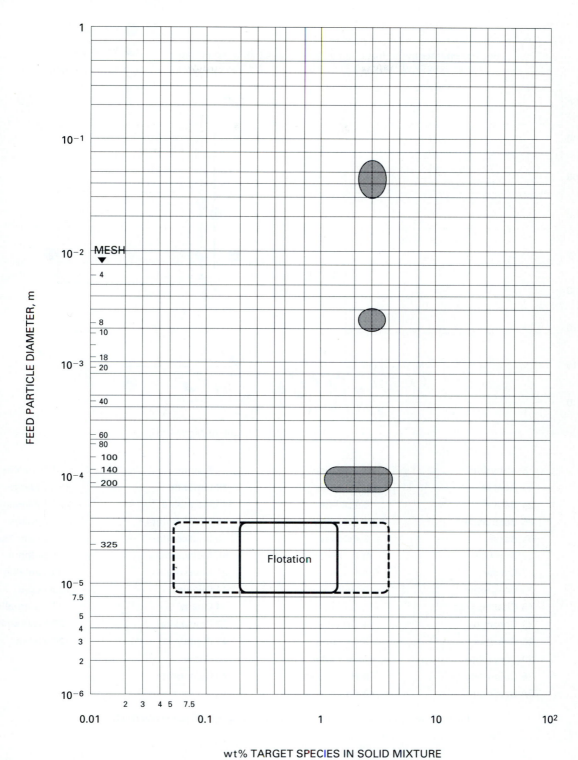

Figure 5–45 General Regions of Applicability for Solid-Solid Separations: Based on Particle Size and Concentration for Leaching and Flotation

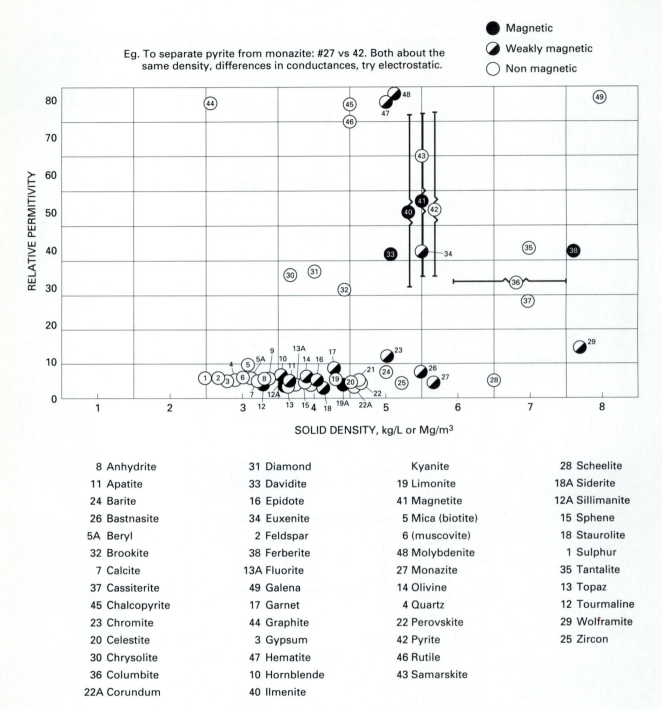

Figure 5–46 Electrical Conductivity, Density (and Magnetic Properties) for Solids

8 Anhydrite	31 Diamond	Kyanite	28 Scheelite
11 Apatite	33 Davidite	19 Limonite	18A Siderite
24 Barite	16 Epidote	41 Magnetite	12A Sillimanite
26 Bastnasite	34 Euxenite	5 Mica (biotite)	15 Sphene
5A Beryl	2 Feldspar	6 (muscovite)	18 Staurolite
32 Brookite	38 Ferberite	48 Molybdenite	1 Sulphur
7 Calcite	13A Fluorite	27 Monazite	35 Tantalite
37 Cassiterite	49 Galena	14 Olivine	13 Topaz
45 Chalcopyrite	17 Garnet	4 Quartz	12 Tourmaline
23 Chromite	44 Graphite	22 Perovskite	29 Wolframite
20 Celestite	3 Gypsum	42 Pyrite	25 Zircon
30 Chrysolite	47 Hematite	46 Rutile	
36 Columbite	10 Hornblende	43 Samarskite	
22A Corundum	40 Ilmenite		

TABLE 5-15. Electrostatic separations

	Exploit Difference in		
	electrical conductivity or permittivity	threshold voltage kV/cm	response to active electrode
Corona-active electrode-drum; HTS	separate good from poor conductors		
Active electrode-drum	separate good conductors from poor conductors	select voltage so that one conducts and other doesn't	if both respond the same, then the polarity of active electrode is important
Triboelectrification	both have similar conductances but different permittivities		adjust sign of active electrode

1. Remove the most hazardous, unstable, and corrosive species early. (Use the data in **Data** Parts C and D to get an idea of the hazard and corrosive rating of the species.)
2. Save the most difficult separations until the last (or do the easy ones first).
3. Separate the most plentiful components early.
4. Leave the high, specific recoveries until last.
5. Avoid extremes in operating conditions (especially pH, since many of the separations involve water).
6. If possible, separate the gas first, then the liquid, and then the solid-solid (except where the liquid is purposefully added for classification concentration to separate solids from solids).

Solid-solid separations seem to have received the most attention and Table 5–16 summarizes these. Since flotation is the workhorse for solid-solid separations, many of the heuristics have been developed for flotation.

5.2 Summary

A starting approach in selecting an option to separate heterogeneous phases is to classify the system according to the **type** of phases: gas-liquid, gas-solid, liquid-liquid, liquid-solid, and solid-solid. Then for each system, use the size-concentration values to make an initial choice.

If more discrimination in options is desired, consider whether:

- clean separation between both phases is required,
- the property difference we ought to exploit is different from "size."

If such factors pertain, then use the details of Section 5.1-2 to make the selection.

When a sequence of options is required, some heuristics are available to guide the sequencing. Most of these have been developed in the context of solid-solid separations. Overviews of the issues for each type of physical separation are given in Figures 5–49 to 5–53.

This minichapter has explored the selection of the equipment for the separation of heterogeneous systems. Details about the sizing and costing are given elsewhere (Woods, 1993a and b). Related to this topic are the selection and sizing of equipment to alter the size of dispersed phases: atomization, sprays, grinding, coagulation, crystallization. These are given in detail by Woods (1993c).

5.3 REFERENCES

BADGER, W. L., and J. T. BANCHERO. 1955. *Introduction to Chemical Engineering.* New York: McGraw-Hill.

BELTER, P. A., et al. 1989. *Bioseparations: Downstream Processing for Biotechnology.* New York: J. Wiley and Sons.

DAHLSTROM, D. A., and C. F. CORNELL. 1971. "Thickening and Clarification." *Chem Eng* **78**,4:63–69.

EPA. 1975. "Process Design Manual for Suspended Solids Removal." EPA 625/1-75-003a, US Environmental Protection Agency: Technology.

HENGLEIN, F. A. 1969. *Chemical Technology.* London: Pergamon Press.

FELSTEAD, J. E. 1954. "Flotation" in vol. 3, *Chemical Engineering Practice,* ed. Cremer and Davies. Stoneham, MA: Butterworths.

KOCATAS, B. M., and D. CORNELL. 1954. "Holdup and Residual Saturation of Hexane in Gravity-Drained Soybean Flake Beds." *Ind Eng Chem* **46**:1219–1224.

MAIER, C. 1982. *Air Classifiers on a Basis of Capacity and Range of Cut Size.* Hamilton, ON: Dept of Chemical Engineering, McMaster University.

MARSHALL, W. R., JR., and S. J. FRIEDMAN. 1954. *Perry's Chemical Engineer's Handbook.* New York: McGraw-Hill.

MICHAELS, E. L., et al. 1975. "Heavy Media Separation of Aluminum from Municipal Solid Waste." *Trans Soc of Mining Engineers* **258**:34.

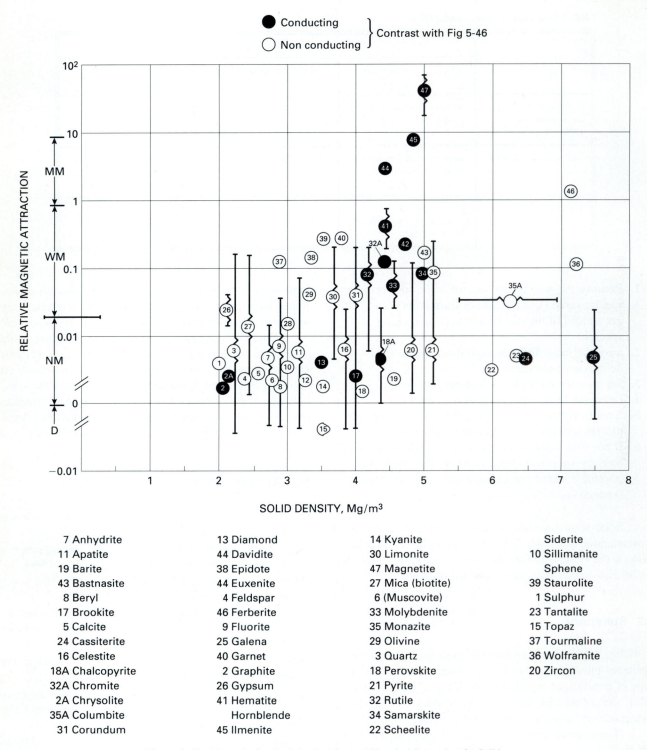

Figure 5–47 Magnetic Conductivity, Density (and Electrical Properties) for Solids

7 Anhydrite	13 Diamond	14 Kyanite	Siderite
11 Apatite	44 Davidite	30 Limonite	10 Sillimanite
19 Barite	38 Epidote	47 Magnetite	Sphene
43 Bastnasite	44 Euxenite	27 Mica (biotite)	39 Staurolite
8 Beryl	4 Feldspar	6 (Muscovite)	1 Sulphur
17 Brookite	46 Ferberite	33 Molybdenite	23 Tantalite
5 Calcite	9 Fluorite	35 Monazite	15 Topaz
24 Cassiterite	25 Galena	29 Olivine	37 Tourmaline
16 Celestite	40 Garnet	3 Quartz	36 Wolframite
18A Chalcopyrite	2 Graphite	18 Perovskite	20 Zircon
32A Chromite	26 Gypsum	21 Pyrite	
2A Chrysolite	41 Hematite	32 Rutile	
35A Columbite	Hornblende	34 Samarskite	
31 Corundum	45 Ilmenite	22 Scheelite	

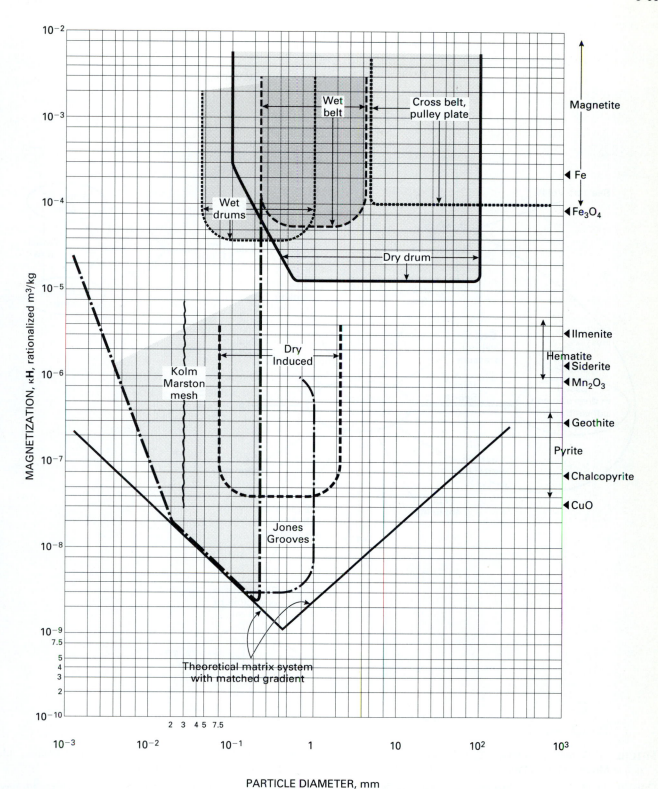

Figure 5–48 General Regions of Applicability for Solid-Solid Separations:
Based on Particle Size and Magnetization for Magnetic Separators
(adapted from Oberteuffer, 1974 Transactions on Magnetics © 1974 IEEE)

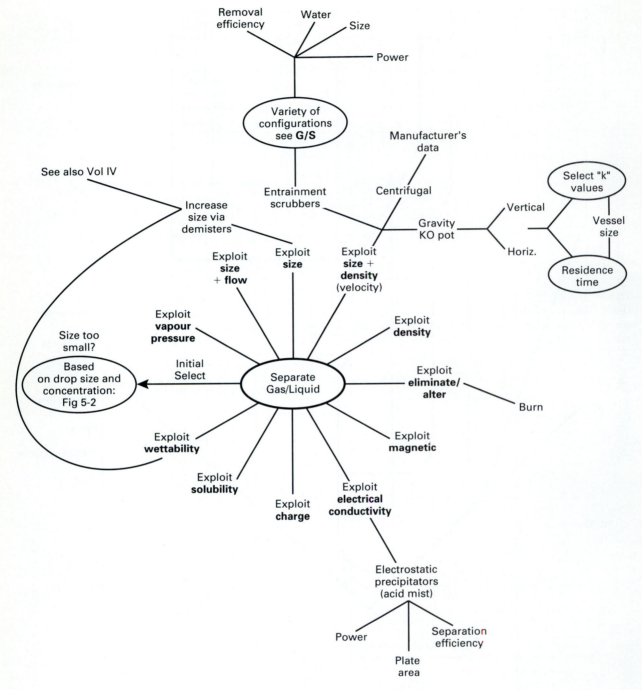

Figure 5–49   An Overview of Gas/Liquid Separations

MITCHEL, F. B., and D. G. OSBORNE. 1975. "Gravity Concentration in Modern Mineral Processing." *Chem Ind* (Jan):58–64.

OBERTEUFFER, J. A. 1974. "Magnetic Separation: A Review of Principles, Devices and Applications." *IEEE Transactions on Magnetics,* **MAG-10,** 2:223–238.

OGLESBY, S., JR., and G. B. NICHOLS. 1978. *Electrostatic Precipitation.* New York: Marcel Dekker, Inc.

PRYOR, E. J. 1965. *Mineral Processing,* 3rd ed. Amsterdam: Elsevier Publishing Co.

TAGGART, A. F. 1954. *Handbook of Mineral Dressing.* New York: J. Wiley and Sons.

WOODS, D. R. 1985. *Surfaces Colloids and Unit Operations,* 4th ed. Hamilton, ON.: McMaster University.

WOODS, D. R. 1993a. *Process Design and Engineering Practice: Selecting and Sizing Physical Separations.* Hamilton, ON: McMaster University.

WOODS, D. R. 1993b. *Process Design and Engineering Practice: Costs.* Hamilton, ON: McMaster University.

WOODS, D. R. 1993c. *Process Design and Engineering Practice: Reactions Mixing and Size Change.* Hamilton, ON: McMaster University.

WOODS, D. R., and E. DIAMADOPOULOS. 1988. "Role of Sur-

TABLE 5-16a. General suggestions about the selection of solid-solid separators

Consider using DMS to preconcentrate <u>before</u> grinding to final liberation size.

Preconcentrate before flotation if the grain size is large and coarse minerals can be removed via density difference before grinding further.

Above 44 μm use gravity concentration/separation techniques instead of flotation.

For feed ore assay <0.3% use a combination of gravity concentration and then flotation.

For feed ore assay 0.5 to 0.7% try flotation (with possible downstream concentration).

For particle sizes <20 μm, neither flotation nor gravity concentration are very effective.

Use flotation only and bypass a preconcentration if the gangue is dense or if the minerals are flaky.

If either "concentration" or "flotation" are possible, then select flotation because it is easier, usually cheaper and gives improved recovery/greater capacity.

factants and Surface Phenomena on the Separation of Dilute Oil/water Emulsions and Suspensions," chapter in *"Surfactants and Chemical Engineering,"* ed. D. T. Wasan, D. O. Shah, and M. E. Ginn. New York: Marcel Dekker, 369–539.

5.4 EXERCISES

5-1 We wish to process 2000 kg/h of inert soybean meal that has associated with it 800 kg of soy oil and 50 kg of hexane. The solvent is 1310 kg/h of hexane that contains initially 20 kg of oil. Our target is to have 120 kg of soy oil in the exit solids. Estimate what we might do.

5-2 Sugar beets are cut into "cossettes" that are 3 to 7 mm wide and about 5 to 8 cm long. The inlet concentration of sugar is 16.5% w/w. The desired exit concentration of sugar in the beet (or marc) is 0.2 to 0.3%. The operating temperature is 75°C. The desired exit solvent concentration is 12% w/w. What might we do to process about 15 kg/s?

5-3 An ore containing 0.6 to 1.2% cobalt arsenides is to be concentrated from its gangue that is primarily quartz, serpentine, and calcite. The density of the mineral is 5.7 to 6.8 Mg/m^3. The ore is available at −8 mm size. What might we do? The density/mass distribution has a Bird number of about 12. The feedrate is 24 Mg/h of ore.

5-4 A serpentine/iron ore contains about 19 to 20% iron. The

TABLE 5-16b. Rules-of-Thumb for Selecting/Developing Units & Networks

General:

- try to float at as coarse a grind as possible (if necessary regrind the concentrate).
- take products (whether they are tails or minerals) out of the main circuit coarse and early; in other words, try a bulk float as first option.
- try to separate early the material making up the majority of the ore (eg. tails).
- for mixed sulfide/oxide ores, float the sulfides first (to prevent depression of the sulfides downstream when sodium sulfide is added to depress quartz for oxide flotation).
- try to work at pH > 7 to minimize corrosion.
- for any recycle stream, try to match the mineral and - chemical composition of the recycle with the steam it joins.
- try to avoid surges to each stage; include buffer storage.
- in any recycle, ensure that all species can escape.

Bulk versus sequential use bulk when:	Flotation Sequencing use selective/sequential differential when:
— all valuable minerals are combined and are coarsely aggregated with respect to the ore as a whole (e.g. sulfide ores),	— all valuable minerals are individual, separately liberated at <200 μm (65 mesh) (eg: lead/zinc; lead/zinc/iron),
— if the concentration of Cu, Pb, Ag in soluble salt forms is high,	— more difficult; the differences in flotability may be less sharp; very careful choice of reagents needed; more delicate.
— it is relatively easy,	
— if depression of one species requires long contact and high reagent concentration.	

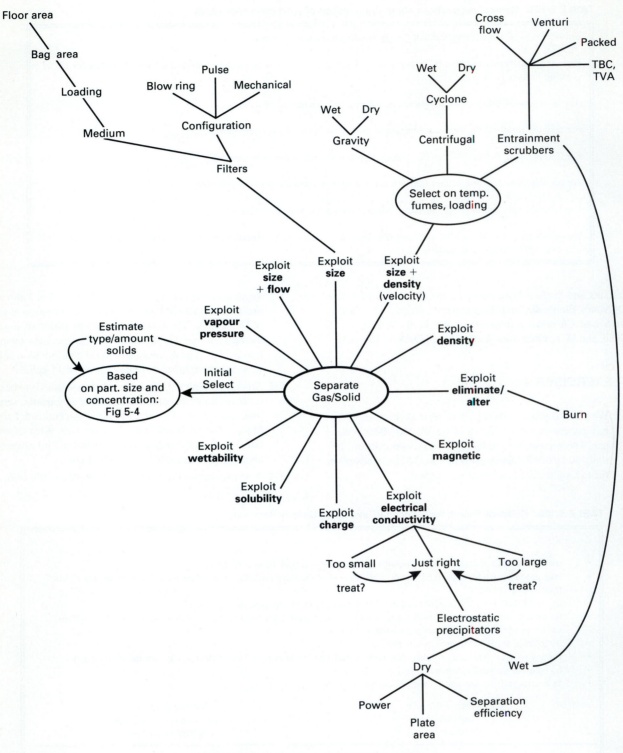

Figure 5–50 An Overview of Gas/Solid Separations

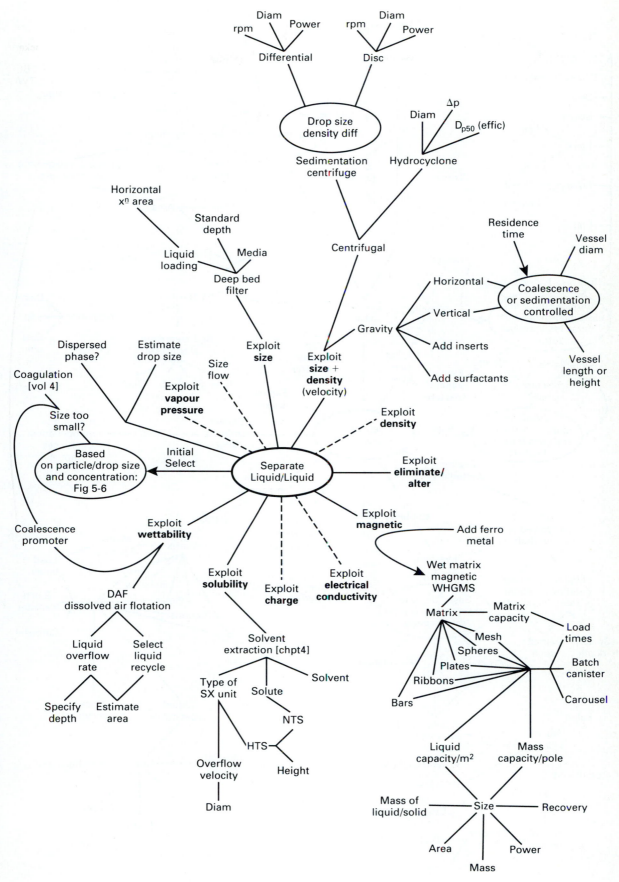

Figure 5–51 An Overview of Liquid/Liquid Separations

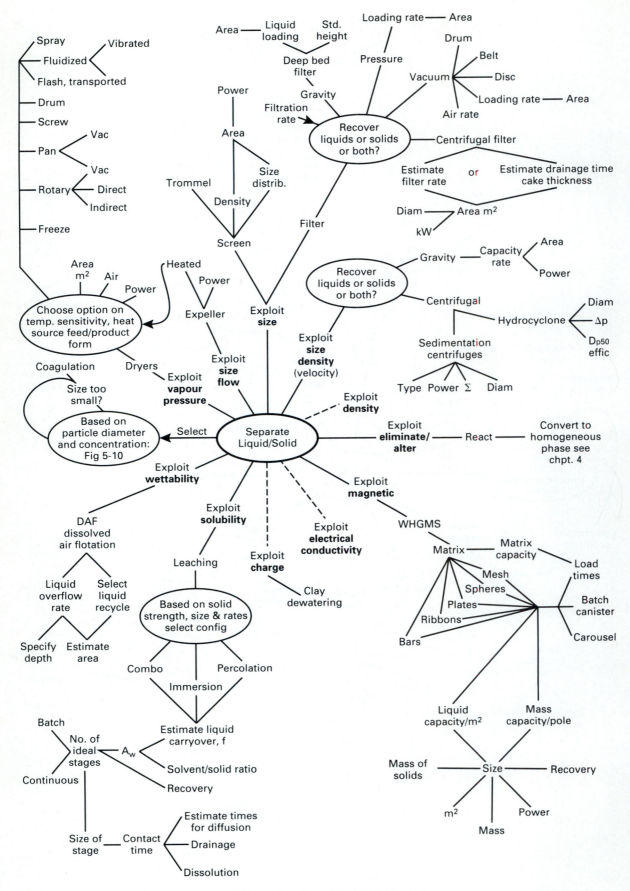

Figure 5–52   An Overview of Liquid/Solid Separations

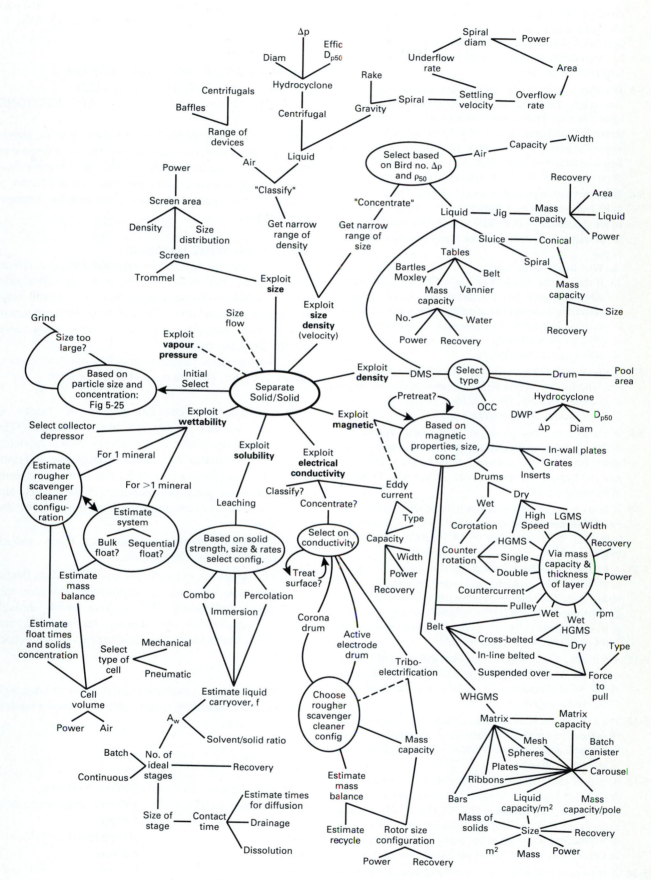

Figure 5–53 An Overview of Solid/Solid Separations

density/mass distribution has a Bird number of about 5. The gangue is silica. To liberate the mineral the ore has been crushed to yield a 0.030 to 0.6 mm size range. The ore has been deslimed to remove all of the particles less than 0.030 mm. What might we do?

5-5   We want to recover the solids from a stream that is 10% w/w solids. The solids are 5 mm diameter, density 1500 kg/m^3, and are fairly robust (as opposed to fragile). The density of the liquid phase is 800 kg/m^3 and is relatively toxic. The total slurry flowrate is 100 L/s. The desired exit liquid content is to be less than 1%. The liquid is to be recovered and recycled. What would you recommend?

5-6   Solids have been ground in a wet ball mill. The product is to be leached. However, before it is, we want to alter the solids concentration from 35% w/w to 75% w/w. The solids density is 2.7 Mg/m^3 and the liquid density is about 1 Mg/m^3.
   a. Why would we do this?
   b. If the particle size is 40% w/w −200 mesh and 100% w/w −28 mesh and the distribution is log normal, sketch the distribution and estimate the geometric mass average diameter and the geometric standard deviation.
   c. Why do we not densify to 0% water?
   d. Recommend some options.

5-7   We wish to separate vermiculite from kyanite and pyroxene. The feed is ground to 6 to 28 mesh and we want to keep the material dry. The capacity is 24 Mg/h. The feed is 32% vermiculite. What might we do?

5-8   We wish to separate phosphate from quartz. The feed is 71.7% phosphate and we would like to remove some of the quartz from this. The material is to be kept dry. The feed-rate is 34 Mg/h. The particles are all less than 35 mesh. What might we do?

5-9   A feed stream of 15% water in Solvent 99 is to be separated. The density difference is 200 kg/m^3, the interfacial tension is 38.5 mN/m. The feed flowrate is 2 L/s. What might we select?

5-10 A frit smelter produces a waste gas containing 2.8 g/m^3 of particulates. The geometric mass median particle size is about 3 μm. The range is from 0.6 μm to about 8 μm. The gas flowrate is about 52,000 dm^3/s and contains about 0.5 g/m^3 of NO_2 and 0.3 g/m^3 of HF. The gas temperature is 200°C to 870°C. What might we do?

5-11 A refractory lined furnace incinerates 490 Mg/d of Municipal Solid Waste. This produces 111,400 dm^3/s of gas at 980°C. What alternatives are available if the dust loading is 3.5 kg/1000 kg gas?

5-12 The exhaust gas from a fluidized bed coal dryer flows at 69,000 dm^3/s at 90°C. The particles are classified as "ultra-fine"; the loading is about 6.4 g/m^3. Some SO_2 is present. What might you do?

5-13 A flow of 43,000 dm^3/s of hot gas exits from a cupola producing 14.5 Mg/h of iron using 140 kg of coke per Mg of iron. The gas temperature is 1150°C. The particle size and loadings are typical of cupola operation. What might you do to produce a waste gas at less than 110°C and containing less than 0.4 kg of particulates per Mg of exhaust gases?

5-14 A mixture of oil/sulfuric acid is to be separated. The temperature is 33°C. The oil is 23% v/v LIX 64N in Escaid

100. The oil/acid ratio is between 1:1 and 1:5. The total flowrate is 6.3 L/s. How might this separation be done?

5.5 PROCESS & INSTRUMENTATION DIAGRAM FOR WASTE WATER TREATMENT ILLUSTRATES APPLICATIONS

Processes to treat waste water must remove both the particulate and the dissolved contaminants. We must dispose of the solids involved in this process. Thus, this process illustrates the application of principles outlined in Chapter 5, with additional application of the separation for homogeneous phases.

PID-5: Waste Water Treatment

Industrial waste waters are contaminated with dissolved organics, solid particulate sludges, and immiscible oils. These combine to provide a liquid that is cloudy and dirty-looking, smelly, and toxic. The water may be a mixture of process water and rain water. The amount of water varies hourly, daily, and monthly.

An illustrative process to treat this industrial waste is shown in PID-5. The process is characterized by:

- duplication of much of the equipment as "standby" so that a working process is "always" available. (The standby equipment is not always shown explicitly in PID-5. Rather it is shown as **D**uplicate or **T**riplicate.) See, for example, **D** for just about every pump and major piece of equipment, **V500, V501,** and **T** for the equalization and holding basins **V502** and **V508.**

- flexibility: with piping to allow us to bypass and adjust the sequence of equipment.

- a lot of storage to handle the hourly and daily fluctuations in flow and concentration. This prevents surges and upsets. This is done mainly through the equalization and holding basins **V502** and **V508.** These former are designed for one-day capacity.

The feed to this example plant is about 30 L/s ± 30% with a variable amount of storm water. The process water is characterized by:

temperature	33°C
pH	10
immiscible light oil	2000 to 3000 ppm
heavy oils and sludges	130 ppm (parts per million)
suspended solids	130 ppm
phenol	50 ppb (parts per billion)
ammonia	8.8 ppm
nitrate	0.1 ppm
sulfides as H_2S	15.7 ppm

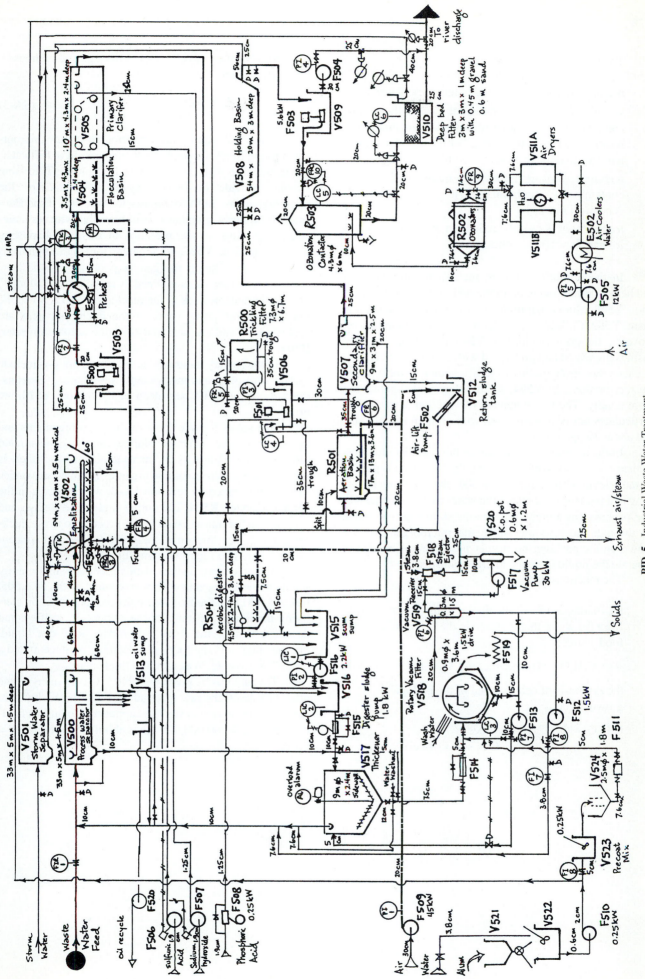

PID–5   Industrial Waste Water Treatment

The process has four sections:

primary treatment: removal of the particulates,

secondary treatment: removal of most of the soluble contaminants; traditionally secondary treatment has meant "biological treatment," or the use of microorganisms to remove targeted soluble species,

tertiary treatment: removal of more of the soluble contaminants; traditionally, this has meant by "chemical means,"

treatment of the solids: reduction of the mass of solids and dewater the solids.

In the primary treatment section, immiscible light and heavy oils are removed in the API, process water separator, **V500.** More separation can occur in the equalization basin, **V502.** Colloidal and micron size particulates are removed in the primary clarifier, **V505.** Feed to this settling basin is preheated and pretreated with 35 ppm of coagulant to cause the tiny particles to join together to form larger clusters or flocs, in **V504,** so that they will separate out more readily. The liquid leaving this section is free of the particulates but is contaminated with 15 ppm soluble oils, and about 30 ppb phenol.

In the secondary treatment section, aerobic microorganisms are set to work to remove the dissolved organics. In the trickling filter reactor, **R500,** the microorganisms cling to a packing over which the water trickles downward. The packing has a high void volume so that air can move up through the packing. Since the microorganisms are grown on the packing, no further processing is needed to separate the liquid from the microorganisms.

The reactions include:

the oxidation of organic matter or the BOD removal from the waste that is stored in the cell as reserved food source:

$$C_xH_yO_z + O_2 \text{ enzymes} \rightarrow CO_2 + H_2O \qquad \text{(5A-1)}$$

the synthesis of cell material or the biological growth of cells:

$$C_xH_yO_z + O_2 + NH_3 \text{ enzymes} \rightarrow \\ CO_2 + H_2O + \text{cells} \qquad \text{(5A-2)}$$

the oxidation of cell material through endogenous respiration:

$$\text{Cells} + O_2 \text{ enzymes} \rightarrow \\ CO_2 + H_2O + H_2O + NH_3 \qquad \text{(5A-3)}$$

In the "activated sludge" reactor, **R501,** the microorganisms are freely suspended and mix around in the water. Air is blown in with diffusers near the bottom. The rising bubbles provide dissolved oxygen, essential for the microorganisms, and mix the reactor contents. Thus flocculated biomass circulates continuously with organic-containing waste in the presence of oxygen. A settling basin is needed, **V507,** to separate the microorganisms from the water. The microorganism population in the "aeration basin reactor" is maintained by recycling the "sludge" of microorganisms back to the reactor via **V512.** Since the microorganisms grow, some of the microorganisms are syphoned off and are disposed of in the aerobic digester, **R504.**

In the tertiary treatment of the liquid, options include carbon adsorption and membranes. In this process, ozonation, in reactor **R503,** and deep bed filtration, **V510,** are used.

The solids include some of the microorganisms and the sludges. The activated sludge can be concentrated through autooxidation by the microorganisms when there is no other food (such as soluble organics in the waste water). This is represented by reaction 5A–3, which occurs in the aerobic digester, **R504.** The output sludge, the sludges and scums from the other vessels in the process go to the thickener, **V517.** The thickened sludge is washed and concentrated further in a rotary vacuum filter **V518.**

An interesting feature of this process is that gravity sedimentation is used in at least five units, **V501, 502, 505, 507** and **V517.** In the first four units the focus was on clarifying the liquid; we called these settlers or clarifiers and designed them on the criteria to produce solid-free liquid. For the latter unit, the focus is on getting as concentrated a "solid" as possible. We call this a thickener.

The unit cost is given in Table 5A–1. In many of the other processes, the raw material cost dominated, but here the raw material cost is $0. The capital cost to build a facility to process 1,000,000,000 L/a is about 6.5 million (MS = 1000).

TABLE 5A-1. Waste water treatment facility: cost contributions to the cost per litre of waste water treated

Contribution from	Approximate Unit Cost breakdown, %
Raw material	0.
Utilities	19
Catalyst and chemicals	1
Labor	29
Credit for byproducts: recycled oil	-31
Maintenance	11
Supervision	3
Depreciation	36
Indirectly attributable costs	32
General expense	0
Total	100

Minichapter 6

Selecting Options for Reactions and Storage

Reactors are the heart of most processes. Here the raw materials are mixed and experience "ideal" conditions for the target reaction to occur to convert all the raw materials into the desired product.

Unfortunately, rarely can this goal be achieved. Because cost of the raw materials usually represents about 50% to 70% of the total processing costs, this means that working out the best compromise to try to achieve the goal is a challenging and vital task. The key issues to consider include:

- most target reactions (from raw material to the desired product) must compete with undesirable side reactions. Our task is to control the mixing, the temperature, the pressure, and often, the type and selectivity/activity of a catalyst so as to optimize the target reaction. We might have to compromise such that we allow an unwanted side reaction to proceed but try to minimize its effect. (For example, steam might be added as a "reactant" to minimize high temperature coke formation through the water gas shift reaction, as illustrated in Table 4–22.)
- reactions occur at different speeds: some are so fast that essentially they are "equilibrium reactions;" some are dominated by the rate of mass transfer to

and/or from the reaction site and so are "mass transfer" controlled; some are controlled by the concentrations of the reactants to different powers: first order, second order, and so forth.

- reactions may be highly exothermic and generate a lot of heat; some may be highly endothermic and require a lot of heat. Thus, we have to evolve options for controlling the temperature.
- some reactions are homogeneous and occur in one phase; others are heterogeneous and occur in either of the bulk phases or at the surface between the phases.
- the optimum target conditions will include some ideal "constant" temperature under which the reactants are kept for a desired time for reaction. However, the temperature is not constant; it varies across the volume of reactants. How much it varies depends on the configuration. For example, for reactants flowing in a tube, the temperature variation across the tube, when it is within a furnace, varies as illustrated in Figure 3–8. Thus, for the minimum variation in temperature a single row of tubes would be used. On the other hand, for the particular reaction we might be able to get away with a 4/1 variation in temperature across a tube cross section and thus use the cheaper furnace configuration with banks of tubes against a

wall. Likewise, the residence time of the reactants in the reactor is not constant for all blobs of reactants entering a reactor; for example, for reactants flowing through a tube, the "plug-flow tubular reactor" or PFTR, the RTD, residence time distribution, is narrow as illustrated in Figure 6–1. For a perfectly mixed, continuous flow stirred tank reactor (CFSTR), the RTD is broad as illustrated in Figure 6–1b. For all reactions, the deviation of the RTD affects the quality of the output. However, for some reactions we can get away with using a broad RTD in the reactor and produce a quality product through the downstream separation scheme. On the other hand, for polymerization reactions, the quality of product is what comes out of the reactor; usually we need a relatively narrow RTD. Thus, the allowable variation in the temperature and in the RTD in the reactor are other considerations in selecting reactors.

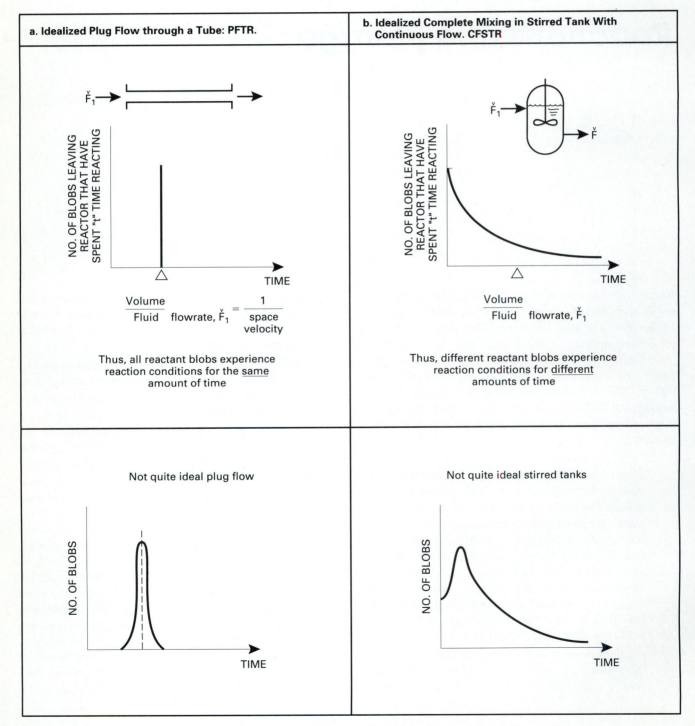

a. Idealized Plug Flow through a Tube: PFTR.

NO. OF BLOBS LEAVING REACTOR THAT HAVE SPENT "t" TIME REACTING

TIME

$$\frac{\text{Volume}}{\text{Fluid flowrate, } \check{F}_1} = \frac{1}{\text{space velocity}}$$

Thus, all reactant blobs experience reaction conditions for the <u>same</u> amount of time

Not quite ideal plug flow

NO. OF BLOBS

TIME

b. Idealized Complete Mixing in Stirred Tank With Continuous Flow. CFSTR

NO. OF BLOBS LEAVING REACTOR THAT HAVE SPENT "t" TIME REACTING

TIME

$$\frac{\text{Volume}}{\text{Fluid flowrate, } \check{F}_1}$$

Thus, different reactant blobs experience reaction conditions for <u>different</u> amounts of time

Not quite ideal stirred tanks

NO. OF BLOBS

TIME

Figure 6–1   Residence Time Distributions for Flow Reactors

To tackle each of these apparent complications, engineers have evolved a range of options. Unfortunately, sometimes these are contradictory and hence we compromise. For complex reactions, the prime criteria is the **selectivity** of the system. Consider first the options developed and used.

6.1 SELECTING A REACTOR CONFIGURATION

In selecting a configuration, first we need to know key fundamentals about the reactions being considered. Then we consider the various issues that affect the choice. How the various equipment options satisfy these issues is described. Then the criteria are applied to select configuration options.

6.1-1 Identifying the Starting Fundamental Information Needed

What fundamental information do you need to know before we can consider selecting a reactor configuration?

Consider first *homogeneous* reactions, in the gas phase or in the liquid phase. For the target and unwanted side reactions we can use Le Chatelier's principle and equilibrium calculations to explore the effect of temperature and pressure on the conversion. Since temperature and pressure excursions away from room conditions are expensive, we would like to operate as close to room conditions as possible. Then we can consider the rate and order of reaction. The rate of reaction doubles with every 10°C increase in temperature. The concentrations of the reactants affect the rate depending on the order of the reaction. If the rate of reaction or the selectivity is poor, we might create a catalyst to alter these. If we do, then usually the catalyst is a different phase and so we move to heterogeneous reaction systems. Once we have identified optimum conditions for homogeneous reactions, we usually have to control the temperature and pressure and provide intimate mixing.

For *heterogeneous* reactions, the important considerations are where the reaction occurs, and to ensure that there is a lot of surface area to promote mass transfer. If a catalyst is used, then we also want to maintain the activity and selectivity of the catalyst. Hence, the conditions for the reaction take into account the reactions and the effect the conditions have on the catalyst. From this (and the same issues explored for homogeneous reactions) we usually arrive at the optimum temperature, pressure, ways of generating surface area, ways of nurturing the catalyst, and mixing conditions.

For any reaction, some of the inevitable side reactions include

- for high temperature reactions with organics (higher than 500°C), the breakdown of the species to form

carbon. This implies that we might want to try to suppress this (through the addition of steam) and/or consider ways to periodically remove the carbon deposits.

reactions with large polymeric molecules and foodstuffs usually result in a fouling of the reactor surfaces (and heat exchange surfaces). We need to consider configurations that will make it easy to remove such deposits.

From all this we eventually arrive at the information needed to select the reactor configuration: the phases present, whether a catalyst is used (and its characteristics), the operating temperature and pressure, the RTD, the need to add/remove heat, and the need to clean the reactor.

For example, select a reactor for:

Case 1 the "dehydrogenation of toluene to make benzene": our optimal conditions, we think, are
phase: gas phase plus solid catalyst: G-S
temperature: 590 to 650°C
pressure: 6 to 11 MPa
rate and order of reaction: first order with respect to toluene; 1/2 order with respect to hydrogen
thermal considerations: exothermic: 50 MJ/kmol toluene
 temperature variation:
catalyst robustness: reasonably robust.
RTD variation:

Case 2 the "cracking of acetic acid to make ketene": our optimal conditions, we think, are
phase: gas phase plus gaseous catalyst: G
temperature: 750°C
pressure: vacuum of 13 kPa absolute
rate and order of reaction:
thermal considerations: endothermic:
150 MJ/kmol acetic acid cracked
 temperature variation: minimize or keep from exceeding 2:1
catalyst robustness: NA
RTD variation: minimize because very rapid reversible reaction and polymerization of ketene could cause problems in downstream processing

Case 3 the "alkyllation of isobutane to produce alkyllate": our optimal conditions, we think, are:
phase: liquid phase plus immiscible sulfuric acid as catalyst: L-L
temperature: 7 to 13°C
pressure: 580 to 650 kPa
rate and order of reaction: very rapid, and complex chain mechanism involving carbonium ions with mass transfer of reactants into the acid drops where reaction occurs and then transfer of the products back into the continuous organic phase

thermal considerations: slightly exothermic: 85 MJ/kmol isobutane reacted
   temperature variation:
catalyst robustness: sensitive to dilution by water
RTD distribution:

Case 4 the "hydrogenation of fish oils": we want to reduce the degree of "unsaturates" (as measured by the **Iodine Value** from an IV of 100 to 10). Our optimal conditions, we think, are:
phase: liquid fish oil, gaseous hydrogen, and solid Nickel based catalyst: G-L-S; a reduction on 1 IV requires 0.0795 kg hydrogen per 1 Mg of oil
temperature: 80°C
pressure: 0.8 MPa
rate and order of reaction: reaction at the catalyst surface between the dissolved hydrogen and the unsaturates in the oil
thermal considerations: slightly exothermic: about 1.5 MJ/ IV change
   temperature variation:
catalyst robustness: reasonably robust
RTD variation:

Case 5 the "suspension polymerization of vinyl chloride to make Polyvinyl chloride": our optimal conditions, we think, are:
phase: liquid monomer drops suspended in water with water soluble "initiator": L-L
temperature: 52°C
pressure: vapor pressure of monomer at this temperature; about 0.85 MPa.
rate and order of reaction: serial, parallel and complex
thermal considerations: exothermic: 115 MJ/kmol monomer converted
   temperature variation: minimize because temperature determines Molar Mass Distribution of the product
catalyst robustness: NA
RTD variation: keep the RTD narrow

6.1-2 Considerations for Mixing, Temperature, Pressure, and a Solid Catalyst

Some of the significant issues are how to mix the reactants, how to obtain the desired reaction temperatures and pressure, and how to handle a catalyst, if one is needed.

For **mixing**, in general, liquids can be mixed relatively easily with mechanical agitators; gases cannot be mixed easily with mechanical agitators. This leads to two basic characteristics,

- no-mixing characteristics that are usually obtained from plug flow through a tube configuration. Here the concentration (and maybe the temperature) changes along the tube length. The RTD for plug flow is single valued and illustrated in Figure 6–1a. This configuration is usually used for reacting gases.

- perfect mixing characteristics that are usually obtained from a mechanically stirred tank configuration or Stirred Tank Reactor, STR. Here the concentration is the same at all locations. The temperature is uniform throughout; for a "batch-operated" STR, the residence time equals the batch time and is narrow. For a continuous flow through a STR or CFSTR, the RTD for perfect mixing is broad and given in Figure 6–1b. This is often used for reacting liquids.

These two ideal options are illustrated in Figure 6–2. We can convert the one configuration into the other mixing characteristics by using a plug flow configuration with a large recycle (to simulate perfect mixing) and by using a series of stirred tank configurations (to simulate no-mixing characteristics). This is illustrated in Figure 6–2. Often mixing options exist, such as motionless mixers in a tube or gaseous/liquid fluidized beds. For G-S fluidized beds the gas goes through the bed in seconds, whereas the solids have a broad RTD and on average remain in the bed a long time. Thus, the configuration and the type of mixing affects the concentrations and temperatures within the reactor and the RTD for the reactants in the reactor.

Very high or very low **pressures** are difficult to maintain if the wall of the vessel is punctured to allow for a rotating shaft for a mechanical mixer. It is difficult to seal the shaft at high pressures. Hence, reacting liquids at very high pressures in a CFSTR is difficult because of the inability to seal around the shaft of the mixer.

Figure 6–3 illustrates how the type of phase present and the pressure offer insight into the usual configuration of the reactor.

For **temperature,** two issues are important: control and level.

The extreme options we have for temperature control are:

- to provide perfect insulation and let the temperature adjust according to the heat of reaction: adiabatic operation; this option has a small capital cost; the equipment is easy and inexpensive to build because heat exchange equipment is not needed.

- to provide perfect heat exchange so that the temperature is constant: isothermal operation. Here, excellent heat exchange is needed if the heat of reaction is significant. Some options for heat exchange for different configurations are given in Figure 6–4.

Some tricks that can be considered to help control temperature include:

1. divide the reaction system in a series of small reactors (small enough so that when each is operated adi-

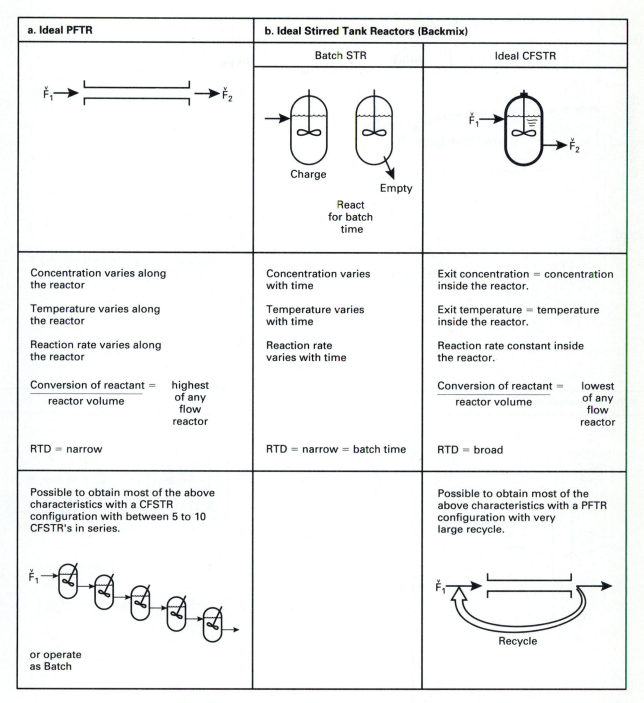

Figure 6–2 Some Options for Reactor Configurations and Their Characteristics
a) Ideal PFTR
b) Ideal Stirred Tank Reactors (Backmix)

abatically, the temperature excursions will not be excessive) and:

a) use heat exchange between each reactor; or

b) inject part of the reactants to each of the small reactors instead of injecting it all at once at the beginning.

These policies are illustrated in Figure 6–5a.

2. add a heat absorber or dispenser such as: an "inert" liquid such as water (as in suspension or emulsion polymerization); an "inert" gaseous heat carrier like steam, an "inert" solid; or a solid catalyst in a fluidized bed operation. (Fluidized operations are considered in detail in Section 3.2-1 c(i) on fluidized bed heat exchange with examples of reactors given in Table 3–11.) These policies are illustrated in Figure 6–5b.

Concerning the temperature level, for temperatures above 500°C, a furnace or kiln configuration is used. For

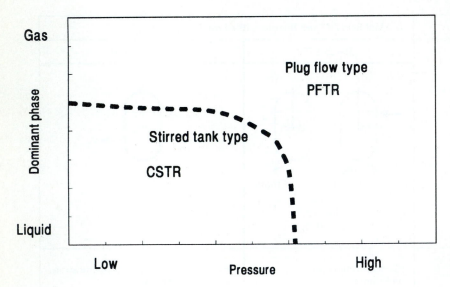

Figure 6–3 Phase and Pressure Suggest Basic Type of Reactor Configuration

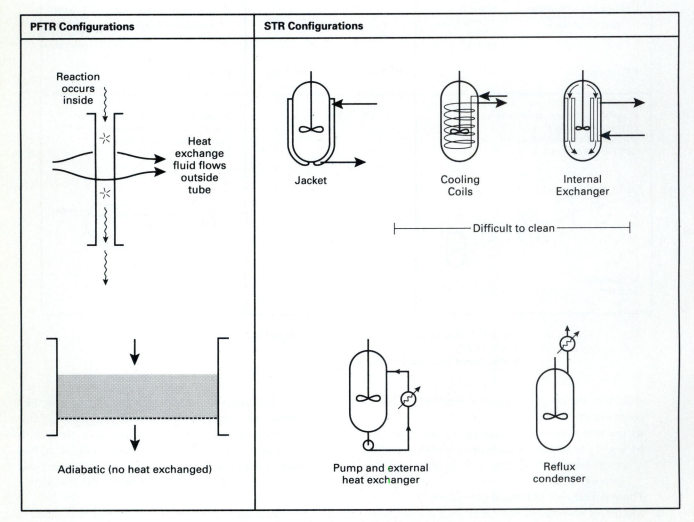

Figure 6–4 Some Options for Heat Exchange

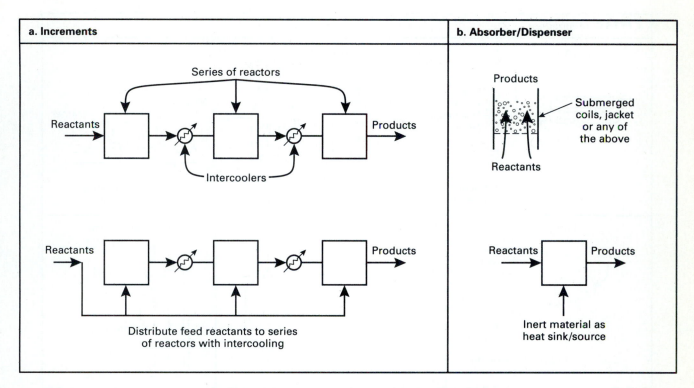

Figure 6–5   Some Policies for Controlling Temperature
a) Using Increments
b) Using Absorber/Dispenser

temperatures between 250°C and 500°C, a high temperature heat exchange medium, such as Dowtherm or those illustrated in Figure 3–12, might be used. The general applicability is shown in Figure 3–5. For low temperatures, the options are discussed in Chapter 3.

6.1-3 Some Optional Configurations

PFTR reactors are basically tubes through which the gases/or liquid flow. The tubes may be the tubes in a shell-and-tube heat exchanger or tubes in a furnace or just long tubes. They could be pressure vessels, long columns, tray configurations. The general types we consider include empty tubes, tubes filled with a "fixed" bed of catalyst, transported bed or slurry reactors, bubble reactors, spray reactors, tray reactors, and thin film reactors. These are illustrated in Figure 6–6. For gas-liquid reactions, the difference surface area and flow characteristics of the different types of contactors are illustrated in Figure 6–7. For gas-solid reactions, the difference in solid residence time and diameter for different configurations is given in Figure 6–8. For these reactors the gas residence time is in fractions of seconds. Figure 6–9 gives the typical residence time as a function of production capacity for different reactor configurations.

STR are basically vessels with some type of mixing supplied. Some of the configurations are illustrated in Figure 6–6. The actual vessel configuration may remain the same but these can be operated *batchwise, semi-batchwise,* or *continuously*. As seen in Figure 6–2, the mode of operation changes the characteristics.

Table 6–1 summarizes which configuration seems to be appropriate considering the **phases** present during the reaction.

Table 6–2 lists some general comments about the applicability of different configurations as different issues become important: reaction rate, selectivity, ability to control the temperature, capacity, and flexibility in operation.

6.1-4 Application of the Criteria to Select a Reactor Configuration.

Table 6–3 summarizes some of the initial considerations and how they affect the choice. Then, use the information in Tables 6–1 and 6–2 and the general background information given in this Chapter to select some options.

Example 6–1:   Select a configuration for Case 1: the dehydrogenation of toluene to make benzene.

An Answer: Based on the criteria in Table 6–3, gas at relatively high pressure and temperature, select a PFTR configuration with catalyst packed in tubes. May operate adiabatically or remove some of the heat.

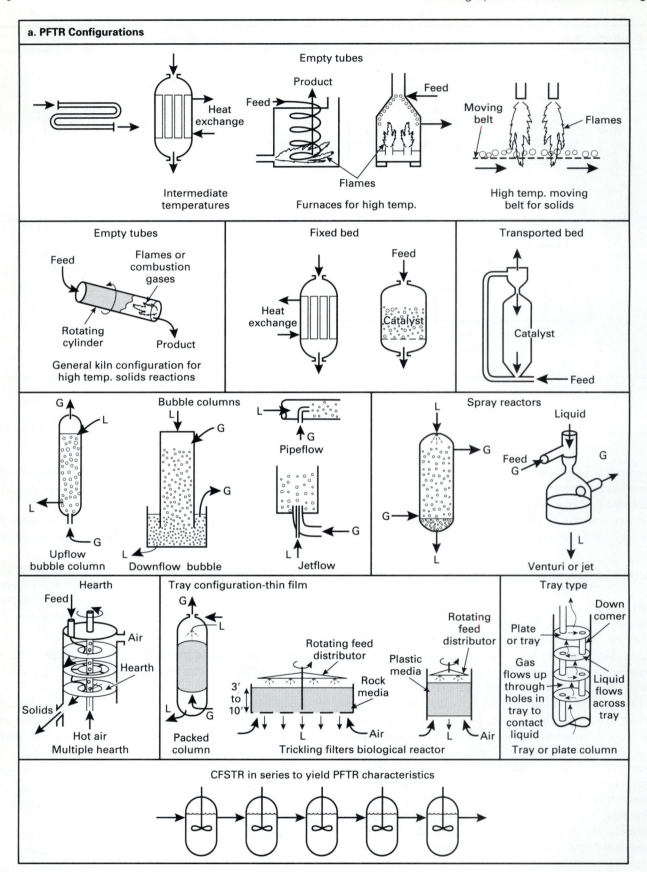

Figure 6–6　Sketches of Reactor Options
a) PFTR Configurations

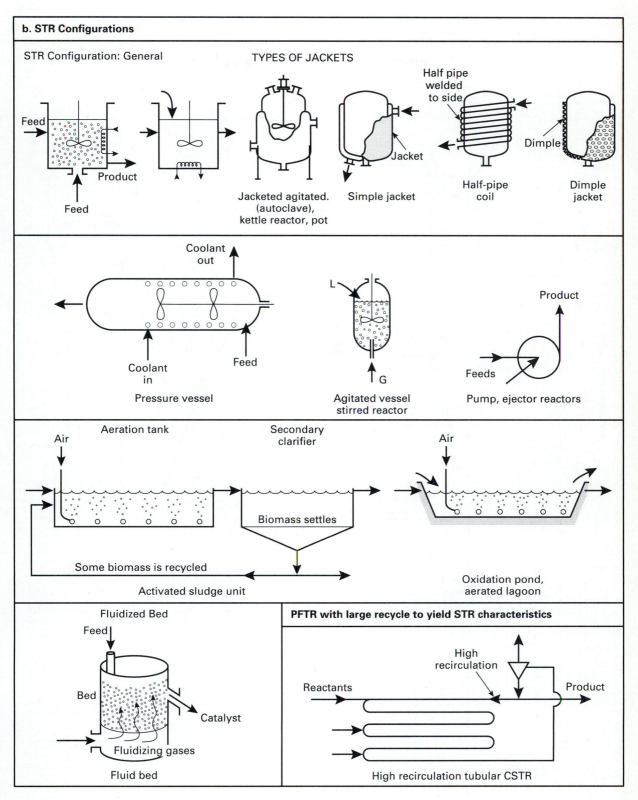

b. STR Configurations

STR Configuration: General

TYPES OF JACKETS

Half pipe welded to side

Dimple

Feed

Product

Feed

Jacket

Jacketed agitated. (autoclave), kettle reactor, pot

Simple jacket

Half-pipe coil

Dimple jacket

Coolant out

Coolant in

Feed

Pressure vessel

L

G

Agitated vessel stirred reactor

Product

Feeds

Pump, ejector reactors

Aeration tank

Secondary clarifier

Air

Biomass settles

Some biomass is recycled

Activated sludge unit

Air

Oxidation pond, aerated lagoon

Fluidized Bed

Feed

Bed

Catalyst

Fluidizing gases

Fluid bed

PFTR with large recycle to yield STR characteristics

High recirculation

Reactants

Product

High recirculation tubular CSTR

Figure 6–6   (Continued)
b) STR Configurations

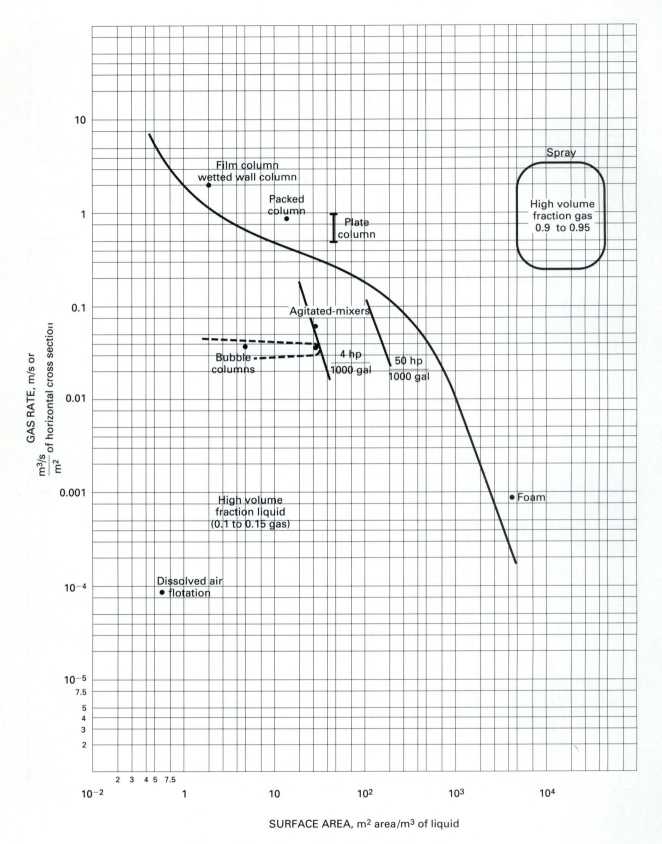

Figure 6–7 Some Characteristics of Gas-Liquid Reactor Systems:
Capacity and Surface Area

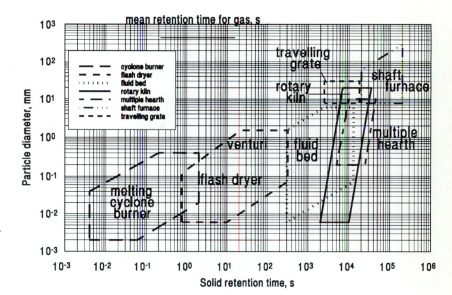

Figure 6–8 Some Characteristics of Gas-Solid Reactors (from Scaleup of Chemical Processes, A. Bisio and R. L. Kabel, eds., 1985 copyright © John Wiley & Sons, Inc. Reprinted by permission of John Wiley & Sons).

Example 6–2: Select a configuration for Case 2: cracking of acetic acid to make ketene.

An Answer: Based on the criteria in Table 6–3, gas at low pressure and high temperature with no catalyst, select an empty PFTR configuration. To maintain a relatively constant temperature and to minimize the pressure drop, use a central coil in a furnace.

Example 6–3: Select a configuration for Case 3: alkyllation of isobutane to produce alkyllate.

An Answer: Based on the Criteria in Table 6–3, liquid at room temperature and high pressure and liquid-liquid system, use a CFSTR configuration. Because this is a relatively clean system, put the tubes inside the reactor.

Example 6–4: Select a configuration for Case 4: the hydrogenation of fish oils.

An Answer: Based on the criteria in Table 6–3, a gas-liquid reactor with suspended solid catalyst with a fair amount of heat generation, use a batch STR configuration. Consider ways of redispersing the hydrogen that bubbles up to the top and fills the vapor space. Include cooling coils but explore ways to clean the coils because there probably will be fouling of the heat exchange surfaces. When the charge is dumped, filter out the solid catalyst.

Example 6–5: Select a configuration for Case 5: suspension polymerization of vinyl chloride to make polyvinylchloride.

An Answer: Based on the criteria in Table 6–3, a liquid-liquid reactor with close temperature control, consider a batch STR or a series of CFSTR. Note that the water was added as a means of trying to control the removal of heat and thus control the temperature.

6.2 SUMMARY

The selection of a reactor configuration draws on fundamental knowledge about the reactions, the desired conditions, and the compromises needed to try to maximize yield, productivity, and quality of the product safely and for a minimum cost.

The issues that affect the choice include the types of phases present, the operating temperature and need for control of the temperature, the operating pressure, the mixing requirements, and the contact times for reaction.

Different equipment options and their characteristics were described. The issues and the equipment characteristics were then used to select possible reactor configurations.

Details about how to **size** reactors and about the selection and sizing of mixers are given by Woods (1993).

6.3 REFERENCES

KABEL, R. L. 1985. "Selection of Reactor Types." Chapter 7 in *Scaleup of Chemical Processes,* ed. A. Bisio and R. L. Kabel. New York: J. Wiley and Sons.

ROSE, L. M. 1981. *Chemical Reactor Design in Practice.* Amsterdam: Elsevier Scientific Pub. Co.

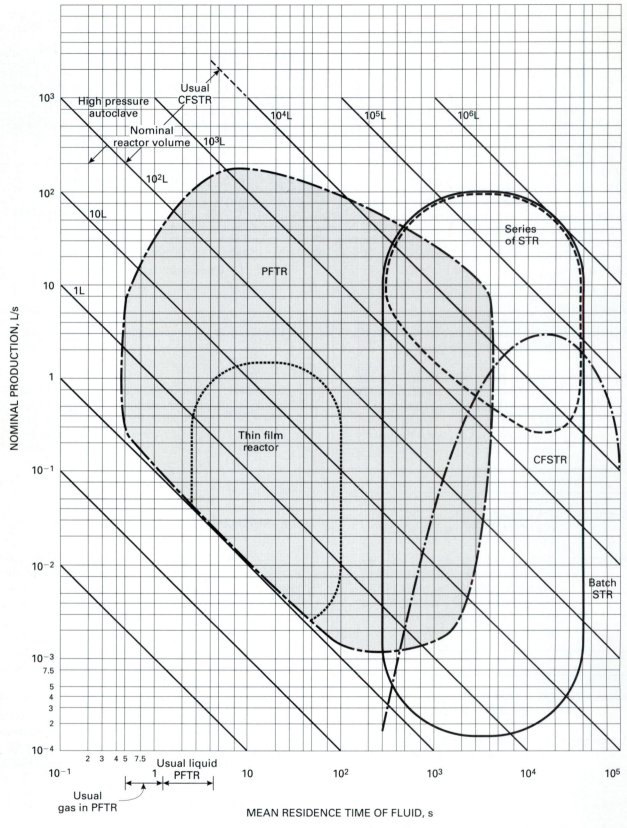

Figure 6–9   Some Characteristics of Fluid Processing Reactors (based on Trambouze et al. (1988) and used with permission from Editions Technip)

TABLE 6-1. How the phases present relate to configuration

Configuration		Homogeneous			Heterogeneous					
		G	L	S	G-L	L-L	G-S	L-S	GLS	
PFTR	empty tube	+++	+							
	fixed catalyst bed						+++	++		
	transport bed/slurry reactor						+	++		especially when solids react
	bubble reactor				++	++				
	spray reactor				++		+	+		
	hearth						++			
	tray config.				+	+			trickle bed	
	thin film				+	+				
STR	batch		+							
	semi-batch		+		+	+		+		especially when product is in a different phase: ppt or gas
	CFSTR		+++		+	+		+		
	fluidized bed						+			

TRAMBOUZE, P., et al. 1988. *Chemical Reactors: design, engineering and operation.* Paris: Editions Technip.

WOODS, D. R. 1993. *Process Design and Engineering Practice: Reactions, Mixing, and Size Change.* Hamilton ON: McMaster University.

6.4 EXERCISES

6-1 For the PIDs given in this book that have a reactor, rationalize the choice or summarize the implications about the reaction.

6-2 Criteria for selecting reactor configurations include the proposed "operating conditions" for the reactor. Table 6–4 gives the conditions for target reactions for some of the processes given in the PIDs for the processes featured in this book.

    a) Apply Le Chatelier's Principle as a possible source of a rationale for the reaction conditions.

    b) Apply thermodynamic equilibrium calculations as a possible source of a rationale for the reaction conditions.

    c) For each, identify possible side reactions that we want to repress. For each, consider first Le Chatelier and then thermodynamics in a search for reasons for the operating conditions.

    d) Consider reaction rate.

    e) Rationalize the choice of reactor conditions.

6-3 Process Flowsheet Development

The process flow diagram for a hydrodealkylation plant to convert toluene to benzene is featured in the next section of this chapter and PID-6. Figures 6–10 and 6–11 represent options shown in the literature for the "same" process. However, Figure 6–10 shows two columns (instead of the three shown in PID–6). These are a stabilizer column, to remove the fuel gas, followed by a column to separate the benzene from the recycle toluene. This flowsheet also shows a lot of energy integration. It also injects "hydrogen" into the reactor to try to control the temperature.

    Figure 6–11 shows only one column to separate benzene from toluene. There is some energy integration.

TABLE 6-2. Characteristics of some configurations

		Selectivity		Heat of reaction	Temp. Control	Reaction Rate	Flexibility	Capacity
		for reaction	of catalyst					
PFTR	Empty tube	for consecutive	Not Applicable	Primarily for Endo-thermic Not Exo-therm	poor to fair	fast	Not very flexible; requires more process integration	
	Fixed catalyst bed	for consecutive	better selectivity & activity but must be stable		usually operate adiabatically; poor to fair	fast		
	Transport bed/slurry react	for consecutive	use if catalyst deactivates rapidly		poor OK for endo	very fast		
	Bubble react				usually good	slow	not for viscous	high
	Spray react					fast		high gas
	Hearth					fast		high gas
	Tray config.				poor to fair	fast		high gas
	Thin film				use for highly exo or endo in viscous liquids			
STR	Batch	for consecutive		Primarily for Exo-thermic	very good	slow	very good	small
	Semi batch	for parallel	backmixing tends to be unfavorable for catalyst selectivity and activity		very good	high or if one component sparingly soluble	very good	
	CFSTR	when need low conc. of reactants			very good	slow to moderate	very good	relatively small for gas-liquid
	Fluidized Spouted	not for consecutive			very good	very fast	poor	

The temperatures, pressures, heat loads, and flowrates in kg/s are shown for various streams.

a) Compare PID–6 (with 3 columns) with Figure 6–10 with only 2 columns. What is the purpose of the third column T-3 in PID–6? In Figure 6–10, column T-3 has been removed. Comment on how Figure 6–10 has been or should be corrected to account for this deletion and the implications.

b) Figure 6–11 shows only **1** column—the benzene-toluene separator. It does not have the stabilizer, column T-1. Comment on how Figure 6–11 has been or should be corrected to account for this deletion and the implications.

6-4 Process Flowsheet Development

For the hydrodealkylation process featured in this chapter, and the process given in Figure 6–11.

a) Consider column T-2 in Figure 6–11. The column pressure is 0.2 MPa and the overhead temperature is 105°C. Quantitatively identify what the species are coming overhead. Quantitatively rationalize the choice of this as the operating pressure for the column.

b) Consider column T-2 in Figure 6–11. Estimate the diameters of the lines on the distillate

  (1) from the top of the column to the overhead condenser, E-6 (the flow in this line is 5.325 kg/s) _____;

TABLE 6-3. Some initial considerations

Feature	Possibility	Implications
State	Gas Gas-solid	usually use PFTR unless extremely exothermic (then consider fluidized bed)
	Liquid	usually use STR unless pressure is high
Pressure	low	state usually dictates
	high	PFTR becomes basic because of sealing around mixer shaft
Temperature	low medium high > 500°C	— previous factors dictate — use heating medium — use furnace configuration or kiln
Type of mixing	PFTR—none	— heat removal more difficult — prefer endothermic reactions — RTD narrow which is preferred for polymerizations
	STR—complete	— heat removal easier; preferred configurations for exothermic reactions — RTD broad — concentration low & uniform
Temperature control	adiabatic—none	— easy to build — used for endothermic (where heat addition may be difficult) or where secondary reactions at high conversions need to be avoided — may need to use inter bed heat exchange or incremental feed or heat sinks
	isothermal—complete	— more difficult to build/operate — use for exothermic
Catalyst	robust	— fixed bed OK

(2) from the reflux drum D-3 to the reflux/product pump P-3 _____;

(3) from the pump discharge to the tee _____;

(4) from the tee back into the column (the flow in this line is 3.023 kg/s) _____;

(5) from the tee to overhead product storage (the flow is 2.302 kg/s) _____.

What is the reflux ratio for this column? Is this reasonable?

c) Consider column T-2 in Figure 6–11. Draw on the diagram the process control schemes you would recommend for the column. Assume that the temperature near the bottom of the column is sensitive to composition. Include detail such as the diameter of valves and associated lines and valves you would recommend.

d) Consider column T-2 in Figure 6–11 and the process control system you recommended in part c. Consider the safety aspects of this column with particular emphasis on start up, shut down, overpressure, "air" failure, and power failure. Mark on the diagram any addi-

tional features you would add, and for all valves indicate whether they should be FO for fail open or FC for fail close.

e) Consider column T-2 in Figure 6–11 and the overhead condenser E-6. Do a four-minute estimate of the required area. Cooling water is on the tube-side.

6-5 Process Flowsheet Development

For the hydrodealkylation process featured in this chapter, and the process given in Figure 6–10.

a) Consider Figure 6–10, where the engineer has recommended extensive "energy integration." To appreciate this, start at the reactor exit at 649°C and note that this stream is used to heat the reboiler on the bottom of column T-1, then to heat the reboiler in the bottom of column T-2, then to preheat the feed to column T-1 in exchanger E-3 and then to preheat the feed to the furnace F-1 (in exchangers E-1).

(1) Why is the heating done in this sequence? Is there another sequence that you think might be preferred?

(2) Note that the columns are controlled by bypassing

TABLE 6-4 Data for reactions for some of the PID processes

Reactions	Temp.°C	Press.MPa	Heat react MJ/kmol	Comments
PID-1A Ethylene: $C_2H_6 <=> C_2H_4 + H_2$ (1)	800 to 860	0.2 to 0.6	144	no catalyst
PID-1B CO_2 removal: $(CO_2)_{aq} + H_2O <=> HCO_3^- + H^+$ (2) $HCO_3^- <=> CO_3^= + H^+$ (3) $H_2O <=> H^+ + OH^-$ (4) amine: $RNH_2^+ <=> H^+ + RNH$ (5) $(CO_2)_{aq} + RNH <=> RNCOO^- + H^+$ (6) Alkali carbonate: $K_2CO_3 + H_2CO_3 <=> 2KHCO_3$ (7)	32	2.8 to 4.5		
PID-3: sulfuric acid $S + O_2 \rightarrow SO_2$ (8) $SO_2 + 1/2\ O_2 <=> SO_3$ (9) $SO_3 + H_2O \rightarrow H_2SO_4$ (10)	970 420 to 600 100	atmos + atmos + atmos +	-298 -98	no cat. vanadium pentoxide catalyst no cat.
PID-4: ammonia primary steam reforming: $CH_4 + H_2O <=> CO + 3H_2$ (11) $CH_4 + 2\ H_2O <=> CO_2 + 4H_2$ (12) secondary steam reforming: $CH_4 + O_2 <=> CO_2 + 2H_2O$ (13) Water gas shift: $CO + H_2O <=> CO_2 + H_2$ (14) $COS + H_2O <=> CO_2 + H_2S$ (15) Methanation: $CO + 3H_2 <=> CH_4 + H_2O$ (16) $CO_2 + 4H_2 <=> CH_4 + 2H_2O$ (17) Synthesis: $3H_2 + N_2 <=> 2NH_3$ (18)	760 to 830 980 390 to 430 220 to 260 300 to 400 400 to 465	1.3 to 4.5 1.3 to 4.5 2.8 to 4.5 2.8 to 4.5 2.8 to 4.5 32	+225 -891 -38 -205 -176 -92	Ni catalyst Ni catalyst iron oxide catalyst Ni catalyst promoted iron oxide catalyst

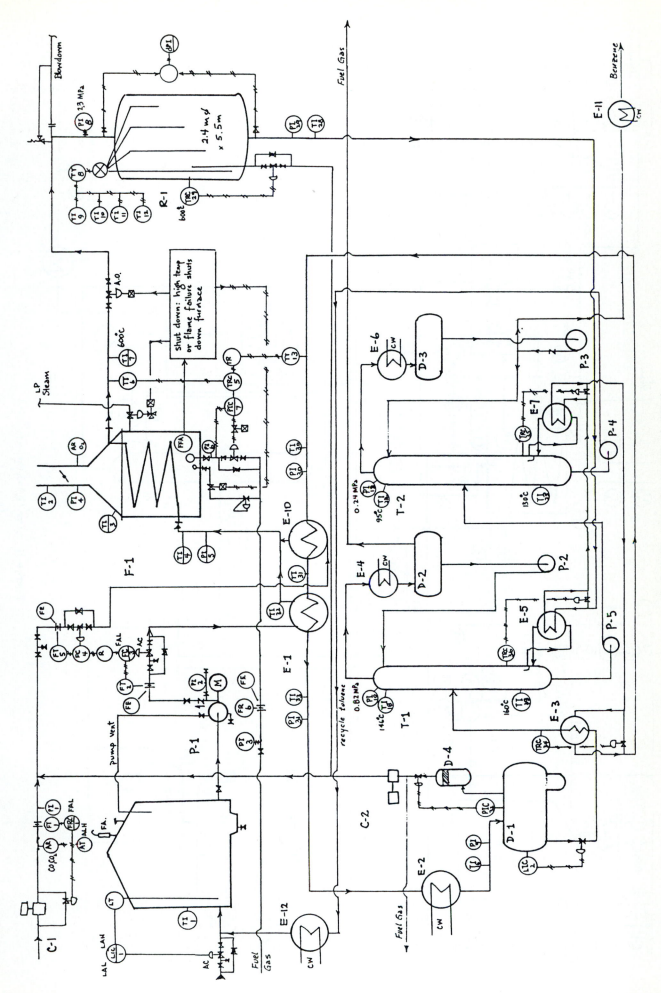

Figure 6–10 A Two-Column Option for PID-6

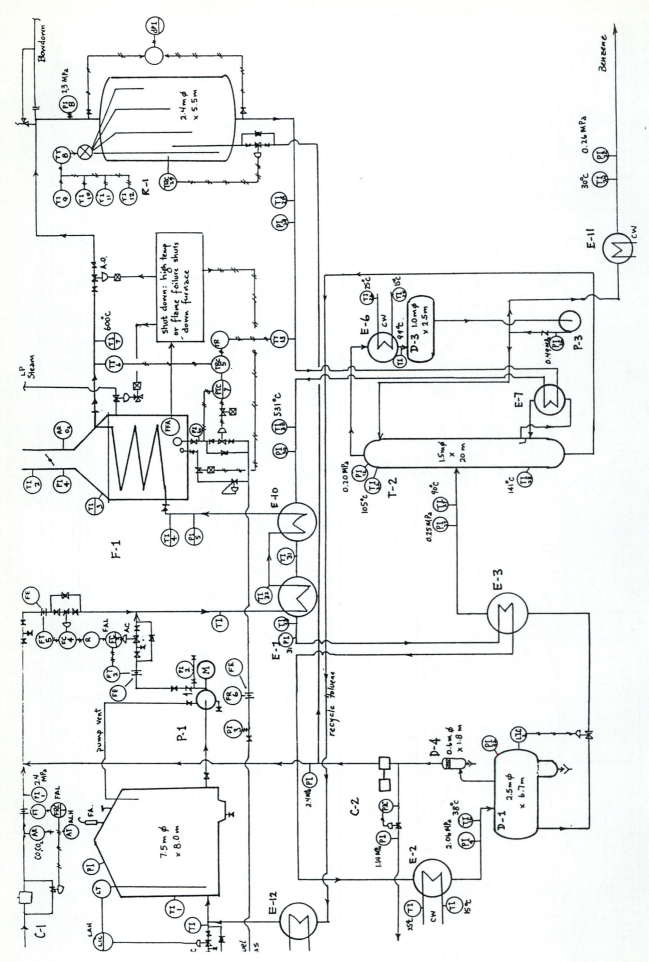

Figure 6–11 A One-Column Option for PID-6

the reboilers; what faults or challenges might this introduce?

(3) For column T-2, the gas stream cools from 609°C to about 481°C. Comment on the boiling characteristics you would predict for the reboiler and the implications. Do you have any recommendations?

(4) For this overall system, comment on how you would start the process up.

b) Consider Figure 6–10 and the pressure drop from the inlet to the reactor (at 2.3 MPa) to the exit after heat exchangers E-1 on the furnace feed preheaters. Assume that the pressure drop is dominated by the pressure drop through the exchangers and **not through the pipe and fittings.** Estimate the pressure in between each piece of equipment

(1) after the reactor _____

(2) after the reboiler in column T-1 _____

(3) after the reboiler in column T-2 _____

(4) after the feed preheater to column T-1 _____

(5) after the feed preheater E-1 to the furnace _____

c) Consider Figure 6–10 and the reactor. The engineer proposes that about 0.32 kg/s of the recycle "hydrogen," is injected into the reactor as "quench" gas. What is its function? Rationalize why you would recommend or **not** recommend this arrangement.

6-6 For the ammonia process, **PID-4**, in the Methane-steam reformer, shown in Figure 6–12, the total feedrate to the reformer is 25% higher than the instrument reads. The readings on the other instruments usually are: inlet reactant gas temperature TI-4 = 454°C (850°F) with the exit CH_4 concentration of 9.5 mol %. The tubewall temperature usually is 960°C (1765°F). The design Δp = 345 kPa (50 psig). Which of the following observations accurately describe the current situation?

A. Nothing, everything works fine because the controllers adjust to give the same exit temperature.

B. The tube wall temperature will be about 30°C higher

because the controllers adjusted to give the same exit temperature. The methane concentration is 7.5 mol %.

C. The tube wall temperature will be about 23.5°C lower than usual because of the cooling effect of the larger mass of gas flowing through.

D. The inlet gas temperature will increase by 38.2°C because of the controller action.

E. With controller action, the methane at the exit will increase to 10.5 mol % and the tubewall temperature will remain at 960°C.

F. With controller action, the pressure drop will be about 500 kPa.

G. With the controller action, the exit product gas temperature will be 20°C lower than the design value.

H. Other.

6-7 For the ammonia process, **PID-4**, in the Methane-steam reformer, shown in Figure 6–13, the total feedrate to the reformer is 25% higher than the instrument reads. The readings on the other instruments usually are: inlet reactant gas temperature TI-4 = 454°C (850°F) with the exit CH_4 concentration of 9.5 mol %. The tubewall temperature usually is 960°C (1765°F). The design Δp = 345 kPa (50 psig). If we operated to keep the exit methane concentration the same, which of the following observations accurately describe the current situation?

A. Nothing, everything works fine because the controllers adjust to give the same exit temperature.

B. The tube wall temperature will be about 30°C higher because the controllers adjusted to give the same concentration.

C. The tube wall temperature will be about 23.5°C lower than usual because of the cooling effect of the larger mass of gas flowing through.

D. The inlet gas temperature will increase by 38.2°C because of the controller action.

E. With controller action, the methane at the exit will increase to 10.5 mol % and the tubewall temperature will remain at 960°C.

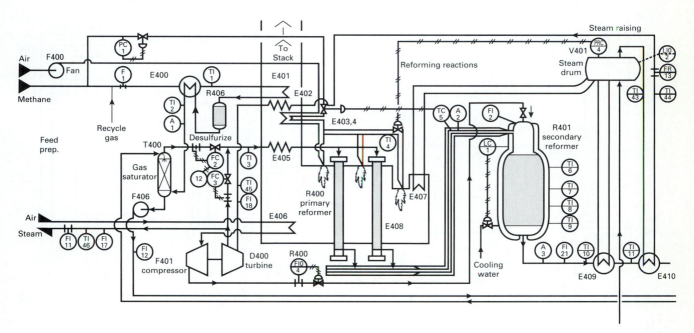

Figure 6–12 Details from PID-4, Ammonia Reformer Control by Temperature

F. With controller action, the pressure drop will be about 500 kPa.

G. With the controller action, the exit product gas temperature will be 20°C lower than the design value.

H. Other.

6-8 For the ammonia process, **PID-4**, in the Methane-steam reformer, shown in Figure 6–12, the steam to gas ratio is 4:1 instead of 3.5:1. The design values are inlet gas temperature TI-4 = 454°C (850°F), the exit CH₄ concentration of 9.5 mol %, the tubewall temperature usually is 960°C (1765°F) and the design Δp = 345 kPa (50 psig). Which of the following observations accurately describe the current situation?

A. Nothing, everything reads the design values.

B. With controller action to yield the control value of the exit gas temperature, the tube wall temperature will be about 8°C higher and the exit methane concentration drops to 7.9 mol %.

C. With controller action to yield the control value of the exit gas temperature, the tube wall temperature remains about the same as the design and the exit methane concentration increases to 9.8 mol %.

D. With controller action to yield the control value of the exit gas temperature, the tube wall temperature will increase by about 3°C and the exit methane concentration remains about 9.5 mol %.

E. With controller action to yield the control value of the exit gas temperature, the tube wall temperature will decrease by 10°C and the exit methane concentration increases to 10.1 mol %.

F. Other

6-9 For the ammonia process, **PID-4**, the reactions in the reformer are:

Steam methane reforming:

$$CH_{4\,(g)} + H_2O_{\,(g)} <=> CO_{(g)} + 3\,H_{2(g)} \quad (1)$$

and the water-gas shift reaction:

$$CO_{(g)} + H_2O_{(g)} <=> CO_{2(g)} + H_{2(g)} \quad (2)$$

The thermocouple for the exit gas temperature TC-5 reads 825°C (1515°F).

Samples are taken of the exit gas and analyzed in the laboratory. The equilibrium constants for the reactions are estimated to be 350 for Eq. 1 and 0.95 for Eq. 2. From Figure 6–14, the temperature corresponding to these equilibrium values are 831°C and 826°C, respectively. Which of the following statements, conclusions, or observations accurately describe the situation?

A. The exit gas thermocouples must be broken, it must be reading low.

B. The data are all consistent.

C. The gas analyses are wrong and new samples should be taken.

D. I can't tell from the data; the evidence is inconclusive.

E. The data are for equilibrium conditions; rate data are needed before I can make any sense out of the evidence.

F. The catalyst is poisoned.

G. Too much steam is present; the steam to hydrocarbon ratio should be reduced from 5.2:1 to 3.5:1.

H. Other.

6.5 PROCESS & INSTRUMENTATION DIAGRAM FOR A HYDRODEALKYLATION PROCESS ILLUSTRATES APPLICATIONS

The process selected for this chapter has interesting implications placed upon it from the choice of "conversion per pass" through the reactor.

PID-6: Hydrodealkylation of Toluene to Produce Benzene:

Toluene is produced from a variety of sources. Unfortunately, we would prefer to have benzene. Thus, an attractive process is to convert the toluene to benzene via hydrodealkylation or the use of hydrogen to break off the

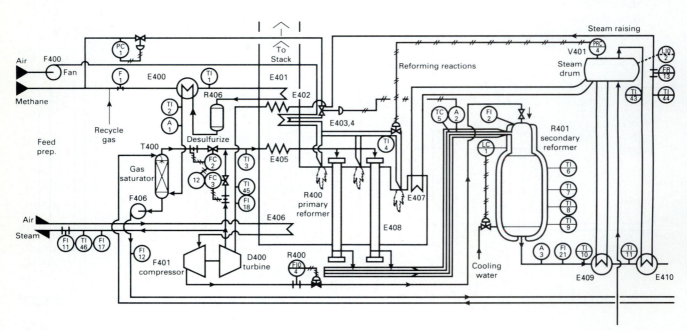

Figure 6–13 Details from PID-4, Ammonia Reformer Control by Composition

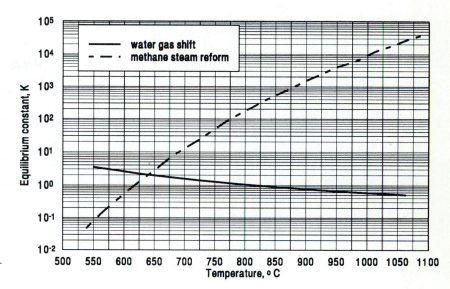

Figure 6–14 Equilibrium Constants for Methane-Steam Reforming and for the Water Gas Shift Reaction

methyl group from the benzene ring. The process is rather straightforward. The target reaction is

$$C_7H_8 + H_2 \longrightarrow C_6H_6 + CH_4 \qquad (6A\text{-}1)$$

Hydrogen and toluene are heated up to reaction temperature and pressure, pass over a catalyst that dealkylates the toluene. The reactor conditions can be changed so that the "conversion per pass" can be 20% to 80%. If the conversion per pass is 50%, then the typical reactor exit gas stream has the following mol % composition:

hydrogen	63.4
methane	25.8
ethane	5.03
propane	1.02
butane	0.52
benzene	1.25
toluene	2.02
xylenes	0.90

The exit conditions are a temperature of 649°C and a pressure of 2.2 MPa. (The inlet conditions were 600°C and 2.4 MPa.)

The feed to the reactor is maintained between 5:1 and 8:1 mol ratio of hydrogen to toluene. The makeup gas feed is, in mol %:

hydrogen	80
methane	10
ethane	5
propane	3
butane	2

The target reaction is first order with respect to toluene and 1/2 order with respect to hydrogen.

Thus, a major function now is to separate and recycle the unreacted hydrogen and toluene. What is interesting is that the two components to be recycled are almost on the extremes, in terms of boiling temperature relative to the desired product, benzene: hydrogen and toluene. **PID–6** illustrates the process.

What makes this process fascinating is that the "optimization" of the process is often done based on the "conversion per pass" through the reactor. The conversion per pass affects the size of all the downstream and recycle processing equipment.

The cost contributions are given in Tables 6A–1 and 6A–2. The raw material cost contribution is interesting in that it dominates. The credits for byproducts comes from the fuel gas from the overhead of tower **T-1**. The value would be increased if this could be upgraded and used as a hydrogen stream, instead of as fuel gas.

For more about determining the optimum conversion per pass and more about this process, see

Amoco-Notre Dame Case Study #8 Their results suggest that 50% conversion is optimum.

"Benzene Design Problem" in "Encyclopedia of Chemical Technology and Design." Vol. 4. ed. J. McKetta. New York: Marcel Dekker. 182 ff. They suggest that 75% conversion is optimum.

G. L. Well's books *Process Engineering with Economic Objectives* and *Process Plant Design* (London: Leonard Hill Publishers, 1973.) The conversion he uses is about 75%.

J. M. Douglas. 1988. *Conceptual Design of Chemical Processes* (New York: McGraw-Hill) provides a detailed look at the design of this process.

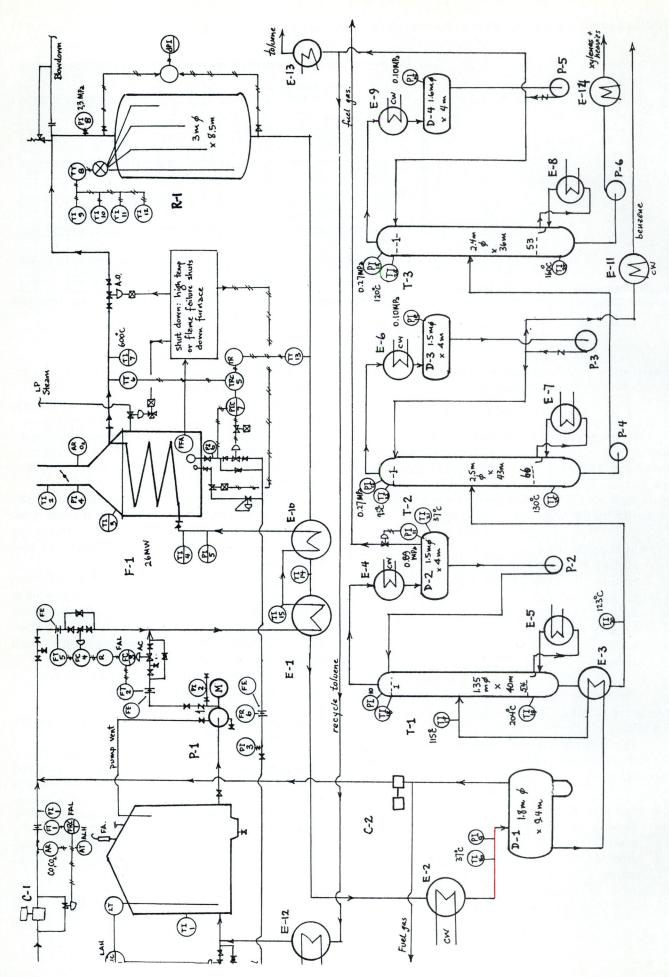

Figure PID–6 Hydrodealkylation of Toluene to Produce Benzene

TABLE 6A-1 Cost Contributions to the Cost per tonne of Benzene product.

Contribution from	Approximate Unit Cost breakdown, %
Raw material	100.3
Utilities	3.0
Catalyst and chemicals	0.1
Labor	1.2
Credit for byproducts	-10.7
Maintenance	0.9
Supervision	incl in labor
Depreciation	2.1
Indirectly attributable costs	3.1
General expense	
Total	100

TABLE 6A-2 Capital Cost Distribution for the Cost to construct a battery limits Facility to produce benzene from toluene. (A 320 m^3/ stream day of toluene feed cost an estimated $16 million ± 30% (MS=1000)*)

Contribution from:	Battery Limits Approximate Capital Cost Breakdown, %
Furnace and effluent heat exchanger	57.6
Reactor	13.5
Compressor and feed storage	15.6
Drums, towers and pumps	13.3
Total	100

* MS is the Marshall and Swift construction cost inflation index. Its value for construction in the process industry was 100 in 1927, was about 300 in 1970 and is 1000 in the early 1990s. Current values of the index can be found in Chemical Engineering magazine. The MS index works well for relating capital costs from one time to another. To estimate the capital cost for the time of interest, multiply the cited cost by the ratio of the MS index at the time of interest/1000. For example, in the first quarter of 1992, the MS index value was 932.9. Thus, the cost of a toluene hydrodealkylation plant at that time would be about 932.9/1000 x $16 million.

Appendices

APPENDIX I: INTRODUCTORY IDEAS ABOUT SELECTING MATERIALS OF CONSTRUCTION

To estimate the cost of the processes we design and the equipment we select, we must specify the materials of construction. As illustrated in Figure 4–7c, this use of alloys can make a dramatic difference in the overall cost of the process and thus the economic viability.

Naturally, we try to select materials that will be inert, that will maintain their strength for the temperature and pressure of the operating system. However, most materials used to construct process equipment react with the materials that are being processed. Hopefully, the rate of reaction is so small that the processed product is not contaminated and the equipment doesn't disintegrate.

Although the rate of reaction or "corrosion" depends on temperature, the amount of vibration, the velocity of the flowing species, the existence of stagnant regions, the conditions at the welds or junctions or joints, the hardness of the water, and trace amounts of specific "spoiler" chemicals, the dominant initial considerations for the choice of materials of construction are:

1. **oxidizing vs. reducing** conditions (some materials of construction *require* oxygen if they are to be corrosive resistant). For example, oxidizing conditions can usually be determined from the oxygen content, or the presence of oxidizing ions or salts (such as ferric or cupric ions or chromate salts), or oxidizing acids (such as nitric or concentrated sulfuric acid). However, care is needed. Figure I–1 shows that sulfuric acid is reducing for concentrations 0 to 70% and oxidizing for concentrations 70 to 100%.
2. the **pH**
3. the presence (or absence) of **chloride ions** and other salts.
4. the juxtaposition of two dissimilar metals so that an electrochemical cell is inadvertently created. The cell operates like a battery and the material that is the anode will react or corrode.

ORGANICS

Many processes process organic fluids. For most organics, carbon steel (mild steel) can be used for the materials of construction. Exceptions are given in the general data listed for each chemical in **Data** Part D a. These data assume that **no** spoiler species are present. For some organics, water is the spoiler chemical, which can have a dramatic effect even if only trace amounts of water are present. In general, be cautious in using carbon steel or stainless steel for organic acids or chemicals that contain an organic group that is an anion for a common acid. This can be especially troublesome at temperatures near the boiling temperature of the liquid. Examples include: acetyl chloride, acetylene tetrachloride, allyl chloride, ammonium citrate, ammonium oxalate, amyl chloride, amyl propionate, benzaldehyde, benzochloride, benzene hexachloride; benzene sulfonic acid, benzyl chloride, butyl chloride, carbon tetrachloride, chloroform, chlorophenol, chlorosulfonic acid, citric acid, dichlorodifluoromethane, diethyl ether, dimethyl sulfate, dinitrochlorobenzene, ethyl chloride, ethylene dichloride, ethyl mercaptan, hexachlorobutadiene, hexachloroethane, hexaethyl tetraphosphate, lead acetate, methylene chloride, nitrophenols, potassium oxalate, n-propyl nitrate, trimethyl phosphite, and vinyl chloride. (from Kirby 1980).

AQUEOUS SYSTEMS

Table I–1 illustrates behavior under three different variable conditions:

- oxidizing, non-oxidizing/reducing, or reducing conditions. This is shown across the top of each of the subtables in Table I–1;
- presence of salts: chloride, nonchloride salts, or no salts. This is shown along the vertical side of each of the subtables in Table I–1;
- the pH: acid, neutral, or base. This is shown as three different subtables in Table I–1.

For each of these 27 options, we note the type of chemicals that might give the conditions, the ratings for different materials of construction. These ratings are patterned after the NFPA ratings and is a scale from 0 to 4.

0 means <0.05 mm/a of corrosion (<2 mils per year) or "excellent" corrosion resistance.

1 means 0.05 to 0.5 mm/a (2 to 19 mpy) or "fair to good" corrosion resistance.

2 means 0.5 to 1.25 mm/a (20 to 50 mpy) or "poor" corrosion resistance.

3 means 1.25 to 3 mm/a (51 to 120 mpy) or "severe" corrosion.

4 means > 3 mm/a (>120 mpy).

Example I–1:   If ferric sulfate is added to solution, what material might be acceptable?

An Answer: Ferric sulfate is a nonchloride salt. Hence, we look in the middle row of Table I–1. We note that this is an "oxidizing" environment and that the pH becomes acid. Under those conditions, carbon steel corrodes at level 3; 316 stainless steel, at level 1. Thus, 316 stainless steel might be acceptable.

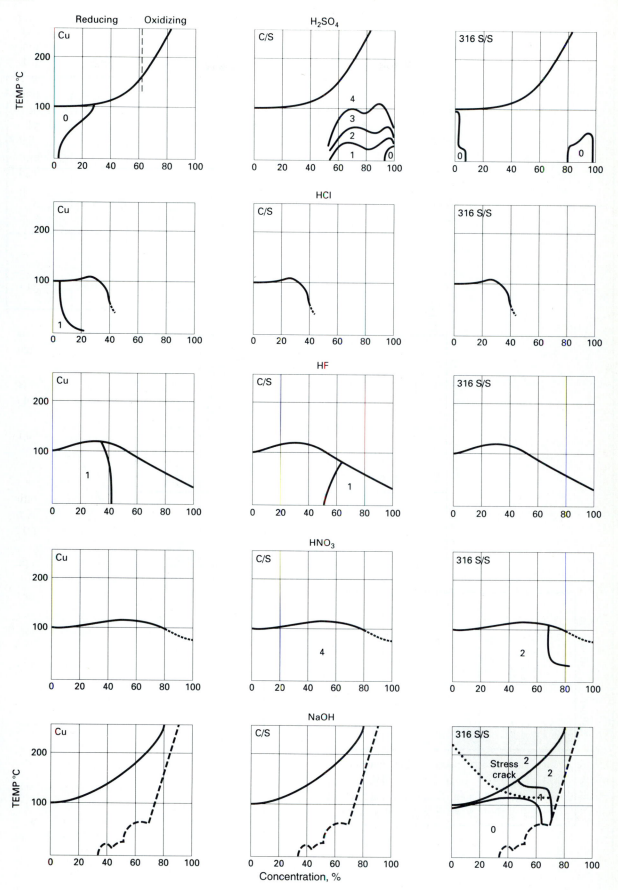

Figure I–1 Corrosion of Copper, Carbon Steel, and 316 Stainless Steel in Strong Acids and Base, As a Function of Concentration and Temperature

TABLE I-1: Environments: Causes and Responses of Copper, Carbon Steel and 316 Stainless Steel.

pH acid < 5

	Oxidizing	Neither	Reducing
Chloride	FeCl$_3$, CuCl$_3$ HgCl$_3$ — Cu c/s 3 s/s 3 — Alloy c Ti, Ta	ZnCl$_2$ NH$_4$Cl — Cu c/s 3 s/s 3	Cu c/s s/s — Ta, Zr, Monel
Non Halide Salts	Fe$_2$SO$_{43}$ AgNO$_3$ — Cu c/s 3 s/s 1	NiSO$_4$ — Cu c/s 3 s/s 1	
No Salts	70 to 100% H$_2$SO$_4$ HNO$_3$ — Cu 3 c/s s/s 0 Zr,	Acetic acid — Cu 1 c/s s/s 1	20 to 70% H$_2$SO$_4$ 10 to 30% HCl 10 to 80% HF H$_2$S — Cu 4 c/s s/s — Monel

pH Neutral 5 to 9

	Oxidizing	Neither	Reducing
Chloride		NaCl — Cu 0 c/s 1 s/s 3	
	NaCrO$_4$ NaNO$_2$ KMnO$_4$ — Cu c/s 0 s/s 0	Na$_2$SO$_4$ NaNO$_3$ — Cu c/s s/s	
No Salts	Cu c/s 1 s/s	Water — Cu c/s 0 s/s 0	0 to 20% H$_2$SO$_4$ 0 to 10% HCl 0 to 10% HF — Cu c/s s/s

pH base > 9

	Oxidizing	Neither	Reducing
Chloride	NaHOCl — Cu c/s 3 s/s 3	Cu c/s 3 s/s 3	
Non Halide Salts		Na$_2$CO$_3$ — Cu c/s 0 s/s 0	
		NaOH	
No Salts	Cu c/s 1 s/s	Cu c/s 0 s/s 0	Cu c/s 0 s/s

STRONG ACIDS-BASE SYSTEMS

Figure I–1 shows the data for the behavior of copper, carbon steel, and 316 stainless steel as a function of temperature and concentration for selected systems. These illustrate that one must be very cautious in trying to generalize. Under some concentrations a metal might be resistant; changing the conditions slightly might alter its behavior dramatically.

Example I–2: We selected 316 stainless steel for a system processing 50% sodium hydroxide at about 90°C. What might happen if the temperature were increased to its boiling point of about 145°C?

An Answer: From Figure I–1, as the temperature increases the system moves into the stress crack corrosion region and the rate of corrosion increases to the 2 to 19 mpy range.

ELECTROCHEMICAL CORROSION

When two dissimilar metals contact one another in a liquid environment, there is the possibility that an electrochemical cell will be created. For such a cell, one of the metals will be the anode (where the "corrosive" reaction occurs); the other becomes the cathode. Table I–2 gives the "electromotive series" for pure elements. This is the thermodynamic, half-cell voltage relative to the hydrogen electrode. The top of the list tends to be anodic relative to those elements farther down the list. The greater the distance apart, the greater the amount of corrosion at the anode. For example, iron is anodic relative to copper. Thus, if copper sheeting is bolted with iron bolts, the iron will corrode and the joint will fall apart. If the reverse were used (copper bolts to hold iron sheets together), the iron would still corrode. However, the impact on the structure would not be as devastating because the material being corroded has the large surface area.

In practice, however, the materials of construction are not pure elements; the environment is not one in thermodynamic equilibrium. Thus, we obtain data for "real" alloys immersed in different conducting liquids. For brine, Table I–3 provides one example environment; the data are called the "Galvanic series." The interpretation is the same; alloys high on the list serve as anodes relative to alloys lower on the list. Some alloys can have their surface "passified" and thus the alloys would appear in Table I–3 as two entries: one where the surface is "normal" or active; the other, where we have treated it to behave "passively." Unfortunately, small cracks and imperfections can cause a "passive" surface to become active. Hence, it is wise to assume that the surface is in its *active* state. Thus if two dissimilar metals or alloys *contact* each other and are both in contact with *a conducting* liquid, then the alloys highest in the table will corrode.

We can use this information two ways. We can be sensitive to the effect of mixing metals and the impact this

---

Example I–3: For a condenser we select Admiralty brass tubes and copper heads with carbon steel shell. What might happen?

An Answer: From Table I–3, carbon steel will serve as the anode and hence will corrode relative to brass. Because the entries are relatively far apart, the electrochemical corrosion can be relatively high.

Comment: We could make the exchanger all out of the same material; we could insulate the materials from each other so that there is no metal-to-metal or connecting contact between the metals. We could try to keep the area of the anodic material large—relative to the area of the cathodic. We could put in material that contacts the carbon steel directly and that is above carbon steel in the table so that this will corrode "sacrificially." Zinc or aluminum might serve this role.

---

would have. Second, we can use this to protect the materials of construction by purposely adding an anodic material that will sacrificially corrode. In this instance, the material used to fabricate the equipment functions as the cathode and is thus protected against corrosion.

ILLUSTRATIVE SPOILER CHEMICALS

Trace quantities of certain chemicals can dramatically affect the corrosiveness of the materials. We have seen already the effects of trace amounts of chloride ion in Table I–1. Others include the following:

- hydrogen, which tends to weaken and embrittle some metals.
- hydrogen sulfide that can alter the corrosion rates.
- water with organic chlorides or sulfates on carbon steel or stainless steel.
- mercury salts on Monel in brine environment.
- oil on rubber lining in acid environments.

TEMPERATURE EFFECTS

Some temperature effects are given in Figure I–1 for strong acids and bases. For carbon steel in neutral pH water with

TABLE I-2: Electromotive series for elements representing the equilibrium half-cell potential relative to the hydrogen electrode.

Anodic	
Potassium	+2.9
Calcium	+2.87
Sodium	+2.7
Magnesium	+2.34
Aluminum	+1.67
Manganese	
Zinc	+0.76
Chromium	+0.71
Iron	+0.44
Cadmium	+0.402
Cobalt	+0.28
Nickel	+0.236
Tin	+0.14
Lead	+0.126
Copper	-0.34
Silver	-0.80
Palladium	-0.83
Mercury	-0.85
Platinum	-1.2
Gold	-1.42
Cathodic	

some oxygen present, the corrosion rate doubles about every 20°C.

OTHER MATERIALS OF CONSTRUCTION

In these notes we have focused on copper, carbon steel, and 316 stainless steel. However, there are a wide variety of options (as has been illustrated in the various charts and tables in this section). Others include:

- concrete
- ceramics
- glass
- fiberglass-reinforced plastic
- plastics
- wood
- linings

Table I–4 lists the names, some of the distinguishing composition, and the relative cost of different alloys.

Some alloys are primarily iron based, others are nickel based, and others are copper based. The costs are very general because these depend on whether the cost is per unit volume or per unit mass and what type of equipment is being discussed. The relative costs are approximate and for piping or plate.

TABLE I-3: Galvanic series relating the relative galvanic corrosion of alloys and metals in seawater environment.

Anodic (corrodes)

Magnesium and magnesium alloys Zinc Galvanized steel Commercially pure aluminum (1100) Cadmium
Aluminum 2024
Carbon steel Wrought iron Cast iron
430 stainless 410 stainless
Ni Resist 316 stainless 304 stainless
Lead-tin solder Lead Tin
Muntz metal Manganese bronze Naval brass
Nickel Alloy 600 Alloy 625 Alloy B Alloy C-276 Alloy 2
Yellow brass Admiralty brass
Red brass Copper

Silicon Bronze
Cupro Nickels
G Bronze
Silver solder
 	
     [nickel passive]
     [Alloy 600 passive]
     [410 stainless passive]
     [304 stainless passive]
     [316 stainless passive]
Alloy C
Alloy 3
Titanium
Alloy 400

Graphite Gold
Palladium **Cathodic**

TABLE I-4: Alloys						
Name	Fe	Cr	Ni	Mo	Cu	Relative Cost, $
carbon steel (mild steel)	°					1.00
Ni-resist (cast iron)	°	2	14		6	
ferritic stainless steels:						
430 stainless steel	°	18				
martensitic stainless steels:						
420 stainless steel	°	12				4
410 stainless steel	°	12				
austenetic stainless steels:						
316 stainless steel	°	16	10	2		6
304 stainless steel	°	18	8			
310 stainless steel	°	25	20			
Alloy 20 Cb3 (Carpenter 20; Nicrofer 3620 Cb)	°	20	37	2	3	17
Worthite (Si 3)	°	20	24	3	2	
Nickel-based alloys						
Alloy 200 (Nickel 200)			99			18
Alloy D (Si 10; Hastelloy D)		1	81	3		
Alloy B-2 (Hastelloy B; Chlorimet 2)			61	28		34
Alloy C (Hastelloy C)	6	16	°	17		
Alloy C-4 (Hastelloy C4)	3	15	59	16		
Alloy C-276	6	16	57	16		34
Alloy G (Hastelloy G)		22	45	6	2	
Alloy 625 (Inconel 625; Nicrofer 6020 HMo)	2	21	61	9		16
Alloy 600 (Inconel 600; Nicrofer 600)	7	15	76			
Alloy 400 (Nickel 400; Monel 400; Nicorros 400)	1		66		32	15
Copper and Copper based alloys						
copper						8
brass: Cu-Zn; Cu-Zn-Sn Admiralty: [Zn 29; Sn 1]				70		
bronze: Cu-Sn; Cu-Sn-P; Cu-Sn-Pb; Cu-Ni						
Aluminum						4
Titanium based						
titanium- grade 7 [0.2 Pd]						50
titanium- grade 2						
titanium- grade 12			0.8	0.3		
Zirconium						90
Tantalum						350

APPENDIX II: IMPLICATIONS OF PROCESS CONTROL ON THE SELECTION OF PROCESS EQUIPMENT AND THE CREATION OF PROCESS FLOW DIAGRAMS

The guidelines given for the selection of process equipment and putting equipment together to create a process are, in general, based on steady-state operation. We assume that after we have selected the equipment we will add "controllers" so that everything operates smoothly and easily.

We know, however, that we must select equipment and systems that

- can be started up, and shut down; where we can accommodate transient changes in feed rate and composition—and not just consider "steady-state" operation;
- are safe and easy to operate;
- can be controlled so as to produce on-specification product.

Thus, **process control and operability** should be considered right from the beginning; they should be accounted for in all the decisions and choices made throughout the whole text. Unfortunately, process control guidelines have not been developed sufficiently that we can include "process control" rules of thumb as early in the selection procedure as we would like.

Here I summarize and try to illustrate some considerations.

II.1-1 COMPONENTS IN PROCESS CONTROL

Control requires a measured variable, a controller/control system, and a manipulated or "control" variable. Thus, to control the temperature of a shower, we

- measure the temperature of the water coming from the shower head by a sensor. The sensor is often your hand or your back. "Temperature" is the measured variable.
- adjust the flowrate of cold water to the shower head; the "flowrate" is the manipulated variable.
- the controller tells us "how we should adjust the flowrate of the cold water." Do we turn the cold water on full blast? Do we increase the flow of water just a bit?

In selecting process equipment, the main concerns are not with the control algorithm or equipment. Rather the seven main concerns are:

1. What are the control objectives?
2. What is the variable to be controlled in any piece of equipment? Can it be measured? Can it be measured reliably and easily? Is the "control variable" the same as the "measured variable"?
3. What is the manipulated variable? Can it be varied over a wide enough range to bring the equipment back into correct operation?
4. What is the delay or dead-time between when the sample is taken, the measurement taken, and corrective action taken by manipulation? How long does it take to measure the effect of the disturbance and return the measurement to its desired value by adjusting a manipulated variable? If the disturbance is infrequent and occurs slowly relative to the control response, the control performance will be good.
5. Can the control system for one piece of equipment or section of the plant be prevented from interfering with the control system of another?
6. Can the hazardous operation condition for the equipment be detected easily? Can the hazard be eliminated quickly and reliably?
7. What is the magnitude and type of disturbance with which the system must cope?

Naturally, the combination of measured variable, manipulated variable, and control system must be such that the steady state and dynamic behavior of the equipment and the system must not change with operating conditions or with time. (Rinard, 1982)

a) Typical Measured Variables

Table II–1 summarizes qualitative data about sensors for different measured variables.

b) Typical Manipulated Variables

The most commonly manipulated variable is the valve position of a control valve. This affects the flowrate of reactants, of heating or cooling media, of fuel.

Other options include altering the power to electric heaters and to motors.

The rotational speed can be manipulated by mechanical pulleys and belts or by changing the power to the motor drive.

Ease	MTBF	Delay	Accuracy
Flowrate (valve position)	8		
Electric Power	10		
Speed	8		
Elevation/position			

TABLE II-1: Qualitative Comparison of Sensors for Measured Variables	Ease	MTBF	Delay	Accuracy
Temperature	10	4	10	10
Pressure	9	4	9	9
Force, Tension, compression	8			
Radioactivity	8			
Fluid flow	8	8	10	8
Liquid level	7	8	7	
Solids level	5			
Weight and weight flowrate	5	4		9
Thickness and Displacement	5			
Velocity	5			
Fluid density/specific gravity	4	4	8	7
Viscosity and consistency	4			7
pH concentration/redox potential	6	5	9	8
Electrical conductivity	7			7
Thermal conductivity	7			
Caloric value	4	5	7	7
Humidity of gases	4			
Moisture content of solids	3			
Optical: colour, reflectance	2			
Turbidity	7			7
Speed: machinery	7	8	9	
Composition: NMR,	1	2		
Mass Spec	1	2		
Chromatography	1	2	3	9
Refractive index	4			10

MTBF Mean times between failure: 10 means long time, example 5 years; 0 means short time, example 30 days. For data see Liptak (1986).
Delay means the time required between the occurrence in the system to when we have a signal representing that occurrence. A 10 means very rapid, with negligible delay; a 0 means a delay of say 24 hours.
Accuracy: 10 means about 1/2%; 6 means about 3%; 0 means about 100%.

II.1-2 IMPLICATIONS FOR TRANSPORTATION EQUIPMENT

Because the most commonly manipulated variable for fluid systems is the flowrate, and because the manipulated variable must be capable of being varied over a significant range, process control has a **major** impact on the selection and sizing of fluid transportation equipment. If the fluid circuit is part of a control system, the *size* of the pump or blower might be 50% to 100% larger than if it is not part of the control loop. It must be sized to meet the maximum demand. The major effect is that the pressure drop across the control valve must be significant. Details of this were discussed in Section 2.3–4. Usually we *select* a centrifugal type of device (as opposed to a reciprocating or rotary device) if it is part of the process control system. (Baumann (1981) outlines when to select a control valve approach vs selecting a variable speed pump.)

Furthermore, there are indirect implications from process control systems that affect the transportation sizing and selection. For example, if we are controlling the temperature in a jacketed reactor, then the "manipulated variable" must be capable of making an adequate and rapid re-

sponse. To minimize the dead-time, the cooling water flowrate might be 3 to 5 times higher than the usual rules of thumb suggest.

In solids transportation, process control, environmental safety, and economics affect the choice of equipment. For dusty or hazardous materials one might prefer to transport by pneumatic or hydraulic rather than by belt conveyors. For more see Goodfellow and Berry (1986).

II.1-3 IMPLICATIONS FOR ENERGY EXCHANGE EQUIPMENT

Process control has a negligible direct impact on the selection and sizing of individual pieces of energy exchange equipment. Temperature and energy input are directly and relatively easily measurable. Usually the control scheme has implications for the transportation equipment (as was discussed in Section II.1–2).

On the other hand, energy integration is a common goal for overall processes. By this we mean that an energy need might be supplied by an energy source elsewhere in the plant. For example, Figure II–1 shows a process where

a. A Process With Little Energy Integration.

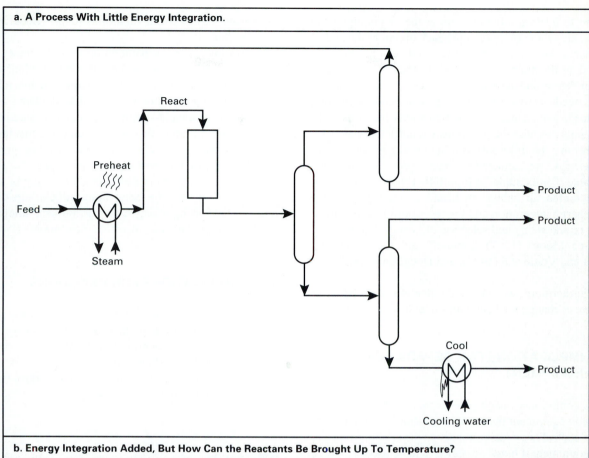

b. Energy Integration Added, But How Can the Reactants Be Brought Up To Temperature?

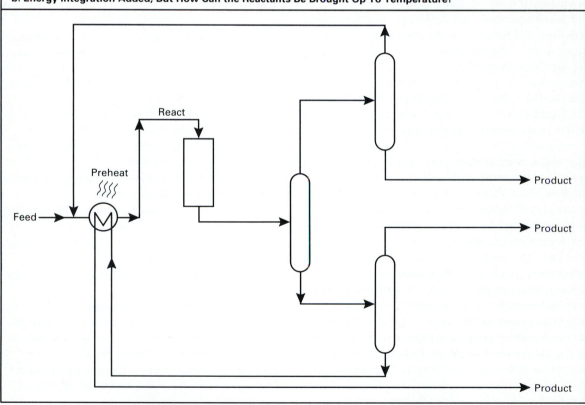

Figure II–1   With Energy Integration, Can the Process Be Started Up?
a) A Process with Little Energy Integration
b) Energy Integration Added, But How Can the Reactants Be Brought Up to Temperature?

we need to heat up a feed stream at the beginning of the process, and yet cool down a product stream at the end of the process. Sometimes these two needs can be mutually satisfied, as illustrated in Figure II–1. Thus, we could save on both steam and cooling water. Thus, considering the energy needs across the whole process and integrating these needs will change the selection and sizing of the individual units (because the heat transfer coefficients and temperatures may be different). But in our enthusiasm to integrate energy we sometimes create processes that are inoperable. For example, in Figure II–1 the processes cannot be started up unless we install a "startup furnace." Thus, energy integration and process control have a direct impact on the sizing and selection of heat exchangers. For more see Moore (1987), Linnhoff and Flower (1978), Linhoff and Vredeveld (1984), and Douglas (1988), Chapter 8.

Furthermore, we must have degrees of freedom so that we can change the heat duties over a wide range.

II.1-4 IMPLICATIONS FOR SEPARATION EQUIPMENT

Process control has a negligible impact on the selection of separation equipment if continuous operation is envisaged. Process control has a major impact on the selection and sizing of equipment if *batch* operation is to be used.

For *batch* operation, the downtime, cleaning time, on-stream time, and process control approaches should be created before or as the equipment is selected and sized. For more see Chowdhury (1988).

For *continuous* operation, the three places of impact are on the deadtime, the relationship between the "control variable" (usually purity) and the measured variable, and on the effect of the control system from one section to another.

The major consideration here is relating the measured variable to the purity of the product, or the quality of the separation. From Table 4–1, for example, we note that only for distillation (where differences in temperature are being exploited) do we have a direct relationship between degree of separation and the measured variable. In other words, we have an "inferential" from which we can infer the concentration. Such inferential variables can rapidly and accurately indicate the condition of difficult-to-measure controlled variables. For all other separation options, we usually have to measure the purity separately by an analyzer and then relate this to other measured variables and hence to the manipulated variables. Hence, based on process control considerations, distillation is the preferred option.

For distillation, not only does the measured variable relate directly to the product purity but temperature is relatively easy to measure and many control systems are available to do this. Indeed, much has been written just on control system design for distillation columns [Buckley et al. (1985), Chin (1979)].

For many other separation options, the purity or degree of separation usually is measured by sampling and analyzing an exit stream. For solvent extraction, this might be after the solvent regeneration stage. The challenge is to relate this measurement—with its associated deadtime—to the operating variables that are measured and manipulated.

Example process control schemes are given by Liptak (1982) for different separation options.

Energy integration also affects the sizing of separation equipment since the temperature and pressure of separation can often be changed to provide heat sources and sinks that might be more attractive. See Moore (1987).

II.1-5 IMPLICATIONS FOR REACTIONS

Control of the reactor is so crucial that usually the process control plays a vital role in the choice of the configuration. Indeed, control of the energy is one of the main criteria for selecting a reactor configuration.

Upsets because of the recycle (and control of those upsets) should be considered as well.

References

BUCKLEY, P. S., W. L. LUYBEN, and J. P. SHUNTA. 1985. *Design of Distillation Column Control Systems*. Research Triangle Park, NC: Instrument Society of America.

BAUMANN, H. D. 1981. "Control Valve vs variable speed pump." *Chem Eng* (June 29):81.

CHIN, T. G. 1979. "Guide to Distillation Pressure Control Systems." Hydrocarbon Process (Oct):145–153.

CHOWDHURY, J. 1988. "Batch Plants Adapt to CPI's Flexible Gameplan." *Chem Eng* (Feb 15):31.

DOUGLAS, J. M. 1988. *Conceptual Design of Chemical Processes*. New York: McGraw-Hill.

GOODFELLOW, H. D., and J. BERRY. 1986. "Clean Plant Design." *Chem Eng* (Jan 6):55–61.

LINNHOFF, B., and J. R. FLOWER. 1978. *AIChE Journal* **24:**633–642.

LINNHOFF, B., and D. R. VREDEVELD. 1984. *Chem Eng Prog* **80,**7:33.

LIPTAK, B. G. 1986. "On-Line Instrumentation and Process Control: Wave of the Future?" *Chem Eng* (March 31):51–71.

LIPTAK, B. G. 1982. *Instrument Engineer's Handbook: Process Measurement Volume*. Radnor, PA: Chilton Book Co.

MOORE, I. J. 1987. "Microcomputer-aided Synthesis of Energy Integrated Distillation Sequences." M. Eng Thesis, McMaster University, Hamilton, ON.

RINARD, I. H. 1982. "A Roadmap to Control-system Design." *Chem Eng* (Nov 29):46–58.

UMEDA, T., et al. 1978. *Chem Eng Prog* **74,** 9:70.

APPENDIX III: ACTUAL PLANT INSTALLATIONS

The specifications for some of the problems are taken from real data where we know from detailed design what equipment and performance actually occurs. This provides useful feedback to us about our quick, order-of-magnitude estimates provided in this text.

2–8: Stoess (1970) p. 58 gives the following options:

a. a vacuum system, 20 cm diameter pipe, and 150 kW. However, this requires a large filter at the exit end.

b. a low-pressure system, 15 cm diameter pipe and 56 kW. The low-pressure system works by using a rotary valve at the base of the storage bin, and screw conveyors to bring the material from each bin to a central input feed hopper and hence to a low-pressure system. Depending on the fluidity of the lime from the screw conveyors, the system may use 20 cm diameter pipe and 53 kW.

2–9: Stoess (1970) p. 61 selects a 10 cm diameter line.

2–10: Stoess (1970) p. 61 selects a 15 cm diameter line and about a 40 kW Rootes blower delivering 650 dm^3/s at 55 kPa.

2–11: On plant trials 31 Mg/h was transported. The line diameter was 10 cm. A key here is not to add molasses to the feed before it was conveyed. When a similar test was done on chick rearing feed containing 5% molasses and with a bulk density of 0.72 Mg/m^3, 21.4 Mg/h was conveyed. However, occasionally there was blocking that stopped the flow, caked material built up on the walls and the throughput was reduced.

2–13: The Black Mesa line is 45 cm diameter with an installed power of 151.83 MW. This line handles this duty.

3–15: Monsanto-Washington Case study 6, provide 4.88 m diameter quenchers that are 6.1 m high for a spray operation. An estimate of the volumetric heat transfer coefficient is 1.4 kW/m^3.°C

3–16: Fair 1972 uses this as a base case example to illustrate calculations for different contacting options. His results are:

- for a packed tower: 1.07 m diameter × 3 m high.
- for a baffle tray tower: 1.07 m diameter with 13 baffles for a total height of 5.79 m.
- for a spray column: 1.07 m diameter with 6.1 m height.

3–17: Bohn (1985) experimentally measured the values to be 2 to 3 × $(F_G)^{1.28}$ kW/m^3.°C

3–18: Jeffreys used a liquid loading of 13.5 kg/s.m^2, a liquid to gas ratio of 8.13:1. The design was a 1.37 m diameter column 2.44 m high packed with stacked packing. An estimate of the volumetric heat transfer coefficient under his design conditions is 2.3 kW/m^3.°C

3–19: A unit that was 3 m diameter and 3.28 m high was installed.

Annotated Subject Index

A

Abrasion, code; definition, 2-39 and **Data for PDEP**, 25; 2-37, -38.

Absorbers: device for gas-liquid contacting so that a soluble species in the gas phase can dissolve in the liquid phase. Many different devices can be devised: columns containing sprays, packings, trays, baffles; liquid-ring pumps or liquid seal reciprocating compressors; or scrubbers where the function may be a combination of removal of particles from the gas and soluble species. If the reverse process of desorbing a species from the liquid into the gas occurs, the device is operating as a Desorber (or stripper).

Absorption, separation method based on dissolving soluble gas species in an "added" absorbing liquid or "entrainer". (see also the opposite process called Desorption). Contrast with adsorption where the species goes to the *surface* and does not dissolve in the bulk phase liquid.

def. 4-33;

pic. 4-36;

selection chart, Fig 4-2, 4-5; Table 4-1, 4-2; Table 4-5, 4-21;

and product purity 4-9;

temperature effects, Fig 4-3, 4-7;

usual size, Fig 4-9, 4-17;

feed-product diagram, 4-36;

characteristics, 4-44;

typical absorption agents, 4-35;

species-agent interaction, data, 4-35;

applications, Fig 4-39, 4-74 and Fig 4-40, 4-77; Table 4-23, 4-75.

physical absorption versus absorption with chemical reaction, 1-21; CO_2 and H_2S removal, 1-21; data for CO_2 absorption, 1-22 to -24;

absorption data of different species by solvents, 1-24 to -25;

related fundamentals: mass transfer and Henry's law.

example applications on P&IDs: PID-1B: CO_2 absorption, T150, T151; PID-3: sulfuric acid, T302, T303; PID-4: ammonia, T401; PID-5: waste water, "ozonator" absorption reactor, R503; aeration basin, R501, 5-77;

Accumulators, see Drums.

Acetaldehyde: [C2H4O] (acetic aldehyde) production capacity, Table 4-4, 4-18;

Acetic acid: [C2H402] (ethanoic acid, vinegar) production capacity, Table 4-4, 4-18;

RO separation Table 4-20, 4-68;

SX, 4-39;

corrosivity, A-4;

molar mass, NFPA, Dow hazard, corrosion rating, freezing temperature, boiling temperature, molar volume, polar and hydrogen bond solubility parameters, pK_a, 4-81; 4-8;

Acetic anhydride: [C4H6O3] production capacity, Table 4-4, 4-18;

Acetone: [C3H60] (2-propanone or dimethyl ketone) production capacity, Table 4-4, 4-18;

RO separation Table 4-20, 4-69;

solubility characteristics, 4-42;

azeotrope with chloroform, 4-26;

Acetonitrile: [C2H3N] (methyl cyanide) adsorption Table 4-9, 4-44

Acetyl chloride: [C2H3ClO] caution about corrosion, A-2;

Acetylene: [C2H2](Ethyne) production capacity, Table 4-4, 4-18;

absorption in liquids, 1-25;

Acidic conditions, and corrosion:

discussion, A-2.

Acids, material of construction for:

application. A-3.

Acid demister, see Demister.

Acid value (AV) of oils:

definition, and data, **Data for PDEP,** 57 and 76.

Acrolein: [C3H4O] (2-propenal) fluidized bed reactor, 3-50;

Acrylic acid: [C3H4O2] (2-propenoic acid or acroleic acid) production capacity, Table 4-4, 4-18;

Acrylics: general drying characteristics, 5-53;

Acrylic fiber: production capacity, Table 4-4, 4-18;

Acryonitrile: [C3H3N] (2-propenenitrile, acrylon or vinyl cyanide) production capacity, Table 4-4, 4-18;

fluidized bed reactor, 3-50;

Activated alumina:

as gas adsorbent, 4-44;

as liquid adsorbent, 4-45;

in the sizing of mass transfer devices. Contrast with the concept "height of a transfer unit". For SX units, HETS is about 0.1 to 1 m except for gravity spray SX units.

Heuristics:
for separation of homogeneous phase: 4-81.
for separation of heterogeneous phases: 5-63.
Hexachlorobutadiene: [C4Cl6] caution corrosion, A-2.
1,2,3,4,5,6-Hexachlorocyclohexane: [C6H6Cl6]
(α-isomer) caution corrosion, A-2.
(γ-isomer, Lindane) caution corrosion, A-2.
Hexachloroethane: [C2Cl6] caution corrosion, A-2.
n-Hexane: [C6H14] (C6) absorption, 1-26;
molar mass, NFPA, Dow, corrosion rating, freezing temperature, boiling temperature, molar volume, polar and hydrogen bond solubility parameters, 4-84;
HGMS: high gradient magnetic separators; see Magnetic separators.
High shear disperser, see Size reduction, liquids.
High tension electrostatic separators, see Electrostatic separators.
Hildebrand solubility parameters: measure of the intermolecular energies and the relative contributions from dispersion, hydrogen bond and polar force contributions. These three parameters, together with the molar volume, are useful in identifying viable options for separating species by solvent extraction, absorption, crystallization, adsorption, and various membrane options. Units of measurement are $(J/cm^3)^{1/2}$ which is equivalent to $(MPa)^{1/2}$.
explanation and definition, 4-55;
initial screening, Table 4-1, 4-2;
calculating the overall and the components: 4-55;
data:
for species, example data, 4-8; **Data for PDEP**, 58 ff;
for membranes, Table 4-17, 4-58; Fig 4-30, 4-59; Fig 4-35, 4-66;
for Reverse Osmosis "rejection", 4-62;
for application, see each of the separation options.
for classes of solvents, 4-42;
for selecting solvents, data: Table 4-8b, 4-40.
Homogeneous: single phase present; as contrasted with heterogeneous, where two or more phases are present, as in gas-liquid mixture. Used to describe
homogeneous phase separation, def., 4-1;
overview of options, 4-88;
homogeneous reactions, def., 6-1.
Homogenizers, see Size reduction, liquid.
Horizontal plate filters, see Filters, LS.
Horizontal ribbon blenders, see Mixing, solids.
HTS, see Electrostatic separation.
HTU: ambiguous acronym because it refers to either "height of a transfer unit" or "heat transfer unit".
"height of a transfer unit"
def., height of packing in a device such that "one phase receives the enrichment (by interphase mass transfer) equal to the *average* driving force in that phase as averaged over the complete device". (see Transfer unit)
data for packed tower distillation, HTU_L is 7 to 20 cm.
"heat transfer unit"
def., 3-18;
data, Fig 3-15, 3-26;
Humidification: related to the process of adding moisture to air but is referred to here as the general process of mass transfer or evaporation of a liquid species into an "inert" or relatively insoluble gas. The gas phase can become saturated and thus equilibrium occurs when

no further change in vapour content occurs. In general, the vapour content is called "humidity" although technically this term refers only to water vapour. The reverse process of removing the vapour from the gas is called dehumidification. Usually we consider the air-water system. The fundamental principles that are important are the simultaneous transfer of mass and heat. For convenience, we usually represent the processes that occur in the gas phase on a "humidity chart". Humidification is important for Dryers, Quenchers, Gas Coolers and Cooling Towers. Humidification is similar to Absorption when there is mass transfer with significant heat transfer. Absorption is Humidification when heat transfer effects are negligible. Contrast with distillation where all the species are in significant and changeable proportions in either gas or liquid phases.
background definitions, 3-58;
enhancement or reduction in heat transfer rate because of simultaneous mass transfer: 3-58;
relating humidification and the direction of heat and mass transfer to various gas-liquid contacting situations, 3-58;
summary chart, 3-60;
predicting volumetric heat transfer coefficient, 3-60;
data and equations, 3-73;
location of the latent heat load in the heat balance, 3-60;
and humidity charts: 3-58 to -62;
adiabatic saturation temperature, def., 3-61 and -59;
air-water chart: 3-59, -70 and -74; and **Data for PDEP**;
air-water-sulfur dioxide chart: 3-63;
benzene-air chart: 3-63;
carbon tetrachloride-air chart: 3-63;
density of mixture, 3-62;
dew point, def., 3-61;
dry bulb temperature: def., 3-61;
enthalpy of the mixture: 3-62;
example use of 3-59 to -62;
humidity: def. 3-61; estimating, 3-61;
humid volume: 3-62;
and Lewis number: 3-62;
psychometric ratio, defined as the ratio of the convective heat transfer coefficient to the gas phase mass transfer coefficient. In practice, the psychometric ratio for the air-water system, is fortuitously about equal to 0.227, the gaseous molar heat capacity. Thus, the wet bulb and adiabatic saturation lines for air-water are coincident.
relative humidity:
def., 3-61;
contrast with "absolute humidity", 3-61;
saturation: def., 3-61;
surface temperature in evaporating drops, 3-61;
toluene-air chart: 3-63;
unsaturation: def., 3-61;
wet bulb temperature, def., 3-61;
o-xylene-air chart: 3-63;
heat transfer coefficients for: 3-44;
Humidity,
def., 3-61;
absolute, 3-61;
relative, 3-61;
Humidity driving force:
for cooling tower design, driving force for mass transfer, use of enthalpy driving force vs humidity driving force for mass transfer, 3-82;
for thermal wheel design, 3-84;
Humphreys spiral sluice: see Sluices.

Ion Exchange Chromatography: ion exchange in which the recovery/regeneration is done by exploiting difference in the isoelectric points of the exchanged species. see also "Packed bed".

    def., 4-49; for details see "Ion Exchange".

    isoelectric point data: Table 4-13, 4-51;

Ionic mobilities: discussion of, 4-64;

Ionic strength:

    important issue for ion flotation, 4-50;

        for foam fractionation, 4-50;

        for bubble and drop coalescence.

    and effect of surface charge, 4-50.

Ion Flotation: separation method based on adding a surfactant with charge opposite to that of the ion, creating a lot of surface area so that surface active materials will concentrate at the surface. Bubbles are commonly used to create the surface area. (related to Foam Fractionation); def., 4-50;

    pic., 4-52;

    selection, Table 4-1, 4-2; Table 4-5, 4-22;

    temperature effects, Fig 4-3, 4-7;

    usual size Fig 4-9, 4-17;

    feed-product diagram, 4-52;

    description and characteristics, 4-50 and 4-22;

Ionized gas, as method of charging. By this mechanism, particles are bombarded by ionized gas particles from a corona. The sign of the charge is the same as the ions. Contrast this with charging by induction where particles move along a grounded drum that move through the region of influence of an active electrode. Thus, a charge is induced on the particles of sign opposite to the active electrode; charging by rubbing (called triboelectrification); or charging by contact where particles touch a charged plate.

    description, 5-63.

Isobutane: (2-Methyl propane) [C4H10] (iC4) absorption, 1-25, -26;

    molar mass, NFPA, Dow hazard, corrosion rating, freezing temperature, boiling temperature, molar volume, polar and hydrogen bond solubility parameters, pK_a, 4-84;

Isoelectric point (related to zero point of charge, zpc): condition (usually the pH) where a particle in a liquid has a zero electrical charge at the surface. This is important for Ion Exchange Chromatography, for "Precipitation" of proteins, and for the coagulation of particulates.

    data: 4-51;

    application to coagulation, 4-49.

    application to IX chromatography, 4-49.

    application to protein precipitation, 4-32.

Isopentane: (2-Methylbutane, or Ethyl methyl methane) [C5H12] (iC5) absorption, 1-25,-26;

    molar mass, NFPA, Dow hazard, corrosion rating, freezing temperature, boiling temperature, molar volume, polar and hydrogen bond solubility parameters, pK_a, 4-84;

IX, see Ion exchange.

J

j-factors:

    for heat transfer, 3-61;

    for mass transfer, 3-62;

Jacketed evaporators, see Evaporation.

Jacketed reactors: see also Heat exchange, jacketed. pic., 6-9.

Jaw crusher, see Size reduction.

Jigs, a separator of Solid-solid systems based on the differential settling velocities of the two species. A layer of mixture sits in water on top of a screen. The water receives an upward pulse that causes the particles to lift and then settle. Repeated jigging or pulsing effects the separation. They may also function with air as the surrounding and pulsed media.

    may also function as a "concentrator" where we encounter a wide range of particle density and a narrow range of particle size: Table 5-1, 5-24.

    Contrast with classifiers; Table 5-1, 5-24;

    pic., Fig 5-19, 5-27;

    select chart, Fig 5-25, 5-33 based on size and concentration;

    general discussion, 5-59;

    exploit property difference and choose recovery, select, Table 5-7, 5-42;

    definition of separation factor, Table 5-2, 5-35;

    separation factor versus concentration, Fig 5-26, 5-34;

    use Bird number to select concentrator type, Table 5-14, 5-63;

    relate density ratio to particle size to select type, Fig 5-44, 5-64;

    types: Batac type; Denver/Harz/ Bendelari/Cooley type; Baum/pulsator type; Hancock type; IHC Holland type; Wemco-Remer type;

    See also devices that do the same function: Concentrators, Tables, Sluices, Dense Media Separators as DSM.

K

Kaolinite: (Clay) hydraulic conveying Table 2-9, 2-60; Fig 2-59, 2-62;

    hardness, 2-39.

Karr column, see Solvent extraction.

Kettle reboiler, see Boilers.

Key component: def., 4-4;

Kilns: rotary cylinder used as a furnace in which the flames directly contact the contents. As an incinerators, reactors and direct contact heat exchangers. (Related to Dryers, rotary; pic., 5-10)

    pic., 3-53, 6-8;

    select, as kilns for solids: Fig 3-5, 3-7;

    size:

        discussion, 3-52;

            chart: diameter and size from capacity: Fig 3-32, 3-55;

        example: E3-15, 3-55;

    as reactors:

        pic., 6-8;

        select:

            phase criteria: Table 6-1, 6-13;

            selectivity, heat of reaction (as "hearths"): 6-14;

            temperature level: 6-15;

    residence time for solids in, 6-11.

    cost of used versus new, 1-6.

Kneader, see Mixing.

Knives, rotary, see Size reduction.

Knock out pots: workhorse for Gas-liquid; and used occasionally for Gas-solid separation.

    for Gas-liquid separation: pic., 5-3;

        chart select, 5-4;

        exploit size and density and base on recovery, select: Table 5-3, 5-38;

    for Gas-solid separation: (settling basin) pic., 5-5;

        chart select, 5-6;

        exploit size and density and base on recovery, select: Table 5-4, 5-39.

    example application P&ID: PID-1A: knock out pots, V102,

in the context of mixing liquids or "blending"/mixing dry solids.

and residence time distribution RTD: 6-2.

importance for reactions: 6-1 ff; details, 6-4;

    operating for reactors, importance of for target reactions: 6-1; as criteria for selection of reactor configuration: Table 6-3, 6-15;

    examples: 6-3 ff; Table 6-4, 6-16;

mixing liquids: usually done by aeration (using columns of rising bubbles) or by mechanical mixers (propellers, turbines, anchors, spirals attached to rotating shafts). Selection depends on size of vessel and fluid viscosity. Power required depends on the function (blending, dispersing, suspending solids, mass transfer, heat transfer, reactions): in range 0.2 to 6 Kw/m^3.

mixing by aeration: surface aerators, 0.02 to 0.04 kW/m^3; by diffused air, 0.15 to 0.25 dm^3/sm^3.

blending dry solids: cone, twin shell ribbon, zig-zag blenders; Muller (or pan or edge roller) mixer; 30 to 100 kW/m^3 working volume for Muller mixer; 2 to 10 kW/m^3 for double cone.

for mixing melts and pastes: extruders, pugmills, rolling mills; for extrusion of polymers 100 to 1000 MJ/Mg; of foodstuffs, soups, pasta and cereals, 75 to 500 Mg/Mg.

Mohs scale of hardness; def., 2-39;

    data of rocks with hardness, 2-39;

    data for solids, **Data for PDEP**, 39 ff.

    conversion to:

        Brinell scale, 2-39;

        Vickers scale, 2-39;

Molality, the number of moles of solute per 1000 g of "pure" solvent.

Molarity, (molar concentration) the number of moles of solute per litre of solution. This value is temperature dependent.

Molar mass, (preferred term for Molecule weight) (item 1 in **ST**)

    and size of molecule, 4-6;

    values of, see **Data for PDEP**. 58 ff.

    and an estimate of the size of the molecule: 4-7, 4-13.

    exploiting differences in: general principles, 4-2. Leads to a wide variety of property differences as illustrated in this table.

    as basis for separation, 4-51;

Molecular geometry, exploiting differences in:

    Table 4-1, 4-2, leads to names of separation options to consider;

    definition of separation factor, 4-6;

    main discussion, 4-50.

    see also Size of molecules.

Molecular kinetic energy: Exploiting differences in for separation:

    for homogeneous phases; Table 4-1 leads to names of separation options to consider, 4-2;

    definition of separation factor, 4-6; leads to gas centrifuge, mass diffusion and thermal diffusion.

Molecular weight, see Molar mass.

Molecular distillation: pic., 4-25;

    description, 4-19;

Molecular sieves, as agents: use Fig 1B-1, 1-21.

Molecular sieves, as a process, see Size exclusion chromatography.

Monoethanolamine: (Ethanolamine, MEA) [C2H7NO] as absorbent Table 1B-2, 1-25; applications Fig 1B-1, 1-21;

    characteristics, Table 1B-1, 1-23.

Motionless mixers, see Mixers.

Motors, electric: pic., 3-3;

    select, 3-4;

    delivery time, 1-8;

    typical speed of rotation, RPM, of, 3-1;

    efficiency, 3-5.

    types: open dripproof, guarded dripproof, TEFC, explosionproof.

Moving bed. (contrast with Fixed bed, Fluidized bed and Chromatographic style for contacting solids with fluids.) Transported beds, moving beds and pneumatically conveyed beds sound very similar. Transported beds and pneumatically conveyed beds are terms used interchangeably. However, a moving bed does not necessarily have particles and fluid moving along together. Hence, "Moving bed" is a separate concept and is noted, especially, in the configurations shown on page 4-13. (Contrast with Pneumatic conveying.)

    contrast "transported" with fixed and fluidized: 3-37;

        fundamentals, 3-37 with illustrative picture, 3-45;

        pic., of equipment configuration, 4-13;

    contrast with fixed and fluidized:

        pic., of equipment configuration, 4-13;

    fluid characteristics for: Fig 3-28; 3-48;

Moyno pump, see Rotary pumps, for liquids.

MTBF, mean times between failure: example data, A-9 and -9.

MSW, municipal solid waste, see Size reduction.

Muller mixer (Edge rolling mill, Pan mixer) see Mixing, solids.

Multicyclone, see Cyclone.

Multijet condenser, see Condensers.

Multiple effect evaporators, see Evaporation.

Multipass heat exchange, see Heat exchange, indirect.

Multistaging:

    principle of, 4-15;

    effect of ease of multistaging on the choice of separation device, 4-20 ff. Details about multistaging are given under each separation option;

    and reactors,

        see "series of STR" to alter mixing, 6-4 with pic., 6-5;

        see "series of small reactors" to control temperature, 6-4 with pic., 6-7.

Multiple hearth furnace: direct contact between the combustion gases and the material. Used for direct contact heat exchange and for gas-solid reactors. pic, 3-53, 6-8;

    select, as kilns for solids: Fig 3-5, 3-7;

    size:

        discussion, 3-56;

        data on solids loadings: 3-56;

        example: E3-17, 3-56;

    example applications, 3-56.

    as reactors: pic., 6-8;

        select:

            phase criteria: Table 6-1, 6-13;

            selectivity, heat of reaction: 6-14;

            temperature level: 6-15;

    residence time for solids in, 6-11.

Myoglobin: zpc Table 4-13, 4-51;

N

N, as in 4N or "4 nines", 99.99; as measure of purity, 4-30;

Nash Hytor vacuum pump: see Liquid Piston.

Nauta-Kosakawa classifier, see Classifiers, air.

Neon: [Ne] adsorption Table 4-9, 4-44;

V

Vacuum: absolute pressure is below usual atmospheric pressure, ie less than 101 kPa.
air leaks into, 2-16 and -21;
air volume-mass relationships under vacuum, 2-20;
illustration and relationship to head, 2-4;
equipment to produce, see Vacuum producing equipment.
creating a vacuum: change the volume of an enclosed gas, for example, by condensation of vapour; or
see Vacuum producing equipment.
use of vacuum:
vacuum distillation;
pneumatic conveying, see Pneumatic conveying.
reactions under vacuum;
vacuum drying, see Dryers.
example hookup to create vacuum in vessel, see vacuum distillation, 4-25;
Vacuum producing equipment: for Details see specific type: Ejectors; Pumps, rotary; Liquid piston and Pumps, reciprocating.
description of options, 2-25 ff.; ejectors are used primarily.
pictures, 2-18;
types: ejectors; rotary sliding vanes, see Pumps, rotary; Liquid piston (Nash Hytor, liquid ring, water ring); mechanical pumps, see Pumps, reciprocating; diffusion pump.
Vacuum distillation: pic., of configuration: 4-25;
Vacuum dryers, see Dryers.
Valves: reliability of, 1-7;
pressure drop across, 2-11 and -12;
star, see Feeders.
control, see Control valves.
Vanner, see Classifiers,
van't Hoff's law: for predicting solidification curve: 4-30.
van der Waals forces, and deep bed filtration, 5-24.
Vapour pressure:
exploiting differences
to separate homogeneous phase:
Table 4-1 leads to names of separation options to consider, 4-2;
definition of separation factor, 4-6;
main description, 4-19;
leads to evaporation, condensation and distillation.
to separate heterogeneous phases:
description, 5-22;
select for liquid-solid; Table 5-6, 5-41; see Drying.
data:
see Cox chart, **Data for PDEP**, 54 and -55;
see boiling temperature, **Data for PDEP**, 58 ff.
Velocity:
of vapour:
in pipes, economic: Fig 2-13, 2-9;
for gas-liquid and gas-solid separation: a commonly-used design criteria is the allowable rise velocity for a vapour so as to allow drops to fall, to prevent foam from forming and to provide contact;
for distillation; see "tower" Fig 2-13, 2-9.
for knock out pots,
for demisters.
of bubbles, drops and particles in air:
data, Fig 5-36, 5-48;
of bubbles, drops and particles in water:
data, Fig 5-36, 5-48;
of liquid:
in pipes, economic: 2-28;
for liquid-liquid separation:

decanters, electrodecanters, 5-49
for liquid-solid separation:
for settlers, hydrocyclones and centrifuges, Fig 5-39, 5-56 and -58;
of sound in dry air at 0°C: 331.5 m/s;
of light: 2.997925 E8 m/s;
Velocity heads, def., 2-9
and pressure drop, 2-11;
Venturi jets, see Scrubbers.
Venturi scrubber: see Scrubbers, gas.
Vessels,
for heat loss from vessels; 3-44;
picture, see types, Reactors or Distillation or Drums.
wall thickness, : for pressure and vacuum vessels; depends on the outer diameter, the internal pressure or vacuum, the temperature. Thus, the required thickness for a 1 m diameter carbon steel cylindrical vessel operating at 8 MPa at temperatures less than 300°C is 3.8 cm. To such thickness estimates a corrosion allowance is added.

If the corrosion is	allow
<0.08 mm/a	1.5 mm
0.09 to 0.3 mm/a	3.0 mm
0.31 to 0.4 mm/a	4.5 mm
0.41 to 0.63 mm/a	6.0 mm

Vibrating incline classifier, see Classifiers, air;
Vibratory feeders, see Feeders.
Vickers scale of hardness,
data 2-39;
conversion to Mohs scale, 2-39;
conversion to Brinell scale, 2-39;
Vinyl acetate: (Ethenyl Ethanoate) [C4H6O2] fluidized bed reactor Table 3-11, 3-50.
Vinyl chloride: (Chloroethylene) [C3H3Cl] fluidized bed reactor Table 3-11, 3-50; caution corrosion, A-2.
Virus: size Fig 4-4a, 4-9; Fig 4-3a, 4-7.
Viscosity,
conversion factor for, Engler, Kinematic, Saybolt and Redwood, **Data for PDEP**, 2;
dimension M/LT;
of gases, liquids: **Data for PDEP**, 83ff.
Vitamin A: decomposition conditions Fig 4-3b, 4-7.
SX, 4-39;
Voidage, see Porosity.
Void volume, in packed beds, see Porosity.
Volume, molar, (item 11 in **ST**)
for gas at 0°C and 101.325 kPa: 22.4 L/mol;
for liquids: molar mass/density; useful in selecting solvents, 4-35;
Volumetric displacement pumps. see Positive displacement pumps.
Vortex breaker, see Vortex.
Vortex, preventing;
importance for NPSH, 2-33.
methods:
vortex breaker: pic., and size, 2-35;
and degree of submergence, 2-35;

W

Wall thickness, see Vessels.
Washing: see Leaching.
Waste water treatment:
PID-5: 5-77;
description of process, 5-76;
reactions, 5-78;
capital and operating costs, 5-79.